TELECOMMUNICATIONS LAW AND PRACTICE

SECOND EDITION

AUSTRALIA
The Law Book Company
Brisbane • Sydney • Melbourne • Perth

CANADA
Carswell
Ottawa • Toronto • Calgary • Montreal • Vancouver

AGENTS:
Steimatzky's Agency Ltd., Tel Aviv;
N.M. Tripathi (Private) Ltd., Bombay;
Eastern Law House (Private) Ltd., Calcutta;
M.P.P. House, Bangalore;
Universal Book Traders, Delhi;
Aditya Books, Delhi;
MacMillan Shuppan KK, Tokyo;
Pakistan Law House, Karachi, Lahore

TELECOMMUNICATIONS LAW AND PRACTICE

By

COLIN D. LONG

LL.B. Partner in Coudert Brothers

SECOND EDITION

LONDON • SWEET & MAXWELL • 1995

First Edition 1988

Published in 1995 by
Sweet & Maxwell Limited
of South Quay Plaza
183 Marsh Wall, London E14 9FT
Computerset by Wyvern Typesetting, Bristol
Printed and bound in Great Britain
by Hartnolls Ltd., Bodmin

No natural forests were destroyed to make this product only farmed
timber was used and re-planted.

A CIP catalogue record for this book is available from The British
Library

ISBN 0 421 50520 6

List of Contributors

John Horrocks, Horrocks Technology, Pirbright, Surrey—Telecommunications Technology (Chapter 1).

Michael Rhodes—Cable and Satellite Broadcasting and other Transmissions (Chapter 4).

Philip Burroughs—Property Rights and the Environment (Chapter 11).

Peter Alexiadis as of Coudert Brothers, Brussels—European Union (Chapter 16).

Peter G. Leonard, Gilbert & Tobin Lawyers, Sydney—Australia (Chapter 17).

Lorne P. Salzman, McCarthy Tétrault, Toronto—Canada (Chapter 18).

Owen Nee, Coudert Brothers, Hong Kong—China (Chapter 19).

Karen Larsen, Berning Schlüter Hald, Copenhagen—Denmark (Chapter 20).

Philippe Shin, Coudert Frères, Paris—France (Chapter 20A).

Dr. Klaus-Jürgen Kraatz, Kraatz & Kraatz, Kronberg—Germany (Chapter 21).

Anne Hurley, Consultant, Minter Ellison, Sydney—Hong Kong (Chapter 22).

Livia Magrone Furlotti, Magrone e Ardito, Rome—Italy (Chapter 23).

Greg L. Pickrell, Coudert Brothers, Tokyo—Japan (Chapter 24).

Masayuki Okamoto, Tanaka & Takahashi, Tokyo—Japan (Chapter 24).

Marjolein J. Geus, Buruma Maris, The Hague—The Netherlands (Chapter 25).

Jim Stevenson, Buddle Findlay, Wellington—New Zealand (Chapter 26).

Almudena Arpón de Mendivil y de Aldama, Gomez-Acebo & Pombo, Madrid—Spain (Chapter 28).

Olof Alffram and Susanne Themptander, Advokatfirman Tisell & Co AB, Stockholm—Sweden (Chapter 29).

Jeffrey Blumenfeld and Christy C. Kunin, Blumenfeld & Cohen, Washington DC—USA (Chapter 30).

Acknowledgements

Once again I am indebted to a number of people for their assistance to me in commenting on and contributing to this book. Quite apart from the sterling efforts of the country contributors, I must thank my colleagues at Coudert Brothers London, Michael Rhodes and Philip Burroughs for their chapter contributions and thanks also go to Alan Harper at Mercury One2One, Chris Athersych at Mercury Communications, Ian Walden of Queen Mary & Westfield College, Sally Davis at Nynex Communications, as well as my colleague Vladimir Khrenov in Coudert Brothers' Moscow office and Tim Slater and Anna McKibbin of Coudert's London office. Last, but not least, my grateful thanks to Annabelle Cox who typed the entire manuscript which, given that she typed the manuscript of the first edition, must be something of a record.

I must also express my general appreciation to all my friends in the telecommunications industry, without whose support and encouragement over the years this book would not have been possible.

Very special thanks go to my wife, Sheila, and my children, who as before have had to extend me patience and tolerance in putting up with my many absences to complete this work.

Foreword to First Edition

The decade of the 1980s will be remembered as a time of remarkable change in British telecommunications. The change resulted from the implementation of a new political philosophy during a period in which technology was developing with unusual rapidity. The pivotal event has been the privatisation of British Telecom. This gave it the freedom from rigid controls and the incentives to operate more efficiently. It also necessitated the establishment of a regulatory regime, which is administered by the Office of Telecommunications (OFTEL). The focus of the regulations is to make things better for customers. As Director General of Telecommunications, I have the job of promoting the interests of customers partly by promoting competition. I regard competition as the regulator's best weapon: a monopolist can afford to neglect the customers, a business facing competition cannot. However, where competition is not yet effective, regulation is needed to prevent the abuse of monopoly power—and regulation is most beneficial if it mimics the effect of competition.

The new regime in telecommunications inevitably involves some complexity. Partly this is transitional. I aim to have as little complexity in regulation as possible and I expect that it can be reduced as time goes by. However, those who have described our policies as "deregulation" have chosen the wrong word. To promote competition, and to protect the interests of customers, in a situation where one supplier has a large degree of dominance, regulation is required and some of the regulations will involve legal—and other—complexity.

The development of a body of expertise about telecommunications law is therefore important to the success of the new policies. Colin Long is one of a group of lawyers who have specialised in this branch of law and, so far as I know, his book is the first in the field. No doubt it, like British Telecom, will have to face competition. I welcome it and wish it success in contributing to—and participating in—effective competition.

June 9, 1988
[SIR] BRYAN CARSBERG
[Former] Director General of Telecommunications

The author is grateful for Sir Bryan Carsberg's kind permission to reproduce the foreword to the first edition.

Preface

When I wrote the first edition of this book in 1987 it was confined to the law and regulation of the U.K. This made sense in a European context at least because at that time a comparative analysis of the telecommunications laws and regulations of the continental European countries would have made a very thin pamphlet indeed. Perhaps only the experience of the U.S. would have been worthy of comparison.

All that has of course changed quite radically. As a consequence of the downward pressure of European Union directives, emanating from the Directorates General for Competition (DG IV) and Telecommunications (DG XIII), and the upward pressure of the forces of international competition, many European Union countries have now developed relatively sophisticated schemes of regulation, although (happily) not rivalling the U.K.'s in complexity.

Elsewhere in the world the name of the game has been infrastructure development. The G7 and similar countries have been seeking to establish national information "highways", whilst the less developed have been hungrily looking to attract international financial institutions, equipment vendors and telecommunication operators (telcos) to invest in and install new digital networks. This dash to re-equip and the evident socio-political importance of extending network penetration at all levels, urban and rural, has in many cases compelled quite conservative countries to liberalise and open up fixed, mobile and satellite services to competition. Recognising that one of the fundamentals for attracting investment is a stable, predictable and transparent regime, these countries are at the same time putting telecommunications laws in place that, with varying degrees of success, create regulatory institutions and licensing prodedures designed to provide a framework for service provision, tariff controls, fair competition and inter-operator issues such as interconnection.

Interconnection (which has its own separate chapter in this book) is gradually surfacing as the key issue for network and services competition throughout the sector globally. Due, primarily, to the incumbents' control of essential facilities and to the resulting imbalance in bargaining power between interconnecting operators, regulation has often intervened and acted in the role of mediator/arbitrator. Not always, however; some governments have preferred not to accord telecommunications special status but to leave market entrants to rely on general competition law for support. This approach is tenable provided the competition law is effective and pragmatic. The New Zealand *Clear* case seems to have unearthed a deficiency of this nature in that particular country's adoption of general competition law as the *entire* solution.

Returning home, U.K. regulation seems to have sailed into a Sargasso Sea of complication. When the 1990 Duopoly Review led to a new open licensing policy, the opportunity should have been taken to revamp the regulatory system to remove the "dominance" elements from non-monopoly operator licences and to reassess in open debate how interconnection charges might best be calculated. Instead, aberrant provisions affecting the sector generally (*e.g.* access deficit contributions—see below) were added into BT's licence, and from 1991 onwards numerous grafts were added to an already limping body of regulation. The result has been burdensome, if not oppressive, for BT's competitors and the cause of liberalisation has been dis-served by arcane rules obscurely drafted, sufficient to confound the regulators just as much as the regulated. More work for lawyers perhaps, but cold comfort for their clients seeking to drive forward business, not regulation.

The hope must be that by the time of the next supplement of this book (volunteers please write in) telecommunications law and practice, not just in the U.K., but around the world, should have matured into a framework which is not bedevilled by accounting mythology or fashionable economic theory. Its successful evolution requires effective supervision while falling short of undue or artificial interference. It also requires the development of a framework which is underpinned by solid, consistent and effective legal principles protecting fair competition and penalising abuses of dominance.

PROPOSALS FOR REFORM IN THE UK

Well, there is now a chance that the regulatory system may be re-oriented in that direction. In his July 1995 statement on the future of interconnection, competition and related issues—"**Effective Competition: Framework for Action**"—the Director General puts forward key proposals for reform of the existing regulations in this area. A more detailed analysis will have to await the result of the consultation procedure on these proposals and the next edition of this book, but the most important changes proposed and other policy principles enunciated are as follows.

Interconnection

- A move to incremental costs, instead of fully allocated costs, as the basis for interconnection charges: this system would be introduced in early 1997 depending on BT's agreement to the Director General's proposed modifications to the price control regime generally, without which the system would have to be referred to the Monopolies and Mergers Commission. In the interim work would need to be carried out in the ICAS (interconnection and accounting separation) working groups to produce an acceptable methodology for determining incremental costs.

- The abolition of access deficit contributions, subject to the removal of the RP1+2 and RP1+5 price caps: again the timing of this change is a little uncertain due to the need to agree the modification with BT, but there is no doubt that the obituary for ADCs has been written. With the removal of the price controls BT's access network costs would be recoverable as BT chooses through its retail prices.
- Call termination interconnection charges will continue to require regulatory oversight: OFTEL invites comments on whether it should set a standard rate (in fact the Director General is rapidly moving towards regular annual determinations of interconnection charges generally).
- As regards other interconnection charges, as an alternative to annual determinations the Director General canvasses the possibility of a "network" price cap (over a basket of wholesale charges) together with a cost floor and ceiling.

Effective Competition

- A new licence condition to be incorporated in BT's licence, and possibly in other operators' licences, prohibiting the abuse of a dominant position and other acts or omissions "unfairly" preventing, restricting or distorting competition in telecommunications; if this can be successfully pulled through the licence modification process then it could be the answer to many practitioners' prayers. The Director General also cautions that if it is found that this new prohibition, together with other controls that he may exercise over BT, are ineffective, there must be a question as to whether all BT's businesses should remain in single ownership (the implication being a threat of a referral to the Monopolies and Mergers Commission to investigate whether it is in the public interest for BT to own and control all these businesses or whether they should be broken up).

BT Tariff Structure

Here a number of licence modifications are proposed, including:

- the removal of the RP1+2 and RP1+5 price caps on residential and "wholesale" prices for exchange lines;
- the introduction of a RP1+0 cap for low user scheme customers;
- amendment of condition 24C of BT's licence in order to widen the range of possible discounts counting towards the RP1–7.5 price cap's discount yield;
- amendment of condition 24D to provide that any increase of revenue

foregone under the low user scheme may count towards this discount yield.

Tariff Flexibility

There is to be no change in the policy of limiting BT's flexibility to target large volume customers. However, the position is to be kept under review in the light of how BT uses its existing flexibility under its licence conditions (17a and 24f), as well as the development of effective competition and the success of OFTEL in using the instruments at its disposal to deal with anti-competitive behaviour. Condition 17A may also be modified to ensure that the flexibility which it and its associated guidelines give to BT do not prevent OFTEL from considering whether discounted tariffs involve undue discrimination or preference under condition 17.

OFTEL proposes to make no change to its policy on BT's obligation to offer a geographically averaged tariff, whilst promising to undertake a study of the "empirical" considerations that should inform future policy on geographic deaveraging of services not forming part of the universal service.

Universal Service

The Director General puts forward a suggested definition of universal service, namely that this should be:

> "Affordable access to basic voice telephony (or its equivalent) for all those reasonably requesting it regardless of where they live."

Comments are invited on the ways in which this service should best be funded. Perhaps most interesting of all, the Director General has concluded from results of studies commissioned by OFTEL, that universal service does not have a significant cost: its net costs (after associated benefits of serving relevant customers) were found to be in the range of £4–40 million and even then the upper limit is suspected to be an over-estimate. Before effective arrangements for setting up universal funding can be introduced, issues such as the arrangements for determining and reviewing universal service costs, the basis for determining who should contribute to these costs and the administration of such a scheme would all need to be resolved.

Generally, what should be made of the Director General's statement? Certainly he could be congratulated for his recognition of fundamental flaws that have existed in the regulatory regime and his determination to put matters right.

However, whilst the Director General is relatively new to his position and can perhaps afford to be comfortable with the steps he is now taking to improve the state of regulation, his Office of Telecommunications (OFTEL) and the Department of Trade and Industry, the department which first formulated the public operator licences and piloted through the governing legislation, cannot feel complacent about these reforms, which are long overdue. In particular, by the time these proposed changes come into effect it will be five years since the Government announced the ending of the BT/Mercury duopoly and the introduction of full competition in telecommunications networks and services[1] yet the regulatory changes now promised should really have been introduced at that stage, when open licensing was first envisaged. In particular, when the duopoly ended in 1991 the opportunity should have been retaken to adopt new legislation, or at least to reformulate the licences which were the bedrock of regulation, so as to recognise the new competitive tensions and balances present in a multi-operator environment.

Thus many years of properly regulated competition, which could have produced a much healthier competitive climate, have been lost but more insidiously a fundamental error in approach to regulatory rule-making was to my mind made in 1991 and has continued to be perpetrated ever since. This is that major issues affecting the telecommunications sector as a whole, in particular its competitiveness *vis à vis* BT, have been regulated through modifications simply to *BT's operating licence*, which is a licence whose conditions cannot be changed without BT's consent or, failing that consent, through a reference to the Monopolies and Mergers Commission and a recommendation by that Commission that the modifications should be made.

Accordingly, the Director General cannot simply make the changes he proposes, even after extensive consultation, simply because he believes they are right. He has to obtain BT's consent which in some respects puts BT in the position of being the second regulator. It must also at times be rather frustrating for him.

This problem of what some might term "regulatory capture" might be soluble with the help of a carrot like the proposal to abolish the exchange line rental price cap, which may well persuade BT to go along with the abolition of ADCs. Why should the regulator, however, be reduced to having to bargain his way to a solution to such a problem? The technical answer is that the requirement for BT's consent to the modification of its licence is a necessary prerequisite to the proper protection of any licensed operator from the arbitrary or capricious amendment of its licence rights and obligations. But the error that was made was that of including, entirely in BT's licence, rules having important economic and competitive effects *across the industry*, effectively locking them in until they could only be changed by consensus of the regulator and BT or through the MMC route, a long-winded process. Since 1991 other licence modifications have been made which will again have cross-sector ramifications and if and when they need to be revised the regulator will find he is in an unenviably weak position.

In his more recent August 1995 report on multi-media, the Director General's consultative document "Beyond the Telephone, Television and the PC", the Director General puts on his virtual reality helmet to envisage the

---

[1] White Paper "Competition and Choice: Telecommunications Policy for the 1990s", March 1991.

impact of what he dubs the broadband switched mass market services and the effect of the arrival of such services on the scheme of regulation. In this statement he admits the strong case for saying that we need only have minimal regulation, leaving the market to decide most issues, but then remarks that regulation may be necessary to promote the development of the market and to protect consumers and competitors.

The Director General's stated aim is to develop a regulatory regime primarily for dominant network owners and to apply a more relaxed system of regulation to operators with no market dominance. Whilst this makes eminent good sense and is consistent with developing best regulatory practice in the more sophisticated telecommunications jurisdictions, if we are to avoid the errors of the early 1990s, a way must be found for creating a free-standing system of regulation that cannot be controlled or manipulated by any of the players. Primary legislation is almost certainly required, but perhaps more than anything, drastic reform of our national competition laws is also long overdue. In Australia, where they are now set to embark on multi-operator liberalisation, the new thrust is to reduce telecommunications regulation and to refocus general competition laws to address the issue of dominance and the control of essential facilities. This could well be the best way forward for UK regulation. It could lead to maintaining regulation for service obligations, consumer protection and technical issues, and developing competition law (and with it the jurisprudence that cuts across industry sector boundaries) to deal with abuses of market power through a prohibition-based system along European (Treaty of Rome) lines. Surely this must be preferable to merely tinkering with an outmoded preliberalisation model that can longer be made to work effectively. The parts on a car can continue to be replaced only until it is more economic to buy a new model: this is what the industry deserves and it must be hoped that the Government will respond. Whether they have the time or the inclination is another matter.

To return to the theme set at the beginning of this preface, telecommunications law and regulation has become internationalised and this is reflected in the many contributions from experts in this field in many of the leading telecommunications economies around the world. However, as one country privatises its telecommunications monopoly, another country moves to open up the sector to competition and there are a number of countries, including some in Europe, Scandinavia, South America and Asia, which have not been covered. The reasons for this are often, but not always, the relatively embryonic nature of the legal system adopted, the absence in some cases of any regulatory system to speak of, the still awaited emergence of the economy of the country concerned or the lack of apparent availability of lawyers specialising in telecommunications work—or any of the above. However, even in the time it takes to take this book from first proof to publication, new countries have surfaced, such as India and Mexico for example, with relatively sophisticated telecommunications laws which would certainly have merited examination; I hope that these countries can be included in forthcoming supplements along with, no doubt, many others which will have emerged in the meantime. These supplements will also cover developments generally in the various countries evolving regulatory regimes.

I also hope in future supplements/editions to cover some of the strictly telecommunication issues arising out of global information systems such as the Internet. At this stage it has proved a little early to discuss these other than in a somewhat speculative manner.

I am extremely grateful to all the contributors to this book (listed on page v) for their diligence and responsiveness. It has been a happy experience for me working with all of them and one which I hope shows that there is a genuine corpus of expertise in an enthusiastic community of lawyers operating worldwide in this fascinating and developing area of law.

My thanks also to the publishers Sweet & Maxwell for their support and encouragement throughout this project.

Preface to First Edition

Ten years ago there was, to all intents and purposes, no such thing as telecommunications law. Even in 1982, with the creation of Mercury Communications as a potential alternative carrier to BT, the scope for competition was being broadened but the mechanisms for its enhancement by regulation had not been developed nor even, perhaps, conceived.

The Government's ideals and its manifesto objectives were finally fully translated into practical and legal form and effect by the 1984 Act, but, even with the panoply of regulation introduced by that Act, the infrastructure for fair competition in both telecommunication service and apparatus supply left some gaps. In the absence of an American style of approach to anti-trust law no privatisation/liberalisation process of this kind can properly be implemented without in some way building on or modifying the pre-existing competition laws of this country. Clearly, in creating a new competition law "overlay" for telecommunications this point was recognised but even with the licensing regime now in place and the powers conferred upon the Director General and OFTEL under the 1984 Act, some weaknesses remain. The recent Government Green Paper on the proposed revision of United Kingdom restrictive trade practices policies shows that our laws regarding anti-competitive agreements will soon be brought into line with the effects-based approach of our continental colleagues and European law, but unfortunately the opportunity is not to be taken by Government to include similar new legislation controlling the abuse of a dominant position. The argument that the 1980 Competition Act effectively serves this purpose is not convincing.

The European Commission's Green Paper on the development of the Common Market for telecommunication services and equipment may have some impact, even in the relatively liberalised environment of the United Kingdom. Two directives are promised for 1988, as to apparatus and services; indeed, the opening up of national barriers to cross-frontier supply promises to be a big challenge to the Community, given the very national nature of each country's telecommunication undertakings, policies and regulations. Even the European Commission itself has not totally avoided the trap of traditional conservative thinking in this respect: the Green Paper acknowledges that network infrastructure, in terms of the physical apparatus necessary to "wire up" a country, should be the exclusive preserve of one undertaking but its suggestion that voice services should be similarly reserved rather cuts across the success of United Kingdom experience in opening up voice service to competition.

This book is designed to serve as an introduction to lawyers and businessmen and as an aid to more detailed consideration of the issues arising out of United Kingdom regulation of telecommunications. It is not intended as an out and out practitioners' textbook; I have therefore kept case law citations to a minimum, not that these are very much help in the area of licence regulations except by way of comparison with other regulated industries, such as electricity, where some of the social service concepts are similarly framed.

The licensing regulations, which form the bulk of the commentary in this book, should be less complicated if and when the restrictions on simple resale are removed next year, but the regulation of competition and therefore of service activity looks well set to continue as an issue into the 1990s, when further network competition may be allowed by the Government. Certainly, so long as BT remains a de facto monopoly, and United Kingdom regulation of competition generally fails to provide all the answers, some industry-specific controls over its activities will continue to be necessary.

In the immediate future there is clearly a need for rationalisation and simplification of the regulations, particularly those in the Branch Systems General Licence and the Value Added and Data Services Licence. Whenever interpretation of the sometimes obtuse language of these regulations becomes necessary, one frequently has the feeling that if one could have been aware of the intention of the draughtsmen, the realisation of that intention in print would have been that much easier to comprehend.

Generally with United Kingdom regulation one is reminded of the story of the traveller who, on asking a bystander the way, received the response from the bystander that he would not have started here. If the opportunity ever arose to do away with the existing basis of regulation and move towards one where the running of telecommunications systems was in fact permitted from the outset and only specific acts or the provision of specific acts or the provision of specific services was prohibited, life would be much simpler.

At the time of finishing this book, there are many issues arising out of United Kingdom telecommunications regulation which remain unresolved or on which further pronouncements are promised. The meaning of many of the terms used in licences has also yet to be tested and it is thus inevitable that this book can, in some cases, only present the problems and possibilities without offering categoric answers. A very good example is the Director General's statement in the very recent PanAmSat* case in which he gives his view of the meaning of "reasonable demand" in typically simple and direct terms. In the coming years it can be expected that this phrase in particular will come under ever closer scrutiny in testing the extent of public telecommunication operators' obligations and responsibilities to their customers. After all telecommunications law is in the end about the rules introduced to promote the interests of customers and not simply to create new business activity.

Contents

PART ONE—UNITED KINGDOM

PART THREE—INTERNATIONAL

CHAPTER 17 AUSTRALIA

CHAPTER 18 CANADA

CHAPTER 19 CHINA

Table of Cases

E.C. Cases

Table of Statutes

Table of Statutory Instruments

E.C. Legislation

National Legislation

International Treaties and Conventions

Glossary to Part One (United Kingdom)

1956 Act	Copyright Act 1956
1973 Act	Fair Trading Act 1973
1980 Act	Competition Act 1980
1981 Act	British Telecommunications Act 1981
1984 Act	Telecommunications Act 1984
1895 Act	Interception of Communications Act 1985
1990 Act	Broadcasting Act 1990
Access Deficit	The amount by which BT's revenue from exchange line connections and line rentals falls short of the fully allocated costs of providing and maintaining customer connection to its network; specifically defined in BT's licence Condition 13.5A.3
Access deficit contribution	A contribution to be made by interconnecting operators to BT's Access Deficit, specifically defined in BT's licence Condition 13.5A3
Calling line identification	A facility whereby a person being called may identify the number from which the call is being made
CEPT	European Conference of Post and Telecommunications Administrations
Digital	A type of telecommunication system communicating in binary digits in the same way as a computer
Director General	Director General of Telecommunications
DTI	Department of Trade and Industry
ERC	European Radiocommunications Committee
ERO	European Radiocommunications Office

ETO	European Telecommunications Office
ETSI	European Telecommunications Standards Institute
Fixed Link	A communications link (*e.g.* by cable, wire or radio)
FRBS	BT's Financial Results By Services, a set of financial statements drawn up by BT on an annual basis and provided to OFTEL for regulatory purposes, especially BT's licence Condition 13
INMARSAT	International Maritime Satellite Organisation
INTELSAT	International Telecommunications Satellite Organisation
Interconnection agreements	Agreements made by operators on the terms on which their networks may be connected to each other, *e.g.* pursuant to BT's licence Condition 13
International Simple Resale (ISR)	Resale over an international leased circuit involving conveyance of messages via the PSTN at both ends of that leased circuit
ISDN	Integrated Services Digital Network
ITU	International Telecommunication Union
NTP	Apparatus at the boundary of a telecommunication system, forming the interface with another system
NTTA	Apparatus of a PTO installed on customer premises, incorporating the NTP of the PTO and providing the means of physical connection and disconnection, as well as testing
Number portability	A facility enabling a customer to transfer from one operator to a second operator and retain the same number
OFTEL	Office of Telecommunications
PCN	Personal Communications Network—a mobile telephone network using cellular radio designed specifically for handheld portable terminals, known in some countries as PCS (personal communications system), in Europe generally working to the DCS 1800 standard (1.8 GHz)

PSN	Public switched network
PSTN	Public switched telephone network
PTO	Public telecommunications operator, licensed under section 7 of the 1984 Act and so designated under section 9 of the Act
RBOCs	Regional Bell Operating Companies; the local exchange carriers created by the divestiture of AT&T
RPOA	Recognised Private Operating Agency
Secretary to State	The Secretary of State for Trade and Industry
SMATV	Satellite Master Antenna Television
SPL	Self Provision Licence, a class licence granted under section 7 of the 1984 Act; discussed in Chapter 6
Telecommunications Code	The Telecommunications Code contained in the 1984 Act. See also Chapter 11
TO	Telecommunications Organisation; a European Commission description of a public telecommunications operator
the Treaty	the Treaty of Rome
TSL	Telecommunication Services Licence, a class licence granted under section 7 of the 1984 Act; discussed in Chapter 6
Video-on-demand	Programmes or films sent independently to customers in response to individual requests
WTA	Wireless Telegraphy Act 1949

PART ONE

United Kingdom

PART ONE

United Kingdom

CHAPTER 1

Telecommunications Technology

By John Horrocks, Horrocks Technology

1. INTRODUCTION

The purpose of this section is to provide an introduction to the technical   1–01
aspects of telecommunications and to explain the main changes that are
taking place in the technology, and how they are expected to interact with
the regulations and markets. It is written for the non-technical reader who
is involved in telecommunications.

What is special about telecommunications? In the public area especially,
the answer is interconnectivity. The benefit from telecommunications to a
large extent depends on the ability to communicate with others in changing
circumstances and in response to a need that may not be foreseeable. This
is known as the any-to-any requirement.

The need for interconnectivity means that communications systems have
to be technically compatible with each other and have to be actually interconnected. This results in there being a major emphasis on international
standards and harmonisation.

With the older analogue technology, the problems of providing satisfactory quality especially for international connections were considerable, and
these technical constraints, coupled with a different general approach to
essential services, resulted in telecommunications becoming a state monopoly. The advent of digital technology has removed many of the technical
constraints and problems in providing satisfactory international services,
and has introduced many new possibilities for telecommunications to be
provided by a multiplicity of different, competing, organisations with interconnected and technically compatible systems.

2. HISTORICAL DEVELOPMENT OF TELECOMMUNICATIONS TECHNOLOGY

Overview

1–02   The telephone was invented just under 120 years ago by Alexander Graham Bell. In those 120 years the telecommunications network has grown to become the largest man-made machine ever made, handling more than 1000 billion calls a year and encompassing the whole globe.

The principal developments during that time were:

1889: Development of first automatic exchange (Strowger).
1895: Development of radio communications by Marconi.
1901: First radio communications across the Atlantic.
1946: First electronic computer.
1948: Invention of the transistor.
1956: First transatlantic telephone cable.
1965: First communications satellite.
       Invention of optical fibres.
1970s: Introduction of digital communications.
       First electronic exchanges.
       Start of major growth in data communications.
1980s: Introduction of ISDN.
       Introduction of cellular radio system.
1990s: Introduction of intelligent networks.
       Multimedia.

The most rapid period of growth is the period from the late 1960s to the present day. This rapid growth is the result of the combination of:

the developments in electronics;

the development of digital communications and the convergence of telecommunications and computing technologies;

the development of optical fibres;

the application of micro-electronics to radio communications.

There are now some 800 million telephones and the number is growing at about four per cent per annum. The growth rate for data terminals is about 20 per cent per annum. This rapid growth has been accompanied by a sharp fall in costs and prices.

1–03   Telecommunications can take two forms—analogue and digital. In an analogue system, the information is represented by a waveform of continually varying amplitude of frequency. In a digital system, the information is represented by a sequence of signals of discrete levels (usually two levels to match the on and off states of a transistor, in which case the representation is binary and each piece on information is called a bit). A fundamentally analogue source signal can be converted into a digital signal through a sampling process.

The earliest systems were telegraph systems which were a form of digital system. Telephony was analogue from its inception up to the early 1970s when developments in computing began to be transferred across into telecommunications. Currently almost all new systems are digital, because such systems offer

(a) The capability of carrying a multiplicity of services in a common form;
(b) greater robustness in the presence of noise or interference;
(c) maximum commonality of components and techniques with computing.

Telecommunications Culture

In order to understand the development of telecommunications it is necessary to understand something about the telecommunications culture, which has been strongly influenced by the way in which telecommunications has been managed and regulated.

1–04

Until the early 1980s, in almost all countries, telecommunications was the subject of monopoly supply with the public network operator normally being a state owned corporation or government department. Subscribers were not allowed to own their own telephones, and there was little choice in the models that could be rented from public network operator. This created a situation where manufacturers supplied only to public network operators, and those operators often became closely involved in the activities of their suppliers by funding research and development and arranging orders to supply a reasonably constant stream of work. This meant that the whole telecommunications culture was dominated by the public network operators.

This monopoly culture was in stark contrast to the competitive culture of the new computing industry which used to a large extent the same technology and saw rapid reductions in costs and improvements in capabilities.

The need for interconnectivity at an international level was met through the ITU (an agency of the United Nations) and especially the CCITT (now ITU-T), which prepared technical recommendations primarily on the interconnection and interoperation of national services. The main participants in the ITU were the public network operators, although other parties could contribute in practice they exerted much less influence. The ITU worked in four year cycles and there was comparatively little pressure to produce recommendations quickly.

The first wave of change to the culture came from the developments in data communications between computers, where the competitive computer culture saw the advantages of standardisation in avoiding customers being locked in to particular vendors, and began to develop Open Systems Interconnection (OSI). The early standards work on OSI was carried out

1–05

through the normal standardisation bodies (ISO), which had a much more open structure than the CCITT with more involvement of manufacturers and academics. Eventually good cooperation between the two groups was established.

The development of competition occurred first in the supply of terminal equipment and is spreading more gradually to the provision of services and infrastructure, starting with data and value added services. Given the need for interconnectivity, and the need to protect the public networks from harm that could be caused by connecting unsatisfactory apparatus, competition created a large new demand for standards for national as well as international use.

In general the recommendations produced by the CCITT, although of major value, were not adequate for the new competitive market because they were not written as standards but more as descriptions; they did not contain clearly identified requirements and consequently there was normally no information on how to carry out testing. They also tended to contain national options.

Thus the process of liberalisation created a major need for technical standards, and this need is being met in Europe to a large extent by the creation of the European Telecommunications Standards Institute (ETSI) out of some of the technical committees of the old European Conference of Posts and Telecommunications (CEPT).

With the digitalisation of telecommunications, the technologies of computing and telecommunications are converging rapidly. As explained above, the backgrounds of the two cultures were profoundly different, having the following characteristics:

Telecommunications:	Closed
	standardised
	monopolistic
Computing:	Open
	unstandardised
	competitive

but liberalisation and the technological convergence is bringing the two cultures together.

3. Technology Drivers

1–06   This section explains the main technological developments or drivers in more detail.

Integrated Circuits and Semiconductors

1–07   Semiconductors have played the key role in technological advances since the mid 1950s. Semiconductors provide both the processing power and much of the information storage power for modern telecommunications.

Integrated circuits, or microchips as they are commonly known, were developed in the late 1960s and the advances in their performance and the reduction in the cost per function have been the central feature of the electronics industry for the last two decades.

Individual semiconductor devices such as transistors are made from germanium and silicon. Silicon is a much better material than germanium for integrated circuits because it is cheaper and easier to use. Gallium Arsenide which is intrinsically faster than silicon can be also be produced but is in general a much more difficult material than silicon to process (and hence more expensive) and finds application only where the very highest speeds are essential.

Integrated circuits consist of a large number of individual components which are all made together on a single die (or chip) cut from a wafer of semiconductor material. There are varying degrees of integration leading to a series of acronyms which in ascending order of complexity are as follows—

MSI—Medium Scale Integration. Several tens of components per chip.

LSI—Large Scale Integration. Several hundred to several tens of thousand components per chip.

VLSI—Very Large Scale Integration. Over 100,000 components per chip to several million.

WSI—Wafer Scale Integration. A number (10–100) of VLSI circuits interconnected on the same wafer.

VLSI is the current state-of-the-art technology and is a major enabling technology for future telecommunications developments. WSI is some way from being an established production possibility and is something of an unknown quantity in the current context. The main emphasis from here on is on VLSI but much of the terminology is common to all levels of integration.

A measure of the technological advance in going from MSI to VLSI is **1–08** the feature size of the devices that form the circuit elements. Feature sizes of these circuit elements are expressed in microns (10^{-6} metres). Currently VLSI using one to five micron circuits are readily available and the trend is towards sub-micron technology, *e.g.* 0.2 micron by the mid nineties. Reduction in feature size obviously means more devices per chip and because chip manufacturing methods are continually being improved, the size of chip that can be produced without defects to give an adequate yield of working circuits is increasing. The combination of smaller feature size and a larger chip means that the number of devices per packaged chip is increasing very rapidly with time and because the cost of the finished product is strongly dependent on the cost of the package, the cost per function is decreasing rapidly with time.

WSI is based on the interconnection of chips on the wafer itself, thus reducing the length of connections, the number of individual packages and the number of bonds joining the integrated circuit to pins on the package. Short interconnections improve speed of operation and the reduced number of bonds is a significant factor in increasing reliability.

The difficulty with WSI is with the yield of integrated circuits during manufacture. Because of localised defects in the silicon wafer, yields are usually well below 100 per cent and the probability is very low that all the circuits connected together on a wafer will work. The solution is to build sufficient redundancy into the wafer and perhaps include a test system and an automatic configuration controller on the wafer as well. In any case, a means must be found, either automatically or manually, to configure a sufficient set of working circuits to make the wafer viable. This is easier if the circuits are of a repetitive nature, and hence memory applications are likely to be the first to be realised.

Is there a limit to the increasing performance of integrated circuits, and if so when is growth going to stop? There are theoretical thermodynamic and quantum limits which suggest that speed cannot be increased by more than three to four orders of magnitude above that currently achieved by the fastest silicon devices. Limits on speed combined with circuit component density are set by the practical problem of heat removal, but it possible to design circuits to minimise dissipation and to circumvent this problem as least to some extent. There are limits on device size but it would be rash to predict what these might be, except to say that the limit of a few times the inter-atomic distance is within the bounds of possibility.

Figure 1 below shows how the complexity of integrated circuits in terms of the number transistors per chip has increased over the 20-year period from 1970.

For the moment all that can be said is that we are a long way from both the theoretical and practical limits, and that progress will continue to be made, albeit perhaps less rapidly, well into the next century.

Fibre Optics

1–09   The development of communications by light transmission along fibres is probably the most significant technological development in communications in the last twenty years. Even though a substantial part of the public networks in many countries use optical fibres, there is still significant scope for further development because the inherent bandwidth, or traffic carrying capacity, of a fibre is some three orders of magnitude greater than that of coaxial cable and some five orders of magnitude greater than that of a copper pair. However only a fraction (less than 0.1 per cent) of optical fibre's inherent capacity is currently being exploited.

The operating point of an optical system is normally expressed in terms of a wavelength measured in nanometres (nm) or microns (μm), and rarely in terms of the frequency which is of the order 10 to the power of 14 or 1014 Hz. Optical fibre communication makes use of transmission windows or bands where, for a particular wavelength, the loss or attenuation in the optically transparent material from which the fibre is made is relatively low.

Figure 1: Growth in Number of Transistors per Integrated Circuit

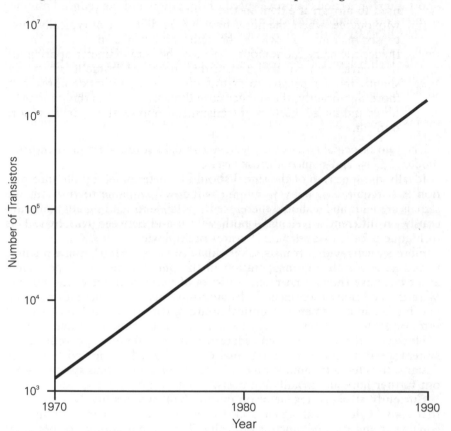

The wavelength at which long distance transmission systems operate has always been dictated by fibre transmission behaviour. Essentially there are three forms of transmission—

(i) Multimode, where the fibres have a core diameter typically of 50 microns or about 50 wavelengths. Light propagates along a multitude of paths being reflected by total internal reflection from the wall, or cladding, of the fibre. Multimode fibres are cheap but limited in performance.

(ii) Graded and Complex Index Profiles where the fibres have a refractive index profile which varies in a complex fashion with distance from the centre of the fibre. In multimode operation, because the velocity of light in the fibre depends directly on the refractive index, the refractive index profile can be chosen to compensate for path length differences. Dispersion performance is not as good as for monomode fibres, and so graded index fibres have not so far been adopted for long distance transmission use and seem unlikely to be used to any great extent in the future.

(iii) Monomode, where the fibres have a core diameter of typically three to six microns and where the light propagates in a single mode (hence the name monomode) because the core diameter approaches the wavelength of the transmitted light. Monomode fibres are slightly more expensive to manufacture than multimode fibres, but, more significantly, the smaller core dimension makes the jointing of fibres and the connection of transmitting and receiving devices more difficult.

1–10   The two principal types of device used as light sources are light emitting diodes (LEDs) and semiconductor lasers.

Ideally the spectrum of the signal should be determined by the information it is required to convey, but the windows available to transmit the signals are wide and it should therefore be possible to send signals simultaneously on different wavelengths both within and between windows. This technique is called wavelength division multiplexing.

Fibre systems with a transmission capacity of 565 Mbit/s using a single light source are already in operation in the public network. Experiments and trials have shown that rates in excess of two Gbit/s are quite feasible. With the addition of wavelength division multiplexing, coherent operation and further improvements in optical sources, detectors and filters, overall rates of 10 to 100 Gbit/s should be obtained within the next five years.

However, the important consideration is that these improvements in system performance can in the main be achieved using existing fibre designs, thus telecommunication operators can upgrade their systems without further fibre and installation costs.

Currently all signal regeneration in long haul systems involves repeaters that convert the optical signal back to the electrical domain, amplify and equalise it and then relaunch it optically. There are a number of potential means of optical amplification under development. There is also considerable interest in the development of optical switches which may replace electronic switches for high bandwidth applications in the longer term.

Software

1–11   The writing of software has developed considerably from the informal unstructured activities of individuals to much more formalised methodology. This development has been necessary as systems have become larger and much more complex. Much of the cost of major telecommunications systems now lies in the development of software and the balance of costs is moving strongly from the recurring costs of hardware to the non-recurring costs of software.

Software development is being helped by the creation of complex tools to assist the specification, writing and testing processes, but one of the main constraints on the pace of development is inevitably the management of teams engaged on a very complex task.

Coding and Processing

Steady progress is being made in the development of new coding and pro-   1–12
cessing techniques. These techniques are aimed at two areas:

(i) Enabling more user information (*e.g.* images, or voice) to be sent
using lower bit rates. This is especially important for radio systems
where the available frequencies and hence bit rates are limited.
(ii) Enabling distortions and errors to be corrected, for example cor-
recting the effects of multiple signal paths in radiocommunications.

Coding is a highly mathematical subject, but it is a vital part of many
radio based networks such as the pan-European GSM network.

4. TECHNOLOGY SYSTEMS

Networks consist of a combination of transmission equipment and switch-   1–13
ing equipment.

Transmission Systems

Transmission systems are concerned with the sending of information via a   1–14
copper or fibre cable or via radio. There are two main issues—the physical
sending of the information in the form suitable for the medium and the
combining or multiplexing together of information into a single complex
form in preparation for it to be sent.
  At the physical level, the transmission systems being installed in public
networks use optical fibres in almost all areas except the local loop, where
fibre is typically at least two to three times as expensive as copper for most
customers. However fibre is being used for larger business customers and
will begin soon to be used for combined TV and telephony systems in
residential areas, at least from the head end to each street.
  At a system level, the main issue is multiplexing, which is the means of
combining a number of different signals onto a single transmission medium.
In digital systems time division multiplexing is used whereby different channels alternately make use of the transmission system. The way in which
the channels alternate may be fixed and regular, or it may vary depending
on the capacity needs of different applications. A combination of the two
techniques may be used.
  The main public telecommunications networks currently use the Plesi-
ochronous Digital Hierarchy which was developed originally to allow the
multiplexing of signals in a situation where the networks were not wholly
digital, and the signals to be combined did not necessarily have precisely
the same rates. A new more complex system has now been developed,
called the Synchronous Digital Hierarchy, which is designed to carry a wide
range of signals and to work at rates typical of those used on advanced

optical fibre systems. This system will be introduced in the public networks in Europe over the next decade.

Multiplexing can also be achieved using packets or frames that are not sent in a regular order and that carry information to identify to which connection they belong. Two new techniques are being introduced:

(i) Frame Relay was developed from concepts for a new ISDN service, but suddenly became the subject of great interest in 1991 when manufacturers began to use the technique for multiplexing together channels from very bursty applications such as local area networks. It has a variable frame length.

(ii) Asynchronous Transfer Mode (ATM) which is a cell technique, of fixed size, designed for the multiplexing and switching of signals from applications with different and varying transmission rates.

Switching

1–15   The current generation of switches are nearly all electronic and digital (some electronic analogue PBXs are still made) and are controlled by software. The main public networks use circuit switches where paths of fixed capacity are set up across a switch and remain for the duration of a connection. Such switches normally communicate with each other using the CCITT Signalling System No. 7.

Dedicated data networks use packet switching where the paths across the switch may vary in capacity, and packets may be queued resulting in variable time delays. The throughput of such switches is low.

Considerable research is going into the development of broadband switches, mainly using ring technologies, that will switch signals in ATM (cell) form and provide very fast packet switching suitable for a combination of voice, image and data.

Intelligent Networks

1–16   The word "intelligence" is used in electronics to describe the ability to access and process stored information. Thus an intelligent network is a network that uses stored information. The public fixed networks are already intelligent to some extent because they use stored information for number translation and routing, but the term "intelligent network" is normally reserved for networks that have more intelligence (stored data and processing) than the current public switched networks, and that have some form of architectural separation between the basic switching and the stored data and processing.

There are already two simple types of intelligent network in use in many countries—the networks that provide premium, freecall and other services, and the cellular radio networks. However a considerable amount of standardisation work is in progress to define intelligent networks, including their architecture and interfaces, in considerable detail.

Intelligent networks are also the subject of growing interest amongst regulators because of their potential for providing a wide range of new services and of separating the provision of services from the provision of the underlying physical network. Regulators are therefore concerned to ensure that there is a competitive supply of services by independent organisations, and also to avoid excessive domination from monopoly network providers.

An intelligent network consists of a number of basic elements. The "ordinary parts" of the network are the transmission and switching facilities that carry the call traffic. These facilities include a common channel signalling system (CCITT No.7) that provides a sophisticated message carrying facility that is independent of the call traffic.   1–17

The additional parts that make the network intelligent are:

additional processing associated with the basic switches to initiate, receive and transfer messages concerned with the handling of calls and the provision of special features;
information resources (*e.g.* databases or message recorders) that are necessary for the provision of special features; and
service control facilities that control the provision of special features and services using the basic network and information resources.

Both the service control facilities and the information resources may be run by an organisation different from the one that runs the basic network.
The Commission is studying the application of Open Network Provision (ONP) principles to intelligent networks.

Satellites

Satellites provide communications by receiving radio signals transmitted by one earth (or ground) station and transmitting them back at a different frequency to another earth station. Communications satellites normally operate in a geostationary orbit which is an equatorial orbit with a period of 24 hours such that the satellite is always in the same position relative to a point on the earth's surface, so that earth stations do not need to move their antennae appreciably to track the satellite.   1–18

Because the signals have to travel to the satellite and back and because the geostationary orbit has a radius some six times that of the earth, transmission via satellite takes appreciably longer than transmission by terrestrial means, and a single satellite hop takes about 250 ms, the exact figure depending on the relative locations of the earth stations and the satellite. This delay is quite noticeable in voice telephony where the speech is interactive, and it also necessitates the use of echo control in speech circuits.

The role of satellites in communications is determined by the differences in their inherent characteristics from those of line communications or terrestrial radio. Because satellites provide radio communications to a large area they are well suited for broadcasting and mobile services, and a single satellite can replace a whole network of terrestrial radio stations. Line communications are less suitable for broadcasting and are unsuitable for mobile services.

Furthermore, because satellites are capable of providing line of sight communications they can use relatively high frequencies and therefore can provide high capacity links over a wide area. In contrast, terrestrial radio systems are limited to either low capacity wide area services or high capacity short distance services.

Satellites can provide communications to any location where there is a ground station, and therefore they are well suited to providing communications across terrain where the political situation or the geography makes the use of cable or microwave links difficult. In addition, ground stations can often be installed quickly since the equipment can be flown in, and this feature makes satellites suitable for providing communications at short notice or for limited periods in the case of special events or in the aftermath of disasters.

Fixed satellite services are however facing increasing competition from cable systems where developments in fibre optics are producing far more substantial reductions in costs than are possible with satellites, thus in the longer term satellites will tend to be used only in situations that exploit their special characteristics. The two main organisations that provide satellite capacity for fixed services are INTELSAT and EUTELSAT.

1–19   The political changes in Eastern Europe are leading to new opportunities for fixed satellite services and VSATs (see para. 1–20 below) as a rapid means of installing a communications infrastructure. Satellites will be used extensively for this purpose in Eastern Germany, and this application will generate additional traffic for EUTELSAT.

Mobile satellite services developed more slowly than fixed services because of technical constraints and because of the high cost of a satellite which can transmit sufficient power for reception by a mobile earth station. The early earth stations used for fixed services had antenna dishes with diameters of the order of 30m which is clearly impracticable for a mobile. However by the early 1970s satellites had become sufficiently large and receiver electronics sufficiently sensitive for telephony communications to ships with antenna diameters of 1 to 2m. These developments led to the formation of INMARSAT the International Mobile Satellite Organisation which provides most mobile satellite services.

Broadcasting satellite services have also developed more slowly than international telephony services because they too have needed small earth stations. The 1977 World Administrative Radio Conference (WARC) prepared a plan for the orbital positions and frequencies to be used for direct broadcasting by satellite (DBS). Each European country was allocated five channels, and the link budgets showed that the larger countries such as France and Germany would need satellite transponders with some 150–200W of power for reception with 90 cm dish antennae and low cost receivers, much greater than the 20 to 30W used for fixed services and difficult to design with the necessary reliability. Such high power systems failed to establish themselves, partly because it became possible to provide coverage of a substantial part of western Europe using the fixed services transponders on the INTELSAT and EUTELSAT satellites. This solution was cheaper and offered the advantage of better international coverage, and so TV broadcasting began in practice by these means, and an independent satellite system called Astra was launched to provide additional capacity.

Satellites are likely to be used increasingly for broadcasting in two roles; they can complement cable systems, by providing programmes to the head ends of cable systems, and they are also well suited to serve homes directly, particularly in areas where cable systems are not available. The development of high definition television and digital television will provide further opportunities for satellites because these applications may require bandwidths that cannot be provided easily within existing terrestrial off-air broadcast networks.   1–20

One application for satellites which first attracted attention during the early 1980s was the provision of business communications within a company using small dish antennae sited on the company's premises. In the USA, these services achieved only very modest success initially; however recently the VSAT (Very Small Aperture Terminal) services have begun to grow more rapidly. The initial services were probably too sophisticated and expensive, whereas VSATs have found their niche in the market by avoiding complexity, concentrating on data and exploiting the broadcasting (point to multipoint) capability of satellites.

The development of satellite services within Europe has been retarded by the existence of PTT monopolies, but with the removal of many of the restrictions as a result of action by the European Commission, innovation in service and further growth in VSAT systems is to be expected.

One important new development is the use of a constellation of low orbiting satellites to provide global coverage to small terminals. Motorola is developing a system of this type called Iridium. There are a number of such systems planned for introduction in the late 1990s, including Iridium, Inmarsat-P, Globalstar and Odyssey. These systems will raise some difficult regulatory issues because they are inherently international and do not fit easily into existing regulatory structures. Developments will be influenced significantly by decisions to be taken by the Federal Communications Commission in the USA.

5. MODERN NETWORKS

The following gives a brief summary and description of the principal newer networks and services.   1–21

ISDN

The Integrated Services Digital Network will provide digital fixed network communications throughout Europe. Its development began in the late   1–22

1970s but it is only recently becoming available. ISDN is not a separate network as it will use some of the same switches and transmission facilities as the PSTN, but it will have its own distinctive digital access standards, its own version of CCITT Signalling System No. 7, and a wide range of supplementary services or features.

There are two types of service. A bearer service is a service for the transfer of information between two network access points, or network to terminal interfaces. For example a switched 64 kbit/s circuit is a bearer service. A teleservice is a description of a complete service provided to a user (rather than to a terminal) by both the network and the terminals. For example voice services and facsimile services are teleservices. It is interesting to note in passing that the distinctions in many licences between voice and data are teleservice distinctions whereas distinctions between circuit switched and packet switched services would be bearer service distinctions.

Within the realms of teleservices and bearer services, the services consist of basic services for straight information transfer, and supplementary services which provide additional functionality, although it is not necessarily true to say that a supplementary service is a value added service in the sense of the legal or licence definition. Basic services may exist on their own, whereas supplementary services exist only in conjunction with basic services.

1–23   Two forms of user access have been defined to date—basic rate access and primary rate access.

Basic rate access is a form of access that is designed to give the user two 64 kbit/s channels, called B channels, and a single 16 kbit/s channel, called a D channel, making a total of 144 kbit/s. The use of the B channels is unrestricted and independent but typically they may be used for circuit switched services (*e.g.* voice), permanently assigned connections and packet switched services, whereas the D channel is used only for packet switched services and signalling between the user and the network. Basic rate access was designed to be capable of being provided over the existing copper pairs in the local network and yet give the user a significantly greater range of services than was possible with the analogue PSTN.

Primary rate access is access at 2 Mbit/s which provides 30 B channels at 64 kbit/s each plus a D channel at 64 kbit/s (higher than the D channel rate for basic rate access). Primary rate access can also provide channels at an intermediate rate of 384 kbit/s called HO or a single channel of 1920 kbit/s called H12 (there is also a USA equivalent called H11 at 1536 kbit/s).

It is important to understand that these channels are access channels over which services (tele-, bearer and supplementary) are provided. The channels are not themselves services and they do not extend unaltered through the public network. The nature of the channels does however limit the services which can be provided on them. For example, whereas the B channel can provide both circuit switched and packet switched services, the D channel can provide only packet switched services because it also carries signalling in packet form.

The Commission is attaching great importance to the provision of ISDN and has issued an ONP recommendation on its introduction, and is funding the development of the necessary approval standards for terminal equipment.

GSM

The pan-European digital cellular radio system is known as GSM after the body called Groupe Speciale Mobile which was initially under CEPT but is now under ETSI, and which was given the task of specifying a European cellular radio system for operation from 1991. (The system is now renamed Global System for Mobiles.) A central requirement for GSM was that it should provide pan-European roaming so that subscribers could use their equipment anywhere in Europe without having to make special arrangements with the local operator. Another requirement was that the efficiency of use of the spectrum should be considerably better than that for analogue cellular radio.   1–24

GSM uses dedicated frequency bands that are common to the different countries in Europe. There are several different power classes for the terminals, ranging from a low power handset to a higher power vehicle installation. The system is designed for a maximum range of 35 km, and like analogue cellular radio the transmitter power is controlled by the base station within the limits of the mobile's class.

In terms of technology, GSM is a most demanding system with the full range of digital techniques, *viz*. equalisation, frequency hopping, sophisticated speech coding, error correction coding, echo cancellation, block interleaving and advanced modulation, being provided to maximise the performance. The degree of processing is such that the battery current drain of the integrated circuits in the mobile is comparable with the current required to provide the RF power for the transmitter.

From the point of view of investment GSM is also most interesting in that by far the main cost has been software and computer aided hardware development, for the system control and the design of the integrated circuits respectively. The cost profile represents the shape of things to come by showing a very high development cost with a relatively low recurring cost for equipment, and with the difference between the two more pronounced than for most earlier systems.

GSM is a two-way mobile telephone system which normally provides contiguous coverage through the use of overlapping radio cells. The available frequency band is divided up between a number of cells which form a repeat pattern, and the band is reused many times as the pattern is repeated across the country. The size of the cells can be varied according to the traffic density with smaller cells and a lower power transmission used in urban areas giving a smaller repeat pattern and greater capacity, and larger cells used in rural areas. Thus in urban areas with dense traffic the cell size is determined by the traffic density, whereas in rural areas it is determined by the propagation limit which in turn is set by the power in the link budget.

Figure 2: Cell Structure (seven-cell repeat pattern)

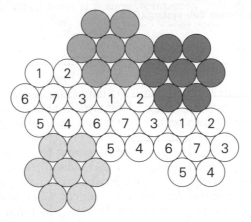

1–25   Communication to the mobiles in each cell takes place through a base station. Base stations can be located centrally in the cells in which case there is one cell per base station, or they can be located at the corner of a cell in which case one base station can serve several cells. The basic cell structure and layout is normally hexagonal and some cells may be divided into three or six sectors. A seven-cell repeat pattern (see Figure 2) is common, although either a four-cell pattern with three-fold sectorisation or a three-cell pattern with six-fold sectorisation may be used in areas of dense traffic. Functionally a sector is very similar to a cell except that the sectors that make up a cell are all fed from the same base station site.

1–26   Figure 3 below shows the structure of the network. Each cell is served by a Base Transceiver Station (BTS) which contains little or no intelligence. A group of BTS are connected to a Base Station Controller (BSC) whose function is to manage the radio frequencies and channels used by the BTS. The BTS are connected to a Mobile Switching Centre, or more accurately a switching centre for mobiles, (MSC). The MSCs are switches with extra intelligence or database facilities having typically five times the processing power that would be found in an ordinary switch. The MSCs together form an intelligent network. Certain MSCs are connected to the PSTN and contain gateway functions for handling calls to and from the PSTN. These MSCs are known as Gateway MSCs (GMSC).

    There are two main types of database concerned with the routing of calls. There is a Home Location Register (HLR) that holds data on all the mobiles that are customers of the network, including information on the services to which the mobile may have access, and the current location of the mobile. Different parts of the HLR may be held at different points in the network, but each mobile has only one file in the HLR.

    The second type of register is the Visitor Location Register (VLR). There are VLRs associated with each MSC and it is common practice for them to be integrated into the switch. Whenever a mobile is active (logged on and able to make or receive a call) most of the data about the mobile that

Figure 3: GSM Network Structure

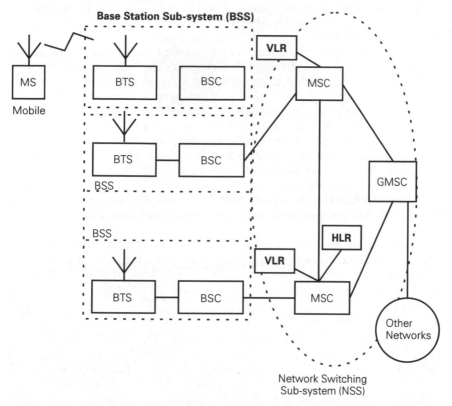

is held in the HLR is downloaded (copied) into the VLR of the MSC in whose area the mobile is. At the same time the HLR is told the exact location of the mobile so that calls can be rerouted to the MSC that is serving the cell where the mobile is actually located.

A feature of GSM that gives great flexibility is that the subscriber's details are held on a smart card and are not stored permanently in the mobile equipment. This means that the subscriber can readily change or borrow equipment. However some networks run an Equipment Identity Register, associated with the HLR, to check automatically that equipment is not stolen.

GSM will provide pan-European roaming enabling a subscriber to be reached or to make calls from his mobile throughout Europe, providing there is local GSM coverage.

Personal Communication Networks (PCN)

As a result of the rapid growth in the use of analogue cellular radio, the U.K. Government chose to stimulate the market by issuing licences for Personal Communication Networks (PCN's) which will bridge the gap in the

1–27

market between GSM networks, as currently envisaged, and the universal mobile telecommunications system (UMTS), which RACE is developing. The standards (known as DCS1800) for these services have subsequently been prepared in ETSI.

Personal communication networks are based on the standards and design work carried out for GSM except that they will use frequencies in the 1.7–1.9 GHz band and employ microcells with sizes ranging from 100 m to 1 km (radius) in areas of dense traffic, and somewhat larger cells elsewhere.

The smaller cell sizes are necessary to ensure that adequate capacity can be achieved from the available spectrum but they also have the advantage that they allow low power handsets to be used, and the handset sizes should be comparable to the early Telepoint sizes (about 200 cm^3). Two classes of handset have been defined with peak powers of 1W and 250 mW.

Many of the developments that will form part of PCN will also be undertaken by the GSM operators at 900 MHz and the services at different frequencies will compete with each other to a significant extent.

Universal Mobile Telecommunications System (UMTS)

1–28   As a part of the European Commission's RACE programme a study is being made of the design of a universal mobile system for introduction towards the end of the 1990s. The aim of UMTS is to produce a personal communicator that will replace both GSM and DECT. The service will probably be in two parts, one aimed at voice and low to medium speed data, and the other at broadband services for which the use of millimetre waves (30 GHz upwards) is being studied. Progress in this area of the RACE programme is slow because the GSM development programme is currently using most of the available research and development resources. It is possible that the objectives of UMTS will be met in practice through further developments of GSM.

DECT

1–29   ETSI has prepared standards for a European digital cordless telephone system known as DECT. DECT will be used for cordless telephones, cordless PBXs and Telepoint applications. It is expected that cordless telephones will be used very widely both by residential and business customers. The potential of the residential market has already been demonstrated by the sales of analogue cordless telephones. The cordless PBX market is only just beginning but is expected to provide rapid growth serving both telephony and data applications.

DECT technology could be used as an alternative to wires for the delivery of services to residential customers. The local operator would provide DECT base stations in the street and the customers would access the network using their DECT handsets.

Universal Personal Telecommunications

Universal Personal Telecommunications (UPT) is a new service that will   **1–30**
enable a subscriber, who is identified by a personal number, to make and
receive calls at any terminal. The service is currently being defined and is
not expected to be offered before 1995 at the earliest, and then only in
certain countries. A pan-European service is not expected until the late
1990s.

In both fixed services and first generation cellular services, the subscriber
is associated with a particular terminal or line. In UPT this association is
broken and the subscriber is identified by a new UPT number which may
be used in a temporary association with any terminal, and thus provides
personal (not terminal) mobility. All billing is carried out with respect to
this UPT number, and so the subscriber is billed at his home address for
all his calls, irrespective of where he makes them. The UPT service may be
provided across multiple networks including networks of different types
(*e.g.* fixed and mobile).

The basic aims and principles of UPT are to:

provide personal mobility by enabling a UPT subscriber to make and
receive calls from any terminal on a global basis (not just within the
same network);
charge on the basis of UPT identity rather than terminal identity;
provide standardised access procedures across multiple networks; and
provide choice and flexibility with regard to the service features selected
by a subscriber.

The concept of the service is as follows. The customer subscribes to the   **1–31**
UPT service of a UPT service provider who allocates him a UPT number.
The subscriber selects the features that he wishes to have, and this selection
is stored in a personal user profile. He also selects the method of access
and authentication that he wishes to use.

The subscriber's UPT number will normally be shown and identified clearly as such on his business card, which should also show his default or
home location.

When the subscriber is away from his normal telephone, he may register
his current location via any terminal of any network, and receive calls or
make calls from that terminal. Registration involves an authentication procedure. Initially this procedure is likely to be limited to the entry of his
UPT number and a secret personal identification number (PIN), using terminals with a tone dialling (DTMF) facility. However the wiping of a magnetic card in a card reader may also be used, and another alternative is to
use a longer and more secure authentication number generated by a small
hand held tone generator. In the longer term, more sophisticated interroga-
tion response systems may be used.

There will be several options for registration at a local terminal. Registration may be set to enable incoming or outgoing calls or both, and registration may be indefinite or for a set period of time. Several UPT subscribers
may be registered at the same terminal simultaneously. However a facility

will be provided to enable the owner of a terminal to reset or annul the registration, so that, for example, he will not receive incoming calls for a visitor who registered at his terminal but has left subsequently without cancelling the registration. Any new registration at any location will automatically cancel a previous registration by the same subscriber. When a UPT subscriber registers at a terminal, he may select whether or not he should give any identification (*e.g.* account number and PIN) when receiving incoming calls; identification could be useful if several people have access to the same terminal.

1–32   The charging and billing arrangements for UPT will need careful planning. The intention is that the subscriber will be billed only by his UPT service provider, not by the operators of the networks that he may use. All billing and charging will be based on the UPT number which should be clearly recognisable so that callers to a UPT number can be aware that they may be paying special tariffs. Callers to a UPT number will be charged only for successful calls.

UPT is expected to be an important service from the late 1990s onwards. Both UPT and radio based mobile communications will provide mobility and in the longer term there will be an element of competition between them. However the two services are fundamentally different in that a mobile service provides mobility between the terminal and the network, whereas UPT provides mobility between the subscriber and the terminal. The subscriber who needs to communicate on the move needs a mobile service, the subscriber who frequently works in different places will probably prefer UPT because it will enable calls to his home fixed number to be redirected cost effectively to his current location. Some subscribers will need both services.

Both UPT and radio based mobile services depend on intelligent networks, so in the long term the relative costs will depend on the costs of radio links and physical exchange lines.

From the point of view of competition, UPT decouples a person's number from the identity of the physical network provider, but this does not solve the barrier to competition of an association with a particular number, because the UPT number is likely to be tied to the identity of the UPT service provider. Thus the problem is moved not solved.

Broadband

1–33   A major part of the Commission's RACE programme is devoted to the development of broadband multiservice networks. These networks will use a combination of new switching technology based on rings and ATM multiplexing techniques in order to provide services with a wide range of fixed and variable bit rates. These new networks will need a new signalling system and work on this is beginning in ETSI.

The traditional distinction between public and private networks is disintegrating and will require a new approach to transmission planning and network quality.

The growth of VSATs and the introduction of low earth orbit systems will make communications increasingly international and it will be difficult to maintain the traditional model of international communications being the interconnection of "national islands". The development of intelligent networks will require detailed regulatory intervention if the full possibilities for competition in the provisions of features and services is to be realised. Increased capabilities of networks, especially intelligent networks with stored information, will raise important new issues of privacy.

CHAPTER 2

Evolution of Government Policy

1. EARLY HISTORY

Regulation of U.K. telecommunications began about the same time as the invention of the telephone itself in 1876. The pace of development of the technology after its invention was remarkable even by today's standards. By the mid-1880s switchboards had evolved to a point where hundreds and even thousands of subscribers could be handled. Private companies in London and in a number of provincial towns introduced telephone exchanges and the Post Office's telegraph "control switchings" in several towns were converted to telephone exchanges.

2–01

As entrepreneurial activities in the establishment of telephone systems began to develop and necessarily impinge on the Post Office's telegraph monopoly, the question arose as to whether such activities amounted to an infringement of the Postmaster-General's "exclusive privilege" of transmitting telegrams, as conferred by the Telegraph Act 1869. The issue came before the High Court, Exchequer Division, in 1880 and the recital of facts in the case gives a fascinating insight to the workings and social impact of the original telephone.[1] The Agreement made by the Edison Telephone Company with its customers (to whom it leased the necessary equipment) provided that the company would:

> "Upon the request made through the said telephone at any time during the continuance of this agreement, between the hours of 9 a.m. and 6 p.m., Sundays excepted, put the lessee in telephonic communications with the telephone of any other subscriber to the said exchange whose wire is free."

The court found that the scope of the 1869 Act was such that any apparatus for transmitting messages by electric signals must be a telegraph whether a wire was used or not. This would extend to "electric signals made, if such a thing were possible, from place to place, through the earth or the air." A somewhat prophetic statement.

2–02

[1] A–G v. Edison Telephone Co. of London (1880) 6 Q.B.D. 244.

The Edison Telephone Company argued unsuccessfully that the telephone and telegraph were significantly different technically and that the Postmaster-General's exclusive privilege should not extend to something not even invented in 1869. The court held that a conversation through a telephone represented a message or communication that was in fact transmitted by a telegraph.

2. END OF THE POST OFFICE "EXCLUSIVE PRIVILEGE"

2–03   The exclusive privilege of the Postmaster-General over the telephone continued for over half a century until 1981. The British Telecommunications Act 1981 made the postal and telecommunications services of the Post Office the responsibility of two separate undertakings. Postal services were retained by the Post Office but the 1981 Act transferred to British Telecommunications the exclusive privilege of running telecommunication systems. Having given with one hand, however, the 1981 Act gave Government the opportunity to take away with the other: this exclusive privilege was said not to be infringed by anything done under a licence granted by the Secretary of State. The political decision had by that time been taken to use this power to authorise one additional operator to compete with BT and a private joint venture company, Mercury Communications Limited (now majority owned by Cable & Wireless plc) was established and licensed to operate a new public voice and data system based entirely on digital technology.

In a move which was to be echoed seven years later by the European Community with its Terminal Equipment Directive,[2] the 1981 Act also empowered the Secretary of State to permit persons other than BT to supply and maintain PABX and other terminal equipment, which he did in 1983.

At the same time the Secretary of State's power to license the running of new telecommunication systems was utilised in order to provide the provision of value added network services (VANs) over the public networks. In 1983 licences were also granted to two operators of the first ever cellular radio mobile telephone systems in the U.K., Racal Vodafone (now Vodafone) and Telecom Securicor (Cellnet).

3. LICENSING OF COMPETITION

2–04   The licences granted under the 1981 Act, notably that to Mercury Communications Limited as a public operator permitted to compete with BT, for example in long-distance services (albeit initially using leased lines—Mercury did not commence switched service until 1986), were the first and at that time the most dramatic moves towards services liberalisation in

---

[2] Directive 88/301 on competition in the markets in telecommunications terminal equipment; [1988] O.J. L131/73.

Europe. It was to be nearly 10 years later (in 1990) that regulations were enacted at an E.C. level[3] liberalising the provision of value added services to the same extent. Yet these stirrings of liberalisation apparently so vital, were with hindsight something of a laboratory attempt at fostering competition. The main problem was that there was no regulatory framework to protect and nurture that competition in the face of a monopoly in both services and equipment supply. It was to be the privatisation of BT that would prove the catalyst for the sophisticated regulatory scheme we have today.

4. Duopoly Policy

By 1983 the Conservative Government had determined to make the privatisation of British Telecom its first "flag ship" privatisation of the nationalised industries. It was accepted that this bold move would necessitate a new code governing activities in the marketplace; U.K. competition laws were weak and ineffective, although this was never officially acknowledged as the reason for developing an industry-specific set of competition rules. In any event the industry, including the public operators, investors and potential investors were all demanding that future policy for competition in the sector be clarified. Therefore on November 17, 1983 the then Minister for Information Technology, Kenneth Baker M.P., made a statement on future competition policy in telecommunications to the Standing Committee on the Telecommunications Bill[4] the fundamentals of which were:

2–05

    (a) The Government did not intend to license operators other than BT and Mercury to provide the basic telecommunication service of conveying messages over fixed links, whether cable, radio or satellite, both domestically and internationally, during the period to November 1990.

    (b) Public systems were to be interconnectable: the Government's stated intention was that any subscriber to one public telecommunication system should be able to call any subscriber to any other public system. The licences of the operators of such systems would provide the rights and obligations for such interconnection and in particular BT's licence would oblige BT to connect its system to any other system where the operator of such system was licensed to make such connection. Where the parties could not agree on the terms of such connection these were to be determined by the Director General. (These provisions now appear in Condition 13 of BT's licence; for a fuller discussion of the scope of Condition 13 and the determinations issued by the Director General under that Condition see para. 2–07 and Chapter 5, paras. 5–43 and 5–52 below.)

---

[3] Directive 90/388 on competition in the markets for telecommunication services; [1990] O.J. L192/10.
[4] November 17, 1983, House of Commons Official Report, Standing Committee A, p. 686.

(c) Simple resale[5] of capacity over circuits leased from BT and Mercury would not (with certain exceptions, eventually in relation to data services) be permitted in the period before July 1989, which corresponded to the original period (which has since been extended) in BT's licence for continued application of the RPI-x formula.[6]

5. NEW TELECOMMUNICATIONS POLICY

2–06   The essential policy aims of the Telecommunications Act 1984 were the privatisation of BT by the sale of just over 50 per cent of its equity, the abolition of its exclusive privilege and its replacement by a public operator's licence, the termination of BT's right to grant licences and further liberalisation by widening the categories of licensable system operators, and reducing the involvement of BT in the approval of apparatus. Perhaps most important of all, the 1984 Act provided for the first of a new breed of privatised utility–style regulators, coupled with a new semi-independent regulatory authority responsible for protecting the interest of consumers and promoting fair and effective competition between operators. This was the first express recognition of the now well-accepted European principle that *operation* should be separated from *regulation* so that those competing in the marketplace should not find they are regulated by the very monopoly organisation with whom they are trying to compete.

At this time (the mid-1980s), this worthy regulatory principle was only half met, essentially because there was so much residual know-how in the incumbent public operator, BT, which could not immediately be replicated by a new regulator. Even today, one very important facet of telecommunications regulation, numbering planning and allocation has only recently been removed from the practical administration of BT and put under the full control of OFTEL. In the early 1980s, whilst OFTEL was ostensibly in charge of approval of telecommunication apparatus for attachment to the public network, in practice much of the important evaluation work, on which the approvals were founded, was carried out by a special BT division called "Teleprove". It was only gradually during this decade that independent testing houses were set up to provide the necessary formal separation of BT as an operator from this regulatory process.

6. REGULATORY AUTHORITY

2–07   The 1984 Act provided that the regulatory authority, OFTEL, should be administered by a new officer, the Director General of Telecommunications, who was first appointed by the Secretary of State for a three-year period from July 1, 1984. The first Director General was Professor (now

---

[5] See para. 2–09.
[6] See para. 5–20.

Sir) Bryan Carsberg, who continued in office until June 30, 1992. His eventual successor, Mr Donald Cruickshank, took up office on April 1, 1993.

The duties assigned to the Director General are set out in section 3 of the Telecommunications Act 1984. Essentially he is to exercise his powers so as to secure the provision of telecommunication services in order to satisfy all reasonable demands and so as to secure that the persons providing such services are able to finance them. More specifically he is bound to promote the interests of consumers and to maintain and promote effective competition in telecommunication activities. For a detailed discussion of the Director General's powers and duties see below.[7]

In 1985 the Director General had his first real opportunity to put his own stamp on U.K. regulation through the referral to him by Mercury of its request for him to determine the terms and conditions applicable to interconnection of Mercury's network with that of BT. In setting the charges that should apply for BT's connection of its systems to those of Mercury and for the conveyance of Mercury calls over its network, the Director General had to perform a delicate balancing act in both promoting competition and at the same time allowing the proper recovery by BT of its costs of providing services, all within the very detailed parameters set out in Condition 13 of BT's licence. This process and the Director General's 1993 re-determination of connection and conveyance charges are discussed in detail below.[8]

7. DEVELOPMENT OF COMPETITION

From 1984 until 1990 the prime areas of active regulation were under-   2–08
standably the relationship between BT and its only public operator competitor, Mercury, the gradual liberalisation of non-voice services (value added and data communications) and the introduction of competition in specialised areas such as cellular mobile, paging and cable television.

Throughout this period the Government remained committed to opening up competition in all but voice services, but was still dogged by the grey distinction between value added services, which involved the computerised manipulation and processing of data, and basic data services, essentially the mere transmission of data from point to point. Gradually, as the operation of managed data networks emerged and the Government consulted the service providers and users on how best to regulate the industry, the day dawned when the Government realised that the regulatory distinction between basic and value added services could no longer be preserved and sensibly enforced. It thus took the policy decision to license the provision by all-comers of everything except simple resale (of voice and data) and basic service, leaving voice telephony and basic telex as the final preserve of BT and Mercury, the PTOs. This was accomplished in the Value Added and Data Services (VADS) Licence, a class licence, issued in 1986. This

---

[7] See para. 3–03 et seq.
[8] See paras. 5–40 to 5–54.

licence, though still extant, is effectively defunct, having been superseded by the more liberal Telecommunication Services Licence.[9]

2–09   At this time all licences, both those governing the running of domestic or branch systems (then known as the Branch Systems General Licence) and class licences like the VADS licence, prohibited the provision of simple resale services. Simple resale, in U.K. regulatory terms, involved the use of a leased line (private circuit) to provide services which passed over the public switched network at each end of the call—in other words both at its origination and its delivery. Mr Baker's 1983 statement had indicated that simple resale would not be permitted until after July 1989 and essentially this had been to enable BT in the interim to rebalance its tariffs for leased lines and long-distance switched services, in order to be able to compete effectively with resellers.

8. ABOLITION OF DOMESTIC RESALE RESTRICTION

2–10   In 1989 therefore the restrictions on simple resale for national traffic were finally removed. Although this was welcome news to the operators, the effects of rebalancing must have been quite significant, for in the period that followed domestic resale of telecommunications never took off. Even today the prime interest in resale is with respect to international telecommunications, where competition is limited to certain routes designated by the DTI, but where the margins are much more attractive.

9. END OF DUOPOLY

2–11   In November 1990 the seven-year duopoly period promised by the Government in 1983 finally expired and in that same month the Secretary of State presented to Parliament a consultative document "Competition and Choice: Telecommunications Policy for the 1990s".[10] This document canvassed a number of options for further liberalisation of the U.K. telecommunications sector and invited comments from the industry, from users and from representative bodies and other organisations. As a result of the consultation the Government announced in a White Paper,[11] published in March 1991, that it had decided to end the duopoly policy and that it would consider on its merits any application for a licence to offer telecommunication services over fixed links within the U.K. Licence applications would be considered against the provisions of the Telecommunications Act 1984: surprisingly, perhaps no specific guidelines were given in the White Paper concerning the manner in which applications should be made nor as

[9] See Chap. 6.
[10] Cm. 1303–13 November 1990.
[11] Competition and Choice: Telecommunications Policy for the 1990s—CM1461.

to their suggested content.[12] According to the Government licence applicants should be left as free as possible to make applications that they judged
best reflected the needs of customers. Above all in granting licences the
Secretary of State was required to exercise his discretion consistent with
the duties imposed on him under section 3 of the 1984 Act and also to
consider advice from the Director General.

Thus in one sentence of the White Paper the Government had wrought a
massive change in U.K. telecommunications policy, though with one major
qualification. This was that with respect to international facilities (satellite
or cable), the Government determined that it would not for the time being
grant international operating licences (such as were enjoyed by BT and
Mercury) "and that it would therefore be premature at [that] stage to invite
applications for such licences". This was in contradistinction to its proposal
in the 1990 consultative document where it stated that it was inclined to
consider applications for international telecommunications licences. The
main reason appears to have been that the Government had become
increasingly sensitised to the difficulties faced by U.K. companies in
obtaining similar authorisations overseas, particularly in the U.S.[13] In the
consultation period doubts had been expressed by commentators as to
whether U.K. regulatory controls alone could ensure the development of
effective competition if U.K. companies were not able to obtain equivalent
licences overseas and if overseas regulators did not share the views of their
U.K. counterparts about what constituted anti-competitive behaviour. Clearly the Government viewed these comments sympathetically and this led
to the conclusion not to invite international operating licence applications
for the time being.

2–12

This issue of reciprocity from other countries with respect to the right
to provide international telecommunication services arose again with
respect to the use of international leased circuits. Here the Government
determined in the White Paper that the existing restrictions on the provision
of all telecommunication services via such circuits (including live speech
and telex) should be removed except that international simple resale[14]
would only be permitted between the U.K. and those countries whose regulatory regimes allowed an equivalent freedom to provide such services in
the reverse direction. Currently, at time of writing, the U.S., Australia,
Canada and Sweden have been so designated.

---

[12] Some guidance can be derived from the 1990 consultative document (see n. 10) which stated that the
Government would take into account the service proposed, the ability of the applicant to fulfil its plans,
the extent to which it might result in more effective competition and the environmental implications of
any request for code powers. See further Chap. 4.

[13] Notably with respect to the application of ss. 214 and 310 of the Communications Act 1934, applicable
to licences for international resale and for carriers utilising radio communications respectively.

[14] International simple resale is the use by a service provider of an international leased circuit to provide
a telecommunication service to an end user whereby the calls passed over that circuit originate on and
terminate through a public switched network at each end. The strict definition is found in the TSL and
various individual operators' licences.

CHAPTER 3

Regulatory Structures, Institutions and Powers

1. The Telecommunications Act 1984: Main Objectives and Functions

The Telecommunications Act 1984 swept away the 1981 Act and estab-   3–01
lished a completely new institutional framework which became the model
for all subsequent U.K. privatisations.

Under the 1984 Act, the Secretary of state appointed an officer, the Director General of Telecommunications, and there were assigned both to the
Secretary of State and the Director General duties to exercise their functions
in a manner best calculated:[1]

"(a) to secure that there are provided throughout the United Kingdom,
     save insofar as the provision thereof is impracticable or not reason-
     ably practicable, such telecommunication services as satisfy all reas-
     onable demands for them including, in particular emergency services,
     public call box services, directory information services, maritime ser-
     vices and services in rural areas; and
 (b) without prejudice to the generality of paragraph (a) above, to secure
     that any person by whom any such services fall to be provided is
     able to finance the provision of those services."

The specific functions and responsibilities assigned to either the Secretary
of State and/or the Director General include—

* Licensing of telecommunication systems: the Secretary of State (the
  Director General may license systems with the consent of the Secretary
  of State, but to date no such consent has been given).
* Modification of licence conditions: the Director General.
* Enforcement of licence conditions: the Director General.

---

[1] s. 3(1) of the Telecommunications Act 1984.

- Approval of contractors and apparatus for the purposes of valid connection to public telecommunication systems: the Director General.

The Director General has other ancillary functions including the responsibility to keep registers of licences of approved contractors and approved apparatus, to review and collect information as to activities connected with telecommunications, and to publish appropriate information and advice for consumers and other interested persons.

In terms of regulating competition the Director General has a responsibility to investigate complaints about the provision of services and the supply of apparatus and, in addition to his powers to enforce licence conditions, has the ability to implement certain provisions of the Fair Trading Act 1973 and the Competition Act 1980 in relation respectively to monopoly situations and anti-competitive practices. These are further discussed below.[2]

2. Interaction of Director General and Secretary of State

3–02   It is not always appreciated that the Director General's functions do not include the formulation and development of policy with respect to competition. This was most graphically illustrated in the 1991 White Paper when the Secretary of State himself announced his intention to grant licences to applicants to provide all manner of telecommunication services nationally (*i.e.* within the U.K.) "on their merits". This is not to say, however, that the Director General cannot influence policy. The Secretary of State turns to him for recommendations regarding telecommunication licensing and much of the White Paper reproduced the Director General's proposals with respect to the control of BT's tariffing, contributions towards the financing of its activities and the terms on which interconnection between public operators should be regulated; his proposals were later subsumed into modifications to BT's licence agreed between the Director General and BT.

The ways in which the Secretary of State and the Director General's powers are circumscribed by law and their duty to follow "due process" are discussed below.[3]

3. Powers and Duties of the Regulator

3–03   The regulatory framework is essentially composed of the various individual licences (the most important of which are those granted to the PTOs) and class licences, the powers of licence modification and enforcement conferred on the Director General, the general powers of the Director General

---

[2] See Chap. 10.
[3] See paras. 3–05 to 3–09.

exercisable under the Fair Trading Act 1973 and Competition Act 1980,[4] and the overlay of the 1984 Act itself. This overlay is vitally important particularly in the context where the Director General is discharging his duties and functions. Many of these confer upon him a considerable degree of discretion but this must be exercised within the ambit of his duties under section 3(2) of the 1984 Act which, although subordinated to section 3(1),[5] provide that both the Secretary of State and the Director General have a duty to exercise their functions:

"in the manner which [each] considers is best calculated:

(a) to promote the interests of consumers, purchasers and other users in the United Kingdom (including in particular those who are disabled or of pensionable age) in respect of the prices charged for and the quality and variety of, telecommunications services provided and telecommunication apparatus supplied;

(b) to maintain and promote effective competition between persons engaged in commercial activities connected with telecommunications in the United Kingdom . . .

(g) to enable persons producing telecommunication services in the United Kingdom to compete effectively in the provision of such services outside the United Kingdom;

(h) to enable persons producing telecommunication apparatus in the United Kingdom to compete effectively in the supply of such apparatus both in and outside the United Kingdom".

The duty on the regulator to maintain and promote effective competition is reflected in the operators' licences themselves. For example Condition 13.5 (f) of BT's licence granted under section 7 of the 1984 Act, dealing with the terms of interconnection of BT's systems with those of another operator, provides that when the Director General comes to make a determination of such terms he must ensure "that the requirements of fair competition are satisfied".

BT's licence is particularly important as one of the building blocks of regulation because it is, by virtue of BT's size and importance and monopoly position, relevant to the community of users and all other operators who must necessarily connect with BT's systems in order to access the generality of customers. Much, therefore, of the U.K.'s regulatory rules are embodied in BT's licence; where they are relevant to non-dominant operators the essential principles first enshrined in this licence are often to be found in other operators' licences.

Policy is also reflected in the licences granted under the 1984 Act, for example the restrictions on international simple resale (ISR)[6]: this is specifically barred from provision under both the Self-Provision and Telecommunication Services Class Licences, the requirement being for the grant of individual ISR licences which, for the time being, will only confer the right

3–04

[4] See Chap. 10.
[5] See para. 3–01.
[6] See para. 5–81.

to convey voice telephony messages to and from certain designated countries (currently the U.S., Canada, Sweden and Australia).

4. The Scope for Regulatory Challenge: Judicial Review and Declaratory Relief

3–05   The exercise of the Director General's powers and duties is governed by
the same laws and procedures as apply generally to government-appointed
officials. If he fails to carry out any of his statutory duties he may incur
civil liability for any damage thus caused. On the other hand his wrongful
failure to exercise his discretion could be challenged by an order of mandamus to compel performance of his duty to act.

Again as with any officer of the Government, a person claiming to be
aggrieved by any act or omission of the Director General would be able
to petition either House of Parliament on the matter. Members of the public
may also address complaints to their M.P. who can in turn refer the matter
to the Minister responsible for it to be pursued by means of a parliamentary
question or referred for investigation by an ad hoc enquiry or tribunal.

Lastly it should not be forgotten that complaints by persons claiming to
have sustained injustice in consequence of maladministration of a Government department such as OFTEL may be forwarded by an M.P. for investigation and report by the Parliamentary Commission for Arbitration—the
Ombudsman.

Judicial Review

3–06   In addition to orders of mandamus, the decisions or determinations of the
Director General could in themselves be challenged in either of two ways,
namely:

(i) prohibition, where an order is sought preventing the particular act
taking place; and
(ii) certiorari, where an order is sought to quash the particular act.

However, the courts are understandably reluctant to interfere in the
decisions of executive authorities given in exercise of their discretionary
powers. Essentially the grounds upon which they are prepared to overturn
such decisions are any of the following, namely:

(a) illegality;
(b) irrationality; or
(c) procedural impropriety.[7]

Illegality would essentially have to be constituted by the Director General
acting *ultra vires* or failing to have regard to relevant considerations or

---

[7] See Lord Diplock in *C.C.S.U. v. Minister for Civil Service* [1985] A.C. 374 at 410.

taking into account irrelevant considerations.[7a] Procedural impropriety would involve the Director General in having failed to follow specified procedures set down in the 1984 Act or in any telecommunication licence, whereas irrationality would equate with "Wednesbury unreasonableness"[8] in other words and in general terms, when he exercises his powers the Director General must act reasonably.

The above three grounds on which the courts can intervene are not neces-   **3–07**
sarily exhaustive: lack of proportionality may be another basis for intervention.[9] Generally the courts are prepared to intervene to correct errors of law by public bodies, including regulatory authorities. A further ground for intervening would be the failure by a regulator to act in accordance with the principles of natural justice.

When the first edition of this book was written none of the decisions or determinations of the Director General had been challenged in any of the above mentioned ways. However since that time we have had the *Chatlines* case (discussed in para. 3–20 below and Chapter 4). Interactive services like, para. 4–09 chatlines have spawned a number of judicial review applications, albeit unsuccessful. In *Maystart Limited v. Director General of Telecommunications* February 17, 1994, unreported, BT had introduced a scheme for customers wishing to access sex lines to opt in for such services by use of a personal identification number; the proposal was that a PIN would be provided free on request to any adult BT customer who would then use it when wishing to gain access to the particular service. One provider of such services, supported by the Telephone Entertainment and Service Providers Association Limited, attempted to persuade the Director General that by so doing BT was in breach of Condition 15.4(a) and Condition 17.1 of its licence. The Director General refused to interfere and somewhat belatedly the service provider applied for judicial review of the decision. In the application it was also alleged that the Director General had failed to exercise his statutory duty under section 3(1) of the Telecommunications Act 1984 so as to exercise his functions to secure services such as would satisfy "all reasonable demands for them".

Maystart failed at the Court of First Instance and in the Court of Appeal   **3–08**
met short shrift. That court considered that as the Director General's predecessor had previously referred these types of services to the MMC and had taken the view, supported by the MMC, that BT customers required some protection from unauthorised access to such services, it was entirely reasonable for the Director General to decide that BT should have increased control over access to such services; there were therefore no grounds to support

---

[7a] See also *R. v. HM Treasury, ex p. British Telecommunications Plc, The Times*, December 2, 1993, C.A., where BT was seeking to quash provisions of The Utilities Supply and Works Contract Regulations 1992 in implementation of Directive 90/531—the Utilities Directive—and Directive 92/13—the Remedies Directive; BT failed in its application to prevent practical application of the regulations pending a preliminary ruling of the ECJ.

[8] *Associated Provincial Picture Houses Ltd v. Wednesbury Corporation* [1947] 2 All E.R. 680. See also *Preston v. IRC* [1985] 2 All E.R. 327 and *Wheeler v. Leicester City Council* [1985] 2 All E.R. 1106.

[9] See *O'Reilly v. Mackman* [1983] 2 A.C. 237.

the view that the Director General had failed in the exercise of his duties and discretions. On the Condition 15 and Condition 17 questions the court declined to find that the Director General could have considered the opt-in scheme unreasonable and/or unduly discriminatory, casting doubt (although *obiter dicta*) on the providers of the services in question even falling within the scope of a "service provider" for the purposes of Condition 15 as well as on the service provided by BT constituting one provided in accordance with an obligation imposed by Condition 17.

Declaratory Relief

3–09   The litigation launched at the end of 1993 by Mercury against the Director General and BT, following his December 2, 1993 determination with respect to the charges payable by Mercury to BT for interconnection, was not an example of Mercury seeking judicial review of an administrative act (*i.e.* the determination), but rather an attempt through the Commercial Court to obtain a judicial ruling on the meaning of BT's licence condition governing interconnection.[10]

These proceedings were instituted by Mercury on December 17, 1993 by its issue of an Originating Summons in the Commercial Court seeking declarations as to the effect of Conditions 13.5 and 13.5A in BT's licence. The declarations sought concerned the manner of determination of "cost" in this specific context, with particular reference to connections, capacity and conveyance of messages, the meaning of the term "relevant overheads" used in Condition 13.5a and 13.5A.3(a), and the way in which the charges payable by an operator to BT should properly reflect the cost incurred in providing interconnection facilities. The Director General and BT first sought to dispose of the case on jurisdictional and procedural grounds, by claiming that the Originating Summons should be struck out essentially on the following bases, namely that:

(a) under the scheme set up by the 1984 Act and the licences granted to BT and Mercury, and more specifically by the existing interconnection agreement between BT and Mercury, the task of fixing the charges to be paid by Mercury had been entrusted to the Director General, and not to the court, and therefore it would not be appropriate for the court to take on itself the task entrusted to the Director General;

(b) it is not the practice of the court to grant relief by way of declaration where the issue sought to be raised is hypothetical or a future question and there is no current dispute; and

(c) because of the position of the Director General, who was only there to exercise powers conferred on him by statute, any issue that arises may be an issue of public law, which can only be taken to the courts—if at all—by an application for judicial review in the Crown

---

[10] Condition 13; see further para. 5–40.

Office list and not by an Originating Summons in the Commercial Court.

At first instance, before Longmore J., he found for Mercury and against the Director General and BT on each of these three main issues. The Director General and BT appealed to the Court of Appeal, where the decision was reversed by a 2 to 1 majority (Hoffmann L.J. dissenting). Mercury appealed to the House of Lords who upheld the First Instance decision, rejecting the defendants arguments on all the above issues. The Lords particularly emphasised the discretion exercised by the trial judge and the fact that his decision should stand unless the arguments were clearly in favour of a different result on appeal. See further para. 5–54 below.

5. Regulatory Foundation: Prohibition of Unlicensed Running of Systems

U.K. regulation of telecommunications essentially starts with a general prohibition. That is in section 5 of the 1984 Act which provides that (subject to certain limited exceptions—see below) a person who runs a telecommunication system within the U.K. shall be guilty of an offence unless he is authorised to run the system by a licence granted under section 7. Even though a system is properly authorised in this way, the applicable licence will not prevent the commission of an offence by the person running the system (variously referred to as the "licensee" or the "operator") if another system or apparatus is connected to the licensed system or telecommunication services are provided by means of that person's system which in either case are not authorised by the licence to be so connected or provided (section 5(2)). This would have the effect that if the licensee connects to its licensed system a system which is licensed to be so connected but through which unauthorized telecommunication services are provided, then although the person running that other system is guilty of an offence under section 7, the licensee is also guilty of an offence, whether or not proceedings are taken against that person. Each offence is subject to the defence that the person charged "took all reasonable steps and exercised all due diligence to avoid" its commission.

3–10

A person guilty of any of these section 7 offences is liable:

(a) on summary conviction, to a fine not exceeding the statutory maximum; and
(b) on conviction on indictment, to a fine.

Currently the statutory maximum is £2000 and the maximum fine on indictment is set according to the offender's capacity to pay.

Since the unauthorised running of a telecommunication system can result in the commission of an offence, it is important to understand what constitutes a telecommunication system. The word "system" is not defined either

3–11

in the Telecommunications Act 1984 or in any of the licences issued under it, but "telecommunication system" is defined as follows, in section 4(1) of the 1984 Act:

> ". . . a system for the conveyance, through the agency of electric, magnetic, electro-magnetic, electro-chemical or electro-mechanical energy, of—
>
> (a) speech, music and other sounds;
> (b) visual images;
> (c) signals serving for the impartation (whether as between persons and persons, things and things or persons and things) of any matter otherwise than in the form of sounds or visual images; or
> (d) signals serving for the actuation or control of machinery or apparatus."

A message could be conveyed by all kinds of means but, as can be seen, a distinguishing characteristic of a telecommunication system under the 1984 Act is that its messages require an input of the type of energy specified in section 4(1). In a typical case it thus seems probable that it would be the person who controls the apparatus permitting such energy to enter the system who would be held to be running that system. Moreover the 1984 Act itself tacitly endorses this approach in one set of circumstances referred to in section 4(2) of the Act.

Although not a general authority on the point, this section provides:

> "For the purposes of this Act telecommunication apparatus which is situated in the United Kingdom and—
>
> (a) is connected to but not comprised in a telecommunication system; or
> (b) is connected to and comprised in a telecommunication system which extends beyond the United Kingdom,
>
> shall be regarded as a telecommunication system and *any person who controls the apparatus shall be regarded as running the system*." (Emphasis added.)

3–12   In general, therefore it seems fair to assume that any person controlling the apparatus comprised in a telecommunication system is running that system, but in any case where the question is crucial, an appraisal of all relevant factors, including not only the person controlling the apparatus, but also its ownership, the premises where it is located and the person sending messages by means of the apparatus, should be carefully made.

OFTEL have given their own interpretation of the term "to run a telecommunication system", in their guidance notes on the SPL and TSL class licences. Acknowledging that the meaning of "run" in this context had not been tested by the courts and so only informal guidance could be given, OFTEL opined that " 'run' does not refer to the day-to-day operation of a system. It refers rather to authority over the system, in particular to control over how the system is made up and how, and for what purposes, it is to be used". It will be noted that OFTEL also acknowledge the relevance of "control" but down-play the significance of day-to-day operation of a system. In my opinion the responsibility for day-to-day operation is a highly relevant, even though not determining, factor.

The specific reason for the inclusion of section 4(2) of the 1984 Act was primarily to prevent circumvention of the licensing requirement by the use of apparatus, such as a satellite ground station, connected purely internationally, for example by wireless telegraphy, to an orbital satellite. The system in which such apparatus would be comprised would extend beyond the U.K. and, but for the wording of section 4(2), would have fallen outside section 5 of the 1984 Act.

As is reasonably evident from the above, even the humble telephone (*i.e.* the instrument itself), certainly with its associated wiring, is a telecommunication system requiring a licence under the 1984 Act. This licence (the most basic) would normally be the SPL, a class licence, discussed in Chapter 6.

Types of Licence

There are three possible types of licence which may be granted under the   3–13
Telecommunications Act 1984 (s. 7(3))—a licence granted to all persons (general), to persons of a particular class (class) or to a particular person (individual). A person running a system in a manner compliant with the terms of a class licence will automatically benefit from its validation by the licence, without further formality; no notification or registration procedure is involved. However where a licensee breaches any condition of the licence the [Director General] may by notice revoke its application to that particular person. (Particular class licences (the TSL and SPL) which are discussed in detail in Chapter 6.)

Logically Separate Systems

The same apparatus operated in different modes under software controls   3–14
may constitute more than one separate system licensable as such. Such systems are known as "logically separate" telecommunication systems.

This situation and the licensing status of logically separate systems is discussed in the Director General's statement of October 26, 1990. This statement gives as an example two networks being run by a company under different licences where the same switch is used to control both. The Director General's opinion here is that the systems would be regarded as being separate provided that "each of the logical systems can be clearly defined as is so organised that any message only flows in a way that is permitted by the terms of the relevant licence in respect of that message". It is not, however, possible to "mix and match" parts of licences so as to achieve communications which are not permitted by either licence taken on its own.

It is difficult to envisage circumstances where more than one entity could run logically separated systems using or sharing the same apparatus. This is because there must be a person "running" the system and such running denotes some control over its practical operation (see above para. 3–12). However as the Director General's statement admits, a way round this would be for one licensee to appoint the other as its agent for the purpose of running the system, provided the appointor had the right to retain control over the use of its system and was responsible for ensuring that its agent acted in accordance with the requirements of the relevant licence.

Agency Relationships

3–15   Confusion sometimes arises over the difference between running a network and running a system. It is important to make a clear distinction: telecommunication regulation is only concerned with the running of systems; a network may comprise different systems run by different persons, for example nodes operated and therefore run by a data service provider linked together by private circuits run by PTOs. Similarly, a system may be run by one person but the services provided over it may be offered by another entity, whether on its own account or as agent of the licensee. In these circumstances compliance with the licence is the responsibility of the person running the system, who must make it his business to ensure that the provider of the services does not cause him to be in breach of the licence.

However, in such circumstances where there is a breach, both the licensee and the service provider might be guilty of an offence; as mentioned in para. 3–10 above, section 5(4) of the Telecommunications Act 1984 provides that where a person commits the offence of running an unlicensed system because of the act or default of some other person, that other person is also guilty of an offence. Under this section proceedings may be taken against either of these offending parties without the necessity for concurrent proceedings against the other.

Contracts for Provision of Unlicensed Services

3–16   Where the telecommunication services to be provided by the operator of a telecommunication system are not covered by a licence granted under the 1984 Act, by virtue of section 5(2) any contract for their provision in this way is liable to be illegal and therefore unenforceable by either party.[11]

Limited Exceptions to Prohibition

3–17   The general prohibition on the running of unlicensed systems does not apply to systems run by broadcasting authorities where the transmissions are either by wireless telegraphy for general reception or closed circuit tele-

---

[11] See, *inter alia*, *Levy v. Yates* (1838) 8 A & E 129; *Dungate v. Lee* [1969] 1 Ch. 545; *Archbolds (Freightage) Limited v. S. Spanglett Limited* [1961] 1 Q.B. 374; *Allan (Merchandising) Limited v. Cloke* [1963] 2 Q.B. 340.

vision on a single set of premises. The prohibition is also not contravened by a system which simply involves the conveyance of light readable by the human eye, nor by the running of a system confined to one set of premises or in a vehicle, vessel, aircraft or hovercraft. There is a further exception covering private systems, such as baby transmitter/receivers. This is in terms of the running, by a single individual, of a telecommunication system unconnected to any other system, whose apparatus is under the control of that individual and where the system is run purely for domestic purposes (Telecommunications Act 1984, s. 6).

Perhaps the most important exception, at least for business, is in sections 6(3) and (4) of the 1984 Act which contain important exemptions where the sounds or visual images conveyed by the system (which must not be connected to any other system) cannot be heard or seen by anyone other than the person carrying on the business or his employees and certain other conditions are satisfied, *e.g.* no services are provided to others by means of the system.

Each licence granted under the 1984 Act has its own integrity: by this I mean that the licensee is required to comply with all its terms and conditions if it is to apply to him and cannot rely in part on one licence and in part on another in order to validate the running of a particular system. Only one licence (class or individual) can therefore be applied to that system at any one time.

It is as a result of the prohibition on unlicensed systems that there has   **3–18** grown up a scheme of licensing to which the Government, in the person of the Secretary of State, has chosen to inject a set of principles and obligations which taken together constitute, for the licensed operators, the regulatory scheme to which they are subject. In this regard it is not just the terms and conditions included in an individual operator's licence which are important to it. Many licences confer obligations on the licensee to provide services and perform functions for other licensees and users and thus create third party rights and benefits, the only qualification being that such rights are not directly enforceable so that this can only be done through the ministration of the Director General. This process, typically British perhaps in its implicit assumption that the regulator knows best, is open to criticism on the grounds that it places a regulatory filter on the enforcement of competition which might have proved rather less effective than the pursuit of properly formulated private remedies through litigation.

6. PROVISION OF TELECOMMUNICATION SERVICES

The reader will already have noted that the activity proscribed by regula-   **3–19** tion is the running of a telecommunication system without a licence, not the provision of telecommunication services. This does not mean however that telecommunication services as such are not regulated, the provision of unlicensed telecommunication services can render the person running the system over which they are provided guilty of an offence under section 5(2)(b) of the 1984 Act. It is important, therefore, to appreciate the nature and extent of *telecommunication* services in comparison to mere *services*. A "telecommunication service" is defined in section 4(3) of the 1984 Act to mean a service consisting in the conveyance by means of a telecommunication system of anything falling within paragraphs (a) to (d) of section 4(1), a directory information service and, broadly speaking, installation and maintenance services. Primary "telecommunication" services therefore involve the conveyance of those things referred to in section 4(1) of the 1984 Act which are signals of different kinds containing sounds, images and other information. For these purposes, "conveyance" includes transmitting, switching and receiving messages by means of a telecommunication system (s. 4(7)).

It therefore seems reasonably clear that a "telecommunication service" does not include the provision of the content of the message conveyed. This seems to be supported by the wording of the definition in section 4(3) as it specifically includes directory information; had this not been the case the provision of such information would not have been a telecommunication service if "conveyance" were to be a necessary constituent.

However, the merest hint of conveyance will probably bring the service within the scope of a "telecommunication service" for these purposes. A service provider who is, for example, providing an information service whereby messages are received, processed and/or responded to in some interactive way will necessarily be returning a message which, if only for a tiny proportion of the distance covered by the message, will be conveyed over the system on which the message originates, *i.e.* that of the service provider or its agent. Nonetheless it would be the act of conveyance and not the provision of information through that conveyance which would be a telecommunication service.

3–20   These distinctions can be important in relation to licensees' obligations under the section 7 licences; for example PTO licences typically require the licensees to satisfy reasonable demands for "telecommunication services" either from the very beginning (as in the case of BT) or at some stage in the development of their networks and share of the market. Likewise the "relevant services" regulated pursuant to the Competition and Service (Utilities) Act 1992[12] are limited to "voice telephony service" and a few other non-communication services and the definition of "voice telephony service" for these purposes is that it means a telecommunication service for the conveyance of speech over exchange lines provided by the designated operator.

The fact that the Telecommunication Act 1984 and the licences granted under it are essentially focused on telecommunication services as opposed to the "content" or information provided using that service became a real issue with respect to the reference of Chatlines' services by the Director General to the MMC on July 19, 1988. (section 13(1) of the 1984 Act

---

[12] See para. 3–26 *et seq.*

requires the MMC to investigate and report on matters relating to "tele-communication services".) In the course of these proceedings it was rightly argued for the service providers that the MMC's brief could not extend to the content of such services as these were not concerned with the convey-ance of messages. Ultimately the MMC's recommendations skirted this issue and required that the public operators, then BT and Mercury, could only provide a "controlled service" if there was in effect a Code of Practice governing the provision of such service and which made adequate provision for compensating those who suffered (*e.g.* financially) as a result of the provision of the service. For these purposes a "controlled service" included services, such as chatline services which supported live telephone conversations.

7. Protection of Consumers and Quality of Service

The Telecommunications Act 1984 makes provision for the interests of consumers to be monitored and protected in a variety of ways.   **3–21**

Interests of sections of the community and minority groups are handled through various committees set up to assist and advise the Director General. There are four advisory committees (ACTs) established in England, Wales, Scotland and Northern Ireland to deal with matters of local con-cern. There is also a committee advising the Director General on telecom-munications for disabled and elderly people (DIEL); there also exists a com-mittee for small businesses (generally referred to as businesses with under 200 employees) known as "BACT".

The ACTs are intended to be OFTEL's prime link with consumers, through whom their concerns and opinions are to be heard, and these com-mittees are also to provide a channel of communication between consumers and the suppliers of services and apparatus. Within each of the four ACT regions there have also been established local telecommunications advisory committees known as "TACS". Each ACT provides an annual report which is published along with the Director General's own report for each calendar year.

Codes of Practice

Another aspect of the Director General's responsibilities of direct import-ance to consumers concerns the terms and conditions (not strictly under customer contracts) or standards (in a non-technical sense) according to which BT and other PTOs would provide their services. In this connection PTO licences require the PTO to consult with the Director General on a Code of Practice for consumer affairs and to publish this code.   **3–22**

The Codes of Practice cover such matters as arrangements for fault repairs, payment of bills and compliance procedures. BT and Mercury's licences also require them to offer a small claims arbitration procedure and details of each operator's scheme are set out in its Code of Practice. The procedure is that if a customer has a claim relating to one of the services covered in the Code and the amount he is claiming is less than a fixed amount (which can be determined from time to time by the Director General), or does not involve a complicated issue of law he may put the dispute to arbitration. The Chartered Institute of Arbitrators administers the schemes and appoints an arbitrator who will come to a decision on the basis of written evidence and written submissions provided by both parties.

Confidentiality of Customer Information

3–23   PTO licences (see BT's licence Conditions 38 and 38A) also require them to prepare and "take all reasonable steps to ensure that" their employees adhere to a Code of Practice in relation to the confidentiality of customer information. This Code restricts the persons to whom customer information may be disclosed and therefore provides a model for all PTOs employees to follow in their dealings with customers. BT's Code of Practice is described in detail in Chapter 5, below.

Quality of Service

3–24   One of the particular responsibilities undertaken by the Director General concerns monitoring and encouraging improvements in the quality of service provided by PTOs, particularly BT. BT and Mercury both publish quality of service statistics annually.

   Surprisingly, PTO licences do not contain specific quality standards and requirements. BT's licence provides that it shall provide telecommunication services on demand "except to the extent the Director is satisfied that any reasonable demand is . . . met by other means", from which it may be implied that BT is obliged to provide a reasonable quality of service to its customers. To the extent that Mercury has an obligation to serve customers within a 10 kilometre radius of its network nodes, a similar implication may perhaps be implied from its licence. The remaining PTO licensees do not have any current obligation to provide service nor, as a consequence, any particular quality requirement.

The Competition and Services (Utilities) Act 1992

3–25   The Competition and Services (Utilities) Act 1992 has fundamentally altered the service quality aspects of telecommunications law (see below); in addition licences of all PTOs do contain minimum requirements regarding interfaces and in particular impose an obligation on the operator to publish the technical characteristics of interfaces in order for customer terminal apparatus to be made compatible with such interfaces. The Director General may also require that PTOs implement "essential interfaces" conforming to appropriate European or other international standards or some other standard specified by the Director General, where necessary to ensure interoperability. These requirements are, as the reference to interoperability suggests, part of the wider context of the requirement on public operators under the E.C. Open Network Provision Directives to provide interoperability with customer equipment and other networks, particularly those of other interconnecting public operators.[13]

"Citizen's Charter" Initiative

Under the Competition and Service (Utilities) Act 1992, which is part of   3–26
the Government's Citizen's Charter initiative, the 1984 Act was amended
to give new duties to the Director General and impose new obligations on
designated operators with respect to certain services supplied to particular
customers. On July 1, 1992 the Secretary of State for Trade & Industry
designated both BT and Kingston Communications as such operators.[14]
Generally he may so designate any public telecommunications operator
which provides at least 25 per cent of the voice telephony services supplied
within its area. Ultimately therefore, PTOs like Mercury and cable operators may fall to be so designated.

The new powers afforded to the Director General enable him, with the
agreement of the Secretary of State, to set standards of service for BT and
Kingston Communications when providing relevant services in individual
cases (individual standards) and also to set overall standards of performance by BT and Kingston Communications (overall standards).

These new powers apply to the designated operators when they are supplying voice telephony services, telephone rental services, directory
information services, directory services and facsimile transmission services
to residential premises, or any premises supplied by a single exchange line,
as well as the public call boxes provided by them.

In addition to the power to set standards of performance for the desig-   3–27
nated operators, the 1992 changes to the 1984 Act also empower him to
require them to give regular (e.g. annual) information to customers about
their performance and require each operator to establish a procedure for
dealing with complaints made by customers in connection with the provision to them of relevant services. He is also empowered to resolve disputes
between the operators and their customers concerning their rights with
respect to individual standards of performance as well as disputes regarding
alleged undue discrimination or preference against or in favour of customers. The Director General may also deal with disputes between BT or
Kingston Communications and their customers regarding bills and has the
power to issue special regulations regarding the procedure for handling
such disputes which includes a panel set up by OFTEL consisting of senior
staff in OFTEL and other persons working on a part-time basis and
appointed throughout the U.K.

---

[13] For discussion of ONP in general see Part Two, para. 16–20 et seq.
[14] Competition and Service (Utilities) Act 1992; (S.I. 1992 No. 1360).

In this connection a designated operator is required to settle with the Director General the criteria by which the operator will require deposits to be made by customers before relevant services are provided to them. In an important blow struck for consumers the 1992 Act goes on to provide that where any person has failed to pay any charge he is not to be disconnected by a designated operator so long as the amount of that charge is "genuinely in dispute". It can perhaps be assumed that once the appropriate body, be it the Director General himself, an arbitrator appointed by him or the court, has adjudged a sum to be due to the designated operator, then it is no longer genuinely in dispute.

The licence enforcement provisions of the 1984 Act (ss. 16 to 18 inclusive) are also applied to designated operators with respect to relevant services so that the achievement of overall standards of performance is deemed to be a condition of the designated operators' licences.

8. Promotion of Fair and Effective Competition

3–28   The Director General is required by the 1984 Act to promote competition as a means essentially of bringing the greatest benefit to consumers. In a move to try to ensure due compliance without his intervention, at an early stage he encouraged BT to produce yet another Code of Practice entitled "Competitive Marketing Guidelines". This particular bible was directed to BT's employees as a way of stressing the importance of fair dealing in their relations with customers and their competitors. The Code does not have any status under BT's licence but does at least reproduce elements of BT's own licence obligations. This Code has now been replaced by a similar code entitled "Competitive Marketing Principles" (1993). This deals with issues of particular sensitivity such as criticism of competitors, pricing rules, prohibition on linked sales, relationships with competitors and use of customer information. Its provisions include such laudable undertakings as that BT will:

- "be honest about our competitors and their products";
- "deal fairly with actual and potential competitors"; and
- "always operate within the legal and regulatory rules".

These principles may be contrasted with published findings of the Director General with respect to his study of competitive practices in the telecommunications market.[15] The study concluded that telephone companies generally had not properly educated their staff about fair competition so that they would know what they could and could not do and would know what they could and could not say about competitors. According to the Director General, "even on a small sample size of staff we talked to, there was evidence of widespread ignorance of their employers' fair trading rules". The Director General added that this was a very worrying feature

---

[15] See press notice 94/27 of August 23, 1994.

and that he would be pressing all the companies to review and strengthen their compliance programmes.

In relation to competition in telecommunications services, as a con-   **3–29**
sequence of the requirements of section 3 of the 1984 Act, the Director General is under a duty to promote fair and effective competition between operators and service providers.

The most direct method of enforcement of competition by the Director General where such competition is secured by particular conditions in operator's licences granted under section 7 of the 1984 Act is through his powers to issue licence compliance orders under sections 16 to 18 of the 1984 Act. This procedure and its effectiveness is discussed below.[16]

In addition to these regulatory powers, the Director General also has at his disposal his concurrent powers (with the Director General of Fair Trading) under the Competition Act 1980 and the Fair Trading Act 1973. These are discussed in detail in Chapter 10.

Each year in his annual report (usually printed in March or April for the previous year) the Director General reports on his actions and initiatives taken to promote and protect competition in the industry.

---

[16] Chap. 4 para. 4–12.

Licensing of Systems

As explored in Chapter 3, one of the main planks of U.K. regulation is the prohibition on the running of telecommunication systems without a licence. This chapter reviews current licensing policy, the procedures for granting licences and their modification, enforcement and revocation.

4–01

1. LICENSING POLICY

Licensing policy is exactly what the title suggests—it is merely policy and is not really to be found in the legislation. Although now nearly four years old, the statements in the White Paper[1] still embody the main elements of Government and OFTEL policy with regard to the telecommunications sector and, in particular, the grant of licences for networks and services. These elements, as so stated, are in relation to the three principal areas of new operators, cable television and BT's tariff structure, as follows.

4–02

New Telecommunications Operators

- The Government will consider on its merits any application for a licence to offer telecommunications services over "fixed links" (*i.e.* fixed network infrastructure) within the U.K.
- In the "short term" (arguably the short term is now coming to an end), the Government is unlikely to grant new international operating (*i.e.* using operators' own facilities) licences, so that for the time being these remain the preserve of BT and Mercury.
- Licences for satellite service providers to interconnect with the public switched network are to be considered on a case-by-case basis taking into account the scope of the interconnection proposed, the range of

4–03

---

[1] Competition and Choice: Telecommunications Policy for the 1990s (Cm. 1461)

services to be provided and such other factors as are judged relevant at the time.

- International simple resale services should only be permitted (on an individually licensed basis) between the U.K. and those countries whose regulatory regimes allow an equivalent freedom to provide such services in the reverse direction.
- Telecommunication circuits may be self-provided on condition that they are not used to provide a telecommunication service to third parties and that the capacity is not re-sold.
- There will not generally be competitions for licences except in cases involving scarce radio spectrum.

Cable Television Services

4–04
- The Government does not intend to remove the "present restriction" on BT, other national PTOs and Kingston Communications from *conveying* entertainment services in their own right until 10 years after the publication of the White Paper (*i.e.* 2001); the Government is prepared to reconsider the position after seven years (*i.e.* 1998) if the Director General then advises that removing the restriction would be likely to promote more effective competition in telecommunications.
- The parents, subsidiaries and associates of national public telecommunications operators will nevertheless be able to tender for any local delivery franchise advertised by the Independent Television Commission, including any over-franchising of existing cable areas.
- *National* PTOs are also now allowed to tender in their own right for a local delivery service licence for any part of the country not at that time covered by a cable or local delivery licence.
- *National* PTOs will not be allowed to *provide* entertainment services *nationally* in their own right for at least the next six years (*i.e.* until 2001); thereafter the Government would expect only to review this policy if advised by the Director General that a change would be likely to lead to more effective competition in telecommunications.

The distinction between "conveyance" and "provision" of entertainment services is further discussed later in this work.[2]

As discussed in Chapter 9 below, notwithstanding the clear statement in the White Paper regarding non-removal of the present "restriction" on BT from "conveying entertainment services", this restriction is far from absolute. It has transpired that its terms merely prevent BT engaging in broadcasting: the restriction, which is to be found in Schedule 3 of BT's licence, is equivalent to preventing BT from conveying messages for the delivery of licensable services (as further described in section 190 of the Broadcasting Act 1990) for simultaneous reception in two or more dwelling houses.[3]

---

[2] See Chap. 9, paras. 9–20 to 9–25.
[3] Contrast with the equivalent but updated service restriction in Sched. 3, para. 3 of new operators' fixed link licences which refers to the prohibition of "conveyance of messages for the delivery of one or more of the services specified in paras. (a) to (c) of section 72(2) of the Broadcasting Act 1990 for simultaneous reception in two or more dwelling houses".

BT's Tariff Structure

- The price cap (see Chapter 5, para. 5–20 below) with respect to BT's   4–05
  general prices should continue (it has most recently been extended
  until July 31, 1997) but with BT being allowed greater tariff flexibility
  in stages over the period to 1996 (licence modifications were introduced in 1991 and 1993 to allow for such flexibility).

2. Grant of Licences

A telecommunication licence may be granted by the Secretary of State, after   4–06
consultation with the Director General, or by the Director General himself,
provided he is so authorised by the Secretary of State (which authority has
not yet been given). As explained in Chapter 3[4] licences may be of one of
three types, general, class or individual.

The 1984 Act confers on its administrators rather greater flexibility in
the administration of individual licences, as opposed to class licences, for
two reasons. First, section 7(6) provides that individual licences may
require the licensee to comply with directions given by the Director General
or prohibit the licensee from doing something unless the Director General
consents otherwise; secondly, individual licences may also provide for the
reference to the Director General for his determination of questions arising
under the licence—the interconnection condition in PTO licences (see
Chapter 4 and Condition 13 of BT's Licence) is the prime example. It
would appear from the language of section 7(6) the 1984 Act that such
discretionary powers cannot be included in class licences.

Although the conditions (but not the provisions of other schedules or   4–07
in the annexes) of a licence may be modified, in the case of an individual
licence, with the consent of the licensee (and even without its consent
where the procedure in sections 13 to 15 is followed), in the case of a
class licence no modification can be made unless no representations or
objections whatsoever are made by persons covered by the licence or
any such representations or objections are withdrawn. Accordingly, a
modification of a class licence in this way is unlikely to be practicable
and is more likely to be accomplished by its revocation and re-issue in
the modified form. Alternatively, the Director General could refer a class
licence modification to the MMC, under section 13 of the Telecommunications Act 1984.[5]

To date, individual licences have been the prerogative in practice of the
Secretary of State and his officials at the Department of Trade and Industry.
As mentioned previously, there are no formal procedures with respect to
applications although the 1990 consultative document[6] mentioned that the

[4] Para. 3–13.
[5] See paras. 4–09 to 4–10.
[6] Cm. 1303 (1990).

Government would consider any proposal on its merits taking into account in particular the service proposed, the ability of the applicant to fulfil its plans, the extent to which it might result in more effective competition and the environmental implications of any request for code powers. Moreover the DTI's Guidance Notes on applications for licences also stipulate that applicants "will need to demonstrate that they have the financial, managerial and technical resources to install and operate the systems proposed. They should therefore supply full details of the source and availability of finance together with details of the forecast financing requirements both in respect of investment and operational and working capital requirements." Obviously such information is or could be commercially sensitive. The 1991 White Paper (Cm. 1461) recognised (para. 2.10) that although the Secretary of State would make public when he was considering an application, he would protect confidentiality by only disclosing a brief and general outline of the services proposed.

Licence Fees

4–08   An initial fee and an annual renewal fee are payable for grant of licences under the 1984 Act. These fees are essentially designed to recover the cost of the work involved in issuing and monitoring the particular licence. OFTEL has therefore divided the different types of licence into five separate categories and the annual fees payable as at April 1, 1995 are as follows—

Major PTO licences (*e.g.* national operators with Telecommunications Code Powers)	£37,000
Minor PTO licences (*e.g.* regional or local operators)	£12,500
Major non-PTO licences (*e.g.* satellite and International Simple Resale Operators)	£6,000
Minor non-PTO licences (*e.g.* narrow band, regional band 3 and SMATV operators)	£1,100
Temporary licences and minor licence extensions	£75

3. MODIFICATION OF LICENCES

4–09   Conditions (*i.e.* those usually contained in Schedule 1) of licences granted under section 7 of the 1984 Act may be modified essentially in two ways, either by agreement or by reference to the MMC. As pointed out above however, class licences are not readily susceptible to modification by agreement, given the representation and objection procedures which must be followed under section 12(4) of the Act.

As regards licence modifications by agreement, the procedure set out in section 12 of the 1984 Act is that the Director General should first publish a notice stating the proposal and the reasons for the modification and specifying a time, of not less than 28 days, within which representations or objections may be made. A copy of such notice is also to be sent to the Secretary of State and if within the specified time the Secretary of State directs the Director General must withdraw the proposal. Section 12(6) sets out the criteria which govern any such direction by the Secretary of State.

There has been at least one instance where the failure by the Director General to give his reasons for a proposed modification led to its successful challenge. This arose in the course of the *Chatlines* case, in which the Director General referred to the MMC the issue of the protection of the public with respect to use of and charges for telephone services involving live multi-party telephone conversations, which led to various consumer-protection modifications (to the licences of BT and Mercury) being recommended by the MMC. However, the Director General's implementation of the recommendations through such modifications with respect to Chatlines was supplemented with changes proposed by him[7] requiring that a Code of Practice be adhered to by providers of "interactive" services (*e.g.* adventure and other entertainment games) where a call to the service could last for more than five minutes. As no prior notice had been given of such a proposal nor of the reasons for it, one service provider challenged the legality of the move through commencement of judicial review proceedings and the proposal was withdrawn following the Director General's agreement to a consent order.

Licence modification references to the MMC arise where the Director   4–10
General has been unable to make the modification by agreement with the licensee. Such references may also be made where the Secretary of State has given a direction under section 12(6)(a) of the 1984 Act or where the Director General himself considers the matter of sufficient importance to refer direct to the MMC. The procedure for a MMC reference is set out in section 13 of the 1984 Act. Under its provisions the reference is to be framed to ask the MMC to investigate and report on whether the particular circumstances giving rise to the proposed modification operate, or may be expected to operate, against the public interest and whether these circumstances, "the effects adverse to the public interest", could be remedied or prevented by the proposed modifications.

Modifications to existing licences through an MMC referral can only be made where the proposals come within the pre-conditions for a reference set out in section 13 of the 1984 Act. It appears to be arguable that a reference must in some way be related to specific conduct rather than to cure, say, general deficiencies in regulation.

[7] Statement of July 27, 1989

As with an agreed modification, the Secretary of State can intervene to direct the MMC not to proceed with the reference (section 13(5)).

Following the MMC Report, which must be provided within six months from the date of the reference, where this concludes that the matters specified in the reference operate, or may be expected to operate, against the public interest, and specifies appropriate modifications to deal with the problem, the Director General must make the necessary modifications. He must, however, first give the same notice as is required in relation to a licence modification by agreement and then consider any representations or objections so made. Again the Secretary of State may then direct the Director General not to make modification, where it appears to him that this is requisite or expedient in the interests of national security or foreign relations.

As discussed below,[8] section 95 of the 1984 Act also empowers the Secretary of State to make licence modifications following a monopoly, merger or competition reference.

4. Revocation of Licences

4–11   Licences may be revoked by the Secretary to State and, in the case of class licences, alternatively by the Director General. The normal grounds for revocation of an individual licence are—

- where the licensee agrees in writing with the Secretary of State that the licence should be revoked;
- where there is a change in a significant proportion of the shares held in the licensee or its parent and the Secretary to State notifies the licensee that he is minded to revoke the licence on the grounds that the change would, in his opinion, "be against the interest of national security or relations with the Government of a country or territory outside the United Kingdom" or the licensee has committed a breach of the condition requiring notification of any such change;
- if the licensee fails to comply with a final order (or a confirmed provisional order)[9] under section 16 of the 1984 Act and the Secretary of State gives 30 days' written notice of revocation; note here therefore that the mere breach of a licence *condition* will not be grounds for revocation—the Director General must first issue an order requiring compliance with the condition and that order must be contravened for revocation to be possible;
- if the licensee becomes insolvent or ceases to provide its telecommunication services or if the licensee takes any action for its voluntary winding-up or dissolution or enters into any scheme of arrangement under the Insolvency Act 1986;

[8] See para. 10–06.
[9] See para. 4–12.

- if an administrator, receiver, trustee or similar officer of the licensee is appointed or an order is made for its compulsory winding-up; or
- if any amount payable by way of licence fee remains unpaid for 30 days after it becomes due and remains unpaid for a further period of 14 days after the Secretary to State has notified the licensee that the payment is overdue.

5. Licence Enforcement

As already pointed out,[10] it is an offence to run a telecommunication system   **4–12**
that is unlicensed; thus any provision contained in the licence grant itself (usually one or two pages at the very beginning of the licence), and in its Annex A and Schedule 3 must, by virtue of sections 5(1) and 5(2) of the 1984 Act, be observed in order to avoid commission of a criminal offence. However, except as mentioned below, a breach of a licence condition (other than non-payment of a licence fee) does not trigger any immediate legal sanction in that for example this cannot *per se* be punished by revocation of the licence. The procedure is in fact that the Director General may issue a licence compliance order,[11] and only if that order is not complied with may the Director General then revoke the licence or request the court to grant a mandatory injunction requiring such compliance, failing which the licensee would be in contempt of court and the court could punish the offender with a fine or imprisonment. Under section 18(6) of the 1984 Act such a failure to comply with the Director General's order would constitute a breach of duty for which the licensee would be liable to any person who suffers loss or damage as a result of the breach.

Only if a licensee fails to comply with a final order (under section 16 of the Act) or a provisional order confirmed under that section (where such provisional order is not subject to review) and such failure is not rectified within three months after the Secretary of State has given notice in writing of such failure to the licensee is the licence finally revocable (on 30 days notice in writing to an individual licensee or any licensee covered by a class licence).

Legal challenge to a compliance order is circumscribed by sections 18(1) and 18(3). Under section 18(1) the court may quash the order but only on the grounds that it was not within the Director General's powers under section 16 or that the requirements of Condition 17 were not complied with. Section 18(3) provides that, aside from proceedings under 18(1), the validity of a final or provisional order is not to be questioned by any legal proceedings whatever. In some cases, for example where a determination has preceded an order (such as under BT licence Condition 17), it may be possible to challenge the determination, not the order, by way of judicial review, and thus avoid any strictures imposed by section 18(3).

---

[10] See para. 3–10.
[11] See also paras. 10–02 to 10–03.

4–13   Breach of a licence condition can have immediate legal consequence in one case. BT and other PTOs' standard conditions of contract contain terms entitling the PTO to terminate service immediately (subject to cure provisions) in the event of the customer contravening his licence (for example, in the case of the generality of customers, the SPL and in the case of most service providers, the TSL). This could have the effect of putting the PTO in the position of licence "policeman"—something which many operators would wish to avoid. It is therefore arguable that PTOs should in practice not exercise their rights to terminate except in circumstances where a licence breach has led to enforcement action by the Director General and even then perhaps only if that enforcement has either failed to prevent the breach or has resulted in the licence being revoked.

In his "watchdog" role the Director General has already received complaints in a number of areas, particularly relating to BT's own licence. The first major matter concerned radio paging where the complaint was made by BT's then four competitors in the provision of radio paging services. The complaint was that BT's practice of joint billing for radio paging services and basic telephone services and its method of charging for radio paging services gave it an unfair advantage and therefore inhibited effective competition. In his report of August 1985 the Director General concluded that BT should be required to introduce separate billing for its radio paging customers as soon as practicable and subsequently BT agreed to a modification of its radio paging licence (a separate licence from its licence as a PTO) establishing this practice as a licence obligation. The Director General was not prepared to go further as requested by the complainants and modify Condition 17 of BT's PTO licence so that any unfair favouring of one of its businesses in relation to any activity would have been prohibited.

4–14   In 1985 the Director General was also called upon to deal with complaints against BT allegedly leaking confidential customer information and cross-subsidising in relation to BT's security services. These services involved monitoring the state of a telephone circuit and the transmission of an alarm to emergency services. Security companies competing with BT claimed that BT had circulated marketing literature to customers whose names would have been obtained in the course of work by other parts of BT providing basic telephone services to such customers at the behest of the security companies. On the basis of BT's code of confidentiality of customer information being in force at the relevant time it was alleged that this activity by BT would have been in breach of the code.

In this case BT argued that its own security service was a basic service rather than a value added service and that therefore the rules in its licence on cross-subsidisation and disclosure of confidential information did not extend to it. However, the Director General appears to have persuaded BT otherwise and BT voluntarily agreed to observe its licence rules on fair trading in relation to such business.

4–15   More recently, there have been a number of cases in which allegations of discriminatory pricing or other activity have been raised with respect to

BT. Towards the end of 1992 BT introduced its Option 2000 tariff which gave discounts on tariffs to large organisations with dispersed sites spending in excess of £1 million per annum on directly dialled calls. In its original form the Director General concluded that the discounts on offer were not necessarily related to possible cost-savings resulting from providing service to a number of sites. This would have made it difficult for other operators lacking BT's geographical coverage to compete and would also have been discriminatory between customers depending on whether they were part of a multi-site organisation. Accordingly, following discussions between the Director General and BT, BT agreed to amend significantly its Option 2000 tariff to remove any discriminatory or other anti-competitive effects.

In 1993, BT introduced its "Sunday special" tariff which involved a special offer for directly dialled national calls made between 3 p.m. and midnight on every Sunday during November and December. The offer meant that directly dialled calls would generally be charged at the local cheap rate rather than BT's published rates.

OFTEL received complaints from competitors of BT about this special offer and concluded that its terms did not cover the costs attributable to the running of BT's network when those costs were assessed on the same basis as that used for setting charges for network usage by BT's competitors. This meant that BT was unfairly favouring its own business to a material extent so as to place other operators at a significant competitive disadvantage. Accordingly the Director General concluded that the offer amounted to undue discrimination against those operators in contravention of Condition 17.2 of BT's licence and made a determination to that effect on August 2, 1993.

6. Licences for Systems using Radio Communications

A telecommunication system which uses radio (wireless telegraphy) requires a licence not only under the Telecommunications Act 1984 but also under the Wireless Telegraphy Act 1949.   4–16

The Wireless Telegraphy Act 1949 (section 19) defines "wireless telegraphy" as meaning:

> "the emitting or receiving, over paths which are not provided by any material substance constructed or arranged for that purpose, of electro-magnetic energy of a frequency not exceeding 3 million megacycles a second, being energy which either—
>
> (a) serves for the conveying of messages, sound or visual images (whether the messages, sound or images are actually received by any person or not), or for the actuation or control of machinery or apparatus; or
> (b) is used in connection with the determination of position, bearing or distance, or for the gaining of information as to the presence,

absence, position or motion of any object or of any objects of any class."

The principal effect of the 1949 Act is to make it an offence for anybody to establish or use any "station for wireless telegraphy" or to install or use any apparatus for wireless telegraphy without a licence for that purpose granted by the Secretary of State. There are certain exemptions from the licence requirement, for example with respect to broadcast relay apparatus, such as SMATV.

4–17   From the point of view simply of telecommunication, the important thing to note is that as in the case of telecommunication systems, wireless telegraphy includes not only emission but also receipt of messages. Accordingly the licence requirements extend to receiving stations as well as transmitters and thus include mobile telephones and radio paging devices. All such apparatus would normally be included in the same wireless telegraphy licence granted to the operator of the particular service.

Notwithstanding the exemption for broadcast relay apparatus, the apparatus used by subscribers to the services of broadcast relay companies, as well as of cable television operators, is also deemed to be wireless telegraphy apparatus by virtue of the proviso to section 19(1) of the Wireless Telegraphy Act 1949. The Secretary of State has issued a general telecommunications licence for such apparatus under section 7 of the 1984 Act.[12]

Wireless telegraphy licences are quite separate from licences required under the Telecommunications Act 1984. Since wireless telegraphy is a means of telecommunication, any station or apparatus for wireless telegraphy would also represent a telecommunication system which, unless it falls within one of the exceptions in section 6 of the Telecommunications Act 1984, would require a licence under section 7 of the 1984 Act.

A wireless telegraphy licence may be revoked by the Secretary of State at any time pursuant to his power under section 1(4) of the Wireless Telegraphy Act 1949; according to the DTI he would exercise this power if the apparatus comprised in the wireless telegraphy station or its performance at any time became "unacceptable" to him. In relation to PTOs this power has been curtailed, in order that licences granted to an operator under both the 1984 Act and the 1949 Act may only be subject to revocation in similar circumstances.[13]

Licences to provide radiocommunication services, particularly mobile telecommunications, are discussed in detail in Chapter 7.

7. Licences for Systems Providing Broadcasting Service

4–18   A telecommunication system delivering television and/or sound services for general reception will need not only a licence under the Telecommunica-

---

[12] Class licence for telecommunication apparatus for reception of services conveyed by means of certain cabled systems November 25, 1986.
[13] s.74 of the Telecommunications Act 1984.

tions Act 1984 but also a licence under the Broadcasting Act 1990. Use of the system capable of delivering entertainment services to more than 1,000 dwelling houses, *e.g.* through SMATV would usually require a local delivery licence from the Independent Television Commission.[14] The regulation of broadcasting is discussed in greater detail in Chapter 9 below.

---

[14] In relation to broadcasting and local delivery licences see further below at paras. 9–08 to 9–10.

Chapter 5

Regulatory Conditions in Individual Licences

PUBLIC TELECOMMUNICATION OPERATORS

Section 7(5) of the Telecommunications Act 1984 provides that a licence   5–01
for the running of a telecommunication system may include "such conditions . . . as appear to the Secretary of State or the Director to be requisite
or expedient having regard to the duties imposed on him by section 3
above". Section 8 of the 1984 Act refers to the conditions which may be
applied to a person running a public telecommunication system, which may
include conditions requiring that person (in essence)—

(a) to provide specified telecommunication services;
(b) to interconnect its telecommunication system with a telecommunication system run by any other person;
(c) to permit the provision of specified telecommunication services by
means of the licensed system;
(d) not to show undue preference or undue discrimination to or against
particular persons as respects its services;
(e) to publish notice of its charges and other terms and conditions.

The terms of section 8 of the 1984 Act are thus very important in her-
alding the most significant conditions which are to be included in the
licences granted to PTOs. This chapter considers the nature of these and
other significant conditions incorporated into PTO licences. Aside from the
1984 Act itself, this represents the body of U.K. regulation of
telecommunications.

Since the licence granted to BT is the most comprehensive and representative of PTO licences, this is taken as the main example for commentary;
however, except where indicated most of its substantive provisions are
incorporated in the licences granted to other PTOs.

1. THE LICENCE OF BRITISH TELECOMMUNICATIONS

5–02   BT's licence came into force on August 5, 1984, the day before BT became a public limited company and a few months before its flotation the same year. The licence is divided into six parts:

    (i) The right to run certain telecommunication systems called "Applicable Systems";

    (ii) conditions regulating how the systems are to be run;

    (iii) arrangements for revoking the licence;

    (iv) authorisation to connect to other systems and to provide telecommunication services;

    (v) exemptions and conditions relating to the application of the Telecommunications Code; and

    (vi) a description of the telecommunication systems BT is allowed to run.

This manner of division of the licence does not particularly lend itself to the easiest explanation and so I propose to begin with a discussion of the licence grant itself and the way in which it could be terminated, followed by the scope of the authority which it confers in the running of the licensed systems and their connection to other systems and then consideration of the all-important conditions (1 to 17 inclusive) set out in Part 2 of Schedule 1, applied pursuant to section 8 of the 1984 Act, as well as the conditions (18 to 53) included in Part 3 pursuant to section 7 of the 1984 Act. The Telecommunications Code provisions are discussed in detail in Chapter 11.

The Licence Grant

5–03   This gives BT permission to run the telecommunication systems—"Applicable Systems"—described in Annex A to the licence and to connect to other telecommunication systems and to provide the telecommunication services as specified in Schedule 3 to the 1984 Act. In running its Applicable Systems, BT must abide by the conditions set out in Schedule 1 to the 1984 Act and the permission can be revoked, or taken away, in the circumstances described in Schedule 2 to the 1984 Act, which, as previously explained,[1] do not include mere breach of a licence condition.

The BT Licence is for a minimum duration of 25 years from June 22, 1984, subject to revocation on at least 10 years' notice, which may not be given before the end of the fifteenth year.

Termination of the licence prior to the end of this 25 year period can only arise in the particular circumstances set out in Schedule 2 which include, *inter alia*, failure of BT to comply with a final order or a provisional order made under section 16 of the 1984 Act (where the order is not subject to proceedings for review and the failure has not been rectified within three months of notice from the Secretary of State), or its insolvency,

---

[1] Chap. 4, para. 4–11.

receivership or liquidation: all contingencies which, it might be said, now seem more theoretical than real.

Licensed Systems

The systems which BT is so licensed to run, set out in Annex A, demon-   5–04
strate the network boundary concept underlying the Act and its regulations. A licensed system is described as one by means of which messages are conveyed or are to be conveyed from one network termination point (NTP) to another such point, for example a normal inland call, or from a NTP to a place which is not a NTP, for example an international call or a call between other places "but in any case not beyond a Network Termination Point". The apparatus comprised in the licensed system must not include terminal apparatus installed in premises occupied by a person to whom BT provides services. This apparatus would form part of the customer's system, typically licensed under the SPL or TSL. Also, the BT system must not be a wireless telegraphy system except where such telegraphy is provided to or from permanent or temporary fixed stations (for example microwave), thus precluding the provision of mobile services (paging, radiotelephone, etc.) under this particular licence.

Connection of Other Systems; Provision of Services

The other systems and apparatus with which BT's Applicable Systems may   5–05
connect are described in Schedule 3, paragraph 1 of the 1984 Act and essentially cover other licensed systems and approved telecommunication apparatus.

The services permitted by the licence are set out in Schedule 3, paragraph 1(b) and comprise:

"the provision by means of the Applicable Systems of telecommunication services consisting in:

(i) the conveyance (not including switching) of messages (not including cable programme services sent under a licence granted under section 58 of the Act[2]) and switching incidental to such conveyance; and
(ii) directory information services

but not any Land Mobile Radio Service."

According to the 1991 White Paper (Cm. 1461, paragraph 4.27), the   5–06
Government intended to maintain the restrictions on fixed operators from providing mobile services under their main licences but asked the Director General to keep the position under review.

The restriction in (i) above on conveyance of cable programme services

---

[2] See now s.190 Broadcasting Act 1990.

now in effect relates to licensable programme services provided as a local delivery service, as defined in the Broadcasting Act 1990, s. 72(1).

According to section 4(7) of the 1984 Act the word "convey" and derived expressions include "transmit, switch and receive." For the purposes of Schedule 3 to the 1984 Act therefore "conveyance" should be construed in this sense.

The explanatory notes to the licence once stated that value added services may not be provided under the licence and that BT would require a separate licence for this. This interpretation of Schedule 3 has been revised in the statement of the Secretary of State on February 25, 1987. The modifications made to conditions of PTO licences, including BT's on May 1, 1987, reflect this revision, by referring in a number of places to the "Supplemental Services Business", itself defined essentially as the provision of a "Relevant Service": this unhelpful nomenclature is simply a rather elliptical description of the types of service which a licensee can provide under the now defunct VADS Class Licence. A PTO can offer the same services under its licence.

Conditions

5–07   Rather than analyse each condition in turn, I shall concentrate on the main substantive elements of BT's licence obligations.

(i) Condition 1: Universal Service

5–08   Condition 1 obliges BT to provide what is ambitiously, but commonly, described as a "universal" telecommunication service. Universal in the limited national sense of BT's "licensed area", which is the U.K. excluding Hull. In particular, BT must provide to every person who so requests both:

"(a) voice telephony services; and
(b) other telecommunication services consisting in the conveyance of messages

. . . except to the extent that the Director is satisfied that any reasonable demand is or is to be met by other means and that accordingly it would not be reasonable in the circumstances to require [BT] to provide the services requested".

There may be other circumstances in which BT could be excused performance of this obligation—see Condition 53.

Condition 1, in referring exceptionally to "reasonable demand" in this way provides an interesting contrast to section 3 of the 1984 Act which itself requires the Director General to ensure that telecommunication services are provided in order to meet "all reasonable demands for them". Note that BT's obligation in Condition 1 is not to provide service in accordance with a reasonable demand but to provide service in whatever circumstances "except to the extent that the Director is satisfied that any reasonable demand is . . . to be met by other means". Accordingly, aside from the exceptions in Conditions 53, BT can only escape liability to provide service under Condition 1 where the demand can be met by other means.

The first occasion on which the Director General was called upon to interpret and apply Condition 1 arose in the *PamAmSat* case. There, the would-be independent satellite operator, PanAmSat, claimed that BT was in breach of its licence Conditions 1, 5, 17 and 35 in refusing to agree to provide service linking customers in the U.K. to PanAmSat's satellite. The Director General gave his view that "reasonable demand is primarily demonstrated in the market: reasonable demand exists if one or more customers will pay a fair price for the service".   **5–09**

Generally, therefore, as a rule of thumb, it would in the author's view be safer to assume "reasonable demand" as meaning a demand which in all the circumstances and after taking account of the exceptions in Condition 53, is reasonable. The matter then becomes one for the Director General to interpret exercising his discretion reasonably.

Condition 1 goes on to oblige BT to install, keep installed and run the Applicable Systems in order to provide these services. Thus BT is obliged to install and (on OFTEL's interpretation at least) to maintain in good running order all the telecommunication apparatus necessary to provide such services, including not only the apparatus (*e.g.* exchange lines) by means of which messages initially are sent, but all the transmission and switching apparatus required to ensure their safe delivery.

It is noteworthy that BT's obligation under Condition 1 relates specifically to voice telephony services and telecommunication services consisting in the conveyance of messages. This is important in relation to other licence conditions, particularly Condition 17 (prohibition on undue preference and undue discrimination) which only applies to those telecommunication services provided by BT "in accordance with an *obligation* imposed by or under" its licence. "Voice telephony" services are not defined, whilst telecommunication services are defined in section 4(3) of the 1984 Act as including the conveyance of messages, directory information service and a service consisting in "the installation, maintenance, adjustment, repair, alteration, moving, removal or replacement of apparatus" for connection to a telecommunication system. The meaning of "telecommunication services" is discussed in detail at Chapter 3, paragraphs 3–11 to 3–12 above.

International services are dealt with separately in Condition 5.

(ii) Conditions 2 and 31 to 33 inclusive: Social Obligations

Condition 2 (rural areas) reiterates the terms of Condition 1 as to universal service, specifically in relation to rural areas. There was sufficient concern expressed in the passage of the Telecommunications Bill regarding rural services and other of BT's social obligations, that pressure was successfully brought to bear to include express mention of such areas.   **5–10**

Special arrangements for the provision of directory information for the blind are referred to in Condition 3 (below). Under Condition 31 BT is required to make arrangements for the provision of telecommunication apparatus which is suitable for the disabled; Condition 32 requires that BT should supply special telephones for those whose hearing is impaired. Under Condition 33, this obligation is extended to require BT to provide apparatus in its public call boxes to assist those people who use hearing aids.

(iii) Condition 3: Directory Information

5–11   Condition 3.1 requires BT to provide directory information services relating to its switched voice telephony services. Such a directory information service is, in the context of "telecommunication service", defined in section 4(3) of the 1984 Act as being a service provided by means of a telecommunication system, thus leaving outside BT's licence obligations any necessity to supply "hard copy" directories.

BT's directory information services are nevertheless at present provided in both written form (in hard copy directories) and, for a standard charge, verbally by communication with an operator (who relies on a computerised data base).

In March 1994 the Director General published findings from an OFTEL enquiry into directory enquiry services. In the short term these were to include:

- finalisation of a code of practice on provision of the services;
- examination of BT's charges for inputting and accessing directory information and supply of printed directories;
- pursuing three options for the provision of printed directories:

    — download of data to enable PTOs to produce own directories;
    — unbranded core directory to be made available to all PTOs;
    — continuation of arrangements for bulk supply of BT printed directories to other operators;

- not to oblige operators to supply or grant direct access to residential information to non-PTOs for the purpose of providing directory information services or producing printed directories.
- to continue to leave PTOs to decide whether to make a separate charge for printed directories.

Longer term proposals included an invitation to PTOs which would like to set up a separate directory information data based, to put proposals to him. Further information is awaited.

There are special provisions regarding information about the customers of other operators contained in Condition 3.2. In practice such other operators, like Mercury and cable operators for instance, pay BT a special charge for input to BT's database of numbers identifying the customers of such operators.

BT must also ensure that its directory information is made available free of charge in a form sufficient to meet the needs of blind and disabled people or, where BT does make a charge for its directory information services and the Director is satisfied that the application of such charges to blind and disabled persons cannot be excluded, BT must pay appropriate reasonable compensation to such persons.

(iv) Conditions 6 to 9 inclusive: Emergency Services

Condition 6 (public emergency call services) requires BT to provide to the public free of charge public emergency call services in order to enable people to reach the police, fire, ambulance and coastguard services.   5–12

Conditions 7 to 10 require BT to continue to provide other emergency services required by the emergency organisations themselves (Condition 7), to provide maritime services such as the Distress Watch Service (Condition 8), and to make special arrangements in the event of major accidents and other emergencies (Condition 9).

(v) Conditions 5, 46 and 48: International Services

Under Condition 5, BT must take all reasonable steps to satisfy reasonable demands for international telecommunication services called—"International Connection Services". This obligation extends not only to the generality of BT's customers but also to other operators, even those competing with BT, such as Mercury. In Mercury's case, in the early days of development of its services, access to BT's international routes for Mercury's telephone and telex customers where Mercury had no direct route of its own was secured by the Director General's two determinations.[3]   5–13

The obligation to provide "International Connection Services" is also significant with respect to other licence conditions, in particular the requirement under Condition 48 to adhere to a Code of Practice on international accounting (see below).

The meaning of "reasonable demand" and its discussion in the PanAmSat decision is looked at under Condition 1, para. 5–09 above.

In order to prevent BT "locking out" competition on international service, Condition 47 prohibits BT from entering into any agreement with an overseas administration which unfairly precludes or restricts similar arrangements by any other PTO. BT is also prohibited from excluding any other PTO from participating in "international arrangements" into which BT proposes to enter in relation to the installation and operation of any submarine cable linking with any foreign telecommunication system.

(vi) Condition 48: Parallel Accounting

Condition 48 (international accounting arrangements) requires BT to try to agree with other PTOs, such as Mercury, a Code of Practice on international accounting. In practice, BT and Mercury failed to agree upon such a Code and on March 4, 1987 the Director General duly issued a Code upon the terms determined by him. Accounting rates are the payments one   5–14

---

[3] See para. 5–52 et seq.

administration or RPOA[3a] makes to another for traffic sent to it by that other administration. There is also a charge which the administration sending the traffic renders to its own customers, known for international accounting purposes as the "collection rate".

The statement issued by the Director General with this particular determination explains government policy on international accounting arrangements. This policy is directed towards ensuring "parallel accounting"; this would require BT and Mercury (and any other licensed provider in the future of international telecommunication services) to adopt the same accounting rates, divisions and methods in their dealings with foreign carriers in order to avoid "whipsawing", the practice whereby a monopoly foreign carrier would be able to try to negotiate cut rates with one U.K. operator to the detriment of another.

Parallel accounting might be viewed in principle as an anti-competitive arrangement, but in the U.K. finds favour with the regulatory authorities because it protects competition between U.K. international operators and is also justified in its protection of the national interest or, as the Director General puts it, "the United Kingdom bargaining position".

In some instances competition will be better served by allowing exceptions to the general rule of parallelism, particularly where whipsawing is unlikely, and the determination therefore permits this in certain cases, for example where there is more than one international operator in the foreign country concerned (this would include the U.S. and Japan). This is subject to the Director General being satisfied that the interests of the relevant operators and of users in the U.K. would not be prejudiced by such an arrangement.[3b]

5–15   An exception to parallel accounting is also made, subject again to the Director General being satisfied, for the situation where an operator has been unable to obtain arrangements for proportionate return with an overseas operator. Proportionate return is intended to secure for the sender of traffic the return to him of a proportionate share of incoming traffic to the U.K. The economics of international service are such that it is vital for the overall profitability of an operator's services to obtain the revenues arising from a share of the accounting rate (sometimes called the accounting rate "in-payment") arising from such return traffic.

Under Condition 48.2, where the Director General is of the opinion that BT is about to enter into or vary an agreement or arrangement with a foreign administration which establishes international accounting methods, rates and divisions in a way which would prejudice the interests of other providers of international telecommunication services, he may issue a direction prohibiting BT from going ahead with the proposal.

On January 9, 1991 the then Director General allowed BT and Mercury to adopt different accounting rates from each other on the U.K.-USA route. They were nonetheless still required to seek the Director General's approval for any change to the rates on this route, whereupon the Director General

---

[3a] See Chap. 15, para. 15–16.
[3b] On July 31, 1995 the Director General approved new accounting rates for BT to apply to calls to and from the US.

has three months to decide whether or not to approve any change. On December 14, 1993 both Mercury and BT were granted the Director General's approval for a reduction in their accounting rates with the USA for various types of calls.

(vii) Conditions 4, 10 and 44: Maintenance and Fault Repair

Maintenance: In support of its universal service obligation, BT must also in certain circumstances provide "maintenance services". These services are closely defined in Schedule 1, Part 1 of the licence and include not only carrying out repairs but also "any activity involving the removal of the outer cover of the apparatus". Excluded are operations incidental to the installation, "bringing into service" (again closely defined) or routine use of the apparatus.

5–16

BT's obligation to provide maintenance services for its customers does not apply in a variety of situations, for example where the system or apparatus to be maintained is beyond repair, or necessary components are not available or where it has not been supplied by BT and it is not a term of the licence for such system or apparatus that BT should provide such services, or where its approval under section 22 of the Telecommunications Act 1984 for connection to BT's systems does not require it to be maintained by BT.

Priority Fault Repair: Any PTO is free to provide such levels of fault repair service, beyond normal repair service, as it finds commercially expedient. However, in BT's licence (Condition 10) emergency services are understandably singled out for special treatment in this respect. Any person or categories of persons in authority:

5–17

(a) engaged in the provision of an emergency service to the public, or the provision or supply of essential services or goods, or in public administration;

(b) whose name is notified by the Director General to BT and who pays BT's charges (unless otherwise determined by the Director General) for the service; and

(c) who has a bona fide need for an urgent repair,

is entitled to such repair by BT in order that disrupted service may be restored "as swiftly as practicable".

This priority fault repair service is to be available for 24 hours a day or such lesser a period as BT may agree with the relevant person. BT also provides to all its customers, by contract, its "premium maintenance services" at varying rates for different response times.

In any case where another party's licence for a telecommunication system or apparatus requires that BT should provide maintenance services (for example in relation to network termination and testing apparatus), Condition 44 obliges BT to publish its charges and other terms and conditions for such maintenance services and to supply such services at those charges and upon the stated terms and conditions, with very limited exception.

(viii) Condition 16 and 24 to 26 inclusive: Charges, Terms and Conditions

5–18   Conditions 16 requires BT to publish its charges (or a method for determining its charges) and other terms and conditions on which it offers to provide its various telecommunication services and not to depart from such charges, terms and conditions. It should be noted that this requirement only extends to telecommunication *services* and ancillary activities which BT is *obliged* by its licence to provide and undertake. Thus the supply of telecommunication apparatus is not regulated in this way. Moreover the condition does envisage the possibility of the Director General excusing BT from compliance with its provisions in certain respects. (It will be noted below with respect to other operators' licences that use of the same language, requiring that charges should be published for services the licensee is "obliged to provide", means that since these operators do not have obligations to provide service until they reach a certain market share there is for the time being no obligation on them to publish charges.)

In the author's view, the obligation not to depart from published terms and conditions should not prevent the application by BT in a particular case of additional terms or conditions which do not conflict with or materially affect the published version, provided these additional provisions are warranted by the special circumstances of the case or of the customer and the customer is not thereby unduly preferred or discriminated against.

5–19   Publication is to be effected by sending a copy of the charges, terms and conditions to the Director General, at least 28 days or, in the case of relevant services (essentially value added and data services) at least one day, before any amendment is made, and by placing a copy in the office of the General Manager of each of BT's telephone areas and sending a copy to any person who so requests. These filing requirements are purely administrative; PTO charges *per se* are not subject to approval by the Director General, or any other form of direct control except, in BT's case, with respect to the services covered by the RPI-x formula (paragraph 5–20 below).

Under the modifications introduced to BT's licence on April 1, 1995 the charges for different interconnection services provided by BT to operators are to be determined by the Director General on the basis of unbundled network components and network parts—so-called "standard services". All operators entitled to purchase these standard services from BT are to pay the same charges.

New Condition 16B.8 allows BT's prices for standard interconnection services to be deregulated by the Director General determining that the market for the provision of a standard service has become "competitive". No criteria are provided for such determination and the situation where the provision of a standard service ceases to be competitive is apparently not dealt with. Generally the notion of a "market" for a particular service (which could be a switching service or a particular network segment) seems somewhat unusual and difficult to assess. If, for example, a particular interconnect service provided by BT has an equivalent and competing offering in certain parts of the country but not in others, query whether the relevant "market" is competitive or not.

Price Controls: RPI-x price cap

The rationale for a "cap" or control over BT's retail prices is that it has a very large market share and is expected to be a dominant force in the telecommunications market for some time. The restriction of a "basket" of BT's prices to an overall limit pegged to the RPI is considered to be preferable to pure rate of return regulation as it controls prices rather than profits and thus encourages efficiency.

5–20

The formula of RPI-x (x being a percentage point figure) ran from the inception of BT's licence in August, 1994, initially for a five-year period but it has been renewed periodically with the consent of BT on a number of occasions since then.

The most recent adjustment was made on March 9, 1993 and the effect of this main price control rule (private circuits being covered by a separate price control—see below) is that for the period August 1, 1993 to July 31, 1997 the rate of increase of the prices of a "basket" of BT's main switched services is limited to the annual rate of inflation minus 7.5 percentage points. The basket covers all BT's "general prices" meaning—

(a) all periodic charges imposed by BT for the use and ordinary maintenance of an exchange line (*i.e.* exchange line rental charges);

(b) all charges for the connection or taking over of an exchange line;

(c) all charges for voice telephony calls using such exchange lines excluding calls from public call boxes and calls from private call boxes wherever the charge to the renter is based on BT's standard public call box charges and excluding also transfer charges in respect of calls from call boxes;

(d) all charges for the facility of transferring, with human operator assistance, charges for voice telephony call conveyance.

Excluded from these charges are BT's charges for domestic and international private circuits, special, priority or emergency maintenance or fault repair services, voice telephony calls forming part of BT's "supplemental services business" (essentially value added and data services) or charges for voice telephony calls conveyed to customers of another operator of a "relevant connectable system" (defined in Condition 13.9 of BT's licence—see below), except where that other operator is a fixed link operator (so that for example such excluded calls would cover calls to and from resellers); lastly, charges for any maritime services provided pursuant to Condition 12. The RPI-7.5 per cent price cap is also subject to the provisions to Condition 24C which excludes from the computation any new benefits arising from BT's special packages covered by the differential charging provisions of Condition 17A. By the same token Condition 24C also stipulates that where benefits in existence on July 31, 1993 are withdrawn there should be compensating reductions in its other prices.

5–21

In addition to the general controls imposed by the RPI-7.5 per cent price cap, Condition 24A.14 provides that individual prices within the basket should not increase by more than the RPI in any 12-month period ending June 30 with the exception of exchange line rentals, for which the limits are RPI + 2 per cent for ordinary exchange lines and RPI + 5 per cent on wholesale lines.

Price controls generally are being reviewed as part of the Director General's on-going review of interconnection and related issues. The latest of his statements, "Effective Competition: Framework for Action" (July 1995), proposes that the RPI + 2 per cent constraint on exchange line rentals should be removed, paving the way for the abolition of access deficit contributions some time in early 1996. Such changes would require modifications to BT's licence.

Price Controls for Private Circuits: Condition 24B

5–22   At the end of 1993 BT's licence was modified with its consent in order to limit increases in the aggregate prices of each of three baskets of private circuit services comprising—

- all inland analogue private circuits;
- all inland digital private circuits;
- all international private circuits, both analogue and digital.

In addition to the aggregate control on each basket, increases in individual private circuit prices are limited as follows—

- RPI + two for all analogue private circuits, whether inland or international and
- RPI + one for all digital private circuits, whether inland or international.

These price controls for private circuits apply for four years until July 31, 1997.

Residential Low User Scheme

5–23   Under Condition 24D BT has introduced a scheme to reduce the aggregate of its charges for line rentals and for calls payable by those of its residential customers who make relatively few telephone calls. This scheme has to comply with guidelines agreed between the Director General and BT. These require a discount on line rental and call unit charges where call units per quarter are below a specified threshold and in such circumstances line rental charges are not to exceed BT's standard published rate and a specified number of call units per quarter is to be allowed free of charge. Provision is made for notification by BT to those persons who are or become eligible to participate in the scheme which is open only to residential customers who have a single line.

(ix) Conditions 17 to 23 inclusive, 35, 36 and 39: Fair Trading

The fair trading conditions in BT's licence are selective in that they are aimed at particular anti-competitive practices, which may not be exhaustive. It is the author's view (as expressed in the first edition) that the competition provisions of the regulatory regime, primarily embodied in PTO licences such as BT's, would have been considerably strengthened by a "catch-all" licence condition along the lines of a domestic equivalent of Articles 85 and 86 of the Treaty of Rome. Indeed such a provision would have been entirely consistent with and justified by the Director General's duty, arising under section 3(2) of the 1984 Act, "to maintain and promote effective competition between persons engaged in commercial activities connected with telecommunications in the United Kingdom".

5–24

For example, predatory pricing which is anti-competitive in its effect is not expressly covered in any of the current fair trading conditions (although the Director General is attempting to do something about this in the context of interconnection—see below). Moreover although Condition 17.2 relates to BT's and other licensed PTOs' competitors insofar as it prohibits the licensee unfairly favouring its own business to the competitive detriment of others, its application is dependent upon there being undue preference or undue discrimination.

Undue Preference and Undue Discrimination

Condition 17 provides that BT shall not show undue preference to, or exercise undue discrimination against, any persons with respect to its various services. Significantly, in the case of its telecommunication services, the condition applies only to those services which BT is by its licence obliged to provide. It is also one of a special category of conditions which, by virtue of section 8 of the 1984 Act, are required to be included in licences granted to operators of public telecommunication systems. In this respect it may be distinguished from later conditions which are included in BT's and other PTO's licences pursuant to sections 7(5) and 3 of the 1984 Act.

5–25

The primary purpose of Condition 17 appears to be to protect consumers and users and only incidentally to maintain effective competition. It may be significant that Condition 17 falls in the part of BT's licence which deals with provisions referred to in section 8 and does not fall within the part of the licence which includes conditions under section 7 (and thus under section 3 of the 1984 Act) which is where the conditions relating to competition are normally found.

Any question as to whether or not BT has indulged in undue preference or undue discrimination is to be determined by the Director General— Condition 17.3.

There is no prohibition on discrimination or preference as such—the test is whether it is "undue". The Oxford English Dictionary meaning for "undue" includes "improper", "unreasonable" and, in particular, "excessive". Further in the House of Lords decision in the *South of Scotland Electricity Board v. The British Oxygen Company* Lord Keith stated that the word "undue" encompassed not only illegitimate reasons but could also mean "excessive". He also considered that it was a matter of fact and degree, in effect a jury question.

5-26   In their further appeal in the same action their Lordships went further into the scope of the phrases "undue discrimination" and "undue preference" and, for example, dismissed the Electricity Board's argument that the complainant would have to show there was another customer paying less in order to begin to prove undue discrimination. The decision is also interesting for its discussion of the relevance of costs to discrimination, Lord Merriman stating, for example, that the phrase "shall not exercise any undue discrimination" had to be considered as a whole and that "a fair distribution of the cost of supplying electricity as between one class of users and another is not to be left out of consideration any more . . . 'than any other circumstance which would affect mens' minds'". This case is also useful for its review of many of the older cases including particularly those under various Railway Acts, where, again, the prohibition on "undue preference" appeared.

The *British Oxygen* cases mentioned above concerned section 37(8) of the Electricity Act 1947, which prohibited area electricity boards, in fixing tariffs, from showing undue preference or exercising undue discrimination against customers. Over the years Electricity Board cases have therefore produced a number of useful decisions on the meaning of these phrases.

The first case involving Condition 17 was where the Director General determined that BT had provided the William Hill Raceline Service at a lower rate than that available to other premium service providers. There have been other much more recent cases involving different tariff packages of BT such as Option 2000 and "Sunday Special" which are discussed in Chapter 4, paragraph 4–15 above.

5-27   The presence of undue discrimination or preference falls to be determined by the Director General in accordance with the facts and by applying relevant legal principles which have emerged from the cases. If and where the Director General establishes that such discrimination or preference is present he should consider whether or not to take action in accordance with his duties in section 3 of the 1984 Act, particularly section 3(2) which deals specifically with competition.

It should be noted that the rule against undue preference and undue discrimination, whether in BT's or any other PTO licence, only applies to those services which are provided in accordance with an obligation imposed by or under the licence.[4] Since BT has a universal service obligation contained in Condition 1 this means that its main telecommunication services are covered by Condition 17. Similarly Mercury, although not having the same universal service obligation, does have extensive service obligations (see below at para. 5–71) and is normally bound by the same rule. However, as discussed below,[5] the other fixed link PTOs recently

---

[4] See *Maystart Limited v. Director General of Telecommunications*, discussed at Chap. 3, para. 3–07.
[5] See para. 5–71.

licensed do not have, initially at least, service obligations and are thus not bound by this rule.

In September 1991 a new provision (Condition 17A) was added to BT's   5–28 licence which allowed for differential charging where this complied with special guidelines agreed between BT and the Director General. These guidelines provide that BT's charges should be set at a level no lower than a level equal to BT's highest standard price for providing the relevant service less 80 per cent of the difference between the highest standard price and the fully allocated costs of providing that service; and any combination of charges offered as a package should:

(a) be designed to appeal to a reasonably broad section of customers;
(b) be available to all customers to meet conditions specified by BT;
(c) not include a condition which establishes criteria for eligibility which unfairly discriminates against a class of customers who might otherwise be eligible for the package; and
(d) achieve greater uniformity between the ratios of the revenues accruing from different classes of customer to the long-run incremental costs of providing service to those classes of customer, and should not be below the long-run incremental cost of providing the relevant services to those classes of customer.

More recently, as part of the Director General's on-going implementation of greater separation in interconnection, he has introduced[6] a new condition in BT's licence.

The new Condition 24F, which came into effect on April 1, 1995, requires BT to give the Director General a "Price Control Notice" whenever it proposes to change the price of certain services. The information required is intended to enable the Director General to verify whether retail or network costs are involved in any retail price reduction and, if they are, to determine whether or not the price change will give rise to undue discrimination or undue preference. Equally, if any price reduction by BT results in that price not covering the fully allocated costs of conveyance, BT must adjust appropriately its interconnection charges ("Standard Service" charges) to other operators.

The effect of Condition 24 should therefore be that where a network com-   5–29 ponent or network part's costs are not fully recovered as a result of a retail price reduction, parallel reductions are made in BT's prices for Standard Services.

Condition 24 represents a mechanism for controlling one form of anti-competitive pricing on which the Director General has chosen to focus but there remains the problem that retail tariffs may also be set on a predatory basis in order to subvert the competitive position of other operators and there is no direct control on such activities; the ability for another operator to

---

[6] See determination with respect to BT and Mercury dated December 2, 1993; determination with respect to ADC waivers granted to ACC, WorldCom and others dated July 1994.

demand interconnection facilities on the same basis may be only part of the story if in the meantime that operator's customers are leaving in droves to sign up with its principal competitor. Given the weaknesses of U.K. competition law it remains a disappointment to many that there is no general prohibition on anti-competitive activity in BT's and other public operators' licences.

Prohibition on cross subsidies

5–30   Condition 18 provides that the Director General may step in to prevent BT unfairly cross-subsidising its apparatus supply business, its production of telecommunication apparatus, its provision of mobile radio services or its supplemental services (value added services) business. Any material transfer between any part of its sub-businesses and these other businesses must be recorded at full cost in its accounting records ("full cost" being defined as including the market rate of interest for the money transferred). "Unfairly cross-subsidising" is, however, not defined and is left to the Director General to interpret. Subsidisation within BT's systems business or from another of its businesses to support the system business is not prohibited.

In the recent *Talkland* matter the Director General had to deal with allegations of cross subsidisation and undue preference. After concluding his investigation the Director General announced (May 17, 1994) his finding that cross-subsidies had taken place but that most of the undue preference allegations were not well-founded.

Under a new licence condition (20B.15) the Director General, where satisfied that BT is unfairly subsidising or cross-subsidising, or has unfairly subsidised or cross-subsidised, any of certain specified businesses (see under "Separate Accounts" below) he may direct BT to remedy the situation.

Access Charges

5–31   Condition 19 empowers BT to impose access charges (a somewhat misleading title in view of the later introduction, in 1991, of access deficit contributions in Condition 13.5A of BT's licence; the two are not to be confused). According to Condition 19.2(c) these charges are intended to cover BT's net costs incurred in providing directory information services, public emergency services, public callbox services and services for those who are disabled or whose hearing is impaired and further to cover BT's losses in fulfilling its universal service obligations under Conditions 1 and 2. As such, certainly in terms of the relationship to universal service, the rationale for access charges, which have never yet been levied, is not dissimilar to that for access deficit contributions. Certainly there appears to be overlap between the two, as the access deficit arises from BT's loss on provision of exchange line access to customers.

Separate Accounts

5–32   Condition 20 obliges BT to maintain separate accounts in relation to its systems business, its apparatus supply business and its supplemental services business so that these can all be assessed and reported on separately. BT's accounting statements must set out the revenue and financial position of each of those businesses and include details of items charged between the various BT businesses or apportioned between them.

On April 1, 1995 a modification of this licence condition came into effect, whereby BT is required to prepare and publish separate financial statements for a number of separate businesses as agreed between the Director General and BT before the new condition came into force. In the first instance, the businesses are the access business, the apparatus supply business, the network business, the residual business, the retail systems business and the supplemental services business as defined. The costs, revenues and assets comprised in each business and the level of disaggregation of each business are as agreed between BT and the Director General before the modification came into force.

Apparatus Production: Structural Separation

Condition 21 required BT to transfer its business of production of telecom-munication apparatus to a subsidiary and to secure that any such subsidiary did not engage in the business of running telecommunication systems. Clearly one implication of such requirement would be that through such structural separation and separate accounting BT should not be able to cross-subsidise apparatus production from its telecommunication services business. Under Condition 21 the Director General can also require BT not to acquire apparatus from its separate apparatus production company without first complying with the open tender procedures specified in this condition.

5–33

Prohibition of Preferential Treatment

Condition 22 requires BT to give other suppliers of apparatus similar facil-ities in any area where a person engaged in BT's systems business delivers or connects apparatus as an incidental of carrying on that business; equality of treatment must also be afforded where a person normally engaged in the apparatus supply business of BT arranges for BT's systems business to install apparatus or provide certain services. Clearly, in its ability to combine the supply of apparatus and services BT is potentially at a competitive advantage and this condition is an attempt to try to restore some balance for BT's competitors.

5–34

Alterations to BT's Systems

Condition 23 is designed to protect both BT's suppliers and its competitors in that if BT is about to change its systems or apparatus in a way which would necessitate modifications to systems or apparatus or even lead to equipment redundancy, BT must first notify the Director General and also follow a procedure for prior consultation and advance warning to suppliers affected. Such a procedure has been settled.

5–35

Prohibition of Linked Sales

Condition 35 prohibits what are normally termed, in competition law par-lance, "tie-ins". Specifically BT is prohibited from requiring a customer for

5–36

any particular service or apparatus to take any other service or apparatus from BT as well, unless there is a good technical reason for this.

The same condition contains a restriction on "packaging" services for the same purpose, where the terms or conditions are more favourable than would apply if the additional service or apparatus were not provided, unless the Director General agrees otherwise. The prohibition is relaxed in Condition 35.3 to enable BT to offer quantity discounts, for example.

Prohibition of Certain Exclusive Dealing Arrangements

5–37   Condition 36 is a strangely conceived and worded condition, preventing BT, without the written consent of the Director General, from imposing extraneous restrictions on its suppliers, with respect both to supply of unrelated apparatus or services and to intellectual property rights.

The aim of this condition is perhaps more accurately concentrated in Condition 36.2, where a requirement by BT for the grant to BT of sole rights in respect of the supply of customer apparatus can be prohibited by the Director General where he is satisfied that the suppliers concerned are not genuinely willing to confer such rights on BT. Unfortunately, Condition 36.4 contains so many exceptions to these rules that much of their good is undone. For example, BT is free to require that "other" telecommunication apparatus be supplied or some other telecommunication service be provided by the supplier where that other apparatus or service "is reasonably related to that supply or provision"; the terms in which this has been drafted are too wide and it would have been preferable had the line normally taken by, for example, the European Commission been followed, reference could then have been made to items which are technically indispensable to the proper exploitation or use of the apparatus or service.

5–38   Again under Condition 36.4 BT is permitted to require the transfer to it of any intellectual property right which the Director General agrees "is necessary or desirable to facilitate the running of any of [BT's] Systems". Admittedly, the Director General's discretion is retained and may give sufficient protection against abuse, but the apparent authority given to any attempted appropriation of a patent or copyright holder's rights seems, on the face, unnecessary and even unjustifiable; possibly the draughtsman meant only to refer to the grant (*i.e.* licence) of intellectual property rights rather than their complete disposal, but the wording is ambiguous. A further paragraph (f) in Condition 36.4 states that BT is free to require the transfer to it of any interest in intellectual property "to the extent that that is reasonably necessary for the purpose of enabling [BT] to secure alternative sources of supply of telecommunication apparatus": again, a term which, although of no direct legal effect on anyone other than BT, might encourage the perception that BT had a right to demand such a transfer. Indeed generally perhaps the best that can be said for these intellectual property provisions is that of themselves they are not enforceable against the legal owners of the rights, who are not bound in any way by the terms of BT's licence.

Intellectual Property

Condition 39 provides that the Director General may give directions to   5–39
prevent BT using intellectual property rights to restrict the availability of
any product where this is liable to prevent the connection of any telecommunication system or apparatus or the provision of any service. The Director General is also empowered to direct BT to grant intellectual property
rights to other persons in order that such connections or services may be
made or provided. There are saving provisions where any requirement on
BT would result in it breaching the terms of any licence or assignment of
intellectual property rights. These provisions can be contrasted with other
industries under monopoly control where compulsory licensing procedures
are not generally available. This condition may therefore be used to prevent
BT from withholding the supply of apparatus, but only where this is contrived by the use of intellectual property rights. A more pro-competitive
licence condition might have extended this principle to any refusal to
supply whether or not the product is, say, patented or the subject of copyright. However, the Competition Act 1980 and/or Article 86 of the Treaty
may be operable in such circumstances.

(x) Conditions 13 to 15 inclusive: Interconnection—Connection of Other Systems

Condition 13 (connection of systems providing connection services) is one   5–40
of the most important conditions in BT's licence affecting liberalisation. It
secures for other PTOs the right to require an interconnection agreement
with BT in order that messages may pass from the other operator's system
into BT's system, typically to reach BT customers and in order also, where
the other PTO is a long-line operator (*e.g.* Mercury) that customers of both
BT and that operator may exercise freedom of choice as to the system by
which their messages are conveyed. Essentially, these principles, to be found
in Condition 13.1(a) and 13.1(b) respectively, represent two of the main
objectives of liberalisation, "any to any" (the ability of any customer of
one public operator to call the customer of any other public operator) and
"customer choice".

The structure of Condition 13.1 is quite important in that the Condition
will apparently not be operable, and therefore recourse to the Director
General for determination of "permitted" terms and conditions (see below)
will not be available unless the agreement being sought by the interconnecting operator is to require BT:

(a) to connect its systems to those of the operator and to establish and
    maintain points of connection for such purpose (of sufficient capa-
    city and in sufficient number to enable messages to be conveyed
    between the systems, etc.);
(b) where the operator is a long line public telecommunications oper-
    ator, to establish such points of connection as will enable customers
    to exercise freedom of choice as to the systems by which their mess-
    ages are conveyed; and

(c) to provide such other telecommunication services as the Director General determines are reasonably required "to secure that Points of Connection are established and maintained" and to enable the operator effectively to provide its connections services (as defined in Condition 13.9).

5–41   The preamble to Condition 13.1 states that unless it is impracticable to do so, BT *must* enter into an agreement with an operator for these purposes if that operator is licensed to run a "Relevant Connectable System" and requires BT to do so.

A relevant connectable system refers to a system which is licensed to be connected with BT's and to provide services for reward to the public but which is *not* run under a class licence for the connection of which BT offers standard terms and conditions satisfying the requirements of Condition 16, although the Director General has the discretion to exclude a system which would otherwise be a relevant connectable system for these purposes.

The procedure under Condition 13 is that the operator should seek agreement with BT on a number of matters specified in Condition 13.4 and if "after a period which appears to the Director to be reasonable" (generally accepted, particularly in the case of a new interconnection agreement, to be in the region of three months) an agreement has not been achieved, either party may apply to the Director General for him to determine the terms and conditions of their proposed agreement. The criteria which the Director General must bear in mind for this purpose are set out in Condition 13.5 and 13.5A, and, in relation specifically to Long-Line PTOs, in Condition 13.6.

5–42   No period is specified in the licence for the giving of the Director General's determination; nor, so far, has any reliable pattern emerged. The 1993 BT/Mercury determination process took 18 months, which is exceptional. Much depends on the complexity of the matter, how earnestly the parties are pushing for issue of the determination and whether there are, and if so how relevant may be, any previous determinations which serve as a precedent or guide. Condition 13 is silent upon how soon after the Director General's determination an agreement should have been entered into, incorporating the terms and conditions so determined, but it can be assumed that, depending on the circumstances, the parties thereafter have a reasonable period of a few weeks or, at the most, months in which to do so and if BT were to default in this (the other operator being willing to sign) that it would be in breach of its licence.

Enforcement of interconnection agreements is to some extent underpinned by Conditions 13.8 and 13.8A. Condition 13.8 was quite limited in its scope, giving the Director General power to direct BT to do certain things only in order to ensure that connections are established or maintained and that messages are conveyed. This inadequacy was recognised and in 1991 a new Condition 13.8A was added. This Condition gave the Director General the additional power to require BT to perform any obligation covered by an interconnection agreement. This applied where the Director General considered that the obligation ought to be performed to achieve the purposes of Condition 13.1 and the operator was not able to enforce the agreement so that the obligation was performed within a reasonable time. In this case the "balance of convenience" required the Director General to take action rather than leave it to the courts, and the operator must have performed all its obligations relevant to the obligations of BT which were not being performed.

Terms and Conditions for Interconnection

Interconnection agreements are freely negotiable in two senses. First of all, interconnecting operators may, if they so chose, contract entirely outside the terms and ambit of Condition 13, but this would be unusual if not imprudent given that the parties would thus forego the benefit of recourse to the Director General in circumstances where they fail to come to an agreement in their negotiations. Secondly, even where negotiations have been quite firmly set within the context of Condition 13, its provisions clearly envisage that the parties may succeed in negotiating the contractual terms and conditions without the need for referral to the Director General.

5–43

Introduction in 1995 of BT's standard terms is likely in practice to "depress' the negotiability of many terms, particularly where the view may be taken that the Director General is likely to support the BT approach.

Condition 13.3 stipulates that BT may require that interconnection agreements should be subject to terms and conditions, but only such terms and conditions as are "permitted" in accordance with Conditions 13.4, 13.5 and 13.6. The latter two sub-conditions deal with the financial aspects, connection and conveyance charges, whilst 13.4 gives a general description of the scope of the matters to be covered.

Condition 13.5A was added to BT's licence as part of the post-duopoly review modifications following the 1991 White Paper, in particular to incorporate access deficit contributions into interconnection payments. There is in fact some overlap between Conditions 13.5A and 13.5. Both these conditions set the basis on which the Director General should determine "permitted" terms and conditions between interconnecting operators where these have not been agreed by them, being terms which, according to Condition 13.5(a)[7] "appear to the Director reasonably necessary (but no more than reasonably necessary) to secure [a]" that the operator pays to BT "the cost of anything done pursuant to or in connection with the agreement including fully allocated costs attributable to the services to be provided and taking into account relevant overheads and a reasonable rate of return on attributable assets". Compare this with Condition 13.5A.3(a)[7] which requires that the charges payable to BT by an operator should cover "(a) [BT's] fully allocated costs of the conveyance calculated on a historic cost basis, including a full contribution to relevant overheads . . .". It is surprising and indeed a cause for some confusion that when BT's licence was modified to include Condition 13.5A, the opportunity was not taken to make clear any distinction between that new Condition and Condition

[7] Amended April 1, 1995.

13.5(a) or even to delete 13.5(a) entirely. We are now, unfortunately, left with two overlapping provisions expressing similar concepts in slightly different ways.

5–44   Condition 13.5(a) was the basis for the first (1985) determination by the Director General, with respect to the terms and conditions for telephony interconnection as between BT and Mercury. The Director General's second determination between these parties made in December 1993 was however also based on 13.5A and included charges to cover the access deficit contribution referred to in Condition 13.5A.3(c).

Condition 13.5A is prefaced by the following—

"Where in pursuance of such an application as is referred to in paragraph 13.5, the Director determines any charge, or the means of calculating any charge, payable to the Licensee by the Operator, he shall do so in accordance with the following provisions".

The structure of the payments, and the cost basis on which they are to be derived, is then set out in the remainder of the condition.

5–45   Condition 13.5A.1 appears to add very little as it simply provides that if the operator interconnecting with BT is collecting the revenue from the calling party then that operator must pay BT a charge in respect of the call. Condition 13.5A.2 then, somewhat out of order, describes the make up of the charges which are to differ according to—

(a) whether the "call" is a local, national or international call;
(b) the portion or portions of BT's systems by means of which the "message" is conveyed; and
(c) the time of day at which the call is made.

Note the intermingling of "call" and "message" which does not particularly aid interpretation. The charging structure is thus set according to BT's retail tariff structure as we know it at present.

Condition 13.5A.3 requires that the charges referred to in 13.5A.2 "shall cover" three elements—

(a) BT's "fully allocated costs of the conveyance calculated on a historic cost basis, including a full contribution to relevant overheads, calculated on the basis of information supplied by [BT] drawn from BT's audited FRBS [financial results by service] figures for financial years ending on or before 31 March 1995 and for subsequent financial years, the audited information in BT's financial statement for the relevant business.";
(b) the applicable rate of return applied to the relevant capital employed (defined in 13.5A.8 as the single rate of return which is notified by the Director General to BT from time to time as reasonable for the systems business); and
(c) until the expiry of the period during which BT is subject to restrictions on increases in residential and single-line business rental charges, an access deficit contribution.

5–46   This contribution or "ADC" as it is generally known, is defined in

Condition 13.5A.3 to mean, in relation to any financial year, the difference between—

(a) BT's aggregate revenue arising from connection charges and periodic charges (*e.g.* rentals) in respect of its exchange lines and

(b) its fully allocated costs incurred in respect of those services on the basis of BT's audited FRBS figures (after 31 March 1995, for the relevant business), and its return on capital employed in providing such services calculated at the applicable rate of return.

Thus it is fundamental to ADCs that they should only last so long as BT is subject to restrictions on increases in its rental charges and that they should be payable only so long as BT makes a loss on its so-called access business, in other words through its provision, use and maintenance of exchange lines to customers. The rationale for ADCs is further explained in various statements of the Director General issued before and at the time of their introduction in 1991.[8]

Condition 13.5A.4 provides that ADCs shall be assessed separately in   5–47 relation to local, national and international calls at a rate of contribution per minute of traffic in respect of each such category, according to a formula set out in that condition. The contribution is only to be paid in respect of a portion of BT's systems, or "segment" as it is normally described in the Director General's determinations, which connects a customer's terminal apparatus with a point on BT's systems at which there is an interconnection with the operator's system. In simple terms therefore the ADC is triggered by conveyance over the portion of BT's system used to collect and deliver calls at their origination and destination.

ADC *"waivers"*

Condition 13.5A.5 was introduced to mitigate the effects of ADCs, ostens-   5–47a ibly for those operators seeking to enter or maintain a presence in a particular market. Its incorporation was a direct result of the backlash which resulted from the Director General's announcement of his intention to introduce ADC's, and is further explained in his statement of July 3, 1991. The Condition provides that the Director General may determine that before July 1, 1997 an operator should make either no or only a partial contribution to the access deficit in respect of its first 10 per cent of market share, subject to the proviso that BT should receive a full contribution—

(i) in respect of calls conveyed using equal access either by preselection or on a call-by-call basis or

(ii) where BT's market share is less than 85 per cent, with respect to the market share of the interconnecting operator and all other operators (taken together) which is in excess of 15 per cent of the total market.

Note that the "total market" for these purposes appears not to be the   5–48 same market as the "market" of the relevant operator to which the 10 per cent limitation (mentioned above) applies. It would be the total national and international call market.

[8] Director General Statements of May 10, July 3 and July 24, 1991.

Under Condition 13.5.A.5 therefore any ADC "waiver" under this condition could be cut back to the extent that the 15 per cent threshold is
breached. This effectively means that waivers granted to all operators could
in those circumstances be reduced on some basis, pro rata to market share
or otherwise, which is not explained in the condition. There is a further
clawback possibility in Condition 13.5A.5(b) which applies where the
interconnecting operator achieves a market share of 25 per cent or more
(again of its own relevant market, not of the total market), so that in those
circumstances the operator should pay a full contribution to BT's access
deficit on all its traffic, not just that in excess of the 25 per cent threshold.
At that stage the operator's entitlement to waivers under 13.5A.5 would
disappear altogether.

The remainder of Condition 13.5A.5 explains the operation of the
waiver provisions and the basis of the grant of waivers, particularly in
sub-paragraph (e). The Director General's policy with respect to waivers
is further explained in his explanatory documents to the determinations so
far made under the condition.

According to these statements the policy of the Director General is that
waivers should be used to encourage:

- "the development of a wide range of services for the consumer to
  choose from; and
- the provision of those services at the lowest sustainable price".[10]

5–49   To those ends the Director General continues, he will have regard to—

  (i) the desirability of granting waivers to those companies likely to
     encourage BT to achieve greater efficiency on a sustainable basis;
 (ii) the desirability of maintaining maximum flexibility to meet the
     demands of changes in the market;
(iii) the extent to which companies will be providing competition to BT
     in the local loop, particularly in providing services to residential
     customers.

As part of the follow-up to the Consultative Document issued by the
Director General in December 1994, "A Framework for Effective Competi-
tion", the Director General proposes in his July 1995 statement "Effective
Competition: Framework for Action" the removal of access deficit contri-
butions from Condition 13.5A (a modification which would require either
BT's consent or a recommendation by the Monopolies and Mergers
Commission) on the basis of the *quid pro quo* of eliminating the RPI + 2
per cent cap on BT's line rental charges.

Condition 13.5B provides that any determination by the Director under
Condition 13.5 may also include a determination as to charges payable by
BT to the interconnecting operator in respect of its conveyance of messages
for BT, including the underlying principles for such charge, as set out in
Conditions 13.5B.1 and 13.5B.3, the latter envisaging a contribution
towards the access deficit of the interconnecting operator.

9–10 See, *e.g.* Annex C of BT/Mercury determination December 3, 1993.

Publication of Interconnection Agreements

Under Condition 16A of BT's licence, BT must publish either an   5–50
adequate description of any interconnection agreement which it enters
into and the precise method of calculation of the interconnection charges,
or the agreement itself, not later than 28 days after entering into the
agreement. In order to protect commercial confidentiality provision is
made for the exclusion of any matter on which the Director General's
consent may be obtained following representations to him by BT, by
the interconnecting operator or by any other person having an interest.
Where BT opts for the first alternative, of providing merely an "adequate
description" of the agreement, then its description of the method of
calculation of the interconnection charges must be "such as to enable
those charges readily to be calculated by a third party". Publication is
effected by sending a copy to the Director General and keeping a list
of the documents, together with details of the persons to whom copies
of the documents or agreements are provided, BT being required to send
such a copy to anyone who so requests on their paying a reasonable
charge.

Note that Condition 16A does not have any effect on agreements
already entered into prior to the introduction of the condition and BT
in particular has a number of these. However, variations to these "old"
agreements are picked up and required to be published (16A.6 and
16A.7).

Standard Terms

Condition 13.8B.1 provides that if the Director General considers there is   5–51
likely to be a category comprising a sufficient number of operators seeking
interconnection determinations and for whom standard terms and conditions should be appropriate, he may require BT to publish such standard
provisions. Condition 13.8B.2 continues that if the Director General is satisfied a standard item or condition proposed by BT is unreasonable and BT
has acted unreasonably in negotiations on that term or condition he may
determine that BT should modify it for the purpose of the particular negotiation or generally in its standard provisions.

Condition 13.5A.3A (inserted by modification of BT's licence effective
from April 1, 1995) refers to interconnection charges for BT's Standard
Services and requires that BT should submit to the Director General
forecasts of its fully allocated costs of conveyance as well as a forecast of
the access deficit contribution, at least two months before the commencement of the relevant financial year. Under the new Condition 13.5A.3
(introduced at the same time) the Director General is to re-calculate and
re-determine interconnection charges for Standard Services at the end of
each financial year; provision is made for adjustments and payments
between BT and the interconnecting operator if these interconnection
charges as finally determined are different from interim charges paid on the
basis of BT's forecasts.

BT has developed standard terms and conditions, including standard price lists, for interconnection. In view of Condition 13.8B and new Condition 13.5A.3 on determination of Standard Prices the frequency and range of requests to the Director General to determine terms and conditions for particular interconnection agreement could reduce significantly over the course of time.

Director General Interconnection Determinations

5–52   To date, the Director General has made a number of different determinations under Condition 13 but without doubt the most important have been the two with respect to the terms and conditions for interconnection of the systems of Mercury Communications Limited to those of BT, first on October 11, 1985 and the second on December 2, 1993. I have set out in the tables below the conveyance charge rates and access deficit contributions set by the Director General in the later of these two determinations, these being the rates currently applicable. However, by its terms the 1993 determination is subject to immediate and annual review, renegotiation and re-determination by the Director General; this could lead to fairly immediate and regular adjustments to these rates.

Segment	Cheap Rate	Standard Rate	Peak Rate
Local Exchange	0.67 pence	1.16 pence	1.53 pence
Tandem Local	0.81 pence	1.41 pence	1.85 pence
Tandem Short National	0.95 pence	1.65 pence	2.17 pence
Tandem Long National	1.20 pence	2.08 pence	2.73 pence

Call Type	Cheap Rate	Standard Rate	Peak Rate
Local	0.27 pence	0.54 pence	0.71 pence
National	0.94 pence	1.48 pence	1.93 pence

International (pence per minute)

Charge Band Outgoing	1	2	3	4	5	6	7
Peak	2.10	3.76	9.26	10.03	11.11	12.51	15.73
Standard	2.10	3.76	9.26	9.24	10.66	12.51	15.73
Cheap	1.71	3.14	7.83	7.94	9.00	8.95	12.62
Incoming	1.96	3.55	8.60	8.76	9.70	11.48	15.13

Charge Band Outgoing	8	9	10	11	12	13
Peak	15.07	26.99	29.29	26.94	28.28	58.01
Standard	15.07	26.88	29.29	26.94	28.28	58.01
Cheap	12.20	23.03	24.83	25.69	27.19	57.08
Incoming	14.37	25.17	27.22	26.68	28.03	57.77

5–53   Many cable television operators have negotiated interconnection agreements with BT, elements of which have been referred to the Director General for his determination. All these determinations are by virtue of section 19(2) of the 1984 Act on the public record, although the Director General has the power under section 19(3) to keep off the register any provision which would be "against the public interest or the commercial interests of any person" and has done so in certain instances.

There is nothing in the 1984 Act which actually requires the Director General to determine the terms and conditions of interconnection agreements; this is a power arrogated to him entirely by virtue of the interconnection licence condition (*e.g.* BT's licence, Condition 13). Moreover, the practice of the Director General, as evidenced in his various interconnection determinations, of requiring that review clauses should be embodied in interconnection agreements and should include provision for recourse to the Director General if the parties fail to reach agreement in the course of such a review, is one conceived by him, as he puts it in the explanatory document to his December 1993 BT/Mercury Determination, "to ensure his powers to make a determination under the terms of the [review] clause" and thus to enable him to make further determinations from time to time at the request of the contracting parties. In practice, parties who succeed in negotiating interconnection without the intervention of the Director General often include their own review clauses which provide for the Director General to make a "determination" on any aspect which the parties fail to agree. When these clauses first began to emerge from operators negotiations and the Director General was asked to consent to taking on such a role he was unwilling to give such consent and preferred to deal with the matter by having the parties request him to "determine" the relevant review clause (under for example BT's licence—Condition 13). In the view of the writer this policy may have been the manifestation of a desire by the Director General, as regulator, not to be drawn into a contractual relationship with the interconnecting parties and therefore to deal with the parties, as a regulator, more remotely by making a determination, ostensibly pursuant to his powers and duties under the relevant interconnection licence condition.

The December 2, 1993 determination made by the Director General, 5–54 pursuant to the review provisions of the interconnection agreement between BT and Mercury Communications Limited, has been the subject of recent litigation. Shortly after the determination was issued, Mercury issued an originating summons asking for a declaration that on the true construction of BT's licence, and in particular Conditions 13.5 and 13.5A, the relevant costs and overheads referred to in those Conditions should bear a particular meaning. The defendants, the Director General of Telecommunications and BT, applied to strike out the summons on the grounds that it was frivolous or vexatious or otherwise an abuse of the process of the court and/or under the inherent jurisdiction of the court; in particular the Director General and BT argued that the relationship between the Director General and both Mercury and BT in connection with his determination was governed solely by public law and that the only procedure open to Mercury to raise the issue was by way of judicial review.[11] The Lords rejected these contentions, as had the judges at the preceding levels.

Rather more problematic was the argument by the detendants that the issues raised for interpretation by the court fell within the exclusive competence of the Director General. On this point the majority of the Court of Appeal had found in their favour. However, in the House of Lords' judgment the Director General's interpretation was open to review by the court as there was no express or implied provision that these matters should be remitted exclusively to him even though in order to carry out his functions he was obliged to interpret them in the first place.

Finally, the defendants had argued that the declarations sought by Mercury were future, academic and hypothetical questions, even though the Director General had already made his determination and was likely to base future determinations on the same interpretation. The Lords found that the issues raised by Mercury were by no means hypothetical. Generally, on the question of striking out for abuse of the process of the court, the Lords emphasised the discretion exercised by the trial judge (which had been in favour of Mercury) and the fact that this should stand unless the arguments were clearly strong in favour of a different result on appeal.

As a result of the Lords' judgment, Mercury's summons for a declaration can now proceed, but at the time of writing the matter has not come to trial on the substantive issues.

Connection of Other Systems and Apparatus

5–55   Condition 14, by contrast with Condition 13, does not apply to relevant connectable systems but imposes upon BT an obligation to connect all other parties' systems to its own system where these systems are licensed to be connected. BT is similarly obliged to connect apparatus to its system where such apparatus is duly approved for this purpose. There are, as usual, exceptions to BT's obligations, in particular where the apparatus did at one time, but does no longer, comply with the relevant standard or standards. There is also an exception where, although conforming to this standard, the apparatus is liable—in BT's opinion—to cause death or personal injury, or damage or materially impair the quality of telecommunications service provided by BT, and the Director General has not expressed a contrary opinion.

Condition 15 permits a person, who is authorised to provide telecommunication services to others, to provide such services whilst its system is lawfully connected to BT's systems.

(xi) Condition 34B: Numbering

5–56   Condition 34 was superseded by 34B as part of the amendments made in 1991.

Condition 34B.5 requires the Director General to determine a specified

---

[11] *O'Reilly v. Mackman* [1983] 2 A.C. 237; also *Roy v. Kensington & Chelsea & Westminster FPC* [1992] 1 A.C. 64, 628.

numbering scheme ("the scheme") in accordance with the national numbering conventions ("the conventions") published in accordance with Condition 34B.9 and to allocate numbers from the scheme to BT in accordance with the conventions.

Under Condition 34B.6 BT is to adopt a numbering plan ("Plan") for the numbers allocated to it by the Director General and to furnish details of and any subsequent material changes to the plan to him. BT is also to furnish details of its plan and any material changes on request to "any other person having a reasonable interest", which should embrace other operators.

Condition 34B.9 explains the conventions. These are to be a set of principles and rules published from time to time by the Director General after consultation with interested parties who are members of the Telecommunications Numbering and Addressing Body and "if deemed appropriate" with end users. Under Condition 34B.9 the conventions are to govern the specification and application of the scheme and the plan of BT and may also include other matters relating to the use and management of numbers, as further detailed in 34B.9.

Condition 34B.10 requires that in setting up and amending the scheme   5–57
and the conventions the Director General is to have regard to certain specified criteria including the need for sufficient numbers to be made available, the need for compatibility of numbering plans, the convenience and preferences of end users and the requirements of effective competition.

In 1994 OFTEL assumed management of the UK numbering scheme. In March 1994, OFTEL published a consultative document containing draft numbering conventions.

Condition 34B.11 requires that BT should provide numbering portability as described in Condition 34B.12.

On February 22, 1995 the Director General issued a notice of his proposal to modify Condition 34.B of BT's licence so as to introduce a power for the Director General to determine the charges to be levied by BT on other operators to provide number portability. BT was unwilling to agree to such a modification and, as a consequence, on April 27, 1995 the matter was referred to the Monopolies and Mergers Commission.

(xii) Conditions 42 and 43: Wiring

Use of wiring is important for liberalisation and the penetration of competi-   5–58
tion. Before 1981 BT's systems embraced wiring and apparatus on customer premises and there was no concept of a network boundary at the interface of the public system with the customer's system, as there is today. An historical overhang from this arrangement was that BT owned much of the wiring in customer premises throughout the country. A customer may wish to make use of this wiring to enhance or reconfigure his system; if it cannot be worked on independently of BT's system, delays and complications could result.

Accordingly Condition 42 restricts BT's installation of integrated wiring and secures unencumbered access to that wiring on reasonable terms where it already exists. Condition 42 provides that BT shall not, except in limited circumstances, install a line on customer premises in such a way as to prevent access to the wires or cabling of any other system on such premises. In relation to existing premises BT is also obliged to install additional apparatus in order that a system on the customer's premises may be run by someone other than BT and so that operations may be carried out on other systems separately from BT's systems.

5–59   Under Condition 43 BT is obliged to make available (*i.e.* typically, rent) to any person, any of its non-system apparatus (*e.g.* wiring) for that person to use in the running of a telecommunication system, typically a customer's system. This is to be done on terms and conditions and subject to charges which are no less favourable to the customer or other user than would apply if BT ran the system itself, provided the maintenance services or supplied telecommunications apparatus comprised in the system. In a case where BT does not retain ownership or control of the apparatus (*i.e.* sells it to the applicant) it is to be made available "at a reasonable capital charge".

Under Condition 43 BT must permit a user of telecommunication apparatus to carry out work on such apparatus except in a case where that apparatus is comprised in BT's systems in such a manner that maintenance services (referred to in Condition 4) cannot be carried out independently of BT's own operations on it systems. The apparatus to which Condition 43 applies is fully described in Condition 43.3.

(xiii) Condition 38: Customer Confidentiality

5–60   Under this condition BT is obliged to operate a Code of Practice restricting the disclosure of information about its customers by its employees engaged in various BT businesses, each of which has a separate code.

The Code mainly deals with the disclosure of information and not with the possibility of unfair advantage arising from the use of information by a person who works for the systems business and obtains information in that capacity and at the same time works for or with some other part of BT's business; such matters are dealt with in separate guidelines established by BT for fair trading practices (see, *e.g.* BT's "Competitive Marketing Principles", latest edition published 1993).

5–61   The main provisions of BT's Code are as follows:

(a) the Code applies to BT's "Systems Business" as well as its "supplemental services business". The Systems Business is defined in Condition 18 of BT's licence and covers its main activity of running a public network together with ancillary activities such as installation, maintenance and repair of apparatus comprised in the network and the bringing into service of other apparatus and systems connected to the network. The supplemental services business, also defined in Condition 18, is essentially BT's value-added and data services business;

(b) the confidential information the subject of the Code is the type of information provided by customers when they order or make enquiries about exchange line service, the installation of additional sockets and the provision of private circuits. This information may be passed on to persons working within the systems business without the customer's consent, but may not be passed to, say, the apparatus business or data services part of BT to help them create a selling opportunity;

(c) those responsible for collecting and storing confidential customer information are obliged to keep it confidential and ensure that it is not disclosed to any BT employees except those who need it in the performance of their duties for the Systems Business;

(d) the exceptions to the prohibition on disclosure of confidential information are where a criminal offence is being investigated or where the law specifically so permits (*e.g.* in the interest of national security) or where details of a customer's account are required for disclosure to specific persons involved in the process of collecting debts or as required by BT's auditors or for disclosure or to other grantors of credit where the customer's account is in default.

The Code is merely what its title suggests, a code of practice; it is not directly enforceable by customers although clearly it would be possible to rely on any breach of the Code in proceedings for breach of confidence or contract by BT. As it specifically states, the Code is without prejudice to any other legal obligations of BT towards its customers.

(xiv) Condition 27: Disputes, etc.: Code of Practice

Under this condition BT is also obliged to issue a Code of Practice giving guidance to BT's customers and employees regarding disputes and complaints. The Code contains sections on service, phone books, bills, operator services, fault repairs, payphones, complaints and arbitration.   5–62

Arbitration

As the Consumer Code of Practice makes clear, BT is obliged to include in its standard terms and conditions an inexpensive independent arbitration procedure for resolving disputes which do not involve a complicated issue of law or a sum greater than a particular sum specified by the Director General from time to time (presently £1,000). The arbitration procedure adopted by BT (and Mercury) is set up under the auspices of the Chartered Institute of Arbitrators.   5–63

(xv) Condition 46: Private Circuits

The current version of this Condition allows for a PTO to obtain private circuits from BT even where the PTO itself provides such circuits, unless the Director General is satisfied that the demand may be met by other means (in the terms of Condition 1.1) or, in the alternative, that the PTO in question would thereby be unduly reliant upon services provided by BT as a means of   5–64 satisfying its own obligations under its licence. In contrast to Condition 13, pursuant to which, *inter alia*, private circuits or quasi-private circuits may be made available for interconnection purposes on cost-based terms and conditions, the provision of private circuits under this Condition would normally be according to retail or wholesale (*i.e.* inter-operator) published charges.

Where BT has published standard charges for private circuits and has published charges for different or similar descriptions of private circuits to other operators, a PTO may apply to the Director General for him to determine whether he is satisfied that the charges in question are reasonable and he may require BT to modify its charge in such a way as to make it reasonable (Condition 46.3). For this purpose the PTO first has to establish a prima facie case that any specific charge is unreasonable.

There is an interesting commentary on priorities in Condition 46.6 which states that nothing in this condition requires BT either to deal with applications from other PTOs in priority to its applications from elsewhere (*i.e.* other customers) or otherwise to discriminate in favour of PTOs, or to act in a way which is likely to seriously reduce the quality of service provided by BT to the generality of its customers. The effect is thus that BT should not be obliged to give any priority or preferential treatment to PTOs to the prejudice of its services to its customers generally.

(xvi) Condition 49: Joint Ventures

5–65   BT is obliged to notify the Director General at least 30 days before its entry into particular agreements or arrangements with third parties as follows—

    (a) for the running of a telecommunication system requiring a licence; for providing telecommunication services or for the production of telecommunication apparatus where that production would lead to a monopoly situation;

    (b) for the establishment of a partnership for such purposes and in such circumstances; or

    (c) for a joint venture to run a telecommunication system requiring a licence or to provide such telecommunications services.

Any such agreement for establishing or controlling a body corporate as referred to in (a) above, or for establishment of a partnership as in (b) above, only applies where BT has or is to have not less than 20 per cent of the voting power in the particular entity.

(xvii) Condition 50: Associates

5–66   This Condition prevents BT avoiding its obligations by using another member of its group ("associate") in that the Director General can intervene to issue directions to BT requiring it to take the necessary steps to ensure that its associate ceases or otherwise remedies the default.

(xviii) Value Added and Data Services

5–67   In providing any service as part of its supplemental services business (equivalent to a value added or data service which can be provided by a non-PTO under the VADS class licence) BT is obliged to abide by the same conditions as apply to major service providers under the VADS class licence.[12]

(xix) Condition 53: Exceptions and Limitations to BT's Obligations

As has been seen, BT's obligations are not absolute. This particular Condi- 5–68
tion excuses BT from any obligation to provide service in the event of *force majeure* circumstances (Condition 53.3) and it is also not obliged "to do anything which is not practicable" (Condition 53.2).

There are further relieving provisions applicable only to voice telephony services. Here Condition 53.4 provides that in specific circumstances BT shall not be under any obligation to provide such services, for example where there is no reasonable demand or where the necessary apparatus is not available. For non-voice services Condition 53.5 provides in addition that BT's obligations may not apply where the provision of service is not economic.

Another potentially significant exception is in Condition 53.6, which excuses BT from being obliged to provide service where the customer is in breach of his contract with BT, or refuses to enter into a contract with BT (unless BT is behaving unreasonably in this respect), or is using apparatus illegally or has dishonestly obtained service from BT.

In the case of BT's supplemental services and in particular the Condition 40A obligation to provide means of access conforming to OSI standards, BT is excused where it is unable to comply because of non-availability of anything necessary for such purpose or, because BT cannot, through no fault of its own, install necessary apparatus. The Director General may also dispense with BT's compliance where interests of consumers would not be promoted to any material degree by the introduction of OSI standards.

The remainder of this Condition preserves for BT a number of rights and discretions which an operator would normally wish to be able to include in its conditions of service.

2. LICENCES GRANTED TO OTHER PUBLIC TELECOMMUNICATIONS OPERATORS

Fixed-Link Operators

Since the duopoly review and the decision of the Government to consider 5–69
on its merits any application for a licence to offer telecommunications ser-
vices over fixed-links within the U.K., a number of licences have been granted to new PTOs. Typical of these is the licence granted by the Secretary of State to Energis Communications Limited on May 24, 1993 which

---

[12] See para. 6–22.

in this section is taken as a representative example of the new licences (referred to herein as the "new-style PTO licence(s)".

Mercury Communications Limited has had a PTO licence in effect since November 8, 1984. In this section I will compare the main areas in which the licences of Mercury Communications and the new operators differ from each other and from that granted to BT.

Scope of Licence Authorisation

5–70   The new-style PTO licences authorise the licensee to provide any telecommunication services except international simple voice resale services (unless such services are provided to countries so designated by the Secretary of State), and international simple data resale services, again unless these services are provided to countries so designated (for example the European Union countries have been designated). The licensees' service authorisation also excludes their ability to convey messages for broadcast in the same way as BT's licence and further excludes mobile radio tails service and international live speech services to and from public switched networks.

Requirement to Provide Telecommunication Services

5–71   In contrast to BT's so called "universal service" obligation,[13] Mercury's licence provides (Condition 1) that it must roll out its network to specific urban areas within precise periods; in practice it well exceeded these targets. As an on-going obligation it is also subject to a licence condition requiring that it must meet demands from customers for service within a 10 kilometre radius of the nearest Mercury node. In practice this seems unlikely to have caused an undue burden because under its general licence conditions Mercury was always permitted to charge at a sensible economic level and there remains (in Condition 50 of Mercury's licence) a built-in safety mechanism, also to be found in BT's licence, relieving Mercury from having to extend its network where this would not be practicable.

The new-style PTO licence puts no responsibility on the licensee to provide telecommunication service until

(a) the licensee has become, in the opinion of the Director General, a "well established" operator in the provision of any telecommunication service of a particular description in the U.K. or within any part or locality thereof; and

(b) the arrangements made by the licensee are inadequate to secure the availability of such a service within the U.K. or that part or locality to any person who may reasonably request it,

and the Director General has directed the licensee to instal and run its systems in a way to secure that such a service is available. For these purposes, a "well established operator" is defined as meaning that the licensee has 25 per cent or more of what is in the opinion of the Director General the relevant market.

[13] See para. 5–08.

Quite apart from the absolution of licensees with the new-style PTO licence from the obligation to provide telecommunication service, this provision is doubly significant in that, as will be seen below, the obligations of such operators as to the publication of charges, terms and conditions and with respect to undue discrimination and/or preference, do not apply in cases where there is no obligation to provide the relevant service.

Public Emergency Call Services

The new-style PTO licences provide that the licensee shall ensure that,   5–72
except to the extent that the Director General determines it is not reasonably practicable, both the numbers 999 and 112 are available as emergency call numbers so that any member of the public by dialling either number may access a public emergency call service.

Mercury's licence (Condition 5) requires Mercury to ensure, to the extent that the Director General determines that it is reasonably practicable, that its voice telephony customers are provided with emergency service as well, enabling them to communicate with the emergency organisations.

Interconnection

In the new-style PTO the licensees are obliged to provide connections and   5–73
message connection services in terms identical to those in the BT licence. The relevant condition (Condition 5) then goes on to detail the "permitted terms and conditions" to be agreed between the licensee and any other applicant operator, again in the same terms as the equivalent part of Condition 13 of BT's licence; Condition 5.5 specifies the basis upon which the Director General should determine any permitted terms and conditions which the parties have failed to agree by negotiation. That Condition is also in terms identical to those in BT's licence but the similarity ends there. The differences are most marked therefore in relation to the financial terms of interconnection.

In this respect Condition 5.6 of the new-style PTO licences provides that: 5–74

(a) costs incurred in the provision of dedicated capacity at a point of connection, but not transmission capacity, should be shared between the parties according to the proportions in which each of them will bill customers originating calls conveyed over the point of connection. These proportions are to be derived from forecasts by each party of the capacity it requires to convey calls for which it will bill the originating customers. Such costs are to be assessed on the basis of the licensee's and the interconnecting operator's respective "fully allocated costs of the establishment of the connection including a reasonable contribution to relevant overhead" and "the application to relevant capital employed of a reasonable rate of return on attributable assets".

(b) Any determination of the charge payable for dedicated capacity may be made subject to the condition that it should not be payable where the party providing such capacity and therefore in whose favour the charge is payable fails to provide the connection within six months of an order, provided it is "reasonable in all the circumstances for it to apply"; this sanction on the supplier of dedicated capacity is to be deemed not reasonable if any of the three following circumstances are operative:

  (i) it was not reasonably practicable for the supplier to provide the connection in time;
 (ii) the other party's request for the connection was unreasonable in quantum having regard to its current and future needs; or
(iii) in order to comply with the time period the licensee would have had to give priority to making the connection beyond the priority given to its own customer generally.

(BT's licence was amended in 1991 to include similar provisions—see Condition 13.5C.2.)

5–75   There are also provisions for extending the six-month deadline to the extent that there may have been delays due to the fault of the interconnecting operator or to *force majeure* and for reimbursement of costs incurred by one party where the other cancels an order requiring connection.

In particular, there is no provision in these licences for the licensee to be paid any access deficit contribution, somewhat surprisingly in view of Condition 13.5B.3 of BT's licence which provides that in determining the access deficit contribution of an operator to BT the Director General may provide that BT should pay a contribution towards the access deficit of that operator.

Mercury's interconnection condition (Condition 12) is in essentially the same terms as the new-style PTO licences. Mercury has interconnection agreements with a number of operators including Vodafone, Cellnet, COLT, Mercury One-2-One and cable operators. Charges applicable under the Vodafone and Cellnet agreements were "determined"[14] by the Director General pursuant to the relevant interconnection licence Condition 12, but have been kept off the public register in accordance with the procedure set out in section 19 of the Telecommunications Act 1984.

Publication of Charges: Terms and Conditions

5–76   On the face of things, this part of new-style PTO licences (Condition 8) is identical to the corresponding BT licence version. However, there is one subtle difference, already touched upon above, which is that the obligation to publish only applies with respect to each description of telecommunication services to be provided "in accordance with an obligation imposed by or under" the licence. Accordingly unless and until those conditions (concerning the operator being "well established", etc.) set out in Condi-

---

[14] Mercury/Vodafone determination—August 1991; Mercury/Cellnet determination—September 1991.

tion 1 of the licence are satisfied there is no such obligation and accordingly these publication requirements remain suspended in their effect.

By contrast not only BT but also Mercury, which has a service requirement condition of very little force, have full obligations to publish charges (on at least 28 days' notice) which gives important prior warning to their competitors who in turn are not required to give equivalent notice with respect to their services. This is an important aspect of fair competition which does not appear to be equitable, certainly as between the non-dominant operators.

Prohibition on Undue Preference and Undue Discrimination

As with Condition 8, although the provisions of this condition mirror the terms in the equivalent prohibition in BT's licence, they are inoperative insofar as they only apply to services which the licensee is *obliged* to provide.   5–77

Reflecting on the above comparison between the licences of BT and Mercury and the other newly licensed operators, it is noteworthy that section 8(1) of the 1984 Act requires that licences granted to operators of public telecommunications systems should include conditions requiring the licensee not to show undue preference or discrimination and to publish its charges, terms and conditions of service. Whilst such obligations are expressed to be included but are in practice inoperative, it is arguable that the requirements of section 8 of the 1984 Act are not satisfied; apparently this argument is not accepted by the DTI. Given that the new-style PTO licences have been published now for over a year, the prospects for a successful challenge to their formulation on these grounds, at least through judicial review proceedings, are probably non-existent.

Cable Operators

Cable operators require to be licensed not only under the 1984 Act but also under the Cable and Broadcasting Act 1984, or its successor the Broadcasting Act 1990, as providers of a local delivery service. The licensing position of cable operators is discussed in detail in Chapter 9, paragraph 9–08.   5–78

The majority of cable operators' licences were originally granted prior to 1991 and therefore preceded the changes wrought by the duopoly review. The new form of cable operators licence, modified to incorporate in particular the right to provide telephony services pursuant to the policy changes set in the duopoly review White Paper (Cm. 1461), has the following main features:

- So-called "must pass" obligations to install the licensee's systems so that a specified number of premises could be served by certain milestone dates; provision is made, however, for relaxation of such requirements where the Director General accedes to a request of the licensee based on the premise that it would be "in the interests of the

sound commercial development of the licensed systems" for the dates or numbers of premises to be modified.
- The obligation to provide telecommunication services (except to the extent the Director General determines otherwise) in accordance with requests in any area where the licensee's systems have been installed.
- The obligation, on the direction of the Director General, to make equal access arrangements available on the request of any other operator, at any time after the licensee has begun to provide 25 per cent of the available exchange lines in any BT local call charge area or in the licensed area; any such Director General direction is conditional on a positive cost-benefit analysis and on the Director General being satisfied on a number of price-related pro-competitive criteria specified in the licence.

5–79
- The obligation, once the licensee has, in the opinion of the Director General become a "well established" operator in the provision of voice telephony services within part of the licensed area, and where the licensee's arrangements are apparently inadequate to secure the availability of such services of any person in such area, and if so directed by the Director General, to take the steps considered appropriate by the Director General to ensure that such services are made so available. (This is equivalent to the obligation on newly licensed PTOs discussed at para. 5–71 above.)
- An obligation to ensure that the licensed systems comply with rules published in 1984 by the Radio Regulatory Division of the DTI with respect to the measurement and limitation of electro-magnetic radiation.

Cable operators are designated as public telecommunication operators and otherwise their licences are in the format generally common to all PTOs. The telecommunications code, with certain exceptions and conditions, also applies to them.

Mobile Operators

5–80   The cellular mobile operators and the newer operators of "personal radio telecommunication services", or PCN or DCS 1800 (the ETSI designation) as they are sometimes known, are also designated as PTOs and have full PTO licences as well as telecommunications code powers. The licences granted to these mobile operators are discussed in detail at para. 7–08 below.

3. Non-PTO Licences

ISR Operators

5–81   Over the past two to three years a number of individual licences have been granted to different organisations to run telecommunication systems to provide international simple resale services. These services are defined in each licence as, in essence, the conveyance of voice or data messages over an international private leased circuit when such messages both originate and terminate on a public switched network in the country at each end. This is sometimes known as "double ended" resale. ISR licensees are subject to licence conditions which in their most significant respects—

(a) require the licensee to adopt a numbering plan;

(b) stipulate that the Secretary of State may require the licensee (where it appears to him to be "requisite or expedient . . . in the interests of maintaining or promoting effective competition in the conveyance of messages to or from the relevant country") to secure that the ratio between its voice and data messages outbound from and inbound to the U.K. is the same as the ratio for the total volume of all messages which respectively are delivered to and sent from the U.K. to and from the relevant countries (Condition 7);

(c) require that the licensee's telecommunication services may only be provided to and from designated countries. For the time being those designated with respect to international simple data resale services are all the countries of the European Union together with the U.S., Australia, Canada and Sweden; international simple voice resale services are only permitted to and from the latter four countries. Negotiations had been ongoing with the U.S. for some time to allow international simple voice resale services between the U.K. and the U.S. and on October 20, 1994 the President of the Board of Trade announced that the U.S. had been added as a designated country so that ISR then became possible directly between the U.K. and the U.S.

Note that for the purposes of interconnection conditions of the fixed link PTOs, an ISR operator is treated as running a "relevant connectable system" which is connectable with those PTO systems under the interconnection condition (*e.g.* in BT's case, Condition 13 of its licence).

CHAPTER 6

Regulatory Conditions in Class Licences

1. THE TSL AND THE SPL

As explained in Chapter 4, in practice, licences under the 1984 Act take   6–01
the form either of individual licences granted to a single licensee or of class
licences which apply to all those running systems in accordance with their
conditions, without the need for individual application. With the exception
of the SMATV licence (see para. 6–23) the class licence system does not
involve any notification, declaration or registration procedure.

The main operative class licences are the Telecommunication Services
Licence (TSL) and the Self-Provision Licence (SPL).

Aside from public telecommunication systems, most systems are run
under the TSL or the SPL. It is indeed a little known fact and certainly not
well understood that even the telephones of residential customers are
required to be licensed and are run under the SPL. Other customer premises
systems based on, for example PABXs, are also run under the SPL, whilst
private systems on which value added and data services are provided are
normally run under the TSL (formerly they would have been run under the
valued added and data services licence—discussed below).

Until September 1992 "branch systems", namely those systems connected to the public telecommunication networks, were licensed by another
class licence, the Branch Systems General Licence. This was replaced by
the first issue of the SPL which has been superseded by the current version.

Applicable Licence Conditions

The first eight conditions of the SPL and the TSL are common to each   6–02
licence. These cover the following matters in much the same way as was
provided by the Branch Systems General Licence.

(i) Condition 1: Approval of Equipment

Apparatus comprised in the licensed system which is to be connected to a   6–03
public telecommunication system must be approved for such connection

under section 22 of the 1984 Act. Approvals framed by reference to branch systems are to be regarded as approvals for connection to the licensed systems.

(ii) Condition 2: Connection Arrangements

6–04   Connection of the licensed systems to a public telecommunication system must be by means of network termination and testing apparatus (NTTA) or by means of wireless telegraphy (*e.g.* mobile apparatus), or by a means which has been specified by the Director General.

The licensee is required not to make or break any tangible connection between the licensed systems and any public telecommunication system, nor to permit any other person to do so (*e.g.* installer/maintainer). This may only occur if the connection to the public system is by plug and socket or the licensee has authority in writing from the relevant public telecommunications operator or the Director General, or that other person or the licensee itself is approved for such purposes (*e.g.* as an installer or maintainer under section 20 of the 1984 Act) and is described in the public list kept by the Director General, or that other person making or breaking the connection is the relevant public telecommunications operator or its agent.

(iii) Condition 3: Technical Requirements

6–05   This requires that the licensed systems be connected to a public telecommunication system only if certain technical requirements for the connection are satisfied; such requirements to be described in a list to be kept by the Director General. In practice the minimum requirements necessary to ensure personal safety and to avoid impairment of the public networks are dealt with in the process for approval of apparatus and the Director has not specified any additional requirements under this condition.

(iv) Condition 4: Relevant Operations on Apparatus Comprised in Public Telecommunications Systems

6–06   This is a corollary to Condition 2 in that it prohibits the licensee from carrying out any operations on any apparatus comprised in a public telecommunication system, including NTTA, unless the licensee is so authorised by the public telecommunications operator or by the Director General; the person carrying out the operations itself is the public telecommunications operator or its duly authorised sub-contractor; or that other person is an approved installer or maintainer under section 20 of the 1984 Act and undertakes only the work which has been specified and described in a list kept for the purpose by the Director General.

(v) Condition 5: Maintenance, Inspection and Bringing into Service of Apparatus

6–07   This condition applies if the licensed system contains call-routing apparatus connected to two or more exchange lines in such a way as to be capable of switching messages consisting of live speech between two or more telephone handsets. In such a case the licensee must have a written contract with a "designated maintainer" for each item of call-routing apparatus.

"Call-routing apparatus" is defined in Annex C as telecommunication apparatus capable of switching messages consisting of two-way live speech telephone calls between two or more items of extension telephone apparatus and two or more circuits forming part of one or more public switched networks.

Serially connected equipment in the licensed system, such as a multiplexer, must also have a designated maintainer unless it is used independently of call routing apparatus. Different items of apparatus may have different designated maintainers. The following operations (defined in Annex C) with respect to call routing apparatus may only be carried out by its designated maintainer:

(a) **Maintenance services**: which include carrying out repairs, verifying or ensuring that the apparatus conforms to its specification or complies with its approval or that terms and conditions regarding the apparatus or its connection or use as stipulated by any public telecommunication operator to whose system it is connected are observed, or any activity involving the removal of its outer cover or the use of any test apparatus or other equipment which does not form part of it.

(b **Inspection**: involving the removal of the outer cover of any item of apparatus where necessary for a public operator or its subcontractor to test the operation of the network or to bring any apparatus into service which is to be connected to it.

(c) **Bringing apparatus into service**: connecting or disconnecting apparatus when the connection is made or broken by means requiring the use of a tool—broadly speaking "hardwired" connections. Call routing apparatus which can be connected or disconnected by use of a plug inserted into a socket is not covered by these provisions and can be connected or disconnected by anyone.

(vi) Condition 6: Emergency Telephones for the Hearing Impaired

This requires that where the licensed system or any of its apparatus is installed   6–08
in a lift to which members of the public have access in circumstances where they may need to use the system for emergency communications, the licensee shall take reasonable steps to ensure that the telephone in question is capable of coupling inductively to hearing aids designed to be so coupled and that there is a notice which indicates that that facility is available together with instructions on its use, affixed on or adjacent to the telephone.

(vii) Condition 7: Connection of Automatic Calling Equipment to a Public Switched Network

This provision polices what might be called "junk" telephone calls. Auto-   6–09
matic calls, not involving live speech or messages transmitted and received

by fax machines, may not be used to send messages to anyone unless that person has consented in writing to receive the message. The licensee is bound to maintain a register of the persons who have so consented, which must be made available for inspection on reasonable notice by the Director General.

(viii) Condition 8: Restrictions on Advertising and Supply Activities

6–10   This Condition is designed to control the making of unsolicited telephone calls for the purpose of advertising or selling goods or services. Where the licensee is making such calls by means of a licensed system and they are conveyed by means of a public telecommunication system and the licensee receives from any customer a request that such calls should cease, the licensee must comply with the request and keep a record of the names and numbers of the customers concerned for inspection by the Director General.

The second part of the Condition also allows the Director General to set up a scheme whereby customers should be allowed to register a general preference not to receive unsolicited calls for the purpose of advertising or selling goods or services. Once the scheme is in operation, no unsolicited sales calls would be allowed to any customer whose number is on a public list set up as part of such scheme.

SPL: Information Obligation

6–11   There is only one further condition in the SPL (Condition 9), requiring the licensee to furnish information to the Director General, typically with respect to the licensee's telecommunication system and the way in which it is run. More significantly, however, the licensee is required to maintain a record of the information described in Annex B to the SPL, which is really quite extensive. It can confidently be expected that the vast majority of SPL licensees are in breach of this obligation, although this could not result in instant revocation because the Director General would first have to go through the licence compliance procedure described in Chapter 4, paragraphs 4–12 to 4–15.

TSL: Other Provisions[1]

6–12   The remaining provisions of the TSL cover the following matters, namely:

- **Display of tariff information**: where the licensee is running telephones which are made available for public use, the licensee may be required to display information relating to the charges and other conditions of use, in accordance with the Director General's specification.
- **Numbering arrangements**: it is possible that, for some services pro-

---

[1] The TSL was amended on November 4, 1994 and with respect to certain Centrex-switched traffic will apparently be treated as if routed over a private circuit.

vided under the TSL, numbering arrangements which are additional to those relating to public networks may be required. For example, the provision of telecommunication services to a single "closed user group" under the TSL may require such arrangements. The licensee is obliged to adopt a numbering arrangement which allows for sufficient numbers to be available having regard to the reasonably foreseeable growth in demand for the services. Furthermore the licensee is obliged to furnish details of the numbering arrangements to the Director General and on request to any other person having a legitimate interest. Any proposals to modify the numbering arrangements must also be notified to the Director General.

- **Privacy, confidentiality and metering systems**: the licensee is obliged to take all reasonable steps to safeguard the privacy and confidentiality of any message conveyed for a consideration by means of its licensed systems and any information acquired by the licensee in relation to such conveyance. In addition the licensee must take all reasonable steps to ensure the accuracy and reliability of any metering system used in connection with its system. Clearly licensees may also need to comply with the provisions of the Data Protection Act 1984 in this regard.

- **Provision of telecommunication services in shared premises**: in order   6–13
to safeguard the rights of occupiers of premises where, for example, a PABX serves different premises in the same building, the licensee is prohibited from refusing to provide service to an occupier except for any of the good reasons specified in the condition such as: impracticability, the need to make a material modification to the licensed systems and refusal by the occupier to enter into a contract with the licensee for the provision of such services. The licensee is also prohibited from imposing a requirement on an occupier to obtain telecommunication services from the licensee and from linking the provision of such services to a binding requirement to obtain other telecommunication services (unless these are indispensable) or apparatus.

- **Wiring**: where there is any wiring or other apparatus owned by the licensee or under its control and installed on premises occupied by any other person wishing to use it then the licensee must secure that such apparatus is available to the user. This requirement is designed to make available for use by occupiers of premises wiring which is either integrated with a public network (typically that of BT) or is owned by a third party such as a landlord.

- **Requirement to furnish information to the Director General**: this is in the same terms as the equivalent provision in the SPL, discussed above.

Authorised Connections and Services

Both the SPL and the TSL authorise licensees to connect their licensed   6–14
system to any public telecommunication system and to various other types of telecommunication system which are themselves authorised for such pur-

pose. The service authorisation provisions in schedule 3 of each licence are quite different, the SPL only authorising the conveyance of messages to the licensee or a member of its group whilst the TSL authorises the provision of telecommunication services of any description other than international simple resale services (voice and data), television programme services and services involving the conveyance of messages for more than 500 metres in lateral distance between points on a single set of premises, where the size of the premises is such that any two points on their boundary are more than five kilometres apart (unless the Secretary of State has specially authorised such services).

Licensed Systems

6–15   The systems licensed under the SPL are simply any systems of any kind, whilst those licensed under the TSL must satisfy the conditions that the apparatus comprised in the system:

   (i) is situated within a single set of premises; or
   (ii) if not so situated, is situated in different sets where none of them is more than 200 metres in lateral distance from any other; or
   (iii) is situated within a single building; or
   (iv) is run by the same licensee and situated on premises specified by the Secretary of State; and
   (v) none of the apparatus consists of mobile radio communications apparatus.

Requirement (ii) above is designed to ensure that operators providing services over an extended network cannot install their own links connecting the various component systems (*e.g.* nodes) together but must lease lines from PTOs for such purpose, or otherwise obtain an individual licence for this purpose.

For these purposes there is no legal definition of "premises" although OFTEL opine that it normally means an area occupied as owner or tenant within a single boundary. Premises could, however, perhaps include an area of land sufficient to affix telecommunication apparatus, *e.g.* a payphone.

2. OTHER CLASS LICENCES

6–16   Besides the TSL and the SPL a few other class licences have been issued as follows:

Outside Broadcast Licence

6–17   This covers systems used to convey messages to or from an outside broadcast site. Understandably therefore the licence is temporary and limits the length of time (60 days in any six-month period) during which licensees can run apparatus at any outside broadcast site, although licensees can apply to OFTEL to have this time limit waived or extended. Licensees are required to register with the DTI.

Satellite Services Class Licence

This licence, issued on August 2, 1991, licenses the running of radio com- 6–18
munication (wireless telegraphy) systems used to transmit messages to or receive messages from satellites. The facilities thus licensed are generally referred to as "ground" or "earth" stations, which may be fixed, mobile or transportable.

The services authorised include any kind of satellite service whether one-way or two-way, point-to-point or point-to-multi-point, including voice, data, video or any other kind of message, for reception within the U.K. or overseas. However there are important restrictions on the services which may be provided under the licence. In particular it does not allow messages to be conveyed between an earth station and a public switched network, either directly or indirectly such as by a private leased circuit. (Any operator wishing to use a satellite earth station for conveying such messages has to obtain an individual licence for such purpose.) Exceptionally licensees are permitted to send messages to a public switched network by means of a satellite under certain specified circumstances.

By the same token the licence does not permit the service of conveying messages from the licensed system to a fixed earth station overseas for transmission into a public switched network, although this restriction does not apply where the earth station is mobile or transportable as defined in the licence.

Radio Alarm Services Licence

This allows the running of radio alarm systems between fixed links, in 6–19
particular the provision of systems linking customer premises to monitoring centres.

Short Range Radio Alarm Systems Licence

This licence covers alarms used in caravan and lorry parks (where an alarm 6–20
fixed to a vehicle is connected to a fixed receiver) and alarms used for neighbourhood watch purposes (where there is a fixed transmitter in a household connected to transportable receiver).

Private Mobile Radio

Private Mobile Radio (PMR) is the regulatory authority's nomenclature for 6–21
privately operated mobile networks intended for use by closed user groups,

such as the police, emergency services and taxi companies. The possibilities for PMR became more interesting when additional radio frequency spectrum was released following the discontinuance of 405-line VHF television broadcasting.

PMR is not a cellular mobile network and is thus "linear" in its configuration. The standard developed in the U.K. for trunked PMR is MPT1327, which has been adopted in a number of other European countries.

There is a class licence for the running of private mobile radio systems, radio paging systems and automatic location systems issued by the Secretary of State on July 28, 1994.

Value Added and Data Services Licence

6–22   Until the introduction of the TSL, a person wishing to run a system in order to provide value added and/or data services, other than BT or Mercury Communications who could provide such services as supplemental services under their PTO licence, would normally have been licensable under the Value Added and Data Services Licence, a class licence. However now that such services may be provided under the TSL the VADS licence has become defunct and it seems unlikely that many companies are choosing to operate under provisions. The VADS licence runs for a minimum 12-year period from April 30, 1987 and is expressed to apply to anyone other than a PTO and its associates.

One good reason perhaps for a person seeking to provide value added and/or data services under the TSL rather than the VADS licence is that the VADS licence subjects what are described as "major service providers" to certain fair trading conditions along very similar lines to those included in PTO licences. A major service provider is a licensee which, in the financial year immediately preceding the year in question, had a group turnover exceeding £1 million in respect of value added and data services or a group turnover exceeding £50 million from sources of any kind. A major service provider must notify OFTEL within 30 days of the date on which the licence conditions first applied to it, of its name and the address of its main office, normally its registered office. Major service providers must also pay an annual fee within 30 days of such notification and annually thereafter in an amount determined by the Director General.

SMATV Licence

6–23   This licence is a class licence, which runs for 10 years from June 10, 1991 covering Satellite Master Antenna TV (SMATV) systems serving no more than 1,000 homes, for the provision of all types of broadcast service; a licensee under this licence is any person specified by the Secretary of State whose name and particulars have been registered by the Director General. See further Chapter 9, paragraph 9–10.

CHAPTER 7

Radio Communications

The general scheme for licensing of systems using radio communications   7–01
is introduced at Chapter 4, paragraphs 4–16 to 4–17 above.

1. GENERAL

The use of wireless telegraphy, otherwise radio communications, for tele-   7–02
communication purposes is regulated under both the Telecommunications
Act 1984 and the Wireless Telegraphy Act 1949. Whilst OFTEL is the
regulatory authority responsible for oversight and enforcement of licences
granted under the 1984 Act, in the case of the Wireless Telegraphy Act
1949 the regulatory body with practical responsibility is the Radiocommunications Agency, an executive agency of the DTI.

    Any "station" run for the purposes of sending messages by means of
wireless telegraphy also constitutes a telecommunication system and, unless
it falls within one of the statutory exemptions, would require to be licensed
under section 7 of the 1984 Act.

    At the same time the Wireless Telegraphy Act 1949 requires that to avoid
the commission of an offence anyone establishing or using any wireless
telegraphy station should also have a licence under the 1949 Act. Its definition of "wireless telegraphy" (see Chapter 4, paragraph 4–16 above) covers
apparatus which both emits and receives signals using radio waves and
thus everything from satellite earth stations to base stations for mobile
cellular systems, mobile telephone handsets and radio paging devices.

2. LICENSING OF SYSTEMS

From the regulatory point of view the licences granted under the 1984 Act   7–03
are rather more important than the 1949 Act licences because the former

contain conditions which ultimately are enforceable by the Director General, whilst the latter tend to be little more than mere permissions to use the allocated frequency.[1] The main areas involving radio communications and the principal 1984 Act licences involved are discussed below.

Mobile

7–04   The main categories of mobile communications licences are as follows:

- cellular licences,
- PCN licences,
- PMR licences,
- radio paging licences,
- mobile data licences, and
- closed user group licences.

I have differentiated cellular from PCN licences as this is the distinction made in OFTEL's list of licences, but PCN licences are also cellular; the main difference between the two main cellular systems, operated by Racal Vodafone Limited (Vodafone) and Telecom Securicor Cellular Radio Limited (Cellnet) and the PCN licensees, Mercury Personal Communications (Mercury One2One) and Microtel (Orange), being that Vodafone and Cellnet offer systems operating to digital (GSM) as well as analogue standards, whilst the PCN licensees operate to a digital standard (DCS 1800, a derivative of GSM) geared to a microcellular technology.

(i) Cellular Licences (Vodafone; Cellnet)

7–05   Vodafone and Cellnet were granted new licences by the Secretary of State on December 9, 1993 and March 22, 1994 respectively. The original licences to these entities restricted the licensed services to mobile radio telecommunication services but in the latest licences this restriction has been removed. This reflects a policy change heralded in the 1991 White Paper[2] which stated that the Government would consider applications from mobile operators to be allowed to offer fixed services using their radio networks. Vodafone and Cellnet are not, however, authorised to provide certain specific services comprising broadcasting, services conveying messages to or from non-U.K. non-GSM systems and international simple voice resale services/simple data resale services except where, in general terms, these are to support the conveyance of messages to or from non-U.K. GSM systems. There are also restrictions *which expired on April 1, 1995* with respect to the use by Vodafone/Cellnet of local fixed radio communication links.

The service availability obligations previously imposed on Vodafone and Cellnet have been preserved in the new licence so that they are each obliged (per Condition 1) to provide to service providers a mobile radio telecommunication service "in an area where 90% of the United Kingdom popula-

---

[1] Wireless Telegraphy Act 1949 licences typically specify the frequency assigned to the licensee, location, antenna height (if applicable) and transmission power.
[2] Competition and Choice: Telecommunications Policy for the 1990s—Cm.1461

tion live". As the new licence puts Vodafone and Cellnet on a par with newly licensed public telecommunication operators of fixed networks, a further provision is included in Condition 1 of Vodafone/Cellnet's licence to the effect that once it becomes a "well established operator" in the provision of any particular telecommunication service, other than mobile service, in the U.K. or any part of it the Director General may (where the licensee's arrangements are inadequate to secure the availability of such a service) require the licensee to make such service available to any person who may reasonably request it.

The other significant change from the original licences to these cellular   7–06 operators is that whereas they were previously prevented from providing telecommunication services direct to users, and had therefore to provide services through service providers, this restriction has been lifted and does not appear in the new licences. This has put Vodafone and Cellnet on a par with their rival PCN licensees. However, in order that service providers should not immediately be put out of business Condition 42 of the licences provides that the licensee should provide mobile radio telecommunication services to a service provider where the service provider is able to demonstrate to the reasonable satisfaction of the licensee that, broadly speaking, 80 per cent of the airtime resold by the service provider is to customers outside of the service provider's group. Vodafone/Cellnet are able to avoid providing services to a service provider, however, where they have reasonable cause to doubt the likelihood of that service provider being able to provide service to others in a "proper and efficient manner" or "financing provision of services" and the Director General has not given a written contrary indication.

Condition 43 preserves the position of end users in the event of a service   7–07 provider ceasing to provide services. In order to allow users the choice of operator/service provider, it requires that during the first three months after a service provider has ceased operations the licensee should not unfairly promote any service provider or its own business so as to place other potential competitors at a significant disadvantage.

Vodafone and Cellnet are PTOs and the Telecommunications Code has been applied to them with certain specified exceptions and conditions.

There is an interconnection condition in the licence which follows the BT/Mercury format, apart from omission of any reference to the access deficit contribution modifications introduced to BT's licence in 1991. The cellular systems are interconnectable with those of other PTOs, without which facility they would not be viable as the vast majority of calls are fixed-to-mobile and mobile-to-fixed rather than mobile-to-mobile, at least for the present. Cellnet and Vodafone have interconnection agreements with BT, Vodafone with Mercury.

(ii) PCN Licences (Mercury One2One; Orange)

These were granted in 1992 and as they are in similar form to the cellular   7–08 licences and thus those of BT and Mercury as well, I will again concentrate on the special provisions.

The main service condition requires that on or following December 31, 1999 the licensee should provide service to service providers in an area where 90 per cent of the U.K. population live and ensure that reasonable demands for the provision of such services are capable of being satisfied, either directly or by inter-system roaming.

As noted above, the PCN licensees are not precluded from selling other than to service providers and may therefore also deal direct with the end-user public.

Handsets connected to cellular and PCN systems are authorised by a class licence "for mobile or portable apparatus connected to cellular tele-communication systems including class licence for mobile or portable apparatus connected to mobile radio or mobile paging systems" published in 1988.

(iii) PMR Class Licence

7–09   Private mobile radio (PMR) systems are systems intended for private use, normally with no access to the PSTN (except in emergency situations). They are used extensively by the emergency authorities, utilities, and organisations which need to keep in contact with vehicles, etc. The PMR class licence[3] authorises the running of private mobile radio systems, radio-paging systems and automatic location systems for the provision of services to third parties. The licence only allows systems to be run and services to be provided under the 1984 Act. The systems also require a licence under the Wireless Telegraphy Act 1949.

(iv) Radiopaging licences

7–10   A number of licences have been granted, most notably to operators of nationwide radiopaging systems. Typically these services involve the ability to communicate a message to the user of a paging device, by the sender of the message first communicating with a computer over the public switched network. A message is then passed from the premises where the computer is located over the fixed links of a PTO, to reach a transmitting station which, by wireless telegraphy, communicates to the paging device.

As with mobile radio licences, typically the licence granted to a radio-paging operator is for 25 years duration and contains obligations with regard to expansion of the coverage of the service nationally. In contrast with most other individual licences, however, the radiopaging licences are relatively brief and omit the fair trading conditions.

Fixed Networks

7–11   Radio communications are extensively used in fixed networks, for example by microwave transmission. The use of wireless telegraphy in this way by public telecommunication operators is normally covered by their existing

---

[3] Issued July 28, 1994.

PTO licences although they will also require Wireless Telegraphy Act licences covering such apparatus.

Cable operators are also eligible in certain areas to be licensed to use microwave video distribution system frequencies for the conveyance of local delivery services. In addition there are a growing number of PTOs[4] who are licensed as fixed link operators specifically for the purpose of using purely radio communications technology to provide a direct-to-home service.

Satellite Services

As these services are provided by means of wireless telegraphy, the earth   7–12
station communicating with the satellite will require a licence both under the 1984 Act and the 1949 Act.

The satellite services class licence under the 1984 Act is discussed in Chapter 6, paragraph 6–18.

Spectrum Management and Frequency Allocation

As in all other countries which are members of the International Telecom-   7–13
munication Union (ITU), the U.K. Government undertakes the task of over-all radio spectrum management, including frequency allocation, in practice through the Radiocommunications Agency.

The Radiocommunications Agency represents the U.K. in ITU discussions of radio issues, including the World Administrative Radio Conferences or WARCs. The Government has not, however, delegated to the Agency the full responsibility for allocating spectrum between different and sometimes competing demands for civil and military applications. This function is performed centrally between the government's spectrum—managing departments. The Ministry of Defence manages spectrum allocated for military use and certain other government departments and bodies (*e.g.* the Home Office, Scottish Office, National Air Traffic Control Service) manage blocks of spectrum for use by services within their areas of responsibility.

The U.K.'s management of the available spectrum is becoming increasingly subject to European Directives and procedures (see Chapter 15, paragraphs 15–22 to 15–27).

3. WIRELESS TELEGRAPHY ACT 1949 LICENCE FEES

Fees are charged for the grant of Wireless Telegraphy Act licences with   7–14
respect to the use of radio transmission equipment. In general fees are set in regulations laid before Parliament which are reviewed annually. Major

---

[4] Such as Ionica.

users of spectrum, such as public telecommunication operators, pay fees determined specifically by the Secretary of State, their fees including an element reflecting the amount of spectrum allocated for their use and the cost of work carried out by the Radiocommunications Agency with respect to their licences.

4. INTERNATIONAL REGULATION

7–15   Radio communications is an area which, if only because of the disrespect of the medium for national frontiers, calls for international cooperation, rule-making and standardisation. The ITU and its agencies involved in such activities, such as frequency coordination, are discussed in detail in Chapter 15.

CHAPTER 8

Terminal Equipment

Terminal equipment, or "telecommunication apparatus" as it is described   8–01
in the Telecommunications Act 1984 (for which reason the term "apparatus" is used in the rest of this chapter) is the subject of regulation at both
a European Union and national (Member State) level. The relevant E.U.
directives are covered in more detail in Part Two.

1. TERMINAL EQUIPMENT APPROVALS

Approval under the Telecommunications Act 1984 and the Telecommunications Terminal Equipment Regulations 1992

Apparatus is a matter for regulation in order that reasonable safety stand-   8–02
ards should be met, that the integrity of public networks should be maintained and that, in particular, using the general criteria set out in Condition
14 of BT's licence and reflected in other PTO licences, such networks
should suffer no "material impairment" as a result of their connection to
any apparatus. For E.U. countries this principle is reinforced by the "essential requirements" specified in the "second phase" Directive on Terminal
Equipment (91/263).[1]

The most common licences applicable to systems which are to be connected to public telecommunication systems, the TSL and the SPL
(discussed in Chapter 6), have a Condition (1) which requires that telecommunication apparatus comprised within the licensed systems and connected
to a public system must be approved for such connection under section 22
of the 1984 Act. However, as discussed below, the Telecommunications
Terminal Equipment Regulations 1992 (TTE Regulations) provide
(regulation 2) that with respect to apparatus to which a Common Technical
Regulation (CTR) applies, these and other licences should be construed as

---

[1] Discussed in more detail in para 16–11 *et seq.*

referring to regulation 8 of the TTE Regulations (essential requirements, etc.) and not to section 22 of the 1984 Act.

There is a further provision (Schedule 3) common to such licences which authorises the connection of the licensed system to other apparatus provided that apparatus is itself comprised in the licensed system or a public telecommunication system, whether or not that apparatus is approved under the 1984 Act.

8–03   Section 22 of the 1984 Act provides that apparatus may be approved either by the Secretary of State or (with the consent of, or in accordance with a general authorisation given by, the Secretary of State) by the Director General. The Director General has this general authorisation. The TTE Regulations provide that section 22 of the 1984 Act shall not apply to "applicable terminal equipment" on and from November 6, 1992. For these purposes "terminal equipment" is equipment intended to be connected to a public telecommunications "network" (*i.e.* PTO applicable systems) or "to interwork with a public telecommunications network being connected directly or indirectly to the termination of a public telecommunications network" (regulation 4(1)). Applicable terminal equipment is terminal equipment for which a CTR is in force. Accordingly section 22 of the 1984 Act will gradually cease to be applicable as CTRs are introduced.

The Director General is required to keep a register of approvals given and designations made under section 22 of the 1984 Act and in practice he also regularly publishes details of recent approvals (see OFTEL News).

In granting section 22 approval to apparatus, the Director General must be satisfied that it would not be liable to cause death or personal injury or damage to property, nor materially to impair the quality of any telecommunication service provided by means of a system to which the apparatus is to be connected. By contrast the TTE Regulations (regulation 8) provide that applicable terminal equipment must satisfy the "essential requirements" *and* comply with both the conformity assessment and connection marking requirements. Essential requirements are defined (regulation 5) essentially to cover the following, namely user safety, safety of employees of PTOs, electromagnetic compatibility requirements, protection of the public networks from harm, effective use of the radio frequency spectrum, where appropriate, and interworking with and via a public network. Conformity assessment requirements are described in regulation 9 and under the E.C. type-examination procedure below. Connection marking requirements are discussed in paragraph 8–13 below.

Categories of Approval

8–04   In U.K. regulatory parlance, there are essentially three different categories of approval:

(a) **general approval**: as the description implies, this covers all apparatus conforming to the description set out in the approval. A manufacturer or supplier of apparatus covered by a general approval need not take any steps to have its products evaluated and approved specifically;

(b) **type approval**: this requires the evaluation process to be followed because once given it applies to every example of the particular type of apparatus. In such a case OFTEL will require to be satisfied not only as to the performance of the samples submitted to it but also on the manufacturing and quality control procedures to be followed by the producers of the apparatus;

(c) **one-off approval**: this is where OFTEL approves apparatus for a specific purpose or for use at a specific site, on a one-off basis.

Since September 1992, the British Approvals Board for Telecommunications (BABT) has had full responsibility for granting type approvals for all types of telecommunication apparatus. However, OFTEL continues to grant general approvals and site specific approval itself. (It has recently issued a general approval for test and development or exhibition purposes in order to reduce the number of site specific approvals needed.) BABT is also the "notified body" (the body tasked with carrying out certification, product checks and associated surveillance tasks) for the purposes of the TTE Regulations.

Under section 23 of the 1984 Act the Director General is required to maintain a public register of all items of telecommunication apparatus approved under the 1984 Act (this is held in the OFTEL Library).

"Grandfathering"

The concept of "grandfathering" was applied in the early days of liberalisation and enabled independent suppliers to obtain an approval to sell apparatus direct to the consumer where they had previously supplied it to BT before its monopoly was removed. The term is also used to refer to all apparatus approvals given under the British Telecommunications Act 1981 to carry forward authorisations for apparatus given by BT before the 1984 Act took effect.   8–05

Approval Standards

Prior to the entry into force of Council Directive 91/263 and the TTE Regulations in November 1992, apparatus was tested against either British Standards or European Standards (NETs) where they existed. Since entry into force of Council Directive 91/263 and the abandonment of NETs, those Common Technical Regulations (CTRs) which are issued from time to time are mandatory on all Member States including the U.K. The CTRs issued so far cover GSM (global system for mobile communications) access and telephony, DECT (digital European cordless telecommunications) and unstructured ONP leased lines. Where apparatus is not covered by a CTR, U.K. regulations and standards continue to apply subject to the requirement (per Articles 3 and 5 of the Directive) that these should not contain   8–06 any requirements going beyond the "essential requirements" specified in Article 4 of the Directive.

Approval Procedures

Mutual Recognition

8–07   In addition to the pan-European approvals pursuant to CTRs, the DTI is in the process of establishing bilateral arrangements with some other European countries for the recognition of approvals to nationals standards. This will allow manufacturers to obtain one-stop approvals for several countries from a single notified body and a single test laboratory.

Applications for approval are usually conducted by reference to existing standards. In the event that there is no existing standard against which to assess an application for apparatus approval, a "special investigation test schedule" procedure is used. Interim approval may be obtained upon submission of a test report from an accredited laboratory which confirms that the apparatus complies with the requirements of the relevant standards. A type of outline approval may be awarded before major capital expenditure is incurred.

E.C. Type-Examination Procedure

8–08   Under the TTE Regulations the E.C. type-examination procedure consists of two parts namely—

- E.C. type-examination and
- E.C. declaration of conformity to type procedure.

The E.C. declaration of conformity to type procedure consists of both conformity to type or production quality assurance and the drawing up by the manufacturer of an E.C. declaration of conformity to type.

E.C. type-examination is described in regulations 16 to 23 of the TTE Regulations, conformity to type in regulations 24 to 26 and production quality assurance in regulations 27 to 35.

For the purposes of E.C. type-examination via a U.K. notified body, that body is to—

(a) examine the technical documentation, verify that the type has been manufactured in conformity with it and identify the elements which have been designed in accordance with the relevant harmonised national standards;

(b) perform or have performed by a recognised test laboratory the appropriate examinations and necessary testing checks with respect to essential requirements; and

(c) perform or have performed by a recognised test laboratory appropriate examinations and test to check that the type meets the relevant CTRs.

For *conformity to type*, which is apparently the system that will be used by most manufacturers, the manufacturer ensures that the manufacturing process assures compliance with the type-examination certificate and declares pursuant to Annex II of Council Directive 91/263 that it does so comply. The notified body (BABT) may then carry out random testing of the product's conformity throughout its lifetime of manufacture. For the *production quality assurance* part of this procedure, described in Annex III of the Directive, the manufacturer must ensure and declare that the applicable terminal equipment fulfils the relevant condition; he must have his quality systems approved by a U.K. notified body, which itself is to carry out surveillance, involving audits at reasonable intervals to make sure that the manufacturer maintains and applies the approved quality system, as well as random visits.

E.C. Declaration of Conformity Procedure

As an alternative to the E.C. type-examination procedure, Annex IV of the Directive envisages that a manufacturer may apply its own quality assurance, both for initial and continuing conformity—a "full quality assurance approved quality system". This is described in Part 4 of the TTE Regulations, regulations 38 to 49. Under this procedure a U.K. notified body must assess the quality system to determine that it satisfies the Directive's requirements with respect to product compliance, systematic and orderly documentation containing an adequate description of the techniques, processes and actions that will be applied and so as to ensure a common understanding of quality policies and procedures. The notified body is also to carry out surveillance, including audits and random visits, in order to inspect the location of the design, manufacture, inspection, testing and storage of the equipment and the manufacturer is bound to provide access for such purpose as well as information, including quality system documentation, facilities and assistance reasonably required by the notified body in carrying out its procedures.

8–09

Withdrawal of Approvals

Approvals may be withdrawn on a number of grounds, namely where:

8–10

- apparatus ceases to meet the requirements against which it was originally approved;
- the stated expiry date has passed;
- the approval holder is in liquidation;
- the applicant has failed to disclose adequately in his original submission the features and facilities of the apparatus or the manner of their operation/interaction;
- unauthorised changes have been made to the apparatus;
- an approval condition is broken;
- annual inspection shows the manufacturers' manufacturing and quality assurance operations fail to meet the required standards;

- exceptionally, an error has been made in the approval process; and
- exceptionally, a significant change occurs in network conditions.

Enforcement

8–11   The TTE Regulations are enforceable in Great Britain by the weights and measures authorities and, in Northern Ireland, by the Department of Economic Development. Where the authorities consider that any applicable terminal equipment is noncompliant, the Secretary of State may serve a prohibition notice on the manufacturer, supplier, user or relevant public operator prohibiting any of them from manufacturing, supplying, putting into service, etc., that equipment or prohibiting the operator from permitting that equipment to remain connected to its system or the supply of telecommunication services by means of that system to any user of the equipment.

Further reading

An excellent series of reference books on European standards and approvals has been produced by John Horrocks (former Deputy Technical Director of Oftel) in his "European Guide to Telecommunications Standards[2]".

8–12   As mentioned in the first edition those interested in the detail of the U.K. approvals processes should also refer to the United Kingdom Telecommunications Approvals Manual.[3]

2. MARKETING AND ADVERTISING

National legislation

8–13   Section 28 of the Telecommunications Act 1984 gives the Secretary of State authority to make orders requiring telecommunication apparatus to be marked with certain information or instructions relating to the apparatus or its connection or use. He has done this under the Telecommunication Apparatus (Marking and Labelling) Order.[4]

Under this Order, which applies only to apparatus which is *not* "applicable terminal equipment" for the purposes of the TTE Regulations, approved apparatus supplied through retail outlets must carry a solid green circle together with details of the approval number if given and a legend referring to the fact that it is prohibited from connection to public networks and again with a legend to this effect. The Order goes on to provide that unapproved apparatus must carry a solid red triangle referring to the fact

[2] Published by Commend.
[3] By Michael Brenton; published by Commed
[4] S.I. 1985 No. 717; amended by S.I. 1985 No. 1031.

that it is prohibited from connection to public networks and again with a legend to this effect.

Section 29 of the 1984 Act authorises the Secretary of State to make orders as to the content of advertisements for telecommunication apparatus. The Telecommunication Apparatus (Advertisements) Order[5] provides (again only with respect to apparatus which is *not* "applicable terminal equipment") that advertisements for telecommunication apparatus subject to the order must, with limited exceptions, display the fact of their approval. The detailed requirements vary according to the advertising medium used.

E.U. Legislation/TTE Regulations

In implementation of Council Directive 91/263, the TTE Regulations pro-   **8–14**
vide that on November 6, 1992 the two marking and advertisements ordered (S.I. 1985 Nos. 717 and 719 mentioned above) cease to apply in respect of "applicable terminal equipment" (see paragraph 8–13 above) and "connection-capable equipment" and "radio connection-capable equipment", the latter broadly speaking covering equipment not intended, adjusted or adapted for connection to a public network.

The connection-marking requirements under the TTE Regulations are that applicable terminal equipment should have affixed to it the following marks in the following order—

- the CE mark (set out in Schedule 2),
- the notified body symbol (Schedule 3),
- the connection symbol (Schedule 3) and
- an inscription identifying the equipment by means of type and batch/ serial number and by the name of the manufacturer or supplier responsible for first supplying it in the Community.

3. ELECTROMAGNETIC COMPATIBILITY

Telecommunications equipment, in common with other electromagnetic   **8–15**
apparatus, can cause interference with other such equipment, sometimes with devastating results. Accordingly, there has always been a need for regulation of electromagnetic emissions and up to the present time the governing legislation was contained in the Wireless Telegraphy Act 1949 and regulations made under section 10 of that Act.

This domestic legislation is now being superseded by the Council Directive 89/336,[6] which has been in force since January 1, 1992. However, as a result of amending Directive 92/31[7] a transitional period has been allowed for full entry into force of the original Directive until December

---

[5] S.I. 1985 No. 719; amended by S.I. 1985 No. 1030.
[6] EMC Directive [1989] O.J. L139.
[7] [1992] O.J. L126.

31, 1995. From January 1, 1996 therefore U.K. domestic legislation will no longer be applicable in this area other than with respect to apparatus excluded from the scope of the Directive.

8–16   The EMC Directive has been implemented in the U.K. by the Electromagnetic Compatibility Regulations (the EMC Regulations).[8] The basic provisions of these Regulations are as follows:

(i) **Electromagnetic disturbance**: the regulations are designed to protect against and regulate electromagnetic disturbance, defined in regulation 4, as "any electromagnetic phenomenon which is liable to degrade the performance of relevant apparatus". Examples are given in the same Regulation and in Schedule 2 of phenomena which may be regarded as electromagnetic disturbance, including electromagnetic noise, unwanted signals and changes in the propagation medium. The performance of apparatus is expressed to be degraded if any of the following types of interference occur, namely permanent, temporary or intermittent total loss of function or significant impairment of function, or otherwise, in a case where the apparatus is for information storage or retrieval, any destruction or corruption of such information.

(ii) **Relevant apparatus**: the Regulations apply to electrical apparatus unless, by exception, it is specified as falling outside their scope or is covered by other directives. For example, electromagnetic compatibility requirements of the Regulations are not to apply to telecommunications terminal equipment if that equipment is required to satisfy such requirements pursuant to Article 4(c) of Council Directive 91/263.

(iii) **General requirements**: supply and taking into service of relevant apparatus is prohibited unless the specified requirements have been complied with, that is to say—

- the relevant apparatus conforms with the "protection requirements" (*i.e.* the requirements necessary to avoid electromagnetic disturbance, per regulation 5);
- the conformity assessment requirements have been complied with;
- the CE mark[9] has been properly affixed by the manufacturer or his authorised representative; and
- the manufacturer or his authorised representative has properly issued an E.C. declaration of conformity for the relevant apparatus.

(iv) **Compliance with conformity assessment**: There are three routes to compliance—

- the standards route, generally known as the self-certification route

[8] S.I. 1992 No. 2372.
[9] Symbol "CE" set out in S.I. 1992 No. 2372, Sched. 4. To be affixed to apparatus, packaging, instructions for use or guarantee certificate.

- the technical construction file route
- the EC-type examination route (see paragraph 8–08 above).

The first two routes apply to apparatus other than radio communication   **8–17**
transmission apparatus and the third applies to radiocommunication
apparatus. There is also a presumption of conformity to the effect that
where apparatus conforms to the applicable EMC standard for that apparatus or the conformity assessment requirements have been complied with
pursuant to the technical construction file route, then there is a presumption that until the contrary is proved the relevant apparatus is compliant
with the protection requirements.

(v) **Enforcement**: enforcement of the EMC Regulations is to be carried
out, in Great Britain, by weights and measures authorities and, in
Northern Ireland, by the Department of Economic Development.
These authorities have powers to purchase and test apparatus as
well as to search for, seize and detain apparatus. Where apparatus
is tested and found to be non-compliant, a suspension notice may
be served to give an opportunity for re-testing by the supplier. If
the product again fails to comply then a prohibition notice may be
served.

The effect of a prohibition notice is to prohibit the manufacturer
from manufacturing, supplying, taking into service or using the
apparatus except with the consent of the Secretary of State.

Fines may be imposed for contravention of the Regulations, on two   **8–18**
levels. For a contravention of the general requirements there is a fine not
exceeding £5,000. If a prohibition or suspension is served and that is contravened, there is a further level of fine of up to £5,000 and three months'
imprisonment. There are other offences such as giving false or misleading
information, the misuse of CE marking, obstruction of enforcement officers
or failure to retain necessary documentation, which may also result in penalties being imposed.

The Department of Trade and Industry have published a guidance document on the Regulations (May 1993) as well as the E.C.'s own guidelines
document (October 1993).

CHAPTER 9

Cable and Satellite Broadcasting and other Transmissions

By Michael Rhodes

1. THE REGULATION OF BROADCASTING IN THE U.K.

In the U.K., sound and television broadcasting has historically been under-   9–01
taken by the British Broadcasting Corporation (BBC) and a number of
licensed independent operators. This background has resulted in the traditional perception of broadcasting as being limited to the transmission of
sound and television programmes by means of wireless telegraphy. However, the transmission media of broadcasting include copper and fibre optic
cable as well what is termed "terrestrial" means, *i.e.* through land-based
transmitters. The similarities between broadcasting and telecommunications (both of which involve the conveyance of information between geographically distinct locations) and the convergence of the telecommunications, broadcasting, publishing, information technology and consumer
electronics industries are leading to the erosion of these traditional boundaries. Broadcasting is now, quite properly, regarded as essentially part of
the same province as telecommunications, particularly in the context of the
cable and satellite sectors.

Broadcasting in the U.K. is regulated not only in respect of the broadcast
itself (including the programme content) but also in respect of any use of
the radio spectrum and the possible ancillary running of a telecommunication system and provision of telecommunications services. The legislative
instruments which address these issues are, respectively:

- the Broadcasting Act 1990;
- the Wireless Telegraphy Acts 1949–1967; and
- the Telecommunications Act 1984.

As terrestrial television broadcasting (that is, the activities of the BBC
and Channels 3 and 4 and the proposed Channel 5) is not the principal

subject matter of this book, this chapter will focus principally on the impact of these three licensing regimes on those sectors of broadcasting which have the greatest relevance to telecommunication, namely cable and satellite broadcasting.

9–02   In order lawfully to engage in a broadcasting activity, a broadcaster will require a licence under one or more of the regimes established under these Acts. Cable television operators require (i) a licence under the 1990 Act (or its predecessor the Cable and Broadcasting Act 1984), which permits the licensee to provide television services within a specified franchise area and (ii) a licence under the 1984 Act which permits the licensee to install and run a telecommunication system over which cable television services and telecommunication services may be provided. Cable television operators only require a licence under the Wireless Telegraphy Acts if they engage in microwave video distribution (MVDS) for part of the transmission of their signals. Satellite broadcasters will require licences under the 1990 Act and the WTA, but should require a licence under the 1984 Act only if they themselves engage in up-linking or down-linking (thus the running of a telecommunication system for such purpose) in the U.K.

In order to simplify matters for prospective licensees for cable television franchises, one application made under the Broadcasting Act 1990 to the Independent Television Commission (ITC) has, since its inception, been treated as an application for all necessary licences.

2. BACKGROUND TO THE BROADCASTING ACT 1990

9–03   The publication in November 1988 of the White Paper entitled "Broadcasting in the '90s: Competition, Choice and Quality" presented the Government's proposals for the re-regulation of broadcasting (and the inherent impact of such re-regulation on the telecommunications sector). In summary, the government's policy was to promote a more open and competitive broadcasting market by implementing, amongst others, the following proposals:

- the replacement of the extant ITV system by a regionally based Channel 3 with positive programming obligations, but also greater freedom to match its programming to market conditions;
- the introduction of a new regime for the development of multi-channel local services through both cable and MVDS;
- the advertisement by the Independent Broadcasting Authority (IBA) of the U.K.'s two remaining direct broadcasting by satellite (DBS) frequencies;
- the grant to all television services (including those of the BBC) of the freedom to raise finance through subscription and sponsorship and, except in respect of the BBC, to carry advertising;
- the creation of a new agency, the Independent Television Commission

(ITC) in place of the IBA and the Cable Authority to licence and supervise all parts of a liberalised commercial television sector; and

- the reform of the transmission arrangements giving scope for greater private sector involvement.

Perhaps the most dramatic change to the broadcasting industry was the introduction of (in the words of the White Paper) a "more commercial element into the award of [television] franchises" by empowering the ITC to operate a financial tendering procedure for licences.   **9–04**

The regulatory framework was modified to convert the 15 ITV franchises that existed before the 1990 Act came into force into Channel 3 franchises. The financial tendering procedure for these Channel 3 licences mirrors the tendering procedure introduced for "cable television" licences under the 1990 Act.[1] The BBC remains outside the scope of this financial tendering procedure. It enjoys its broadcasting rights by virtue of its Royal Charter which came into force on August 1, 1981 and is for a period of 15 years. In a White Paper published in July 1994 entitled "The Future of the BBC: Serving the Nation Competing World-Wide", the Government confirmed that the BBC should continue to be the U.K.'s main public service broadcaster with its primary role to make and broadcast programmes for audiences throughout the country. The Government proposed that the BBC should be granted a further Royal Charter for a period of 10 years.

3. REGULATORY AUTHORITY—THE ITC

Section 1 of the 1990 Act provides for the establishment of the ITC consisting of a chairman and a deputy chairman appointed by the Secretary of State for Trade and Industry together with such number of other members appointed by the Secretary of State, not being less than eight nor more than 10, as he may from time to time determine. The constitution of the ITC as a statutory body corporate is detailed in Schedule 1 to the 1990 Act.   **9–05**

The functions of the ITC, specified in paragraphs 9–03 to 9–04 above are to regulate television programme services which are provided from places in the U.K. by persons other than the BBC and the Welsh Authority and to regulate the provision of additional services from places in the U.K. As the expression "television programme service" is defined to include not only the broadcasting of television programmes for general reception in the U.K. (or any part thereof) but also non-domestic satellite service[2] and licensable programme services[3] (that is, the provision of content of essentially any type), the jurisdiction of the ITC is drawn particularly wide. This jurisdiction extends to the regulation of telecommunication signals sent by wireless telegraphy by means of the use of spare capacity within the signals

[1] See paras. 9–10—9–14.
[2] See paras. 9–27—9–29.
[3] See paras. 9–35—9–40.

carrying any television or radio broadcast service on a frequency notified by the Secretary of State. Additionally, the ITC is responsible for the regulation of local delivery services[4] (which are synonymous with cable television services licensed under the Cable and Broadcasting Act 1984).

9–06   Section 2(2) of the 1990 Act provides that it is the duty of the ITC to ensure that a wide range of services is available throughout the U.K., to ensure fair and effective competition in the provision of such services and to ensure that such services are of a high quality and offer a wide range of programmes calculated to appeal to a variety of tastes and interests.

Section 13 of the 1990 Act expressly stipulates that the provision of a service regulated by the ITC without being authorised to do so amounts to a criminal offence. The intervening sections within Chapter 1 of the Broadcasting Act 1990 specify certain general provisions about licences and licensed services including certain restrictions on the holding of licences.

Section 5 of the 1990 Act obliges the ITC to do all that it can to secure that certain entities, including local authorities, religious bodies, political bodies and advertising agencies do not own or participate in entities holding certain licences including local delivery services licences.[5] The detailed provision of such restrictions is specified in Schedule 2 which also specifies restrictions which prevent accumulations of interests in licensed services and restrictions on the cross-ownership of different licensed services which may be amended at the discretion of the Secretary of State.[6]

9–07   The ITC is able to enforce the restrictions on the holding of licences by revoking licences if any change in the nature or characteristics of the licensee, or any change in the persons having control over or interests in it, occurs, such that had the change occurred before the licence had been granted, the ITC would have refrained from granting the licence.[7]

The ITC can revoke a licence and, pursuant to section 41 of the 1990 Act, reduce the term of the licence and impose a fine on the licensee if it fails to comply with the conditions of the licence or with any direction of the ITC and the ITC considers revocation to be in the public interest.

4. CABLE TELEVISION—LOCAL DELIVERY SERVICES

9–08   In February 1982, the Government-commissioned report of the Information Technology Advisory Panel recommended the development of a U.K. cable network carrying information and other services as well as entertainment. This recommendation was followed by the publication in 1983 of a White Paper[8] on the "Development of Cable Systems and Services" which

---

[4] See paras. 9–08—9–09.
[5] Broadcasting Act 1990, Pt. 2, Sched. 2 lists the categories of disqualified persons.
[6] The provisions of Sched. 2 to the Broadcasting Act are supplemented by The Broadcasting (Restrictions on the Holding of Licences) Order 1991 (S.I, 1991 No. 1176) and The Broadcasting (Restrictions on the Holding of Licences) (Amendment) Order 1993 (S.I. 1993 No. 3199).
[7] s. 5(5) of the Broadcasting Act 1990.
[8] Cmnd. 8866 (1983).

adopted the proposals of the Hunt Committee in October 1982 concerning the establishment of a privately financed cable television industry. Subsequently, 11 pilot cable franchises were awarded under the Broadcasting Act 1981 (which was repealed by the Cable and Broadcasting Act 1984).

Prior to the enactment of the 1990 Act, cable television operators were regulated by the Cable and Broadcasting Act 1984 (together with the 1984 Act as the definition of telecommunication system is wide enough to catch a cable television system). Under Part 1 of the Cable and Broadcasting Act 1984, cable television operators were granted prescribed diffusion services licences which were colloquially referred to as "cable licences". Since 1991, new cable television franchises have been granted local delivery operator (LDO)[9] licences under Part 2 of the Broadcasting Act 1990 rather than "cable licences" under the Cable and Broadcasting Act 1984. Although the 1984 Act was repealed in its entirety by the 1990 Act,[10] Schedule 12 to the 1990 Act provides for the continuation in force of "cable licences" granted under the 1984 Act and Part 2, Schedule 12, paragraph 2 provided for a six-month period from January 1991 within which a licensee under a "cable licence" could request the ITC to grant a local delivery services licence in substitution for that "cable licence".

In contrast with the Telecommunications Act 1984, the 1990 Act regulates the provision of services rather than the running of systems; LDO licences permit the provision of local delivery services. Section 72 of the 1990 Act defines local delivery services as being services consisting of the use of a telecommunication system (whether run by the licensee or any other person) for the purpose of the delivery of one or more specified services for simultaneous reception in two or more dwelling houses in the U.K. The specified services are: **9–09**

- any television broadcasting service;
- any non-domestic satellite service;
- any licensable programme service (that is, the provision of television programmes);
- any sound broadcasting service which is regulated by the Radio Authority; and
- any licensable sound programme service (that is, the provision of sound as opposed to television programmes).

5. SMALL MASTER ANTENNA TELEVISION (SMATV) SYSTEMS

As mentioned in Chapter 6 above, there exists a 10-year class licence for SMATV systems covering fewer than 1,000 homes. However, the benefits of the licence are not applicable to any person unless that person has been   **9–10**

---

[9] The acronym "LDO" will be used in this chapter to refer to Local Delivery Operators; that is, operators under licences for the provision of local delivery services. Although this acronym has, perhaps, not entered into common parlance, its use can be traced back to the Government's Consultative Document prior to the Duopoly Review (Cm. 1303).
[10] Sched. 21 to the Broadcasting Act 1990.

specified by the Secretary of State and such specification will not be made with respect to a system to be installed in an area where a licence has already been granted to a broadband cable or local delivery operator unless a right of first refusal has been given to the holder of such licence and has not been exercised. A cable or local delivery operator will be given the right to provide an alternative service to that proposed unless it is in default of its build timetable or has failed to honour a previous commitment to provide an alternative service elsewhere in the cable/LDO area. Any operator so qualifying is to be given 20 working days in which to offer to provide an alternative service which must provide a similar or superior service to that proposed by the SMATV applicant, at a reasonable price, and in particular to include at least the same major satellite and terrestrial channels and a similar range of alternative channels. If the cable operator/ local delivery operator waives the right of first refusal or fails to honour his commitment to provide an alternative service, the SMATV operator will be specified as a licensee under the SMATV class licence.

6. The Grant of LDO Franchises

9–11   It is the ITC's current policy not to grant more than one LDO licence in any franchise area in order to encourage the development and construction of local "alternative infrastructure". Furthermore, the ITC refrains from granting LDO licences in respect of areas which are already served by licensees operating under "cable licences" granted under the 1984 Act.

While a large number of cable television licences were granted under the old regime, blanket coverage of the U.K. had not been achieved by 1991. As the stated policy of the ITC is to grant only one "cable licence" or LDO licence for each franchise area, since its formation, the ITC has merely been plugging the gaps in the existing coverage.

9–12   As of June 5, 1995, 132 franchises had been awarded. Although these franchises are, by and large, located in the most densely populated areas of the country, they only cover about 64 per cent of U.K. homes. A report from the Trade and Industry Committee on Optical Fibre Networks (July, 1994, H.C.) confirmed that the then Telecommunications Minister believed that further franchises would be offered and taken up and that the eventual coverage could be 80 to 85 per cent of homes; however, no date was specified either for the franchising or actual construction.

Section 73 of the 1990 Act provides that the ITC has a discretion as to which licences are granted. Where the Commission proposes to grant a licence to provide local delivery services, the ITC must publish a notice stating this intention, specifying the franchise area and inviting applications for the licence. The ITC's published policy, with regard to its discretion as to which franchises are advertised, is that it "will be guided by the interest expressed to it in new local delivery franchises and the areas they might cover". As a purely practical matter, the ITC will be unlikely to advertise an area in respect of which no interest has been expressed by a potential applicant.

Having determined that a particular area will be the subject of a potential new franchise, the ITC advertises that area and invites applications for the licence. It is the ITC's practice to specify, in the advertisement, available MVDS frequencies for use in connection with the conveyance of local delivery services in that franchise area.

Licence applications in response to the advertisement must be accompanied by: **9–13**

- details of the applicant;
- details of the availability of funding necessary to construct the cable television network in the franchise area, together with independent verification of such details;
- a business plan detailing the projected business development throughout the term of the franchise;
- a technical plan indicating the parts of the franchise area which would receive service, the timetable within which that service would become available, the technical means by which such service would be conveyed and the extent to which the applicant intends to engage subcontractors; and
- a statement of the applicant's cash bid.

The ITC follows a two-stage procedure when considering applications   **9–14**
for franchises. Section 75 of the 1990 Act provides that the ITC shall not proceed to consider an applicant's cash bid unless any telecommunication system proposed to be used by the applicant satisfies the requirements of the Director General and Secretary of State and the ITC is satisfied that the applicant would be able to maintain the licensed service throughout the period for which the licence would be in force.

Section 76 of the 1990 Act provides that the ITC shall award the licence to the applicant submitting the highest bid (that is, the highest annual sum during the continuance of the licence) unless it appears to the ITC that there are exceptional circumstances which make it appropriate to award the licence to another applicant. Such exceptional circumstances are specifically deemed to be in existence where it appears to the ITC that the coverage proposed to be achieved by an applicant not submitting the highest bid is substantially greater than that proposed to be achieved by the applicants submitting the highest bid. Accordingly, it would appear that the priority for the ITC is to achieve the greatest coverage for local delivery services even at the expense of revenue generated by the grant of licences.

The ITC charges an application fee which is calculated according to a   **9–15**
sliding scale based on the size of the franchise area advertised. In addition to the initial licence fee (that is, the cash bid of the successful applicant) LDOs must pay an annual licence fee. The annual licence fee is also calculated by reference to a sliding scale based on the size of the franchise area

but which also takes account of the number of homes "passed" by the franchisee.

LDO licences are granted for a term of 15 years and may be renewed on one or more occasions for further periods of 15 years. Applications for renewal must be made within the last five years of the licence but not later than the date on which the ITC publishes a notice inviting applications for a new licence in respect of the relevant franchise area.

The grounds on which the ITC may refuse to renew a licence are limited. These grounds include (i) the decision by the ITC to grant a "fresh" LDO licence for a different area to that previously licensed and (ii) the failure by the licensee to achieve the coverage specified in the technical plan submitted with the licence application within the period specified in the technical plan.[11]

7. CABLE TELEVISION—TELECOMMUNICATIONS ACT LICENCE

9–16   In summary, section 4 of the 1984 Act defines a "telecommunication system" as being a system for the conveyance of sounds and images by various technical means. The ITC takes the view[12] that any system used to deliver a local delivery service will be a "telecommunication system" for the purposes of the 1984 Act. Accordingly, a licence will be required under the 1984 Act in order to run the system. This licence will be required by the firm which is running the system over which a local delivery service is conveyed, which may be a firm other than the holder of an LDO licence if such holder has contracted for another firm to run the system.

The 1984 Act licence required in connection with an LDO franchise is granted by the Secretary of State for Trade and Industry under section 7. It endures for the same period or until the same expiry date as the relevant LDO licence (typically, initially for 15 years). This licence may be granted earlier than the LDO licence in order to permit the licensee to establish its network infrastructure before the period of the LDO licence starts to run. Clearly, time spent building a cable network is "dead time" in terms of service delivery and in order to make maximum use of the finite period of the LDO licence it is usual, after award of a franchise, for the franchisee to postpone requesting the ITC to grant the licence until the network has, at least to some extent, been constructed and a revenue stream is available.

9–17   The 1984 Act licence will specify the authorisation to run a system which is determined by the requirements of the particular franchise; for example, an authorisation may include systems to provide inter-active broadband services, MVDS services or one-way services such as those provided by a SMATV[13] licensee.

---

[11] Broadcasting Act 1990, s. 78(4).
[12] See the General Notes for the Guidance of Franchise Applicants published by the ITC on May 11, 1992.
[13] Satellite Master Antenna TV systems are small cable systems providing satellite and terrestrial TV channels, typically to a block of flats or a discrete housing development.

The terms and conditions of each 1984 Act licence, including the extent   9–18
of services authorised and the existence, or not, of powers under the Telecommunications Code, vary according to the nature of the local delivery
system proposed. Telecommunications Code powers are only granted in
respect of a system if it could not reasonably be built without them. In
order to be eligible for powers under the Telecommunications Code,
licences for running two-way broadband cable systems contain conditions
derived from section 8 of the 1984 Act and such systems have been designated as public telecommunications systems. The most important of these
section 8 conditions are certain requirements to connect to other systems
and to allow services carried on those connected systems to be provided
via the licensee's own system if a customer so requests.[14] The licensee is
also subject to the standard PTO licence obligations to publish charges,
terms and conditions and not to show undue preference towards nor undue
discrimination against any persons or class of persons. A licensee may also
be subject to a service obligation to satisfy reasonable demands for certain
telecommunication services.

The licence may include appropriate build milestones which are based
upon the project timetable included in the technical plan submitted by the
licensee with its original franchise application. It is OFTEL's responsibility
to enforce the build milestones in the licence. The ultimate sanction for a
failure to achieve the milestones is revocation of the licence.

Licensed systems must meet specified technical and safety requirements,   9–19
including those related to radio interference. Broadband cable systems must
also meet certain additional BSI and other standards.

The ITC's current guidelines state that the fee for issue of a licence conferring Telecommunications Code powers is likely to be of the order of
£10,000. Licences not conferring Code powers are likely to cost about
£2,000. The higher fee for licences with Code powers reflects the much
greater costs, including in particular the costs of the statutory consultation
on the terms of the licence required under the 1984 Act. Annual renewal
fees are also charged. Initially these will be £2,500 in the case of licences
with Code powers, and £1,250 for SMATV or MVDS-only licences, or, in
each case, a sum not exceeding 0.08 per cent of the annual turnover of the
licensee's system business where that sum is the greater.

Licences under the 1984 Act may be revoked with the agreement of the
licensee or if the licensee persistently breaks licence conditions, becomes
insolvent or otherwise ceases to trade or no longer holds a local delivery
service licence from the ITC. They may also be revoked if there is a change
of ownership or control of the licensee which in the opinion of the Secretary
of State is against the interests of national security or relations with a foreign government. The licensee is not entitled to transfer its 1984 Act
licence.

To encourage further the provision of competitive local loop telecommu-   9–20
nications services, the Duopoly Review removed the restriction on cable

---

[14] See Chap. 5, paras. 5–73—5–75 on interconnection generally.

television operators providing voice telephony in their own right. Until this restriction was removed, cable television operators could only provide such services as agents of BT or Mercury. In addition to this liberalisation, cable television operators are now permitted to provide switched telephony and other telecommunication services to their customers. The Duopoly Review also permitted cable television operators to interconnect directly their network infrastructure with that of other cable television operators in contiguous franchises, enabling the provision of service over a larger area without the need to make use of the network of a national PTO. Cable television operators could also agree on indirect interconnection using the transit facilities of other operators, either PTOs or cable operators.

8. COMPETITION FROM PTOS AND VIDEO-ON-DEMAND

9–21   LDOs are not only protected from competition by the ITC's current policy of only granting one licence per franchise, they are also protected from competition by the restriction on national PTOs prohibiting national PTOs conveying national broadcast entertainment services in their own right until 2001 at the earliest.

This restriction is (for example, in the context of BT) contained in paragraph 1(b)(i) of Schedule 3 to BT's PTO licence which expressly states that the licence authorises "the conveyance . . . of Messages (not including cable programme services sent under a licence granted under Section 58 of the [1984 Act]) . . ." By virtue of section 190 of the 1990 Act, the expression "cable programme services sent under a licence granted under Section 58" should now be construed as a reference to services sent under a local delivery services licence. Accordingly, PTO licences do not authorise the *conveyance* of local delivery services.

9–22   Part 5 of The Broadcasting (Restrictions on the Holding of Licences) Order 1991 (S.I. 1991 No. 1176) now provides for a limited right for national PTOs to apply for LDO licences which are advertised after March 31, 1994 and, in respect of which, no part of the franchise area lay within an existing LDO (or "cable television licence") franchise immediately prior to the invitation for application for the licence in question.

The rationale for the restriction on PTOs providing (as compared with conveying) national broadcast entertainment services was to provide the embryonic cable industry with time to develop and become recognised as a real alternative to terrestrial wireless telegraphy broadcasting networks and at the same time to encourage the development of competitive telecommunications infrastructure in the "local loop". It was appreciated that otherwise BT with its economies of scale and scope could combine entertainment with telecommunication services in a way that could quickly undermine the economic position of its fledgling cable competitors. Regulatory constraints on BT, for example on cross-subsidy between telecommunications and entertainment services, would have been inadequate to

deal with the adverse competitive impact for the cable industry. The duopoly White Paper stated that this restriction would be reconsidered by the
Government in 1998 if the Director General were to advise that its removal
would be likely to promote more effective competition in telecommunications.

In the meantime, it was becoming apparent that notwithstanding the   9–23
restriction on conveyance of local delivery services BT was planning nevertheless to send video entertainment programmes to customers via BT's
normal infrastructure including local loop exchange lines. The question was
whether such activity would breach the restriction. On September 29, 1993,
the ITC issued a statement confirming that BT and other national PTOs
may, under their current licences, convey "video-on-demand" services
which consist of the provision of television programmes over the respective
networks of national PTOs in response to specific requests (made over the
telephone) for such programmes. The key distinction between "video-on-
demand" services and broadcast entertainment services (which remain
restricted) is that "video-on-demand" is not provided for simultaneous
reception in two or more dwelling-houses in the U.K. Since section 72 of
the 1990 Act defines local delivery services as consisting of the use of a
telecommunication system for the purpose of delivering a specified service
for simultaneous reception in two or more dwelling houses in the U.K.,
"video-on-demand" would certainly seem to be outside the scope of the
regulation of local delivery services. Thus the grant of an LDO licence is
not a prerequisite to the conveyance of such "narrow-cast" services and
national PTOs are therefore able to convey "video-on-demand" services
under their existing licences.

On July 19, 1994, the Trade and Industry Committee published a report   9–24
on "Optical Fibre Networks" having received evidence during six hearings
from PTOs, industry associations, unions, equipment manufacturers, regulators and representatives of the administration of the U.S. which is heavily
promoting the "information superhighway" or "national information
infrastructure".

In summary, the reason for the investigation by the Trade and Industry   9–25
Committee into the importance of developing a national optical fibre network, was the concern that government policies could be hindering or not
sufficiently encouraging the development of the most advanced infrastructure and services and that this could result in the U.K. falling behind other
countries, with damaging consequences.
One of the principal recommendations of the Committee was that:

"The Government should reduce the uncertainty concerning the restrictions on PTOs by directing OFTEL and the ITC to review the licences
for current franchise areas, taking account of the build obligations
contained in the licences, with a view to allowing competition into
franchise areas by providing for the lifting of the restrictions on PTOs

on a franchise-by-franchise basis at specified future dates, subject to the principle that all cable franchises should be exclusive for seven years from the granting of the original licences; and the Government should make clear that all restrictions on PTOs conveying or providing entertainment will be lifted by the end of 2002, provided that the PTOs permit fair and open access to their networks."

9–26   In November 1994, the DTI published a document entitled "Creating the Superhighways of the Future: Developing Broadband Communications in the UK".[15] This document considers the opportunities and challenges arising from the convergence of communications, information technology, media technologies and services and the development of multimedia. It also incorporates the Government's response to the recommendations of the Trade and Industry Committee in its report on "Optical Fibre Networks".

In response to the specific recommendation that the restrictions on PTOs concerning broadcast entertainment services be lifted, the DTI stated that it:

" ... considers that the present regulatory framework continues to be the best way of providing a suitable climate for the development of communications technology, infrastructure and services. That regulatory environment will undoubtedly evolve as more infrastructure is established, as technology advances and as the market for broadband services and applications becomes more competitive. To move away now from the policies in the 1991 White Paper would adversely affect ... existing or prospective franchises. ..."

Accordingly, the Government squarely declined to accept the Trade and Industry Committee's recommendation and reaffirmed its commitment to the restrictions on national PTOs conveying and/or providing broadcast entertainment services which were detailed in the 1991 White Paper.

9. SATELLITE BROADCASTING SERVICES

9–27   The role of satellites in telecommunication has been well established for a large number of years, particularly in the context of international communications. Two types of satellite are used for broadcasting: telecommunication satellites which are usually low or medium-powered and are known as "FSS" (fixed satellite service) satellites and broadcasting satellites which are high-powered and known as "BSS" (broadcasting satellite service) satellites. Through its World Administrative Radio Conferences (WARCs), the ITU co-ordinates the available radio frequencies and allocates geostationary orbital positions or "slots", for both types of satellite.

The ITU defines BSS as "a radio communication service in which signals transmitted or retransmitted by space stations are intended for direct reception by the general public". It was indeed originally intended that BSS should be the basis for direct-to-home (DTH) satellite broadcasting and to

[15] Cm. 2734 (1994).

that end the U.K. was allocated various BSS frequencies which ultimately were awarded to British Satellite Broadcasting, now subsumed into British Sky Broadcasting (BSkyB).

However, technological developments were such that soon medium-powered FSS satellites became capable of transmitting direct-to-home; the ITU definition of FSS as "a telecommunications service between fixed stations using one or more satellites" did not preclude this. However, initially the use of FSS satellites in this way would have been encumbered by the need to obtain licences under the Wireless Telegraphy Act 1949 and the Telecommunications Act 1984 for the necessary receiving equipment. This situation was resolved in January 1989 when the Secretary of State issued a regulation[16] exempting the use of wireless telegraphy apparatus capable only of receiving transmissions; in the following months the government issued a general class licence under the Telecommunications Act 1984 authorising all persons (with certain limited exceptions) to run a telecommunication system to receive messages from orbital apparatus. Thus, since that time it has been possible for transmissions via FSS satellites (such as Astra) to be lawfully received by the end user in the U.K. without the need for an individual licence.

The Broadcasting Act 1990 provides for two types of satellite television services, domestic and non-domestic.

Section 43 defines a domestic satellite service as a television broadcasting   9–28
service where the television programmes included in the service are transmitted by satellite from a place in the U.K. on an allocated frequency and for general reception in the U.K.

Section 43 of the 1990 Act defines non-domestic satellite services which consist of the transmission of television programmes by satellite other than on a frequency allocated to the U.K. for broadcasting by satellite but for general reception in the U.K. or in a country prescribed by the Secretary of State by order. Such non-domestic satellite services are deemed to be subject to U.K. regulation where the programmes are transmitted from a place in the U.K. or the programmes consist in material provided by someone in the U.K. for inclusion in the satellite service.

The ITC has a discretion regarding the grant of licences to provide   9–29
domestic satellite services. Such a licence may authorise the provision of the service consisting in simultaneous transmission of different programmes on different frequencies. The procedure for granting such domestic satellite services corresponds to the new competitive procedure for granting Channel 3 licences and LDO licences.[17] An important distinction between on the one hand domestic satellite services and on the other local delivery services and non-domestic satellite services is that Schedule 2 to the 1990 Act provides that non-E.C. nationals are disqualified from holding licences for domestic satellite services.

---

[16] S.I. 1989 No. 123.
[17.] See paras. 9–11—9–15.

Section 45 of the 1990 Act provides that an application for a licence to provide a non-domestic satellite service shall be made in such manner as the ITC may determine and shall be accompanied by such fee (if any) that the ITC may determine. When considering an application for such licence, the ITC may only refuse to grant the licence if the proposed non-domestic satellite service would not comply with the general requirements specified in paragraphs 9–16 to 9–20.

10. PROSCRIPTION of UNACCEPTABLE FOREIGN SATELLITE SERVICES

9–30   Sections 177 and 178 of the 1990 Act extend U.K. regulation further to encompass foreign satellite services; that is, services consisting wholly or mainly in the transmission by satellite from a place outside the U.K. of television or sound programmes which are capable of being received in the U.K.

Section 177 provides that the Secretary of State may, by order, proscribe a foreign satellite service.

The ITC and the Radio Authority may notify a particular service to the Secretary of State if they are satisfied that there is repeatedly contained in the programmes included in the service matter which offends against good taste or decency or is likely to encourage or incite to crime or to lead to disorder or to be offensive to public feeling.

An order proscribing a foreign satellite service may only be made by the Secretary of State if he is satisfied that such an order is in the public interest and that it is compatible with any international obligations of the U.K.

9–31   Once a foreign satellite service has been so proscribed, any person committing the following acts shall be guilty of an offence:

- supplying any equipment or other goods for use in connection with the operation or day-to-day running of the proscribed service;
- supplying, or offering to supply, programme material to be included in any programme transmitted in the provision of a proscribed service;
- arranging for, or inviting, any other person to supply programme material to be so included;
- advertising, by means of a programme transmitted in the provision of a proscribed service, goods supplied by him or services provided by him;
- publishing the times or other details of any programmes which are to be transmitted in the provision of a proscribed service or (otherwise than by publishing such details) publishing an advertisement of matter calculated to promote a proscribed service, (whether directly or indirectly); and
- supplying or offering to supply any decoding equipment which is designed or adapted to be used primarily for the purposes of enabling

the reception of programmes transmitted in the provision of a proscribed service.

11. Television Without Frontiers

Notwithstanding the adoption of the "Television without Frontiers" Dir-   9–32
ective, the U.K. regulatory authorities still purport to retain a right to control, on the grounds of morality and public policy, the content of broadcasts (which can be received in the U.K.) The stance of the U.K. regulators
has been established by the courts in a case colloquially known as "Red
Hot TV".[18]

CT, a Dutch television company, transmitted by satellite between midnight and 4 a.m. to subscribers who had purchased or hired a special
decoder. The transmission comprised hard-core pornographic programmes
originally from the Netherlands (known as "Red Hot Dutch") and subsequently from Denmark (known as "Red Hot Television"). These programmes were capable of being received in many European countries
including Britain.

Acting on a report from the Broadcasting Standards Council, on April
6, 1993 the U.K. Government issued an order under section 177 of the
Broadcasting Act 1990 proscribing the programmes.[19] This order amongst
other things criminalised:

- publication of the programme times and other programme details; and
- the supply of decoding equipment enabling the programme to be
  received.

In accordance with Article 2(2) of the "Television without Frontiers"
Directive, Member States are obliged to ensure freedom of reception on
their territory of television broadcasts from other Member States. However,
Member States may provisionally suspend re-transmissions of television
broadcasts if such broadcasts manifestly, seriously and gravely infringe the
provisions of the Directive relating to the protection of minors from television programmes which contain pornography.

CT applied for judicial review of the proscription order on the basis that
"Red Hot Television" was received in Britain by direct transmission via
satellite and not by re-transmission by an intermediary in the U.K. Accordingly, Article 2(2) of the "Television without Frontiers" Directive required
the U.K. Government to ensure the freedom of reception of the programme
as Article 2(2) only permitted Member States to block "re-transmissions".

In the proceedings, it was reported to the High Court that a senior Commission Official had opined that the expression "re-transmission" in Article
2(2) included direct reception. The court referred to the European Court
of Justice under Article 177 of the EEC Treaty the question of interpreta-

---

[18] *R. v. Secretary of State for National Heritage, ex parte Continental Television B.V.* (1993) 2 C.M.L.R.
333 D.C.; (1993) 3 C.M.L.R. 387, C.A. (Red Hot Television).
[19] The Foreign Satellite Proscription Order 1993 (S.I. 1993 No. 1024).

tion of Article 2(2) of the "Television without Frontiers" Directive. The decision of the European Court of Justice is still awaited.

12. RADIO BROADCASTING

9–33   Satellite sound broadcasting services provided from places in the U.K. are regulated and licensed by the Radio Authority. For these purposes a satellite sound broadcasting service is any sound broadcasting service (other than one provided by the BBC) which consists—

- of the transmission of sound programmes by satellite from a place in the U.K. for general reception there or
- the transmission of such programmes by satellite from a place outside the U.K. for general reception there, if and to the extent that the programmes included in the service consist of material provided by a person in the U.K. who is in a position to determine what is to be included in the service (so far as it consists of programme material provided by him).

Such a licence would be for eight years and the Radio Authority can only refuse to grant a licence to a person who is not a fit and proper person to hold it. The Radio Authority has the power to draw up a general code for programming and advertisements with which all services licensed by the Radio Authority must comply.

9–34   Any cable broadcast which uses the radio spectrum to distribute its services in the franchise area and any satellite broadcast will require a licence under the WTA for the establishment and use of wireless telegraphy "stations". Wireless telegraphy licences are issued by the Radiocommunications Agency to users of such stations. In the context of a cable television operator, this may be the franchisee under the LDO licence or, as with the requisite 1984 Act licence, the WTA licence may be issued to another firm which has contracted to provide and use wireless telegraphy stations in connection with the particular franchise (for example, National Transcommunications Limited which acquired the broadcasting infrastructure of the former IBA).

The Wireless Telegraphy Apparatus (Receivers) (Exemption) Regulations 1989 (S.I. 1989 No. 123) removed from the scope of this licence requirement wireless telegraphy apparatus "which is inherently incapable of transmission". Thus, the possession, installation and use of a dish for receipt of satellite broadcasts does not require a licence under the WTA.

Prior to the award of a WTA licence, the Radiocommunications Agency must be satisfied that the requisite part of the radio spectrum is available and can be used without causing undue interference to others and without itself being unduly interfered with. Thus applicants for a WTA licence must provide with their respective applications details of the proposed sites of all wireless telegraphy transmitters, the height of the proposed sites and proposed antenna heights, together with the proposed frequencies, power levels, polarisation and proposed emissions and the radiation patterns envisaged from each antenna.

Section 75(2)(a) of the 1990 Act provides that the Secretary of State must   9–35 confirm that the proposed telecommunication system is acceptable in the context of wireless telegraphy regulation before the ITC can consider a particular applicant for the award of an LDO licence. In this context, the Radiocommunications Agency will award the WTA licence authorising the use of specified frequencies from specified sites at specified power levels once the ITC has awarded the cable franchise to a particular applicant.

As with 1984 Act licences in the context of cable television franchises, the WTA licence will run for the same period or until the same expiration date as the LDO licence, unless it lapses or is otherwise revoked in advance of such date. The Secretary of State may revoke a WTA licence in accordance with section 1(4) of that Act.

The WTA licence specifies the technical parameters within which the licensed radio communication must be conducted. It also specifies that the wireless telegraphy apparatus concerned should be designed, constructed, maintained and used in such a way as not to cause any undue interference to any other users of the radio spectrum.

A fee is payable annually in advance in respect of the grant of a WTA licence. This fee is intended to cover the costs of issuing and enforcing the licence. The licence terminates automatically in the event that the fee is not paid by the due date. The fee is determined by reference to a sliding scale based upon the number of homes in the franchise area.

13. CROSS-OWNERSHIP MEDIA RESTRICTIONS

Legislative Background

Schedule 2 to the 1990 Act contains restrictions on the holding of licences   9–36 which may be awarded by the ITC and the Radio Authority (a section of the Department of Trade and Industry which is responsible for granting licences to provide sound broadcasting services).

Television Cross-Ownership

Part 2 of Schedule 2 to the 1990 Act relates to television. The limit con-   9–37 tained in Schedule 2 to the 1990 Act to the effect that no person may hold more than two licences for regional Channel 3 services is supplemented by two statutory instruments, so that a person may not hold two licences for London, or until such licences have been granted, for two specified contiguous areas. In addition, the holder of a regional Channel 3 licence who has a non-controlling interest in a second Channel 3 licensee company may not have more than a 20 per cent interest in a third such company, nor more than a five per cent interest in a fourth.

The holder of a national Channel 3 licence may not have more than a 20 per cent interest in a second national Channel 3 licensee company, nor more than a five per cent interest in a third. Similar limits are applied to Channel 5. The holder of a licence for national or regional Channel 3 or Channel 5 may not have more than a 20 per cent interest in a company which holds a licence to provide a service falling into another of those categories, nor more than a five per cent interest in a third such company. Those who invest in, but do not control, Channel 3 and Channel 5 licensee companies are limited to a 20 per cent interest in a third regional Channel 3 company and a 20 per cent interest in a second Channel 3 or Channel 5 company if one company provides a national Channel 3 service or Channel 5 service. Further investment in Channel 3 and Channel 5 is limited to five per cent; and those with more than a five per cent interest in three Channel 3 or Channel 5 companies are limited to a five per cent investment in any other such companies.

In its proposals on Media Ownership of May 1995 the Government proposed relaxations in these limits: see para. 9–42 below.

Radio Cross-Ownership

9–38   The rules contained in Part 3 of Schedule 2 to the 1990 Act concern radio. They provide for a points scheme, by virtue of which various categories of radio licence are allocated points according to the size of the population in the areas for which they are provided. Subject to minor exceptions, no person may hold radio licences such that the number of points attributable to them exceeds 15 per cent of the points allocated to the radio system as a whole. In addition, one person is limited to two licences for the largest type of local radio service and six licences for the next size down (and four licences if a national radio licence is also held). Subject to limited exceptions, a local radio licensee may not hold, nor hold more than a 20 per cent interest in a company which holds, a licence to provide a local radio service which overlaps substantially with its service.

Newspaper Cross-Ownership in Broadcasting

9–39   Part 4 of Schedule 2 to the 1990 Act contains supplementary limits for newspaper ownership. The proprietor of a national newspaper may not have more than a 20 per cent interest in a company which holds a licence to provide a local radio service and vice versa. The proprietor of a national or local newspaper may not have more than a 20 per cent interest in a company licensed to provide a *domestic* satellite service,[20] nor more than a five per cent interest in a second such company, and vice versa.

[20] The expression "domestic satellite service" is defined in s. 43 of the Broadcasting Act 1990 as being a television broadcasting service where the television programmes included in the service are transmitted by satellite from a place in the U.K. (on a frequency allocated to the U.K. for broadcasting by satellite) for general reception in the U.K.

Participation of PTOs

Part 5 of Schedule 2 to the 1990 Act concerns national PTOs. It provides   9–40
that a national PTO with an annual turnover of more than £2 billion, its
associate, a person who controls either of them or a body controlled by
such an operator or its associate, may not hold a licence to provide a Channel 3 or Channel 5 service, a domestic satellite service or a national radio
service. Subject to a limited exception, a national PTO may not hold a
licence to provide a local delivery service. Since April 1, 1994 such an
operator may now bid for such a licence if it is for an area not at that time
covered by a licence authorising the provision of a local delivery service or
a prescribed diffusion service.

Participation by Foreign and Other Entities

In certain sectors of the broadcasting industry, for example cable television,   9–41
the U.K. Government has essentially encouraged foreign investment, particularly from the U.S. Thus no restrictions exist in respect of foreign participation in cable television (that is, foreign ownership of local delivery services licensees). Similarly no such restrictions exist in respect of the
provision of what are termed "non-domestic satellite services". These are
services which are either transmitted from outside the U.K. or are transmitted from inside the U.K. on a frequency other than one which has been
allocated for general reception in the U.K. The rationale for permitting
unrestricted foreign participation in non-domestic satellite services is clearly
the fact that the introduction of restrictions on foreign ownership would
result merely in the migration of such services abroad. Bodies which are
not nationals of or ordinarily resident in E.C. Member States are disqualified from holding all other broadcasting licences granted by the ITC (or
the Radio Authority in respect of radio broadcasts) as are bodies which
are controlled by non-E.C. nationals or bodies not ordinarily resident in
E.C. Member States. In addition, the 1990 Act prohibits political and
religious bodies owning broadcasters.

Future policy

There is growing pressure in the U.K. for a substantial review of this form-   9–42
alistic approach to cross-media ownership. The convergence of the telecommunications, broadcasting, publishing, information technology and consumer electronics industries is generating a greater need for flexibility in
this area. On August 3, 1994 the *Financial Times* reported that the High
Court of England had upheld a decision of the Radio Authority to approve
a scheme allowing Emap, a media group, to have a stake in eight large
metropolitan radio licences, two more than are allowed under the 1990
Act. Under the scheme, Emap created a new corporate structure enabling
two of the licences to be "warehoused" in a company which is jointly
owned (on a deadlock basis) by Emap and Schroders, its merchant bankers.

The success of this scheme is likely to precipitate similar schemes until the points system of regulation of cross-media ownership is simplified.

9–43   In May 1995 the Government published its proposals on Media Ownership; in summary these suggested a number of measures to restructure media regulation in order to allow media companies more flexibility to exploit the opportunities offered by new technologies, as follows:

- establishing an independent regulator with powers to govern media ownership;
- defining media markets and establishing their overall size through the measurement of audience or revenue share;
- expressing the value of shares in one media sector in terms of shares in another;
- determining thresholds of ownership in the national, sectoral and regional/local media markets beyond which acquisitions would have to be referred to the media regulator; and
- providing a set of public interest criteria against which the regulator would have to assess existing holdings above the thresholds, as well as any proposals for merger or acquisition which brought holdings above them.

In particular, the Government proposed to simplify the rules concerning the control of *television* licensees by:

(i) maintaining the current two licence limit on regional Channel 3 licences, but enabling broadcasters to expand to up to 15 per cent of total television audience share thereafter, including the control of up to two regional Channel 3 licences (but not both London licences), or one regional Channel 3 licence and the Channel 5 licence, or one regional Channel 3 licence and the national Channel 3 licence;

(ii) abolishing the rules which limit ownership between the various means of delivery (terrestrial, cable and satellite), subject to certain percentage audience share, market share and licence holding limits.

In addition, the Government decided to exclude from these relaxations those satellite and cable companies with more than 20 per cent ownership by newspaper groups with a national circulation share of more than 20 per cent, so that such companies will continue to be restricted to a 20 per cent holding in one Channel 3 or the Channel 5 licence, and 5 per cent in any further such licences.

Merger Control

In addition to the specific rules on media concentration, the U.K's general regulation of mergers and monopolies can restrict the establishment of media/broadcasting and other concentrations. The Monopolies and Mergers Commission may block or reverse proposed or implemented mergers if they are liable to distort competition.

CHAPTER 10
Competition Law

This area of telecommunications regulation is a hybrid of specially formu- 10–01
lated legislation and quasi-subordinate legislation and existing fair trading/
competition laws. It is characterised, as is all U.K. competition law, by a
lack of direct enforceability and the requirement, under our domestic rules,
for administrative intervention. By contrast, E.U. competition rules, discussed in full in Chapter 16, are directly applicable and give rise to rights and
remedies in the U.K. national courts as well as before the Commission and
the European Court. E.U. competition rules appear set to be of increasing
relevance so long as U.K. domestic competition law remains relatively weak
and ineffective.

1. REGULATION UNDER THE TELECOMMUNICATIONS ACT 1984

As has been seen,[1] the licences of public telecommunication operators con- 10–02
tain conditions imposing reasonably extensive, but selective, prohibitions
on certain aspects of anti-competitive behaviour. These notably include
undue discrimination, linked sales (otherwise known as "tie-ins"), exclusive
dealing arrangements and restrictive use of intellectual property rights.
Enforcement of these licence conditions is primarily a matter for the Director General under sections 16 to 19 of the 1984 Act. Study of these
sections will show that the procedure for bringing a recalcitrant operator
to heel is somewhat long-winded and may in certain circumstances prove
to be insufficiently speedy to provide an effective remedy. To summarise
this procedure, where the Director General has investigated a possible
breach of a licence condition by an operator (*i.e.* person licensed to run a
system) and is satisfied that the operator is in breach, the Director General
can do one of two things. If he fears that the breach will continue and, in

---

[1] See Chap. 5, paras. 5–30 to 5–39.

particular, may cause loss or damage to any person, he may make a provisional order to secure the operator's compliance with the particular licence condition. Clearly a provisional order is intended to be temporary and by virtue of section 16(7) of the 1984 Act it ceases to have effect at the end of a maximum period of three months unless it has been previously confirmed.

10–03   Otherwise, and instead of a provisional order, the Director General may proceed directly to the making of a final order. The procedure for such an order, which also applies to confirmation of a provisional order, is that the Director General should first give notice to the operator concerned that he proposes to make or confirm the order and setting out its effect. Such notice should state the relevant conditions of the licence which the operator is alleged to have contravened and give a time, at least 28 days from the date of publication of the notice, within which the operator may make representations on, or objections to the proposed order.

If notwithstanding these representations or objections the order is confirmed or made, the operator may under section 18(1) of the 1984 Act apply to the court for determination of the question whether it was within the powers of the Director General to make the order or whether he complied with the procedural requirements mentioned above. If the court finds that the order was not within the Director General's powers or that he did not follow the correct procedure, it may quash the order or any relevant provision which it contains. Section 18 goes on to state that except as it provides, "the validity of a final or provisional order shall not be questioned by any legal proceedings whatever".

All this may appear relatively unremarkable. However, an order once made can have severe implications: the obligation to comply with the order is a duty owed to any person who may be affected by its contravention. Any such person is able to take action not only against a defaulting operator breaching an order but also against any person guilty of an act which induces the operator's breach or interferes with its performance of the order. In any such proceedings pursuant to section 18 it will be a defence to prove that the defendant "took all reasonable steps and exercised all due diligence to avoid contravening the order".

10–04   These particular provisions giving rise to liability for inducement of a licence breach caused particular consternation to BT's employees' trade unions during the passage of the Telecommunications Bill. Prior to its enactment, the prevention of transmission of telecommunication messages had been a criminal offence; in the event this criminal liability was removed and substituted by the civil liability imposed by section 18(6) of the 1984 Act. This sub-section was said at the time to have been drafted with particular care so that a person inducing a breach is only liable if his act is done wholly or partly for the purpose of achieving the result of causing the breach. Accordingly, it will be for the aggrieved person seeking a remedy to prove that that was indeed the purpose for which the act was done.

The Director General is required to keep a register not only of licences granted but also of final and provisional orders made under the relevant sections of the 1984 Act.

Breach of a licence condition should be distinguished from failure to adhere to the provisions of the licence grant itself, for example the permitted services in Schedule 3. Such a failure would put the operator's activities beyond the protection of the licence so that, in therefore running an unlicensed system, the operator would become guilty of a criminal offence.

2. INHERITED FUNCTIONS—FAIR TRADING ACT 1973 AND COMPETITION ACT 1980

Whatever his rights and powers in any particular case, the Director General   10–05 has a general duty under section 49 of the 1984 Act to investigate complaints in relation both to the supply of telecommunication services and of telecommunication apparatus.

Section 50 of the 1984 Act specifically transfers to the Director General the functions which were conferred upon the Director General of Fair Trading under the Fair Trading Act 1973 ("1973 Act") in relation to the investigation and control of monopoly situations in the telecommunications industry. Similarly, the functions of the Director General of Fair Trading under sections 2 to 10 ("Control of Anti-Competitive Practices") of the Competition Act 1980 ("1980 Act") have been transferred to the Director General in respect of telecommunication services and apparatus.

These functions are in both cases, under the 1973 and the 1980 Act, to be exercised "concurrently" with the Director General of Fair Trading and if either of the two Directors wishes to take any action under either piece of legislation he is first to consult with the other Director. Neither Director is to exercise such functions if the same functions have been exercised by the other Director in relation to the same matter.[2]

Under section 50 of the 1984 Act, the Director General must exercise the Director General of Fair Trading's functions as to courses of conduct detrimental to consumers of telecommunication services or apparatus, if and when requested to do so by that Director General.

Licence Modification: Under section 95 of the 1984 Act, where, following   10–06 a monopoly, merger or competition reference, the Secretary of State exercises any of his powers under Parts 1 and 2 of Schedule 8 to the Fair Trading Act 1973, he may also provide for the revocation or modification of a licence granted under section 7 of the Act. This would enable him, for example, to amend the telecommunication licence of a PTO which had been the subject of a particular reference of this type and this power would appear to extend to all provisions of such a licence, not just its conditions. By contrast, licence modifications would of course be more usually effected

---

[2] See s. 50(4) of the 1984 Act.

pursuant to sections 12 to 15 of the 1984 Act,[3] when such modifications are limited to the licence conditions.

3. MONOPOLY AND MERGER REFERENCES UNDER THE 1973 ACT

(i) Monopoly references

10–07   A monopoly reference can be made by the Director General, on the basis and in accordance with the procedure set out in sections 47 to 56 of the 1973 Act, but only in respect of the supply of telecommunication apparatus, not in relation to the running of a telecommunication system.[4] A reference in respect of telecommunication services would therefore have to be made by the Secretary of State.

10–08   There are two possible types of monopoly reference, one "limited to the facts" and the other "not limited to the facts". There have been very few references limited to the facts. As regards those not so limited the Monopolies and Mergers Commission (MMC) may only investigate and report on such a reference with a view to determining whether or not a monopoly situation exists and whether any act or omission complained of in the reference operates, or may be expected to operate, against the public interest. Such matters would have to relate to prices or recommendations or suggestions as to prices, refusal to supply and preference "whether by way of discrimination in respect of prices or in respect of priority of supply or otherwise" (this latter concept is broader than the undue discrimination provisions in BT's licence) or to any matter which may be regarded as "a step taken for the purpose of exploiting or maintaining, or attributable to the existence of the monopoly situation" (see section 49(3)).

Under section 55(1) of the 1973 Act the MMC must normally report on a monopoly reference within the period specified in the reference. (The period can be extended more than once and there is no overall limit). If the report concludes that any matters do operate against the public interest the Secretary of State may by order made by statutory instrument exercise any of the powers specified in Parts 1 and 2 of Schedule 8 to the 1973 Act. These powers are wide-ranging but essentially enable the Secretary to proscribe the offending activities in relation to prices, refusals to supply, preference and discrimination. Part 2 powers are only exercisable with the approval of Parliament.

In passing, it should be noted that under section 78 of the 1973 Act the Secretary of State may at any time require the MMC to report on the general effect, on the public interest, of practices which are "commonly adopted" in order to preserve monopoly situations or on practices which

[3] See Chap 4, para. 4–09.
[4] 1973 Act, s. 50(2)

appear to be "uncompetitive practices". He may also require the MMC to report on appropriate remedial action required.

(ii) Merger references

Merger references remain the preserve of the 1973 Act and the Secretary   **10–09**
of State. He may refer to the MMC a "merger situation" qualifying for investigation, being a situation where two or more enterprises, one of which is carried on in the U.K. have "ceased to be distinct" and either:

(a) as a result, at least a quarter of the relevant goods or services are supplied by or to, one and the same person or such a previously existing market share has been increased; or

(b) the value of assets taken over exceeds (currently) £70 million.

4. COMPETITION REFERENCES—COMPETITION ACT 1980

The transferred functions of the Director General under sections 2 to 10   **10–10**
of the 1980 Act empower him to investigate anti-competitive practices, to report on the results of such investigation, to accept undertakings in relation to any anti-competitive practices the subject of such report, and to refer any such practice to the MMC. The Deregulation and Contracting Out Act 1994 extends the power granted to the Director General by section 9 of the 1980 Act to seek undertakings to remedy any situation found by the MMC to operate against the public interest. Under the 1994 Act, the Director General may request undertakings in respect of conduct which he has reasonable grounds for believing may be anti-competitive and may operate against the public interest, providing he has first issued a notice of his intention to do so and considered any representations made to him.[5] If an undertaking is accepted by the Director General at this stage, he will be precluded from referring the matter to the MMC while that undertaking is in place.

Under section 2 of the 1980 Act, and substituting the 1984 Act's amendments, a person is said to engage in an anti-competitive practice:

> "If in the course of business that person pursues a course of conduct which, of itself or when taken together with a course of conduct pursued by persons associated with him, has or is intended to have or is likely to have the effect of restricting, distorting or preventing competition in connection with the production, supply or acquisition of *telecommunication apparatus* in the United Kingdom or any part of it or the supply or securing of *telecommunication* services in the United Kingdom or any part of it."

Such anti-competitive practices have their own near-equivalent to the   **10–11**
Article 85, minor agreements exemption. Under the Anti-Competitive Prac-

---

[5] s. 12.

tices (Exclusions) Order 1980,[6] a course of conduct is excluded from constituting an anti-competitive practice if it is:

(a) a course of conduct described in Schedule 1 to the Order; this includes contracts for exports (where of course any such conditions affecting EEC trade could still merit review in relation to Article 85); or

(b) the course of conduct of a person whose (or whose group's) annual turnover in the U.K. is less than £5 million and who (individually or whose group) enjoys less than one-quarter of a relevant market (*i.e.* in respect of the particular goods or services involved).

Although the language of section 2 of the 1980 Act, in describing an anti-competitive practice, is similar to its Euro-equivalents, the similarity becomes less apparent on examination and comparison of the procedure adopted by the 1980 Act to deal with such practices. Whereas, for example, the European Commission can investigate and issue legally binding pronouncements as to the validity of agreements and practices within the bounds of Article 85, in some cases giving rise to directly enforceable claims by persons affected, the Director General can undertake investigation but cannot take any executive action himself. Once the matter has been referred by him to the MMC it is out of his hands unless and until he is to seek undertakings (if any) arising out of the MMC report.

10–12   The MMC's terms of reference and its ability to make suitable recommendations to bring an end to an anti-competitive practice are heavily circumscribed. In particular, on both a monopoly reference under the 1973 Act and a competition reference under the 1980 Act, the MMC's objective is to discover whether or not there are effects adverse to the public interest. If the public interest is not adversely affected, the MMC has no locus' to make any other determination of the matter. This emphasis purely on the public interest criterion appears to be one peculiar to U.K. anti-trust and it would have been preferable in terms of potential harmonisation of European competition laws and certainly preferable for those seeking to prevent anti-competitive practices or who might be prejudiced by them, for the test simply to be whether or not the particular situation gave rise to a significant or appreciable restriction on competition or represented the abuse of a dominant position.

Thus it may come as a surprise, at least to foreign lawyers, that under the 1980 Act an anti-competitive practice is not *per se* illegal and does not invite legal sanction until the whole procedure of Director General investigation, MMC report and Secretary of State order has been completed.

---

[6] S.I. 1980 No. 979.

5. Restrictive Trade Practices

Broadcasting Legislation: Channel 3 Networking Arrangements

These are arrangements under which the 15 regional Channel 3 licensees   **10–13** in the U.K. organise themselves in order to obtain and make available programmes for simultaneous broadcasting in all the U.K. regions so as to provide a nationwide service to compete with other nationwide broadcasters. The arrangements are essentially concerned with the commissioning and acquisition of programmes to be shown on the licensees' channels. The Broadcasting Act 1990 introduced measures to make the networking arrangements more competitive and thus prevent larger companies abusing their position. A Network Centre has been formed by the 15 regional Channel 3 licensees to select programmes suitable for the network and acquire on behalf of the licensees broadcast rights in those programmes.

The 1990 Act provides that the Restrictive Trade Practices Act 1976 shall   **10–14** not apply to the networking arrangements provided they are as specified in an order made by the Secretary of State having consulted the Independent Television Commission (ITC) and the Director General of Fair Trading. The arrangements were examined by the Director General of Fair Trading to determine whether they satisfied a competition test under the Broadcasting Act 1990 concerning the effect on competition in any business activity. It was as a result of this requirement that the networking arrangements first formulated by the licensees were referred to the ITC and subsequently to the Director–General of Fair Trading who concluded that the arrangements failed the competition test for two main reasons:

  (i) the requirement that independent producers should contract with a licensee and not with the Network Centre was likely to restrict or distort competition in the production of television programmes;
 (ii) the standard terms of contracts, which were effectively non-negotiable, gave exclusive U.K. transmission rights to the Network for years with an option to extend for a further five; in addition the standard terms for acquisition of programmes by licensees for independent producers, where there is a 100 per cent funding, sought rights in perpetuity.

The Director General of Fair Trading believed that those terms would restrict the ability of independent programme makers to compete in the production of programmes and restricted competition between broadcasters.

The Director General of Fair Trading also concluded that the arrange-   **10–15** ments did not satisfy the criteria set out in the Treaty of Rome, Article 85, paragraph 3, because although they contributed to improving the production of services and gave consumers a fair share of the resulting benefits, they imposed restrictions which were not indispensable to the attainment of such benefits. The Director General of Fair Trading considered that it was possible to arrange matters so that contracts could be made by independent producers with the Network Centre.

As a result the networking arrangements proposed by the licensees were referred to the MMC whose report essentially agreed with the Director–General of Fair Trading findings and concluded that (a) provisions of the networking arrangements which precluded direct contracting between an independent producer and the Network Centre, and (b) provisions in the networking arrangements dealing with the acquisitions of rights failed in each case to satisfy the competition test.

10–16   The current networking arrangements comprise four publicly available documents which provide guidance on the practices which will be adopted in relation to access to and the selection of programmes by the Network Centre and include the form of contract to be used. As a result of the MMC report, both licensees and independents have equal access to the Network Centre for submitting programme proposals and the terms on which rights are to be acquired are subject to free negotiation between the Network Centre and the producer of a programme. The Network Centre Code of Practice provides that the Centre will not seek to acquire rights greater than are necessary for the broadcast interest of the licensees and rights will not normally be acquired for longer than an initial period of five years with an option to extend for a further two years. These modifications which were recommended by the MMC mean that the current networking arrangement in the opinion of the MMC now satisfy the competition test.

The restrictive trade practices corollary to the MMC recommendations is that under the Broadcasting (Restrictive Trade Practices Act 1976) (Exemption for Networking Arrangements) Order 1994[7] the Restrictive Trade Practices Act 1976 is deemed not to apply to these networking arrangements (set out in Schedule 2 to the Order).

---

[7] S.I. 1994 No. 2540.

CHAPTER 11

Property Rights and the Environment

By Philip Burroughs

1. BACKGROUND

Privatisation of previously Government run bodies has resulted in the **11–01** former statutory undertakers losing many of the freedoms they had hitherto enjoyed when dealing with land. In their guise as statutory undertakers they benefited from very limited statutory controls on their activities with respect to land and a new framework of rules and regulations was necessary in order to control the operations of the "new" privately owned companies.

The new framework needed to deal not only with the impact of statutory controls where previously the former statutory undertakers had little or none, but also to deal with the inter-relationship of the new companies with private land owners.

Whilst the usual principles of common law, landlord and tenant law and planning law apply in the usual way in relation to telecommunications activities, this new framework can be considered in three major areas:

(i) the PTO's licence;
(ii) the 1984 Act (including the Code); and
(iii) planning.

The PTO's licence amongst other things contains the mechanism whereby firstly Schedule 2 of the 1984 Act is applied to the PTO and secondly further conditions and restrictions are applied to the PTO in relation to land.

The 1984 Act governs the framework with regard to the telecommunica- **11–02** tions industry and Schedule 2 of the Act sets out a statutory code which has particular relevance to the relationship between private landowners and the rights of PTOs. The New Roads and Street Works Act 1991 which repealed the Public Utilities Streetworks Act 1950 inter-relates closely with the 1984 Act.

Planning legislation has developed since the passing of the 1984 Act to deal specifically with planning and environmental issues arising in relation to telecommunications installations and apparatus. Particular reference will be made to the Town and Country Planning General Development Order 1988 (S.I. 1988 No. 1813) (as amended) (the GDO) where Schedule 2, Parts 24 and 25 deal with areas of permitted development directly related to telecommunications.

In this Chapter, consideration is given to the areas of statutory "interference" with PTOs and the rights of landowners. Consideration is not given to the general operation of the common law and landlord and tenant issues which, as mentioned earlier, apply in the ordinary way.

2. The Licence

11–03   A PTO is someone who has been granted a licence to operate a telecommunications system under section 7 of the 1984 Act and has been designated as a PTO under section 8. The licences granted to BT and other PTOs contain, in Schedule 4 of each licence, detailed provisions relating to land and planning.

Clause 2 of the licence granted to BT dated August 5, 1984 states:

"The telecommunications code contained in Schedule 2 to the Act shall apply to [BT] for all purposes except those *not* relating to the Applicable Systems and subject to the other exceptions and Conditions set out in Schedule 4 for so long as this Licence is one to which section 8 of the Act applies."

The expression "Applicable Systems" means the telecommunication systems which are identified in Annex A to the BT licence and which BT is thereby licensed to operate. It should be remembered that the Applicable Systems of each PTO can be very different in nature. For example the Applicable Systems of BT, Mercury and others require an ability to lay cables in roads, cross private land, fly lines over land and to lop trees which interfere with signals. However, the operators of cellular mobile networks do not have such requirements, their operation being a network of cell antennae. Accordingly, several paragraphs of the Code may well be inapplicable to the specific PTO's Applicable Systems.

Schedule 4 of each PTO licence renders the relevant PTO subject to additional controls. Whilst these relate to both planning controls and the provisions of the Code, some industry practitioners have expressed the view that the relevant statutory provisions are more stringent, potentially rendering Schedule 4 superfluous in practice. Nevertheless, non-compliance with Schedule 4 would technically mean that the PTO is not operating in accordance with its licence, with all the consequences that potentially flow from that. The Schedule contains some 18 conditions ranging from obligations (Condition 1) to instal lines underground in conservation areas (subject to certain limited exceptions) to (Condition 18) furnishing the Secretary of

State and the Director General of Telecommunications with descriptions of the telecommunications apparatus it is installing and proposing to install above the surface of the ground pursuant to the Code.

Two of the conditions set out in Schedule 4 are worthy of particular attention; although for more general reference it should be appreciated that the licences were granted in 1984 and the statutory references which are quoted verbatim will now have been superseded. For example, Schedule 4 of the licence granted to AT&T Communications (UK) Limited on December 20, 1994 contains updated references to The New Roads and Street Works Act 1991 and the Town and Country Planning Act 1990.

(i) Condition 2: Listed Buildings and Ancient Monuments

"2.1 Except in the case of emergency works, the Licensee shall before **11–04** installing lines, poles or other telecommunication apparatus in proximity to a building shown as Grade I or, as the case may be, Category A in the Statutory List give notice to the Planning Authority. Where the installation would detrimentally affect the character or appearance of the building, or its setting, and the Planning Authority indicates within 28 days of the giving of the notice that the installation should not take place, the Licensee may install the apparatus only if the Secretary of State (having consulted the Planning Authority) so directs in writing, or with the agreement of the Planning Authority.

2.2 For the avoidance of doubt it is hereby declared that nothing in this Licence affects: (a) the statutory requirement that the consent of the Secretary of State shall be obtained before any work is carried out which will affect the site of an ancient monument scheduled under sections 1 and 2 of the Ancient Monuments and Archaeological Areas Act 1979 or section 7 of the Historic Monuments (Northern Ireland) Act 1971; or (b) the obligation imposed on the Licensee by virtue of section 55 of the Town and Country Planning Act 1971 (or by section 53 of the Town and Country Planning (Scotland) Act 1972 or by Article 32 of the Planning (Northern Ireland) Order 1972) to obtain listed building consent for any works which affect the character of a building in the Statutory List or involve the demolition of any part of such a building."

In the context of listed buildings it is of course an offence under the Planning (Listed Buildings and Conservation Areas) Act 1990 to demolish, or to alter or to extend so as to affect its character as a building of special architectural or historical interest any listed building without the written consent of the local planning authority (or the Secretary of State for the Environment). It is no doubt arguable that the erection of satellite dishes on a listed building is an alteration for these purposes. The possibility of substantial fines (not exceeding £20,000) and imprisonment mean that great care must be exercised in this area.

Condition 2.2 (b) of the licence granted to AT&T refers to section 7 of the Planning (Listed Buildings & Conservation Area) Act 1990 and Article 44 of the Planning (Northern Ireland) Order 1991.

(ii) Condition 6: Highways

11–05   With respect to maintainable highways BT's licence provides that:

"6.1 For the avoidance of doubt it is hereby declared that paragraph 6.2 applies in addition to any obligations of the licensee under the Public Utilities Street Works Act 1950 and any order made under Section 11(1) of the Act.

6.2 Except in the case of emergency works, before executing any works involving the breaking up of a maintainable highway in connection with the installation of any telecommunication apparatus in that highway the Licensee shall give to the Highway Authority written notice of its intention to do so describing the proposed works and shall consider any written representations made by that Highway Authority within 8 days of the giving of the notice by the Licensee in the case of an overhead line or any underground Service Line and within 29 days of the giving of the notice by the Licensee in other cases."

In the AT&T licence there is no equivalent of paragraph 6.1. The obligations in that form of licence are as follows—

"6.1 Except in the case of Emergency Works or Urgent Works, before executing any works involving the breaking up of a maintainable Highway or, in Scotland, a Public Road in connection with the installation, inspection, maintenance, adjustment, repair or alteration of any Telecommunication Apparatus in that Highway or that Road the Licensee shall:

(a) in the case of an overhead line or another underground Service Line consider any written representations made by the Highway Authority or, in Scotland, the Road Works Authority within 7 working days after the giving of any such notice as is required to be given to the Highway Authority under section 55 of the New Roads and Street Works Act 1991 or, in Scotland, to the Road Works Authority under section 114 of the New Roads and Street Works Act 1991 or, in Northern Ireland, paragraphs 1 (3) and 3 (2) (a) of Schedule 3 to the Electricity Supply (NI) Order 1972 as amended by the Telecommunications (Street Works) (NI) Order 1984;

(b) in all other cases, consider any such written representations made within 29 days of the giving of any such notice; and

(c) unless the Highway Authority or, in Scotland, the Road Works Authority consents otherwise, shall not commence those works until the expiry of seven working days or 29 days as the case may be".

"Emergency" is defined as "emergency of any kind, including any circumstance whatever resulting from major accidents, natural disasters and incidents involving toxic or radio-active materials".

Clearly, whilst the provisions of the relevant licence have importance insofar as the relevant PTO is concerned and provide an insight into the obligations likely to be imposed on any potential PTO, consideration should now be given to the more general application of statute.

3. THE TOWN AND COUNTRY PLANNING GENERAL DEVELOPMENT ORDER 1988 (GDO)

The PTOs are subject to the planning legislation in the ordinary way. However, a considerable degree of freedom is afforded to the PTOs and in relation to telecommunications generally through the GDO. The Town & Country Planning General Development Order 1977 has now been replaced by the GDO and Classes XXIV and XXV of the 1977 Order are now found in Schedule 2, Parts 24 and 25 of the GDO. Under Part 24 the permitted development relates to land controlled by the PTO or carried out in accordance with the PTO's licence. Development permitted under Part 25 relates to other land perhaps where a third party is to benefit from telecommunications services whether by microwave antenna or satellite antenna. 11–06

Schedule 2, Part 24, "Development by Telecommunications Code System Operators" sets out a class of permitted development in the following terms:

"Development by or on behalf of a telecommunications code system operator for the purpose of the operator's telecommunication system in, on, over or under *land controlled by that operator or in accordance with his licence* consisting of (a) the installation, alteration or replacement of any telecommunication apparatus, (b) the use of land in an emergency for a period not exceeding 6 months to station and operate moveable telecommunication apparatus required for the replacement of unserviceable telecommunication apparatus, including the provision of moveable structures on the land for the purposes of that use, or (c) development ancillary to radio equipment housing."

It is thought that the expression *in accordance with his licence* refers to the exercise of rights under the Code, and that the consent of the occupier of land is required for such development under Paragraph 2 of the Code. It seems likely that every customer order will fall within Part 24 (unless it falls within other exceptions) and accordingly, in practice, customer order forms contain an appropriate form of consent for these purposes.

11–07   For these purposes "land controlled by that operator" means land which is freehold owned by the operator or held under a lease *granted* for a term of not less than 10 years. However, paragraphs A.1 and A.2 of Part 24 go on to provide *exceptions* to the levels of permitted development and *conditions* subject to which the development is permitted.

As to *exceptions*, there are two important categories, the first relating to apparatus which *is* on a building, the second relating to apparatus which is not. In the former case development is not permitted if the *installation* of apparatus would exceed 15 metres above ground level or *alteration or replacement of installed apparatus* would *exceed* the height of the existing apparatus or 15 metres above ground floor level whichever is the greater.

In the later case, development is not permitted in respect of the *installation, alteration or replacement* of apparatus firstly if the height of the apparatus itself would exceed 15 metres on a building which is 30 metres or more in height or 10 metres in any other case, and secondly, if the highest part of the apparatus would exceed the highest part of the relevant building by more than:

(i) 10 metres in respect of a building which is 30 metres or more in height;

(ii) eight metres in respect of a building which is more than 15 but less than 30 metres in height; or

(iii) six metres in any other case.

11–08   The other exceptions to permitted development are set out in paragraphs A(1)(e) to (k) as amended by S.I. 1991 No. 1536. These range from consideration of the ground or base areas of certain structures to the dimensions of microwave antennae.

As to the *conditions* pursuant to which development is permitted, essentially where the development consists of the installation, alteration or replacement of telecommunications apparatus then any antennae or supporting apparatus so installed, altered or replaced *on a building* shall so far as is practicable be sited so as to minimise its effects on the external appearance of the building. Further, where the development is permitted under Class A(b) (emergency period not exceeding six months) then the development is subject to the condition that any apparatus or structure so provided must be removed from the land and the land restored to its condition before the development took place.

Schedule 2, Part 25 "Other Telecommunications Development" sets out two further categories of permitted development. These categories are not limited to land controlled by a PTO but to telecommunications development in its widest sense. The first deals with microwave antennae and is expressed in the following terms:

"The installation, alteration or replacement on any building or other structure of a height of 15 metres or more of a microwave antenna and any structure intended for the support of a microwave antenna."

11–09   As with developments permitted by Schedule 2, Part 24, such development is subject to conditions and exceptions. The exceptions apply if, *inter alia*,

  (i) the building is a dwelling house or within the curtilage of a dwelling house;
 (ii) the development is by PTO under Schedule 2, Part 24;
(iii) the development would result in the presence on the building of two or more microwave antennas;
 (iv) in the case of a satellite antenna the size of the antenna including its supporting structure but excluding any projecting feed element would exceed 90 centimetres;
  (v) in the case of a terrestrial microwave antenna (i) the size of the antenna when measured in any dimension but excluding any projecting feed would exceed 1.3 metres and (ii) the highest part of the antenna or its supporting structure would be more than three metres than the highest part of the building or structure on which it is installed or to be installed;
 (vi) it is on article 1(5) land and
(vii) it would consist of the installation, alteration or replacement of system apparatus within the meaning of section 8(6) of the Road Traffic (Driver Licensing and Information Systems) Act 1989.

The second category of permitted development under Part 25 specifically **11–10** deals with satellite antennae and is expressed in the following terms:

> "the installation, alteration or replacement on any building or other structure of a height of less than 15 metres of a satellite antenna."

The antenna shall so far as is practicable be sited so as to minimise its effect on the external appearance of the building or structure on which it is installed and when no longer needed for the reception or transmission of microwave radio energy is to be removed as soon as reasonably practicable. However, this development is not permitted if (i) the building is a dwelling house or is within the curtilage of a dwelling house, (ii) the development is by a PTO under Schedule 2, Part 24 (iii) it would consist of an installation, alteration or replacement of system apparatus within the meaning of section 8(6) of the Road Traffic (Driver Licensing and Information Systems) Act 1989 (iv) the size of the antenna (excluding any projecting feed element, reinforcing remounting or brackets) when measured in any dimension would exceed (a) 90 centimetres in the case of an antenna to be installed on a building or structure on article 1(7) land, (b) 70 cm in any other case, (v) the highest part of the antenna to be installed on a roof would, when installed, exceed in height the highest part of the roof (vi) there is any other satellite antenna on the building or other structure on which the antenna is installed (vii) it would consist of the installation of an antenna on a chimney (viii) it would consist of the installation of an antenna on a wall or roof slope which fronts a waterway in the broads or highway elsewhere.

When considering the operation of the GDO in relation to any potential aspect of "permitted development" consideration needs to be given as to whether the relevant local planning authority has exercised its rights to restrict or entirely remove the deemed planning consents afforded by the GDO through the operation of the rights afforded to planning authorities in Article 4 of the GDO. However it should be appreciated that some guidance has been given in relation to planning policy.

11–11   The Department of the Environment issued a Planning Policy Guidance Note (PPG8) in 1988 which was subsequently reissued in 1992. Paragraph 5 states that the Government's general policy on telecommunications is to facilitate the growth of new and existing systems, the Government also being fully committed to environmental objectives including well-established national policies for the protection of countryside and urban areas. Local planning authorities are directed "to respond positively to telecommunications development proposals, especially where the proposed location is constrained by technical considerations, while taking account of the advice on the protection of urban or rural areas in other planning policy guidance notes. They should bear in mind the wider environmental benefits—for example, if driver information systems ensure a better use of existing roads infrastructure or if the application of telecommunications technology reduces the need to travel—that may in particular cases outweigh such direct adverse effects on the visual impacts of new marks on the area."

4. THE TELECOMMUNICATIONS ACT 1984

11–12   Whilst, arguably, the most important aspect of the 1984 Act in so far as property is concerned is the *impact* of the provisions of Schedule 2 (considered in paragraph 6, below) there are two sections of the 1984 Act which are worthy of separate mention.

Firstly, section 34 provides that with the consent of the Director General the Secretary of State may authorise a PTO to purchase compulsorily land which is required by the PTO for or in connection with the establishment or running of the particular PTO's system or as to which it can be reasonably foreseen that it will be so required. Land acquired under these provisions cannot be disposed of (nor any interest or right in or over it) except with the Director General's consent. The Acquisition of Land Act 1981 applies to any such compulsory purchase as if the operator were a local authority within the meaning of that Act. Certain provisions of the planning legislation have effect in relation to land acquired compulsorily by a PTO as though the PTO were a statutory undertaker for the purposes of that legislation.

The power to acquire land by compulsory purchase extends to a right to acquire easements or other rights over land. There is also a power under the Act, s.37 at any reasonable time, to enter upon land for the purpose of ascertaining whether the land would be suitable for use by the PTO for or in connection with the running of its system. In order to protect the land owner there are obligations to make good damage and pay compensation in respect of damage caused.

Secondly, section 96 of the 1984 Act, albeit yet to be brought into legal effect,[1] has the potential to interfere with leases in so far as the lease might restrict or prohibit a lessee's desire to install telecommunications apparatus or systems. Essentially it enables the tenant to have the relevant prohibition or restriction treated as if it were subject to a proviso that the landlord's consent would not be unreasonably withheld. Further, under s.96(2) of the 1984 Act the question of whether consent is unreasonably withheld shall be determined having regard to all the circumstances and to the principle that no person shall be unreasonably denied access to telecommunications systems. It is the author's understanding that there are no proposals to bring this section into force.

5. THE TELECOMMUNICATIONS CODE

In this section consideration is given to the Telecommunications Code   **11–13** which is set out in Schedule 2 to the 1984 Act. The Code has effect where it is applied to a person who is authorised by a licence granted by the Secretary of State to run a telecommunications system. When applied to the PTO the Code operates subject to such exceptions and conditions as may be imposed in the operator's licence. We have already observed the nature and extent of those conditions which have been imposed in respect of the licence to BT.

It should be noted that the Code will not prejudice any rights or liabilities which arise at any time under any agreement entered into before the Code comes into force and which relates to the installation, maintenance, adjustment, repair, alteration or inspection of any telecommunications apparatus or the keeping of any such apparatus installed on, under or over any land.

The exercise of a right conferred by paragraphs 2 or 3 of this Code is pursuant to a statutory power and will override private restrictive covenants, that is those not entered into pursuant to statute. Thus a covenant with a local authority, for instance a planning obligation entered into pursuant to section 106 of the Town & County Planning Act 1990, will continue to bind a PTO.

The written agreement of the *occupier* of land is required for conferring   **11–14** on the PTO a right for the statutory purposes of—

  (i) executing any works on that land for or in connection with the installation, maintenance, adjustment, repair or alteration of telecommunications apparatus; or
  (ii) keeping telecommunications apparatus installed on, under or over that land; or

---

[1] The Secretary of State is authorised under s.110(5) to make an order to bring this section into legal effect.

(iii) entering that land to inspect any apparatus kept installed (whether on under or over that land or elsewhere) for the purpose of the PTO system.

In relation to a customer placing an order with a PTO, the customer will be the "occupier" for these purposes and as noted earlier will usually be giving his consent on signing the order form.

Owners of the land are not bound by such rights granted by the *occupier* unless:

  (i) he conferred the right himself as occupier of the land;
 (ii) he has agreed in writing to be bound by the right; or
(iii) he is for the time being treated by virtue of sub-paragraph (3) of paragraph 2 of the Code as having so agreed or he is bound by the right by virtue of the impact of sub-paragraph (4) of paragraph 2 of the Code.

A tenant in a multi-let office building requiring telecommunications apparatus to be installed will be the occupier in respect of his demised premises but faces difficulties. It may be necessary to bring the apparatus cables through common parts of the building in respect of which he is not the "occupier". The protection of the Code will not help in relation to those common parts. Further, such tenant will need to have regard to the terms of his lease because covenants prohibiting certain types of alterations may well impact on the tenant's proposed installation or the works required to install it.

11–15   In the event that a particular owner or occupier is unco-operative and fails to agree that one of the rights should be conferred on the PTO or that his interest in the land should be bound there is power in paragraph 5 to apply to the Court for an order conferring the proposed right or providing for it to bind any person or any interest in the land and (in either case) dispensing with the need for the agreement of the person to whom notice was given (28 days prior notice is required with the application for the order).

Further, for the statutory purposes the PTO has the right in order to cross any land which is used wholly or mainly either as a railway, canal or tramway or in connection therewith to install the line and other telecommunication apparatus on, under or over that land. It should be remembered that the Applicable System of some PTOs will not require such rights, for instance in respect of a cellular mobile network. There are additional rights to carry out works and inspections. Telecommunications apparatus so installed must not interfere with traffic on the railway, canal or tramway. There are provisions for the giving of notice and reference to arbitration.

The court in such circumstances must make an order under these provisions if it is satisfied that any prejudice caused by its order is capable of being adequately compensated for by money or is outweighed by the benefit accruing from the order to the persons whose access to a telecommunication system will be secured by the order. In making such a determination the court is to have regard to all the circumstances and to the principle that no person should be unreasonably denied access to a telecommunication system. The relevant court, having jurisdiction in these matters in England and Wales, is the County Court.

Where the court makes an order dispensing with the need for a person's agreement the order must include such terms with respect to the payment of consideration as would have been fair and reasonable if the agreement had been willingly given, and further, such terms as appear to the court appropriate for ensuring that the persons bound by the matters to which the order relates are adequately compensated whether by the payment of such consideration or for any loss or damage sustained by them as a consequence of the exercise of those rights.

Paragraphs 10 and 12 of the Code deal with the right of the PTO to fly   **11–16**
lines over property without consent of the person owning or occupying the property above which the line passes subject to an objection procedure set out in paragraph 17 and providing that it does not interfere with the carrying on of any business on that land.

Paragraph 19 of the Code deals with the situation where any tree overhangs a street and in doing so interferes or obstructs the working of any telecommunication apparatus used for purposes of the PTO system or will obstruct or interfere with the working of any telecommunication apparatus which is about to be installed for those purposes. The operator may by giving notice to the occupier of land on which the tree is growing require the tree to be lopped. If, within 28 days of the notice, the occupier of the land gives a council notice objecting to the lopping the notice will only have effect if confirmed by a court order. If the operator's notice is not complied with and the 28-day period has expired without counter notice being given or a court order confirming the notice has come into force, the operator may then cause the tree to be lopped himself. There is a provision for the court on application from any person who has sustained loss or damage in consequence of the lopping to make an order to the operator to pay compensation in respect of the loss damage or expenses as it sees fit. The provisions of the Code specifically apply to crown land as well as land in private ownership.

6. UTILITIES

PTOs are subject to the provisions of the New Roads and Street Works   **11–17**
Act 1991. Reference is made in paragraph 9 of the Code to "a street" and or "maintainable highway" as those terms fall within the meaning set out in Part 3 of the 1991 Act.

Under the 1991 Act there is a new mechanism for the giving of notice and co-operation with the relevant street authority. However, it is anticipated that on streets other than traffic sensitive streets, urgent works will be able to take place without advance notice subject to a requirement to give notice within two hours. It will be recalled that the terms of the licence granted to the PTO contain specific obligations for the giving of notice to the relevant Highway Authority in relation to emergency works.

The rights afforded to the PTO under paragraph 9 of the Code and indeed those rights given under paragraphs 2, 10 and 11 are subject to the overriding provisions of paragraph 3 of the Code to the effect that the rights should not be exercisable so as to interfere with or obstruct any means of entering or leaving any other land unless the occupier for the time being of the other land *conferred*, or is bound by a right to interfere or obstruct that means of entering or leaving the land.

Paragraph 23 of the Code applies to work to be executed by other "relevant undertakers" which are likely to involve a temporary or permanent alteration of any telecommunication apparatus kept on under or over any land for the purposes of the operator's system. Relevant undertakers for these purposes are essentially statutory authorities or other PTOs. The paragraph provides for the giving of notice by the relevant undertaker and a procedure for the service of a counter notice. Service of a counter notice by the PTO to the relevant undertaker must state whether i) the PTO intends himself to make any alteration made necessary or expedient by the relevant undertaker's works, or ii) that he requires a relevant undertaker in making such alterations to do so under the supervision and to the satisfaction of the operator. The PTO has a right to execute the works itself following the giving of an appropriate counter notice and the expenses incurred or sustained are recoverable from the relevant undertaker.

7. REGISTRATION

11–18   Paragraph 2(7) of the Code provides that the rights granted to the operator for statutory purposes are not subject to the provisions of any other statutory requirement for the registration of interests in, charges on or other obligations affecting land.

8. SATELLITE ANTENNAE: DWELLING HOUSES

11–19   It should be noted that dwelling houses are specifically excluded from the operation of Schedule 2, Parts 24 and 25 of the GDO. For the sake of completeness under Schedule 2, Part 1 of the GDO, permission is given for Class H development comprising the installation, alteration or replacement of a single satellite antenna upon a private dwelling house or within the curtilage of the house. This is subject to the proviso to the effect that the size of the antenna should not exceed the following limits when measured in any direction (but excluding any projecting feed element, reinforcing rim, mountings and bracket). In the case of an antenna mounted on a chimney, 45 centimetres; on or within the curtilage of a dwelling house other than on a chimney, 90 centimetres; and 70 centimetres in any other case. The highest part of the antenna *should not be higher* than the highest part of the roof or chimney on which it is erected. The antenna should be sited so as to minimise its effect on the external appearance on the building and an antenna no longer in use should be removed as soon as reasonably practicable.

CHAPTER 12

Intellectual Property in Telecommunications

Leaving aside the question of content, which is not within the scope of this **12–01** work, there are two main areas in which intellectual property rights are relevant to telecommunications—transmission and technology. The principal intellectual property rights of relevance here are copyrights and patents.

There is nothing new in the proposition that proprietary rights not only confer an effective monopoly, but are also liable to give the proprietor a competitive edge over its rivals: this has been demonstrated as much in the telecommunications industry as elsewhere and recently has given rise to real tensions leading, in the field of standards, to allegations of breach of E.U. competition rules (see paragraphs 12–09 to 12–12 below).

The phenomenon of the inability of regulation to keep pace with technology has proved to be particularly apposite to intellectual property protection, which the digital/computer age has shown to be wanting in a number of areas: numerous grafts onto basic legislation have proved necessary and will continue to be needed to cope with innovation in telecommunications.

1. COPYRIGHT

Protection of broadcasts and cable programmes

This aspect is dealt with in the Copyright, Designs and Patents Act 1988. **12–02** Under section 16 of the 1988 Act, copyright in the relevant work of authorship, as well as in a broadcast or cable programme, restricts broadcasting of the work (and of the original broadcast or cable programme) and inclusion of the work (and the broadcast/cable programme) in a cable programme service. Copyright in a broadcast or a broadcast work is not infringed by its reception and immediate retransmission in a cable programme service where such re-transmission is in pursuance of a "must carry"[1] obligation, or where the broadcast is made for reception in the area in which the cable programme service is provided—see section 73 of the 1988 Act. The free public showing or playing of a broadcast cable programme does not, however, amount to copyright infringement, by virtue of section 72 of the 1988 Act.

The performance of a literary, dramatic or musical work in public is an act restricted by the copyright in the work and for this purpose "performance" includes, according to section 19(2) of the 1988 Act, "any mode of visual or acoustic presentation, including presentation by means of a sound recording, film, *broadcast or cable programme* of the work." In this connection, lest any telecommunication operator involved in facilitating such presentation might be deemed an infringer, section 19(4) exonerates the person by whom visual images or sounds conveyed by electronic means are sent.

Definition of "broadcast" and "cable programme"

Broadcast

12–03   Section 6 of the 1988 Act defines a "broadcast" as a transmission by wireless telegraphy of visual images, sounds or other information which—

(a) is capable of being lawfully received by members of the public, or
(b) is transmitted for presentation to members of the public.

For these purposes, an encrypted transmission is to be regarded as capable of being lawfully received by members of the public only if decoding equipment has been made available to members of the public by or with the authority either of the person making the transmission or the person providing the contents of the transmission (section 6(2)). The broadcaster is, according to section 6(3) of the 1988 Act, the person transmitting the programme, if he has any responsibility to any extent for its contents, *and* any person providing the programme who makes, with the person transmitting it, the arrangements necessary for its transmission.

Satellite broadcasts

12–04   The vexed question of where an infringing act perpetrated by a satellite broadcast actually takes place is addressed in section 6(4) of the 1988 Act which provides—

"a place from which a broadcast is made is, in the case of a satellite transmission, the place from which the signals carrying the broadcast are transmitted to satellite".

This is not an entirely satisfactory or helpful explanation. Take the

---

[1] "Must carry" programmes are the BBC and Independent Television broadcasts which a cable operator must include in its services by virtue of its licence.

example of a broadcast where the originating signal is first transmitted over a land line, which could obviously be from one country to another, and then up-linked from the second, or a subsequent, country to a satellite: in such a case is the signal carrying the broadcast transmitted from the first country where it originated or from the country of up-link? Clearly, this will depend upon the circumstances, but the question is at its most difficult when the signal is uninterrupted at all stages and simply passes through the earth station in "real time". On the other hand, if the signal is received at the earth station and then stored or re-processed prior to transmission to the satellite, it is clearly much more likely that the broadcast would be deemed to take place at that final point of transmission.

This issue is now addressed and dealt with somewhat differently in the Directive on copyright and satellite broadcasting.[2] Article 1 of this Directive defines the restricted act of "communication to the public by satellite" as being the act of introducing, under the control and responsibility of the broadcasting organisation, the programme-carrying signals intended for reception by the public into an uninterrupted chain of communication leading to the satellite and down towards earth. Thus the Directive establishes that the restricted act takes place where the signal is introduced into an essentially automated and open path leading to the broadcast. This Directive requires that Member States should bring into force the requisite laws and Regulations by January 1, 1995.

Cable Programmes

Section 7 of the 1988 Act defines a "cable programme" as any item **12–05** included in a cable programme service, being a service consisting wholly or mainly in sending visual images, sounds or other information by means of a telecommunications system, otherwise than by wireless telegraphy, for reception either—

(a) at two or more places (whether for simultaneous reception or at different times in response to requests by different users), or

(b) for presentation to members of the public, unless it is a service covered by any other exceptions in section 7(2). (These exceptions cover interactive services allowing two-way flow of messages in real time between the recipient of the service and the provider, as well as a service run exclusively for the internal purposes of a business.)

Note here the key difference between the 1988 Act definition of a "cable programme service" and the Broadcasting Act 1990, s. 72(1) definition of "local delivery service", which refers to a service consisting in the use of a telecommunication system for the purpose of delivery of (amongst other things) a licensable programme service *for simultaneous reception in two or more dwelling houses in the United Kingdom.* Whilst therefore section 72 of the 1990 Act excludes programmes which can be "down-loaded"

---

[2] E.C. Council Directive 93/83 on the co-ordination of certain rules concerning copyright and rights related to copyright applicable to satellite broadcasting and cable retransmission. See also Chap. 16, para. 16–50.

172   Intellectual Property in Telecommunications

electronically in response to a specific request,[3] thus precluding the need for a local delivery licence for such a service, a cable programme service of such a kind nevertheless retains the benefit of copyright protection by virtue of the wider language used in section 7 of the 1988 Act.

Duration of copyright

12–06   Under section 14 of the Copyright, Designs and Patents Act 1988, copyright in a broadcast or cable programme endures for a period of 50 years from the end of the calendar year in which the broadcast was made, or the programme was included in a cable programme service; repeat broadcasts or cable programmes are ignored for this purpose.

The Directive on the harmonisation of the term of copyright protection[4] provides in Article 3 that the rights of broadcasters will expire 50 years after the first transmission of a broadcast, whether this broadcast is transmitted by wire or over the air, including by cable or satellite. This Directive is due to be implemented by Member States before July 1, 1995.

Competition law

Ownership of intellectual property rights confers a form of monopoly in the specific subject matter of the rights. Whilst the *enjoyment* of such a monopoly is thus (typically) validated by statute under E.U. competition rules the *abuse* of such a monopoly may amount to the abuse of a dominant position and thus be illegal under Article 86 of the Treaty.

In an important recent case, this issue came before the European Court of Justice.[4a] This case, generally known as *Magill* after the name of the Irish publisher which was the original complainant, concerned the publication of TV listings in the U.K. and Ireland. The companies involved, the BBC, Independent Television Publishing (the publishing arm of Independent Television) and the Irish broadcaster, Radio Telefis Eireann, each owned copyright in their respective advanced weekly TV listings which they licensed to a restricted number of publishers for publication on a daily basis. Magill wished to produce a comprehensive weekly guide but was refused the necessary licences to reproduce the listings and made a complaint to the European Commission alleging the refusal to be an abuse contrary to Article 86.

The European Commission found in favour of Magill, its decision was upheld on slightly different grounds by the European Court of First Instance and was again upheld on appeal to the European Court of Justice. Applying principles which it had already recognised in *Volvo v. Veng*[4b], the Court held that the broadcasters had abused their dominant position by preventing the introduction of a new product, namely a comprehensive weekly television guide. The Court confirmed that refusal to license will

[3] For a discussion of video on demand see Chap. 9, paras. 9–21—9–26.
[4] EC Council Directive 93/98.
[4a] *Radio Telefis Eireann and another v. European Commission* [1995] All E.R. 415.
[4b] Case 238/87.

not in itself constitute an abuse; the act must be considered in the context of the relevant market(s) concerned and the circumstances of the refusal itself, such as where it prevented the emergence of a new product for which there was substantial demand. The Court also noted that in this case there was no justification for the refusal, thus indicating that there might be situations in which potentially abusive conduct might be excused, although without giving any guidance as to what constitutes justification.

Given the importance of copyright and other intellectual property rights in telecommunications, in particular with respect to standards and essential facilities and resources, the *Magill* case, although not really creating new law, is of considerable relevance and can be expected to colour the approach of parties to the whole question of access to proprietary technology.

2. INTELLECTUAL PROPERTY RIGHTS IN STANDARDS

According to the European Commission, a standard is a technical speci-   **12–07**
fication relating to a product or an operation which is recognised by a large number of manufacturers and users.[5] Council Directive 83/189[6] lays down the following definition in its Article 1(2), namely "standards shall mean a technical specification approved by a recognised standardising body for repeated and continuous application compliance with which is in principle not compulsory". The intellectual property rights relevant to standards are normally patents and copyrights, but could also extend to design rights (including rights in semi-conductor topographies), trade secrets and trade marks. As such, the general principles applicable to such intellectual property rights are equally applicable in the telecommunications sector.

The European Telecommunications Standards Institute (ETSI) (comprising telecommunication administrations, network operators, users, manufacturers, research institutions and private service providers) has the task of drafting European telecommunication standards. The role and activities of the ETSI are described in more detail in Chapter 15.

At the international level, the relevant standards-making body is the ITU operating through its Telecommunications Standards Bureau which produces recommendations now known as ITU-T recommendations; previously these were known as CCITT recommendations.

Under the Agreement on Technical Barriers to Trade (TBTA), concluded   **12–08**
under the auspices of the General Agreement on Tariffs and Trade (GATT), the E.U. has obligations to third countries with respect to the preparation, adoption and application of technical regulations and standards. In particular, the E.U. is obliged to ensure that standards are not prepared, adopted

---

[5] Communication to the Commission on intellectual property rights and standardisation—COM (92) 445 final.
[6] Laying down a procedure for the provision of information in the field of technical standards and regulations; [1983] O.J. L109.

or applied where these will create obstacles to international trade. The TBTA also requires the E.U. to ensure that products imported from the territory of any party to the TBTA should be accorded treatment no less favourable than that accorded to like products of national origin and to like products originating in any other country. In a case where European standards encompass intellectual property rights, the E.U. is also obliged to ensure that the importer from a country which is a party to the TBTA should be able to obtain licences from the holder of the intellectual property rights for importation, marketing, sale and use in the E.U. on fair, reasonable and non-discriminatory terms.

ETSI standards-making

12–09   It has been the tradition until recently for standards-making bodies (*e.g.* the ISO) to act on a voluntary basis, so that their codes and guidelines would normally not be binding but merely advisory. Moreover, the principle of the freedom of holders of intellectual property rights to exploit them, or not, at their discretion, is respected: there is normally no binding obligation on such holders to make their proprietary information available or to grant licences over rights which are embodied in standards. This can have one major drawback in that there is a risk, which can, of course, be mitigated by appropriate searches, that standards may be prepared and even adopted either in ignorance of relevant intellectual property rights (IPRs) or in circumstances where the holder(s) of such rights is or are unwilling to grant appropriate licences on fair and reasonable terms.

The ETSI determined to try to reduce this risk by embarking on a course of adopting a policy which in principle would compel each of its members to give an undertaking to grant licences of any "essential" property rights which it owned or controlled. For these purposes essential property rights are defined as meaning—

> "that it is not possible on technical but not commercial grounds, taking into account normal technical practice and the state of the art generally available at the time of standardisation, to make, sell, lease, otherwise dispose of, repair, use or operate equipment or methods which comply with a standard without infringing that IPR".[7]

12–10   Given that ETSI membership is considered desirable for all those entities and organisations wishing to be involved in, if not to influence, standards-making, this linkage of ETSI membership to the granting of intellectual property licences, notwithstanding the many qualifications and conditions embodied in the undertaking, was a cause of considerable anxiety and concern for the main telecommunications equipment manufacturers. Accordingly, although procedures were built into the ETSI policy enabling IPR holders to withhold licences, many manufacturers continued to oppose the

---

[7] The term "equipment" here means any system or device fully conforming to a standard and the term "method" means any method or operation fully conforming to a standard; standard being any European telecommunications standard.

policy and the undertaking, culminating in the U.S. Computer and Business Equipment Manufacturers Association (CBEMA), bringing to the European Commission a complaint and request for interim measures against the ETSI in June 1993. This complaint had two main allegations, the first that the ETSI IPR policy violated Article 85 of the EEC Treaty and the second that the ETSI and its members, in pursuing the policy were acting in violation of Article 86 of the Treaty.

Eventually the complaint was dropped after ETSI determined to intro-   **12–11** duce a new, albeit interim, intellectual property rights policy: in November, 1994 the ETSI General Assembly approved a two-year interim policy requiring its members to use reasonable endeavours to inform ETSI if they became aware of "essential" intellectual property rights, whereupon ETSI would request the holder of these rights to undertake to grant irrevocable licences on fair and non-discriminatory terms. If a member were to refuse to license its intellectual property rights, and alternative technology were not available, a member might be required to provide a written explanation and the matter might be referred to the ETSI Council. One of the main differences in the new policy compared with the controversial previous proposal is that there are no longer to be sanctions for non-disclosure of essential intellectual property rights.

In this connection, it is interesting to note the statement in the Commis-   **12–12** sion's communication "Intellectual Property Rights and Standardisation"[8] that if special rules for the co-existence of intellectual property rights and standardisation were developed on an industry-specific basis, any resulting lessening of intellectual property rights could lead to a shift in production by manufacturers away from that industry, and could disadvantage, rather than stimulate, European production.[9] Generally, this communication is required reading for anyone interested in the competition policy considerations arising with respect to standards-making. Throughout this document it is apparent that the fundamental principle which guides the Commission's approach to standardisation is that standards-making bodies should ensure that the standards are available for use on fair, reasonable and non-discriminatory terms.

3. REGULATION IN TELECOMMUNICATION LICENCES

Condition 39 (Intellectual Property) of the BT licence provides that the   **12–13** Director General may give directions to prevent BT using intellectual property to restrict the availability of any product where this is liable to prevent the connection of any telecommunication system or apparatus or the provision of any service. The Director General is also empowered to direct BT to grant intellectual property rights to other persons in order that such connections or services may be made or provided. There are saving provisions where any requirement on BT would result in it breaching the terms

[8] COM (92) 445 final.
[9] *ibid.*, para. 4.8.12.

of any licence or assignment of intellectual property rights. This Condition may therefore be used to prevent BT from withholding the supply of apparatus, but only where this is contrived by the use of intellectual property rights. A more procompetitive licence condition might have extended this principle to any refusal to supply whether or not the product is, say, patented or then subject of copyright. However, the Competition Act 1980 may be operable in such circumstances.

In the context of copyright licensing, the general powers of the competition authorities to regulate anti-competitive activities under the Fair Trading Act 1973 and the Competition Act 1980 have been enhanced by section 144 of the 1988 Act. If, following a reference to the MMC under either Act, the MMC specifies in its report matters which may be expected to operate against the public interest, and which include conditions in licences granted by the owner of copyright in a work restricting the use of the work by the licensee or the right of the copyright owner to grant other licences or a refusal of a copyright owner to grant licences on reasonable terms, then the powers conferred by the Fair Trading Act 1973 on the Secretary of State include the power to cancel or modify those licence conditions and/or to provide that copyright licences shall be available as of right.

In exercising the powers under this section, regard must had for the U.K.'s obligations under international copyright conventions.

CHAPTER 13

General Legal Issues

1. CRIMINAL LIABILITY

Interception of Communications

The interception of messages sent by telecommunication is governed by 13–01
two main pieces of legislation, the Interception of Communications Act
1985 (the "1985 Act") and the Wireless Telegraphy Act 1949.

The 1985 Act is primarily intended to protect the integrity of the public
networks and to provide an appropriate statutory framework for legitimate
phone tapping by Government security agencies. It creates an offence of
intercepting communications without authorisation as well as a process for
the official authorisation of interceptions of communications sent through
public telecommunication systems.

Section 1 of the 1985 Act creates the offence of intentionally intercepting 13–02
communications in the course of their transmission by means of a public
telecommunication system, subject to four exceptions, these being:

   (i) where the communication is intercepted "in obedience to a warrant"
      issued under section 2;
  (ii) where the person intercepting has reasonable grounds for believing
      that the person to whom, or the person by whom, the communica-
      tion is sent has consented to the interception;
 (iii) where the communication is intercepted "for purposes connected
      with the provision of . . . public telecommunication services";
 (iv) where the communication is by wireless telegraphy and is inter-
      cepted, with the authority of the Secretary of State, "for purposes
      in connection with the issue of licences" under the Wireless Tele-
      graphy Act 1949 or the prevention or detection of interference with
      wireless telegraphy.

The important point to note is that the criminal activity is that of *inter-* 13–03
ception of the communication in the course of its transmission by a *public*

telecommunication system. The public telecommunication systems in the U.K. are those designated under section 8 of the 1984 Act and include the fixed link operators such as BT, Mercury and others more recently licensed, broadband cable operators and the cellular mobile, including PCN, operators. For this purpose, therefore, public telecommunication systems must be clearly distinguished from all other systems, particularly those operated under the SPL or TSL class licence. The importance of the distinction has been illustrated in a number of cases, quite a few of which have involved police surveillance of criminals. In a recent decision *R. v. Ahmed & Others* (1994), unreported, C.A., Criminal Division, Evans L. J. stated the salient principles as follows:

"... first, we held that the interception of a communication takes place when, and at a place where, the electrical impulse or signal which is passing along the telephone line is intercepted in fact. Secondly, if there is an interception of a private system, the communication which is intercepted is not at that time passing through the public system ... Thirdly, the fact that later or earlier signals either have formed part of, or will form part of, the same communication or message does not mean that the interception takes place at some other place or time. Finally, 'communication' in our judgment does not refer to the whole of a transmission or message; it refers to the telephonic communication which is intercepted in fact ... that consists of what has been variously described as the electrical impulse or signal which is affected by the interception that is made".[1]

13–04   For these purposes therefore it is important to understand where and at what point a public telecommunication system begins and ends—its interface with a private system. This is at the network termination point (NTP), which is normally to be found at the distribution frame set up to connect incoming public system lines with internal wiring (for example, in case of offices) or at the master socket or block terminal installed for the internal wiring to be connected to it (as in residential systems).

Unfortunately, the 1985 Act does not contain a definition of interception. A dictionary definition of the word "intercept" is to prevent, check, or hinder. It might thus be thought arguable that monitoring or eavesdropping on telephonic conversations, even over a public telecommunications system, would not be interception but this somewhat narrow view is not supported by authority: see for example *R. v. Preston*[2] where the Court of Appeal held that there was interception where the police monitored a telephone conversation over a public system.

Where wireless telegraphy apparatus is used to intercept messages the Wireless Telegraphy Act 1949 would be relevant.[3] The 1949 Act, s. 5 provides that anyone who uses wireless telegraphy apparatus "with intent to obtain information as to the contents, sender or addressee of any message

---

[1] Cited with approval in the House of Lords *R. v. Effik*; *R. v. Mitchell, The Times,* July 22, 1994.
[2] (1992) 95 Cr. App. R 355.
[3] See *Francome v. Mirror Group Newspapers* [1984] 2 All E.R. 408.

(whether sent by means of wireless telegraphy or not)" without authority from the Secretary of State is guilty of an offence.[3a]

Offences by PTO Employees

Under the 1984 Act (as modified by the 1985 Act) there is a separate   13–05
offence concerning interception of messages by employees of public telecommunication operators. A person engaged in the running of a public
telecommunication system who "otherwise and in the course of his duty
intentionally modifies or interferes with the contents of a message sent by
means of that system" is guilty of an offence (section 44). Likewise, a
person who, again otherwise than in the course of his duty, intentionally
discloses to any person the contents of any message which has been intercepted in the course of its transmission by means of that system or discloses
any information concerning the use made of telecommunications services
provided for any other person by means of that system is also guilty of an
offence (section 45). Under section 45 of the 1984 Act there are exceptions
provided for disclosures made in connection with the prevention or detec-
tion of crime or criminal proceedings, for disclosures under a warrant
issued by the Secretary of State under the 1985 Act and for any disclosure
as to the use made of telecommunications service where this is in the interest of national security or in pursuance of an order of the court. Section 45
will not apply to catch a PTO employee who wrongfully *uses* information
comprised in a message which has been intercepted in the course of its
transmission.

Under the Wireless Telegraphy Act 1949, s. 5 any person who by means
of wireless telegraphy sends any message which is to his knowledge false
and misleading and likely to "prejudice the efficiency of any safety of life
service or endanger the safety of any person or of any vessel, aircraft or
vehicle" is guilty of an offence. Similarly, any person who otherwise than
with the authority of the Secretary of State or in the course of his duty as
a Crown servant and except in the course of legal proceedings discloses
any information as to the contents, sender or addressee of any such message, being information which would not have come to this knowledge but
for the use of wireless telegraphy apparatus by that person (or by another)
is guilty of an offence.

Fraudulent and Improper Use of Telecommunication Systems

Under section 42 of the 1984 Act a person who "dishonestly obtains a   13–06
service provided by means of a licensed telecommunication system with
intent to avoid payment of a charge applicable to the provision of that
service" is guilty of an offence. This could extend to cover the position of

---

[3a] This appears applicable to the criminal use of scanning devices to "clone" mobile 'phones: however, there is no power of arrest under the WTA.

a "hacker" wrongfully using telecommunications in order to obtain data from a computer. Hacking is the subject of a specific offence under the Computer Misuse Act 1990 in that a person is guilty of an offence under section 1 of that Act if he knowingly causes a computer to perform any functions with intent to secure unauthorised access to any program or data held in any computer.

Control over pornographic and other offensive messages sent via telecommunication systems is provided by section 43 of the 1984 Act, whereby any person who sends a message which is grossly offensive or of an indecent, obscene or menacing character, or sends a false message for the purpose of causing annoyance, inconvenience or needless anxiety to someone is also guilty of an offence.

Fraudulent Reception of Broadcasts and Cable Programmes

13–07   As satellite broadcasts have become more proliferate, so has the ingenuity of unscrupulous dealers seeking to profit from making available the necessary equipment to access and, where necessary, de-code the broadcast signal. This problem was addressed in sections 297 to 229 of the Copyright, Designs and Patents Act 1988 (the "1988 Act") which created both a criminal offence and a civil remedy.

The criminal offence, contained in section 297(1) of the 1988 Act, involves dishonestly receiving a programme included in a broadcasting or cable programme service provided from a place in the U.K. with intent to avoid payment of any charge applicable to the reception of the programme. This is an attempt to catch the individual who is simply trying to avoid payment of a charge for a pay-television service and, in practice, is likely to prove to be difficult to police.

Civil remedies are created for certain broadcasters, in section 298 of the 1988 Act, against a person who makes, imports or sells or lets for hire any apparatus or device designed or adapted to enable or assist persons to receive programmes or other transmissions "when they are not entitled to do so" and against the person who publishes any information which is calculated to enable or assist persons to receive programmes or other transmissions "when they are not entitled to do so". The section only applies to transmissions made by a broadcaster who charges for the reception of programmes included in a broadcasting or cable programme service provided from a place in the U.K. or who sends encrypted transmissions of any other description from a place in the U.K.

13–08   The case of *BBC Enterprises v. Hi-Tech Xtravision*[4] highlighted the ambiguity in the drafting of section 298 of the 1988 Act. This case concerned the transmission throughout western Europe (excluding the U.K.) by the BBC of the BBC Europe Service via satellite. The service was encrypted and transmitted from the U.K. for reception via a decoder. The

---

[4] [1993]3 ALL E.R. 257.

BBC had an agreement with a company called Sat-Tel under which the BBC would only authorise the use of decoders which were designed and made by Sat-Tel. The BBC derived its revenues for the service through the sale of decoders. Hi-Tech Xtravision made decoders capable of decoding the encrypted BBC Europe Service and sold them in western Europe at a price lower than that charged by Sat-Tel. Thus any unauthorised reception using the "pirate" decoders would occur outside the U.K.

The BBC issued a writ against Hi-Tech relying on section 298 of the 1988 Act. Hi-Tech moved to strike out the statement of claim as disclosing no course of action. Interlocutory proceedings hinged on the meaning of a person receiving programmes "when they are not entitled to do so" in section 298(2)(a). The House of Lords stated that the key to the interpretation of this expression is to be found in appreciating that the whole of section 298 is concerned with the legal relationship between one class of persons, the providers of programmes and the senders of other encrypted transmissions on the one side and another class of persons, those capable by the use of the necessary decoders of receiving such programmes and other encrypted transmissions on the other side. Thus the House of Lords determined that it was not only justifiable but necessary, in order to achieve the purpose of the section, to interpret the expression "not entitled to do so" as having the meaning "not authorised by the provider of the programmes or the sender of the other encrypted transmission to do".

The House of Lords confirmed that "a person who seeks to charge for programmes or sends encrypted transmissions has a right not to have other persons making apparatus or devices designed to be of use to persons not authorised by him to receive the programmes". At the same time the House of Lords acknowledged that if providers of satellite programmes emanating from the U.K. are to be protected in respect of the collection of charges made by them for the reception of such programmes, that protection must, so far as is practicable, have effect, albeit indirectly, in relation to reception by persons in other countries.

In light of the ambiguity exposed by *BBC Enterprises v. Hi-Tech Xtravision*, a second criminal offence was introduced by section 179 of the Broadcasting Act 1990. This offence involves making, importing, selling or letting for hire any "unauthorised decoder". A "decoder" means any "apparatus which is designed or adapted to enable (whether on its own or with any other apparatus) an encrypted transmission to be decoded". The word "decoder" is here being used in its widest sense and would include a smart-card as well as integrated decoding/decrypting device. An "unauthorised decoder" is one which will enable encrypted transmissions to be viewed in decoded form without payment of the fee (how so ever imposed) which the person making the transmissions charges for viewing those transmissions, or viewing any service of which they form part. In this context, a "transmission" means any programme included in a broadcasting or cable programme service which is provided from a place in the U.K. It should be noted that the definition of "transmission" may not be wide enough to embrace the delivery of data (for example a computer game) as opposed   **13–09** to audio or audio-visual works and may therefore be inadequate to cover the anticipated uses to which broadcasting is likely to be put in the future.

Section 299 of the 1988 Act gives the Secretary of State the power to extend sections 297 and 298 to programmes and encrypted transmissions emanating from a country outside the U.K., subject to that country offering reciprocal protection to U.K. broadcasters. The Home Office issued a press release on May 2, 1990 stating that section 298 of the 1988 Act would be extended to services from abroad regardless of whether the country of origin offers reciprocal protection to U.K. broadcasters if it appears to the Secretary of State that the reception of those services in the U.K. is materially effecting U.K. broadcasters. Section 179 of the Broadcasting Act 1990 deleted the requirement for reciprocity altogether. At the time of writing sections 297 and 298 of the 1988 Act had been extended to apply to programmes and encrypted transmissions emanating from Guernsey.[5]

2. Contractual Liability

Background

13–10   It is only since 1984 that BT (as successor to British Telecommunications and the Post Office) has been fully liable in tort for failures of performance and in contract for breach of the terms and conditions of its services.

All PTOs under a current obligation to provide service to customers must now publish their contractual terms and conditions by filing these with OFTEL although OFTEL does not have any responsibility to, nor does it in practice, approve them in any respect. As explained above[6] there are many new PTOs that have no obligation to publish their charges, terms and conditions and this obligation will not be activated until they become, in the context of their licences, "well established".

In a recent case[7] the Master of the Rolls has suggested that BT's powers to terminate its customer contracts without express cause or to rely on an "oppressive" contract term may possibly be subject to some constraints implied by virtue of BT's powerful position and its obligations under licence conditions to make telecommunication services available to customers.

Service Standards: Liability and Limitations

13–11   In the absence of anything contained in a contract to the contrary, the supply of telecommunication services falls squarely within the Supply of Goods and Services Act 1982. Under section 13 of that Act there is implied

---

[5] The Fraudulent Reception of Transmissions (Guernsey) Order 1989, S.I. 1989 No. 2003.
[6] See Chap. 5, under "Publication of Charges: Terms and Conditions".
[7] *Timeload v. BT plc* (1993) unreported, C.A.

a term that the supplier shall exercise reasonable care and skill in the supply of the relevant services. This is obviously not a warranty which is particularly apt for or susceptible to the vagaries of modern telecommunications.

In the absence of any exclusion and limitation clauses, the liability of the PTOs with respect to any failure in performance of their services could be quite open-ended. The principles in *Hadley v. Baxendale*[8] apply to a telecommunication service contract as to any other contract. Under the "first rule" in this case all such damages as are the natural consequence of the breach are recoverable; under the "second rule" extraordinary losses are recoverable where these were in the contemplation of the parties as a probable result of the breach: *Victoria Laundry (Windsor) Ltd. v. Newman Industries Ltd.*[9] Accordingly, a PTO could incur such liability, whether consequential or direct. Not surprisingly, therefore, as in many other industries, PTOs have adopted the practice in their standard conditions of service of excluding and limiting their liability.

BT have embodied their terms in two main documents, their "Conditions for Telephone Service"[10] and their "Customer Service Guarantee". The Customer Service Guarantee, which originally was introduced after considerable pressure by OFTEL, provides fixed compensation payments for delays in the installation and repair of telephone lines and network services, for missed appointments and for disconnections made in error. Delays in repairs and installations are calculated by reference to a daily payment equivalent to one month of the customer's network rental for each day of delay beyond the due date up to a maximum of £1,000 for each residential line. In additional, according to the same guarantee, if the customer can prove he has lost money as a direct result of BT not being able to provide or restore service within the agreed time, the customer may be entitled to compensation up to a maximum of £1,000 per residential line.

With respect to business or payphone lines this maximum is increased to £5,000 with a maximum aggregate liability of £20,000.

Having conferred these service guarantees BT in its terms and conditions also excludes liability and applies the general limitation of £1 million for any one event and £2 million for any 12-month period. Although such limitations are subject to the "reasonable test" in the Unfair Contract Terms Act 1977, as is well known it is not possible at all to exclude or limit liability for negligence, personal injury or death by reference to any contract term. Limitations and exclusions are also subject to the test of "fairness" in the Unfair Terms in Consumer Contracts Regulations; any term "which contrary to the requirement of good faith causes a significant imbalance in the parties' rights and obligations under the contract to the detriment of the consumer" is deemed unfair and not binding on the consumer.[10a]

13–12

[8]   (1854) 9 Exch. 341.
[9]   [1949] 2 K.B. 528.
[10]   Issue for February 1994.
[10a]   S.I. 1994 No. 3159, regs. 4 and 5.

Customer Insolvency, etc.

13–13   In any case where a company, which is a customer of a PTO, is the subject of an administration order or goes into liquidation or suffers any similar action as listed in section 233 of the Insolvency Act 1986, the PTO cannot make it a condition of continuing service that any outstanding charges are paid, provided the administrator or liquidator has personally guaranteed payment for such service. This requirement also extends to the PTO not doing anything (for example threatening to cut off) to the same effect as such a condition.

3. SUPPLY OF APPARATUS

13–14   Apparatus are goods and are therefore assimilated with general law in relation to sale of goods, particularly the Sale of Goods Act 1979, the Supply of Goods and Services Act 1982 and the Consumer Protection Act 1987.

Conditions implied under the Sale of Goods Act 1976 are well known and understood and supported by a considerable body of case law; readers are therefore referred to other works on this specific subject. The Supply of Goods and Services Act 1982 discussed above in relation to liability for the supply of telecommunication services, also deals with the hiring of goods. This is important for suppliers of telecommunications services, particularly the PTOs, as much of their equipment is hired to consumers. The 1982 Act implies certain conditions as to the right of possession of the goods and, where the owner (bailor) is acting in the course of a business, as to their quality and fitness.

The Consumer Credit Act 1974 regulates "consumer hire agreements" being agreements made by a person with an individual ("the hirer") for the bailment of goods to the hirer. Such an agreement must be capable of subsisting for more than three months and must not require the hirer to make payments exceeding £15,000. If these conjunctive elements are not present the agreement will not be regulated. For these purposes an individual includes a partnership. Moreover, a telecommunications operator carrying on a consumer hire business of this kind must have a licence under the Consumer Credit Act 1974, obtained by application to the Office of Fair Trading. Public telecommunication operators may apply to the Secretary of State for exemption from the requirements of the Act in relation to consumer hire agreements.

13–15   The form and content of regulated consumer hire agreements is dictated by regulations laid down under the 1974 Act. These requirements are detailed and comprehensive and careful attention has to be given to them; if the regulations are not complied with the agreement will only be enforceable against the hirer by court order.

Under the Consumer Protection Act 1987 strict liability is, with certain qualifications, imposed for death or personal injury or loss or damage to "private" property "caused wholly or partly by a defect in the product."[11] The persons liable for such damage can include the producer manufacturer, including any assembler, a person who puts his trade mark or name on a product, and any importer into the E.U. Such liability cannot be excluded by contract.

It is also important to note that under Part 2 of the 1987 Act a person is guilty of an offence if he supplies goods to or for *consumers* (*i.e.* not to industry or business) which fail to comply with the general safety requirement (*i.e.* which are not "reasonably safe having regard to all the circumstances"). It will also be a criminal offence to supply goods prohibited by safety regulations. All these safety requirements are further explained in the legislation.

4. CONTRACTUAL FORMATION BY TELECOMMUNICATION

The general principle of law applicable to the formation of a contract by offer and acceptance is that the acceptance of the offer by the offeree must be notified to the offeror before a contract can be regarded as concluded: *Carlill v. Carbolic Smoke Balls.*[12]   **13–16**

As an exception to this general rule, where a contract is made by letter or telegram the acceptance is deemed to be completed as soon as the letter is put in the post box and that is the place where the contract is made. Identifying the place where the contract is made may be uninteresting for the parties, but in litigation over international contracts it is important: one of the grounds on which the English courts will give leave to serve a writ outside their jurisdiction under R.S.C., Order 11 is if the contract has been made in England.

This so called "postal rule" is based on considerations of practical convenience and commercial expedience, arising from the delay which is inevitable between delivery of a latter or telegram and receipt of it.

However, where a contract is made by instantaneous communications, for example by telephone, the contract is complete only when the acceptance is received by the offeror, and the contract is made at the place where the acceptance is received. The Court of Appeal in *Entores Limited v. Miles Far East Corporation*[13] decided that an offer accepted by telex was effectively instantaneous and therefore fell within the general rule and did not fall within the exception covering letters and telegrams.   **13–17**

The *Entores* case was considered by the House of Lords in *Brinkibon v. Stanag Stahl.*[14] The House of Lords unanimously approved the *Entores* principle as a general, though not necessarily a universal, rule. Lord Wilberforce said:

[11] Consumer Protection Act 1987, s.2.
[12] [1893] 1 Q.B. 256 at 262.
[13] [1955]2 All E.R. 493.
[14] [1983]2 A.C. 34.

"Since 1955 the use of telex communication has been greatly expanded and there are many variants on it. The senders and recipients may not be the principals to the contemplated contract. They may be servants or agents with limited authority. The message may not reach, or be intended to reach, the designated recipient immediately: messages may be sent out of office hours, or at night, with the intention, or on the assumption, that they will be read at a later time. There may be some error or default at the recipient's end which prevents receipt at the time contemplated and believed in by the sender. The message may have been sent and/or received through machines operated by third person. And many other variants may occur. No universal rule can cover all such cases; they must be resolved by reference to the intentions of the parties, by sound business practice and in some cases by judgement where the risks should lie . . ."

Accordingly where an electronic communication is not instantaneous for any reason, analogies to the postal rule may be apposite.

13–18   Regarding communication by facsimile or other forms of communications, the following principles can be drawn from these authorities:

(a) Acceptance of a contract by facsimile will, as a general rule, take effect when the fax is received on the offeror's fax machine; this will also be the place where the contract is made. The fax is, however, analogous with telex in that although this can be stated as a general rule, there may be certain situations, for example when the machines are operated by third parties, where this general rule will not apply. On the whole, however, fax communication should be treated as "instantaneous communication", similar to telex.

(b) Communications using electronic message-handling system, such as electronic mail is not so straightforward. In some electronic mail systems a message may be added to a recipient's mailbox immediately on being sent. In others the message may be delayed and stored in the course of transmission. Clearly when the sender gives the command "send" to the system he has, in ordinary mail language, "posted" the message. However he could delete the message before the addressee could read it or send a further message cancelling the previous message, which could be "collected" by the recipient at the same time as the first message.

At the distant end matters are even more capable of variation. The addressee will not "collect" his message until he "logs in" to the host computer. Even if he does not then "call up" the message he is in the position of having knowingly received mail without, as it were, having opened it.

13–19   The essential question is, however, where and when the message is then delivered to the recipient. The prudent approach would be to apply the principles laid down by Lord Wilberforce in *Brinkibon*, namely that no universal rule can cover all cases and that the problem should be resolved by reference to the parties' intentions, sound business practice and a judgment of where the risks should lie.

With electronic mail systems, therefore, in the author's view it is probably safest to assume that delivery would normally take place when the message is entered into the recipient's mail box even though the recipient may then be unaware of its arrival. Unfortunately though this does not deal satisfactorily with the situation where the sender of the message is able subsequently to delete the message or cancel it by a further message. Accordingly, the courts may have to develop special rules for that particular situation and where, for example, the sender is aware that his original message has not been read by the recipient judges may well take the view that in the meantime senders should be able successfully to cancel such messages by subsequent communication.

Since there have been no judicial pronouncements on the latest forms of electronic communication the safe course is for parties actually to specify their proposed modes of communication and, in their contracts, how and when notices are to be taken as received. The practice of "registered" electronic mail, whereby this action is notified to the sender, would reinforce this.

Faults could arise within a telecommunication system which prevent the message reaching the addressee's mail box or otherwise corrupt the message received. Where any message, for example by telex or facsimile or electronic mail, is garbled, or in error in some way, the basic legal principle is that the sender of the message is not responsible for the faults of the operator of the communications system or the system itself. In the case *Henkel v. Page*,[15] where the offer was to buy 50 rifles and the defendant, not wanting this number, telegraphed "send 3 rifles", the telegram in fact reached the seller in the form "send the rifles" and the seller therefore despatched 50 rifles. It was held that the buyer was not bound to accept more than three rifles.

13–20

In the *Entores* case Lord Denning discussed failures in what would now be the recipient's (*e.g.* SPL) system. He said:

13–21

"In all instances I have taken so far, the man who sends the message of acceptance knows that it has not been received, or he has reason to know it. So he must repeat it. But suppose that he does not know that his message did not get home. He thinks it has. This may happen if the listener on the telephone does not catch the words or acceptance, but nevertheless does not trouble to ask for them to be repeated: or if the ink on the teleprinter fails at the receiving end, but the clerk does not ask for the message to be repeated: so that the man who sends an acceptance reasonably believes that his message has been received. The offeror in such circumstances is clearly bound, because he will be estopped from saying that he did not receive the message of acceptance.

[15] (1870) C.R. 6 Exch. 7.

It is his own fault that he did not get it. But if there should be a case where the offeror without any fault on his part does not receive the message of acceptance, yet the sender of it reasonably believes it has got home, when it has not, then I think there is no contract."

The point to note here is therefore that if the recipient's system is at fault he is nevertheless to be deemed to have received the message as sent. The same principle would apply to bind the sender where it is his system which is faulty.

13–22 The other important area in relation to contracts is that of service of notices and other documents, by way of electronic mail. It is desirable, in particular, where electronic mail is a permitted method of service, to stipulate when the communication is deemed to be made. It would seem apt that this should be the date it is first stored in the other party's electronic mailbox, to avoid the risk of delayed accessing.

Increasingly electronic mail service providers are offering the facility of automatic forwarding of messages from the host computer to customer (*i.e.* addressee) terminals. Where this is the case the legal effects of such a system of conveying messages become very much assimilated with telex, assuming the forwarding is in fact done in "real time".

CHAPTER 14

Network Interconnection Agreements

1. INTRODUCTION

Interconnection is a topical subject and one of increasing interest to oper-   14–01
ators and service providers as liberalisation advances and networks proliferate. However, it is not really a new issue; a hundred years ago, for
example, in the U.K., many different local networks existed which were
not interconnected and this was eventually one of the reasons they became
unviable. It took a unifying entity, ultimately the Post Office, to provide
the capability to bring them together into one cohesive national network.

This chapter examines the strategic, legal and regulatory issues affecting
the necessary arrangements for interconnection, both of networks and the
services provided over them, as embodied in contractual form—interconnection agreements. I shall be looking at these agreements in the general
and international context and not only therefore with reference to the U.K.
regulations governing interconnection (which were discussed in detail in
Chapter 5, paras. 5–40 to 5–54). Although most of the discussion is geared
to fixed or mobile network interconnection and the provision of telephony
services through such interconnection, the principles and issues raised are
just as relevant to non-network (or non-facilities-based), interconnection,
e.g. of data communication services.

There are, as this book should demonstrate, many different approaches
around the world as to the way in which efficient and fair terms and conditions of interconnection should be achieved between operators. On the one
hand there are those who believe that the dominance of "national" public
operators dictates that market entry support must be given by providing
for regulatory intervention where terms cannot be agreed between the new
operator and the monopoly provider. On the other hand, there are those
countries who believe commercial agreements between undertakings should
be left to be freely negotiated without the "distortion" of a regulator imposing such terms; very often, nonetheless, there is a recognition (for example
as in New Zealand) that competition law should apply to the monopolist
to ensure that its negotiating stance does not represent an abuse of its
market power.

The reason a chapter like this can be written at all is that notwithstanding this pattern of different regulatory bases for interconnection, there is consistency in terms of the issues and principles involved. I will turn to these after a brief review of the development of interconnection regulation in some of the leading telecommunications economies.

14-02   To date, internationally, no consistent set of guidelines has emerged for this subject. Each country has tackled the issue of interconnection and evolved its own approach in keeping with its own legal principles and regulatory framework. In the U.S., beginning with landmark cases on connection of terminal equipment, progressing through the establishment of the rights of specialised carriers to interconnect with AT&T, the industry had arrived by the end of the 1970s at the stage where basic rights of switched services interconnection had been granted to all carriers. Attention has since turned mainly to the issue of price, the appropriate level of charge for the right to access the local facilities of the Bell operating companies, and to the matter of the quality of interconnection including its enhanced forms (co-location of facilities, local interconnection, etc.).[1]

In the U.S., the driving force behind the achievement of interconnection was private litigation initiated by prospective competitors of the established carriers as well as regulatory action; acceptance was ultimately won for the notion that the competition in the provision of inter-state services which interconnection would make feasible would be in the public interest. In Canada, the process which led to the introduction of competition followed a similar course.[2]

Outside North America, the pattern has been somewhat different. There has not been the same propensity to introduce competition through litigation and regulatory intervention. Rather, reforming governments have tended to create the appropriate liberalisation framework (with the honourable exception of New Zealand) and then proceed to issue licences to compete with the established operator on the basis of a new set of rules and conditions created specifically for the new competitive environment. These rules have also recognised that new operators competing with the incumbent telecommunications organisation (TO), whether or not government-controlled, will not always have an easy time reaching negotiated arrangements with the monopolist. Typically therefore they have reserved that disputes between competing operators on issues subject to regulation—such as interconnect and local access—should be resolved by the regulators themselves.

New fledgling networks cannot hope to replicate established networks developed over decades, if not from the late nineteenth century, nor are they likely to find the investment to take on such a task. Accordingly much of regulators' time and attention in the future is likely to be focused on this issue of the relationship of the new operators and service providers to

---

[1] See "Expanded Interconnection with Local Telephone Company Facilities" (1992) 7 FCC Rcd 7369 (special access) and (1993) FCC Rcd (switched transport services).
[2] See Ryan, *Canadian Telecommunications Law & Regulation* (1993), § 705.

the TO, particularly the terms on which their networks should be interconnected.

In the U.S. the competitive long distance carriers have long since achieved   **14–03** interconnection with the monopoly RBOCs (and their monopoly itself is now under attack) and in the U.K. facilities-based competition has existed since 1984, giving rise to interconnection agreements between BT and Mercury Communications (the first licensed "fixed-link" or facilities-based operator), as well as between various locally franchised cable operators and BT/Mercury. In Europe the thrust of interconnection policy has generally been determined by the direction of liberalisation, which in most countries has relied so far solely on the mobile sector to open up public service competition. Now the European Commission is taking a much more active interest. Interconnection emerged as a priority issue for European telecommunications policy in the Bangemann Report on *Europe and the Information Society*:

> "Two features are essential to the deployment of the information infrastructure needed by the Information Society; one is a seamless interconnection of networks and the other that services and applications which build on them should be able to work together (interoperability)".

The Commission is now pursuing this objective and in its Green Paper on the "liberalisation of telecommunications infrastructure and cable television networks" (Part II) identifies the goal as an open interconnected environment where there are no *a priori* restrictions on network interconnection, and where telecommunications-based services can operate seamlessly over interconnected networks. The Commission notes that interconnection agreements are subject to screening under Articles 85 and 86 of the Treaty and proposes a directive specifically on the subject of interconnecting public networks. Shortly before this book went to final print the Commission published its proposal in the form of a draft ONP Interconnection Directive, the main elements of which were that—

- all parties should have the right to enter into commercial and technical agreements to interconnect;
- conditions for interconnection should be based on the established principles of Open Network Provision;
- a negotiating framework should be set ensuring that commercial negotiations result in a fair and timely agreement, with the regulatory authorities responsible for preventing any abuse of negotiating power;
- interconnection charges should satisfy certain specific principles, namely they should—

    (a) be a matter for commercial agreement, subject to supervision and, if necessary, timely intervention by the National Regulatory Authority, and subject to the competition rules;
    (b) encourage efficient and sustainable market entry and be based on the underlying costs of an efficient operator;
    (c) ensure that the cost of inefficiencies is not passed on to intercon-

necting operators, nor threaten the financial ability of an operator to fulfil its licence obligations;

(d) be transparent, non-discriminatory and sufficiently unbundled;

(e) identify separately charges related to recovering losses due to
tariff imbalancees resulting from regulatory obligations;

(f) be separately identified where they relate to the provision of
uneconomic service obligations; and

(g) be based on approved cost-accounting systems implemented
under the supervision of the National Regulatory Authority in
order to ensure transparency and non-discrimination.

Outside of North America and Europe and putting aside Australasia and
Japan, few countries have a huge array of different networks let alone competing services and many still only have one public network providing basic
telephony to the public. However, a large number are actively considering
liberalisation, if not privatisation, and those that have already begun the
process will have identified how essential it is to ensure that new entrants
must be given the opportunity to compete on a fair and equal basis. The
single most important asset in ensuring such equality of treatment is a
properly balanced interconnection agreement.

2. THE REGULATORY CONTEXT

14–04   It can readily be appreciated that in newly-liberalised markets the prospects
for a fledgling operator achieving fair and reasonable terms for an interconnection agreement with its monopoly competitor are dim if not non-
existent. It therefore makes a great deal of sense for regulations to underpin
the negotiating process by providing recourse to a telecommunications
regulatory authority or some other dispute resolution mechanism perhaps
not quite specific to telecommunications. Exactly how this may be achieved
varies from country to country, the U.K. being at one detailed and interventionist end of the spectrum and New Zealand, with its reliance purely on
general market dominance legislation, more or less at the other.

In attempting to regulate the interconnection aspects of the industry and
putting aside the effects of supranational regulation such as that increasingly being introduced at the initiative of the E.C. Commission, individual
countries can therefore be expected to tailor their own rules according to
their traditional legal approach and in the context of their own existing
monopoly control and competition laws. It is therefore not surprising that
those countries with "thin" anti-trust or competition law backgrounds (e.g.
U.K.) have adopted detailed telecoms-specific regulatory regimes, whilst the
countries of continental Europe and elsewhere that prefer a less rigorous
and more broad-brush approach, but nonetheless have reasonably substantive domestic competition laws, have tended towards the adoption of
general interconnection charging principles (at best) or merely administrative recourse procedures, particularly concerning conditions of access, leaving these to be underpinned, where necessary, by existing national competi-

tion legislation. For example, most European continental regimes (*e.g.* France, Germany, Netherlands) simply lay down broad guidelines, leaving it to the parties to negotiate as best they can and giving the appropriate Minister or other regulatory authority the discretionary power to resolve crucial differences on regulated issues.

In all cases, prudent planning preparatory for interconnection negotiations will involve an assessment not just of the other contracting party's background but of the legal position of the regulatory authority concerned and its likely approach to the key issue of interconnect charging, particularly in the light of sector policy generally and of any precedents set in prior rulings. Such an assessment must also necessarily take into account the general mix of local (as well as E.U. where appropriate) anti-trust, competition and restrictive practices laws, particularly those controlling monopolistic activities such as those of the relevant TO. Many domestic laws already contain prohibitions on abusive conduct, on discrimination between customers, whether commercial or end-users, and on the prevention of exclusive or exclusionary arrangements. All these need to be included as background material in the negotiator's "manual".

3. Negotiating Interconnection

Before embarking on interconnection negotiations, a thorough awareness   **14–05**
of the business and regulatory context is required. This will entail:

- Understanding the parameters within which the other party will be inclined to negotiate;
- Where the other contracting party is a TO, reviewing how it is constituted and whether it is privatised or state-owned;
- Determining how it is funded and reviewing its accounts and accounting policies, particularly as to costs and their allocation;
- Ascertaining its cost structure so far as possible (many TOs do not know their own cost bases);
- Looking at its rates of return with respect to different parts of its business and whether these are controlled or there are other tariff policies or constraints within which it must operate.

Consider here also the other arrangements which the TO has which may be analogous or provide important strategic comparables. At best these may include interconnection agreements with other operators (*e.g.* mobile) and even with its own functionally distinct operations, for example those offering mobile services. Even in the absence of such direct comparisons, indicators of equivalent transactions may provide important clues and arguments in support of negotiations, particularly over interconnect payments and rates: for example, the TO's prices for leased lines for large or bulk customers, prices charged for the provision of capacity for resale, and connection costs charged to private network operators for public network connections.

Where possible, examples of interconnection arrangements, charges and principles adopted in other countries, particularly those with comparable states of telecommunications development should be examined. Comparisons of interconnect payments agreed or determined by regulatory authorities, of leased line rates for equivalent situations, of routing handover arrangements and of the approach to other critical interconnect issues (see 4 below) may be impressive to an adjudicating regulator if not to the other negotiating party.

4. Key Principles and Issues in Interconnection

14–06   There are at least six key areas of interconnection:

- the charges to be paid,
- the timely provision of interconnect capacity,
- access to appropriate facilities to minimise duplication of resources,
- access to services and functionality needed to support advanced as well as basic services for customers,
- "equal" access for customers to interconnected networks, and
- commitment between the parties to the provision of good quality grade of service to customers.

Considering each of these in turn—

Interconnection Charges

(i) Connection charges

14–07   A number of questions arise here: are there to be charges for the physical connection work involved in conjoining two networks given that each party might be expected to derive some benefit? Are regulatory principles clear in requiring a TO not only to seek such interconnection in order that its customers may reach the customers of the other operator but also that at a minimum there should be an arrangement for offsetting the resultant costs incurred by each party? How are these costs to be determined and will they be based on strictly cost-related principles? Finally, looking at the "downstream" impact of interconnection, are there costs indirectly incurred by the TO as a consequence of adjustments necessary to deal with interconnection traffic and should these be built into interconnection (therefore usage) payments or in some more direct way such as a one-off charge?

(ii) Usage (call) charges

14–08   Various models exist around the world for determining interconnection call charges, that is the charges which one operator should pay or effectively bear for the service provided by the other operator in collecting or delivering its calls.

In Australia there is the DAIC (Directly Attributable Incremental Cost) system, in the U.S. the concept of access charges reflecting traffic-sensitive and non-traffic-sensitive costs, in Canada "contribution charges"[3] and "traffic switching and aggregation charges"[4] (plus a special transitional charge designed to allow the telephone companies to recoup a part of the "start-up costs" associated with the modification of their networks necessary to permit interconnection),[5] and in the U.K. (for the moment) fully allocated costs plus access deficit contributions, although the latter are now likely to be abolished. All these charging regimes are attempting in different ways to recover the incumbent's relevant costs and overheads according to different economic models and taking account of their own particular competition and regulatory policies, with special reference to universal service commitments and tariff rebalancing.[6]

Besides the more obvious arguments regarding the appropriate economic and accounting basis for determining cost, there is considerable potential for argument over whether charges should be fixed or recurrent and how these relate to costs actually incurred. Mercury's dispute with the Director General of Telecommunications and BT[7] has centred on the argument (by Mercury) that the real costs actually incurred by BT in providing interconnection facilities and services are predominantly for the provision of the standing capacity required to support Mercury's calls (a fixed cost) and, to a much less significant degree, for the variable costs involved in conveying calls to and fro over the network. By contrast until very recently in the U.K. the general assumption had been that call (i.e. conveyance) charges should be set by reference to time and distance; this is now highly questionable as in a digital world calls become increasingly independent of time and distance. Moreover, gearing interconnection pricing to per minute charging is liable to force an interconnecting operator into the same retail charging matrix as the monopoly operator; clearly the risk is that this is unlikely to produce pro-competitive and pro-consumer benefits.

The fact that charges, in order accurately to reflect true underlying cost, should take into account that some costs are traffic-sensitive whilst others are not, has been given clear regulatory acknowledgment in the FCC decision on access charges.[8]

A related issue is the impact of a variation by, say, a TO in its own interconnect charging structures and rates. This will be felt immediately where there is some direct relationship between such public rates and the interconnection payments; even where this is not the case such fundamental

14–09

---

[3] Represent the interconnecting companies' contribution to the non-traffic sensitive costs of running the local network.
[4] These allow the telephone companies to recoup the traffic sensitive costs they incur in providing interconnection.
[5] See Ryan, *supra*, § 706.
[6] See for example the FCC's Access Charge Order and its Universal Service Fund.
[7] See above at para. 5–54.
[8] "MTS and WATS Market Structure" (1983) 93 FCC.

changes by a TO might well merit review of the interconnect payments with respect to their structure and level, although this may depend on whether most favoured nation or similar non-discriminatory controls apply (*e.g.* by contract or by regulation) as between the operators.

Provision of Capacity

14–10   An interconnection agreement has elements of a supply contract. The commitment to delivery times which the TO is prepared to give and the compensation, if any, available for failure to meet such a commitment will be matters to be chiselled out of the parties negotiating blocks. Nor is this a one-sided affair for each operator will typically be looking to the other to forecast capacity requirements in order that proper provision can be made in due time and that contractually binding delivery commitments can be met. Yet a new incoming operator with no previous track record will find it very difficult to give sensitive and accurate forecasts in the early years.

Here the issue of priority of service is liable to cause debate, if not problems. Monopoly public operators with public service obligations tend, to hold to the view that when supply is at all constrained the general public should come first, quite neatly sometimes leaving their competing operators holding the baby. Again, if regulatory principles do not address this then the matter should be thrashed out in negotiations, at the very least giving the applicant operator equal rights with other connecting operators.

Access to facilities, exchanges, etc.

14–11   The place at which and the manner in which networks are connected are not simply technical issues to be resolved by engineers. Operators seeking interconnection should approach these matters on the basis that they must be afforded the possibility of physical connections at any point in the other operator's network at which such connections are technically feasible. In practice, assuming no capacity constraints, there are rarely any such technical obstacles, the matter being essentially a function of the work, materials and costs involved. Yet the points of connection and the levels in the other operator's network hierarchy at which connections are to be achieved can affect such strategically sensitive aspects as the quality of calls using interconnection, the extent to which the applicant operator will be able to carry calls himself and thus potentially save costs, post-dialling delays and redundancy/diversity of routing for security against network malfunctions.

In other countries the strategic importance of the location of interconnection points has been well recognised. The Modification of Final Judgment[9] found that AT&T had for some years raised tenuous technical objections and generally exercised restrictive policies against connection being provided to its local exchange facilities. Judge Greene therefore ordered that

[9] *United States v. AT&T*, 552F. Supp. 131 (1983) affirmed *sub nom. Maryland v. United States*, 460 U.S. 1001 (1983).

as a consequence of the divestiture the RBOCs should provide all inter-exchange carriers with access to their local exchanges on a basis and on terms equivalent to those provided by AT&T. Since that time decisions by the FCC have further expanded the range of facilities to which access is to be provided by the local exchange carriers (LECs), culminating in expanded interconnection requiring that certain alternative access providers be given actual or virtual co-location with a LEC's central office or otherwise at least on economic and technical terms comparable to those available within the LEC's own "office" (*i.e.* exchange). Local regulatory authorities have followed an equally pro-competitive line.[10]

The question of the appropriate facilities for direct interconnection is inextricably linked to the provision of interconnect links. These are physical links to be established between the two networks, involving the connection of one to the other at a particular point of connection. This is a matter of perhaps surprising strategic importance raising a number of questions. Will the applicant operator for example, be allowed access to the premises of the TO to install his cables and terminating equipment? Who will be responsible for maintenance of the link and what level of priority and quality of care will be provided for the link? If the configuration or location of an interconnected exchange is to be changed, what impact will this have on the links and how will any attendant costs be borne?

Access to services and functionality

Beyond the rather technical matter of physical interconnection lurks one of altogether more sensitive commercial and competitive character: service and functionality interconnection. This relates to the level of functionality or "intelligence" at which networks will be enabled to interconnect, and the types of service, particularly advanced or enhanced (*e.g.* 0800 numbered services), that will be supported by interconnect (particularly through use of proprietary software) and thus available across the networks to the customers of both parties. This is an area which is of key significance for the future development of networks which, gradually, national regulatory authorities and supra-national agencies like the European Commission are beginning to recognise.   **14–12**

Equal Access

The U.S. experience has shown very clearly the importance of equal access. As implemented in the U.S. this enabled customers to pre-arrange to have their calls routed through competing carriers by dialling "1 +" as well as allowing indirect exchange carriers to receive equivalent transmission quality. It has also enabled competitive inter-exchange carriers to identify the   **14–13**

[10] See Comparably Efficient Interconnection Arrangements and Instituting Proceeding, Opinion 91–24, NYPSC, where the New York Public Service Commission ordered NY Telephone to interconnect with local service competitors.

originating party's phone number for billing purposes and to determine when a call has been completed, thus facilitating the preparation of detailed billing statements. In the U.K. there was little impetus behind the introduction of equal access during the early years of liberalisation in the 1980s but this was probably mainly due to the relative lack of digital exchanges and the costs of conversion to support equal access. Now this technical problem has been overcome, there is a further inhibition in that, at the time of writing at least, any competing operator seeking equal access from BT is bound automatically to pay access deficit contributions[11] on equal access calls. Accordingly for the time being, companies like Mercury prefer to rely on what has been termed "easy access" simply allowing access by the use of a three digit number, with the selected operator relying on calling line identity to validate the customers' status. This is not of course "equal" but rather "improved" (in relation to the previous situation) access.

Real equivalence requires that an equal number of digits be used to place a call on any network. Such "equal access" is provided for in the U.S., Canada and Australia. Where the implementation of such a scheme is not feasible, users must be able to key-in an identifier before dialling their call. Even if they must accept a dissimilar dialling scheme, competitors should strive to obtain arrangements which ensure similar post-dialling delays, with the carriers paying comparable, if not equal, tariffs for access to the local loop.

In Japan equal access has been discussed but not yet implemented; this is in part tied up with the restrictions on NTT's freedom to rebalance its prices and the lack of any access charge regime. By contrast, in Australia, easy/equal access has been available to the second carrier (OPTUS) since its inception, in a first phase by the use of CLI and a simple access code (akin to the U.K.'s easy access) and in a second phase by "preselection" in a manner similar to the way equal access can work in the U.S.

Quality of Service

14–14   Interconnection does not instantly create a unified network comprising the two interconnected systems. For some time to come, networks will be disparate, based on differing standards and approaches to quality control and customer care. It is critical therefore to make provision in an interconnection agreement for how system failures, emergencies and disruptions are to be dealt with as well as how routine maintenance is to be carried out. In some cases, to give flexibility, updateable codes of practice are seen as the best mechanism for dealing with these matters. The inclusion of compensation schemes can also be considered, particularly in jurisdictions where such schemes already exist between public operators and their ordinary customers.

---

[11] See Condition 13.5A.5 of BT's Licence and paras. 5–47 to 5–49 above. However, access deficit contributions may soon be abolished.

Besides the above key areas there are other matters of concern in the context of ensuring reasonably durable but flexible arrangements for interconnection:

(i) Numbering

Numbering is a major competitive issue. Most countries which are contemplating a liberalised regime intend to move control of numbering from the incumbent TO to a separate independent regulator. This regulator needs to ensure that there is an adequate and equal supply of numbers available to meet the needs of both parties. Further, to counteract the prospect of customers having to change their numbers in order to access competing operators—which presents a significant barrier to entry—there should be an early examination of the possibility of local number portability, so that customers can change operators without changing their numbers. **14–15**

(ii) Mid-term reviews

It is inconceivable that an interconnection agreement negotiated by parties in the early stages of liberalisation should last for very long without needing to be overhauled in the light of experience, if not changes in regulation. It is essential for such mechanisms to be included and for a means to be found whereby failures to agree can be resolved, typically by regulatory authority on the basis of principles guiding entry into the interconnection agreement in the first instance. Where such regulatory intervention is not possible, the parties must consider some alternative means of impasse removal. Whilst experts and arbitrators may spring to mind, the role of an arbitrator is more typically equivalent to that of a judge dealing with disputes concerning alleged failures of performance rather than inability to agree. Experts cannot operate in a vacuum and need clear guidelines if they are not to reach the wrong decisions for the wrong reasons. Equally rules of procedure should be clearly laid out. **14–16**

Aside from wholesale reviews of interconnection agreements, the parties may also consider it worthwhile reviewing the specific element of interconnection payments on a more regular basis, perhaps annually and gearing changes to specific benchmarks whilst never losing sight of basic cost-related principles. Whilst these benchmarks may include public network tariffs for a TO's customers generally, this is dangerous for the applicant operator. Linking its own tariff packages and adjustments to those of the TO may severely reduce its flexibility in offering innovative tariff packages and options.

Telecommunication networks are not "fossilised" and will always tend to run ahead not only of regulation but also of contractual agreements. As mentioned with respect to the structure of interconnection, imagination and some foresight is required as to the potential for technological improvements and their impact on the inter-working of the two systems. Specifically, in these situations the parties might usefully address how changes to systems are to be notified as between the parties, how necessary adjustments are to be effected and how relevant consequential costs are to be borne.

(iii) Liability and Compensation

14–17   Aside from the possibility of compensation for delays in delivery of capacity and failures in network performance, the parties will need to address the more general issue of default, when more general contractual principles will apply (including the application of *force majeure* relief), inclusion of liquidated damages wherever practicable and the imposition of limitations and exclusions of liability albeit subject to consumer protection (if applicable) and other relevant statutory restrictions on such limitations.

(iv) Dispute resolution

14–18   Failures to agree pursuant to review mechanisms may as stated above be preferable to relying on regulatory authority for resolution. Equally such regulatory authority may be prepared or able to play a role in solving contractual disputes, but this would be the exception rather than the rule. In practice, most regulatory authorities have their resources stretched quite enough by dealing with specific tasks assigned to them under local legislation. The contracting parties may therefore need to have recourse either to the courts or to other more traditional dispute resolution procedures such as arbitration.[12] Specifically with respect to technical and financial matters, however, there may well be scope for either interim or final adjudication by experts in these more narrow areas.

5. Impact of Competition Laws

14–19   As recognised by the European Commission in its recent Green Paper on the liberalisation of telecommunications infrastructure[13] interconnection agreements are capable of having restrictive effects or of being affected by the abuse of a dominant position. Instead of applying Article 85 of the Treaty on a case-by-case basis, and recognising the calls from sector players for a stable and predictable regulatory environment for interconnection, the Commission has been considering measures which would set out the conditions under which interconnection agreements are compatible with E.U. competition rules. In particular, at the time of writing, it has proposed an ONP Interconnection Directive for this purpose.

The types of issues capable of raising competition concerns under Articles 85 and/or 86 include: number and location of points of connection, equal access, the costs/charges of providing interconnection, service delivery and quality, numbering and number portability (especially where such matters

---

[12] In September 1991, the FCC adopted its Policy Statement and Orders with respect to "Use of Alternative Dispute Resolution Procedures in Commission Proceedings", etc. (1991) 6 FCC Rcd 5669 exhorting the role of mediation, arbitration, settlement negotiation and negotiated rulemakings.
[13] See para. 14–03 above.

have not been fully removed from the control or interference of the TO), access to premises or equipment, directory services, intellectual property and the exchange of technical/commercial information.

So far as the application of Article 85 is concerned, generally interconnection agreements tend to be predominantly pro-competitive in nature, in that they support the communication of messages from a customer of one network to the customer of another and, where infrastructure competition exists, promote customer choice of network operator. On the other hand, depending on the symmetry of the relationship of the interconnecting parties (relative market position and bargaining power) there exists the possibility for abusive practices and collusive arrangements which might taint the agreement.

With respect to Article 86, there is a question whether it is capable of **14–20** imposing an obligation on TOs to meet requests for interconnection by other operators. In turn this will depend on issues such as the definition of the relevant market and the economic strength of the TO concerned. Certainly Article 86 appears very capable of application to refusals to provide interconnection facilities, either at all or on reasonable terms. To date, however, there has been no specific case before the European Commission or the European Court dealing with interconnection.

Outside of the E.U. practices vary widely in the application of competition/anti-trust laws to interconnection. In the U.S. in the 1979 anti-trust suit brought by the Department of Justice against AT&T (in which it was alleged that AT&T had illegally monopolised the U.S. market for long-distance services) the Department of Justice's position was that AT&T had a continuing obligation under the anti-trust laws to permit interconnection and that compliance with the requirements of the Communications Act was not in itself sufficient. Consideration of that issue was deferred to trial and the case was subsequently settled.

In Australia, so far, interconnection agreements have been regulated under the Telecommunications Act but, if enacted, the Competition Policy Reform Bill will introduce a statutory "essential facilities" regime to apply to industry generally and this would have considerable relevance to interconnection arrangements. Under this proposal the Minister would be able to declare services to be subject to the access regime where the owner of the underlying facility had a "bottleneck" control over that facility and access to it was required by other parties in order to compete. Once the services had been so declared the parties would negotiate with each other and if they could not reach agreement the Australian Competition Commission would have appropriate arbitration powers.

By contrast in New Zealand there is no regulatory regime specific to **14–21** telecommunications and interconnection relationships are subject only to the general prohibition, contained in section 36 of the Commerce Act 1986, on the use of a dominant position in a market for the purpose of preventing or deterring competition. The application of this section to interconnection came dramatically before the New Zealand Courts and ultimately the Privy

Council in the *Clear* case.[14] Clear Communications Limited, a recognised public network operator for the purposes of the Telecommunications Act 1987 had applied for interconnection with Telecom New Zealand's network in order to support its toll bypass service and proposed local services. Negotiations broke down and Clear brought an action alleging breach of the section and claiming damages. After Clear lost in the High Court and won in the Court of Appeal, the Privy Council finally decided that Telecom had *not* misused its dominant position. In particular it was held that application of what has become known as the "Baumol-Willig Rule", (as propounded by the economics professors Baumol and Willig) enabled Telecom to continue to secure monopoly rents but did not involve use of its dominant position. Moreover, their Lordships concluded, it was unjust and impracticable to construe section 36 to extend its scope to produce a quasi-regulatory system when introduction of such a system was expressly provided for, with necessary powers and safeguards, in another part of the Act.

Notwithstanding their decision, the Privy Council were realistic enough to recognise that given the relative bargaining position of the parties there was bound to be fundamental disagreement between them as to the principles applicable to interconnection and that in the absence of guidance as to such principles the parties were as the High Court had said "negotiating in a fog". The Privy Council expressed regret that their decision could only decide whether or not Telecom misused its dominant position and could not decide whether or not Telecom misused its dominant position and could not decide whether Clear's stance in negotiations was reasonable, let alone fix the terms for interconnection.

The *Clear* case is thus a prime example of the difficulties and delays faced by applicants for interconnection in the absence of regulatory underpinning for terms and conditions to be negotiated with monopolists. The reverse side of this coin is best represented by the U.K. where an extremely complex set of rules governing interconnection has built up over the years rather like one skin graft on top of another, with the unfortunate result that the rules have become more and more arcane, more controversial and even litigable.

Check-List of Interconnection Terms and Conditions

Set out below is a summary of the main provisions which would normally be expected to find their way into network interconnection agreements:

- **Points of connection:** the parts of the two networks (*e.g.* switches) at which they are to be interconnected.
- **Interconnection links:** the transmission facilities between networks to connect to points of connection.
- **Provision of capacity:** the installation of lines at points of connection.

---

[14] Telecom Corporation of New Zealand Limited and others v. Clear Communications Limited (1994) unreported, May 26, 1994.

- **Quality and standards:** standards for ensuring the technical and operational quality of interconnection is properly maintained.
- **International:** securing arrangements for access to international facilities and services.
- **Ancillary services:** arrangements for access to operator assistance, directory information and other supporting services.
- **Payments:** the charges payable to each operator to the other for conveying calls.
- **Billing:** invoicing for interconnection charges.
- **Numbering:** arrangements necessary to ensure that calls are properly routed and conveyed through the networks to their destination.
- *Force majeure*: exclusions of liability for *force majeure* circumstances.
- **Liability:** agreed levels of responsibility for errors and omissions.
- **Confidentiality:** provisions securing confidentiality with respect to information flowing between the operators.
- **Review:** mechanisms for review and renegotiation of the agreement and resolution of disputes, possibly involving recourse to an independent expert or regulatory authority.
- **Duration:** the life of the agreement and any notice of termination.
- **Default:** provisions of premature termination in the event of default or insolvency, etc.

Equal access and number portability are two separate isues which, whilst not in themselves absolute prerequisites to efficient interconnection, are a vital factor in enhancing the comprehensiveness of interconnecting operators and promoting choice.

CHAPTER 15

International Regulatory and Satellite Organisations

1. INTERNATIONAL REGULATORY ORGANISATIONS

The International Telecommunication Union (ITU)

Introduction

The International Telecommunication Union (ITU) was established under   15–01
the International Telecommunication Convention in 1932 with the merger
of the International Telegraph Union and the International Radio Telegraphic Union. In 1947 the ITU became a specialised agency of the United
Nations. Today, the ITU comprises some 160 member countries.

The role of the ITU is to regulate the development of telecommunications
on a worldwide basis. The underlying principle behind the organisation is
that a coherent set of international regulations is essential to ensure efficient
communications systems both between regions of the world, and within
those regions. Today the ITU derives its principles and authority from its
Constitution and its Convention (of Nice, 1989, and since amended in
Geneva, 1992) as well as from its Administrative Regulations which
"complement" both the Constitution and Convention. The Regulations
comprise the International Telecommunication Regulations (replacing the
old Telegraph and Telephone Regulations) and the Radio Regulations.

Structure

The ITU is made up of several executive organs which reflect the variety   15–02
of tasks with which the organisation as a whole is concerned. This structure
was reorganised with effect from March 1, 1993, as a result of the Additional Plenipotentiary Conference held in Geneva at the end of 1992. The
main changes affected the International Telegraphic and Telephone Consultative Committee (abbreviated to CCITT from its French title) and the

International Radio Consultative Committee (CCIR from its French title). The present structure of the ITU is now as follows:

The Plenipotentiary Conference

15–03   The Plenipotentiary Conference is the supreme organ of the ITU. The Conference is now to meet once every four years and consists of all ITU member countries. It determines the general policies of the ITU in the light of the overall principles as set out in the Convention. These principles are:

(a) to maintain and extend international co-operation between all members for the improvement and the rational use of telecommunications of all kinds, as well as to promote and to offer technical assistance to developing countries in the field of telecommunications;

(b) to promote the development of technical facilities and their most efficient operation, with a view to improving the efficiency of telecommunication services, increasing their usefulness and making them, as far as possible, generally available to the public;

(c) to promote the use of telecommunication services with the objective of facilitating peaceful relations; and

(d) to harmonise the actions of nations in the attainment of those ends.

The Council

15–04   The 36 members of the Council are elected by the Plenipotentiary Conference. The Council acts for the Plenipotentiary Conference in between the latter's quadrennial meetings. It is responsible for the implementation of the provisions of the Constitution, Convention and Administrative Regulations, the decisions of the Plenipotentiary Conferences and the decisions of other conferences and meetings of the ITU.

The General Secretariat

15–05   The General Secretariat, elected by the Plenipotentiary Conference, is the Chief Executive and legal representative of the ITU. His role is to give legal advice to the various organs of the ITU; to deal with all instruments of ratification and accession to the Constitution and Convention; and to publish the texts of all legal instruments of the ITU.

Administrative Conferences

15–06   Administrative conferences meet at irregular intervals to consider special telecommunications matters. Such conferences aim to oversee the implementation of the Constitution and Convention, ITU policy and the Administrative Regulations. There are World Administrative Conferences and Regional Administrative Conferences. The former can enact a partial or complete revision of the Administrative Regulations, whilst the latter can deal only with a specific issue of a regional nature.

The Telecommunications Standardisation Sector (ITU-T)

The Telecommunications Standardisation Sector (ITU-T) was established   **15–07**
with effect from March 1993 at the Additional Plenipotentiary Conference
of December 1992. It essentially consists of the former CCITT, but has
also taken on board the standards work of the former CCIR. It is therefore
concerned with technical, operating and tariff issues. The ITU-T issues
recommendations with a view to standardising telecommunications on a
world-wide basis. The system of study groups, as established within the
framework of the old CCITT, is maintained under the new body. World
Telecommunications Standardisation Conferences, held every four years,
produce a list of technical telecommunications subjects or questions for
investigation by the study groups. The study groups then submit their
reports and recommendations to the next conference. If adopted by the
conference, these recommendations become highly influential. Such advice
is almost always followed as it represents the collective wisdom of operating administrations and companies, manufacturers and designers of equipment throughout the world, all of whom participate in the study groups.

In recent years the CCITT, and now subsequently the ITU-T conferences,
have adopted an accelerated procedure for approving the recommendations
of study groups. Thus ITU-T recommendations can be approved by correspondence, provided the study group proposing them supports them unanimously. Equally, recommendations now no longer have to be published
as complete sets every four years, but can be published individually as soon
as they are approved and thus can take immediate effect.

An advisory group has also been set up within the new ITU-T framework
to review priorities and strategies; to review progress on the work programmes; to provide guidelines for the study groups; and to recommend
ways of improving cooperation and coordination with other standards-
setting bodies.

The Radiocommunication Sector (ITU-R)

The Radiocommunication Sector, abbreviated to ITU-R, is basically con-   **15–08**
cerned with the remainder of the CCIR activities not taken on by the
ITU-T. The new Radio Regulation Board has assumed all the functions of
the former International Frequency Registration Board (IFRB). It is therefore concerned principally with the efficient management of the radio spectrum, the examination and registration of frequency assignments and ensuring that the Radio Regulations are applied.

The structure and *modus operandi* of the ITU-R closely reflect those of
the ITU-T. The sector operates through World and Regional Radiocommunication Conferences. World Conferences meet every two years to
review and to revise the Radio Regulations; Regional Conferences meet as
and when necessary to discuss regional issues.

The smallest unit of the ITU-R is similarly the study group. The technical
work carried out by the study groups is managed by the Radiocommunication Assembly, which also approves the study group recommendations.

In the past recommendations could only be approved at Plenary Assemblies, which were held every four years. Since 1988, however, where a recommendation is supported without serious dissent by the members of a study group, a procedure similar to that used by the ITU-T of approval by correspondence has been permitted. This has reduced the approval process from a potential maximum of four years to a mere four to six months. Equally, recommendations could previously only be published together in four-year cycles. Now they can be published individually, or in sets, as soon as they have been approved.

The Telecommunications Development Sector (TDS)

15–09   The Telecommunications Development Sector is essentially the old Telecommunication Development Bureau (BDT), unchanged. It usually operates within the framework of UNDP, in giving advice on developing telecommunications systems. The main forum for this are its conferences, both world and regional.

The Scope and Authority of ITU Rules and Regulations

15–10   Article 4 of the Constitution of the ITU confirms that the main instruments of the Union are its Constitution, Convention and Administrative Regulations. The Constitution states that the provisions of these instruments together shall be binding on all members of the Union. In the case of any inconsistency between these various instruments, the Constitution shall prevail over the Convention which, in turn, shall prevail over the Administrative Regulations.

The Constitution and Convention have from time to time been amended at Plenipotentiary Conferences, each new version being subsequently ratified anew by ITU members. The latest versions were drawn up at the Additional Plenipotentiary Conference held in Geneva in 1992, although as at December 31, 1993, many members were yet to deposit instruments of ratification with the General Secretariat.

Scope

15–11   **The Constitution (Geneva, 1992):** this is stated to be the "basic instrument of the ITU". Its provisions set out the purposes of the ITU, its composition, the rights and obligations of its members, and its structure.

The Constitution also contains a section entitled "General Provisions Relating to Telecommunications". These provisions define various rights and obligations of a very general nature which should form the basis of the provision of international telecommunications services. Examples of such provisions include the right of the public to use the international telecommunications service, the right for members to stop or suspend such services in certain circumstances, and the principle of the secrecy of telecommunications.

15–12   **The Convention (Geneva, 1992):** much of the Convention is concerned with procedural issues regarding the functioning of the ITU. It also contains provisions relating to the operation of telecommunication services, for example concerning charges and free services, the rendering and settlement of accounts and inter-communication. The financing of the ITU is also dealt with in the Convention, as is the procedure for arbitration in the event of a dispute between members.

The International Telecommunication Regulations (Melbourne, 1988): The   15–13
preamble to the International Telecommunication Regulations states that their main purpose is to establish general principles relating to the provision and operation of international telecommunication services offered to the public, as well as to the underlying international telecommunication transport means used to provide such services. They also set rules applicable to "Administrations" and "Recognised (Private) Operating Agencies" (R(P)OAs).

The Regulations thus contain numerous statements of general principle, all underpinned by the basic aim of "promoting the development of tele-communication services and their most efficient operation, while harmonising the development of facilities for world-wide telecommunications". More specific provisions establish rules for the routing of international traffic and for charging and accounting.

Regulations and Recommendations: The Regulations clearly delineate the   15–14
difference in their authority from that of the Recommendations made by the CCITT (now the ITU-T). Article 1.4 states that any references to recommendations in the Regulations are not to be taken as giving to the former the same legal status as the Regulations. Nevertheless, the Regulations frequently seek to encourage compliance with any relevant recommendations to the greatest extent practicable by both "Administrations" and "R(P)OAs".

Authority

Under Article 6 of the Constitution, the ITU instruments are stated to be   15–15
binding on members with regard to "all telecommunication offices and stations established or operated by them which engage in international services or which are capable of causing harmful interference to radio services of other countries" (unless such services are exempt).

The same Article also places an obligation on members "to take the necessary steps to impose the observance of the provisions of [the ITU instruments] upon operating agencies authorised by them to establish and operate telecommunications and which engage in international services or which operate stations capable of causing harmful interference to the radio services of other countries".

Recognised Operating Agency (ROA) Status

Two concepts central to the issue of ascertaining which operators are sub-   15–16
ject to the authority of the ITU instruments are thus "Operating Agency"

and "Recognised Operating Agency". The former is defined in the Annex to the Constitution as:

"Any individual, company, corporation or governmental agency which operates a telecommunication installation intended for an international telecommunication service or capable of causing harmful interference with such a service".

Recognised Operating Agency (ROA) is defined in the following clause thus:

"Any operating agency, as defined above, which operates a public corres-pondence or broadcasting service and upon which the obligations pro-vided for in Article 6 of this Constitution are imposed by the Member in whose territory the head office of the agency is situated, or by the Member which has authorised this operating agency to establish and operate a telecommunication service on its territory".

It therefore follows that an operator providing international services is only directly bound by the rules and regulations of the ITU if the national government which licences him has fulfilled its obligation under Article 6, either by use of suitable provisions in its licence or by some other means, to impose the burden of compliance with the provisions of ITU instru-ments. Thus, the ROA "tag" appears merely to denote that an operator providing an international "public correspondence or broadcasting service" has been made directly bound by the ITU rules. In theory, all such operators should be bound (Article 6 (2)); the title "recognised" merely confirms that this has been done.

Although it seems to be standard practice for members to notify the ITU of those operating agencies which it has "recognised", and thus to assist the ITU in the maintenance of its records of ROAs, there appears to be no express obligation on members to do this.

Settlement of Disputes

15–17   The procedure for the settlement of disputes between members is dealt with in Article 56 of the Constitution. Where negotiation fails, the Article pro-vides for recourse to arbitration in accordance with the procedure set out in the Convention. Alternatively, a dispute could be settled by reference to the Optional Protocol on the Compulsory Settlement of Disputes where both member parties to the dispute are also parties to that Protocol.

The European Telecommunications Standards Institute (ETSI)

15–18   The European Telecommunications Standards Institute (ETSI) was estab-lished in 1988 pursuant to a recommendation that a new European stand-ardisation body be established outside the existing framework of Commun-ity standards-setting bodies. The needs for such a body became evident in the light of the special nature of the telecommunications industry and the special regime for mutual recognition of terminal equipment which the European Commission aimed to introduce.

The role of ETSI is to formulate standards in the fields of telecommunications, information technology and broadcasting. Its members are drawn directly from national administrations, public telecommunications operators, manufacturers, users and research bodies. Thus there is direct participation of interested parties at European level rather than through their national delegations headed by a national standards body. At present, over 20 countries are represented within ETSI, although non-European companies and organisations are often invited to participate as observers at meetings of ETSI's technical assembly.

Standards approved by ETSI have the status of voluntary standards. It is then up to the national governments and Commission to transform them into mandatory standards. This can be done either by transforming them into so-called NETS (*Normes Européenes de Télécommunications*), pursuant to the procedure agreed by CEPT (Conference of European Postal and Telecommunications Administration) member countries, or by the Commission itself pursuant to its new "global approach".

Although initially formed so as to operate completely independently of existing Community standards-setting bodies, co-operation and coordination with those bodies continues to be necessary, particularly with respect to the grey areas which overlap with the telecommunications sector, and in order to maintain a unified standardisation policy.

ETSI has recently come under attack for a number of alleged **15–19** shortcomings:

(a) Intellectual Property Rights: ETSI had attempted to adopt a policy on the application and licensing of intellectual property rights with respect to standards which was challenged by a number of manufacturers. This is discussed in Chapter 12, (paragraphs 12–07 to 12–12) above.

(b) Slowness of the Process: ETSI typically takes between two and five years to define and approve a standard. A major concern is that by the time these standards have been set, they may be irrelevant to the needs of the market. Brussels is also concerned at the scope of ETSI's standardisation work. A current estimate is that there are 3,500 standards in progress. Many overlap, while some are not strictly relevant to the telecommunications market. Both the Commission and the ETSI members are also worried that the ETSI standards-making process is failing to keep up with the speed of development in telecommunications where product cycles, particularly for software, are becoming increasingly short.

(c) Parochialism: A number of ETSI members have also recently been frustrated by ETSI's apparent reluctance to act on developments outside the E.U. This is the case particularly in the GSM digital cellular industry. Here, a standard was drawn up for Europe, but has received wide acceptance in Asia and Africa. It has the opportunity to become a genuine worldwide standard, but political and commercial issues in different countries may stand in its way.

The European Radiocommunications Committee and the European Radiocommunications Office

The European Radiocommunications Committee (ERC)

15–20   The European Radiocommunications Committee (ERC) is one of the three Committees which report to the Plenary Assembly of the European Conference of Postal and Telecommunications Administrations (abbreviated CEPT from its French title). It provides a forum for the radio regulatory administrations of the CEPT member countries.

The ERC's main tasks are to develop radiocommunications policies and to coordinate frequency, regulatory and technical matters in this field. The Committee also develops guidelines in preparation for ITU activities and conferences. ERC proposals in these areas generally take the form of decisions, recommendations or reports. Decisions are binding agreements between the regulatory administrations represented on the Committee and can therefore play a major role in the harmonisation of radio regulatory regimes within the CEPT countries.

15–21   **Working groups:** The Proposals for harmonisation measures are prepared by the working groups of the ERC. Currently, the ERC has four permanent working groups: the Conference Preparatory Group, the Working Group FM (Frequency Management), Working Group RR (Radio Regulatory), and Working Group SE (Spectrum Engineering). Ad hoc groups are also established for specific tasks.

15–22   **The ERC Steering Groups:** The Steering Group is, as its name suggests, the co-ordinating body for the activities of the ERC, the working groups and the European Radiocommunications Office (ERO). Its membership comprises the Chairmen and Vice-Chairmen of the ERC and its working groups and the head of the ERO.

The ERC formally co-ordinates its activities with certain other organisations, notably the European Commission and the European Technological Standards Institute (ETSI), on the basis of Memoranda of Understanding (MoU). Under the MoU with ETSI procedures are agreed for these bodies to work together closely in both the development of standards for systems or equipment and the harmonisation of frequency bands and regulatory requirements necessary for the use of such systems or equipment.

The MoU with the Commission addresses the exchange of information, the coordination of activities and the possibility for the Commission to assign to the ERO specific research tasks in the field of telecommunications.

The European Radiocommunications Office (ERO)

The European Radiocommunications Office (ERO), located in Copen-   **15–23**
hagen, was established in May 1991 in order to support the work of the
ERC, the ERC Steering Group and the working groups. It provides a focal
point for consultation and undertakes studies on the use of the radio spectrum and on the harmonisation of spectrum for new systems. It is funded
by the radio regulatory administrations of the CEPT.

The ERO was established on the basis of a MoU, currently signed by
over half of the CEPT member administrations. The MoU was to be
replaced by the "Convention for the Establishment of the European Radiocommunications Office" which will, *inter alia*, see the creation of the Council of the ERO, consisting of representatives from the contracting administrations, which will take over the management of the ERO itself.

Tasks

Besides generally supporting the ERC and the working groups, the ERO   **15–24**
has more specific functions:

- to provide a centre of expertise for the identification of problem areas
  and new possibilities in the radiocommunications field;
- to draft long-term plans for the future use of the radio frequency spectrum in Europe;
- to support and work together with the national frequency management authorities;
- to co-ordinate actions and provide guidance for research studies;
- to conduct consultations on specific topics or parts of the frequency
  spectrum;
- to maintain a record of important ERC actions and of the implementation of CEPT recommendations and decisions;
- to provide the ERC with status reports at regular intervals; and
- to perform studies for the European Commission on the basis of the
  MoU and the Framework Contract concluded between the E.U. and
  the ERC/ERO.

The ERO also produces information documents on various subjects and
is involved in the development of software for a frequency information
system database.

Consultation

Various consultation mechanisms have been integrated into the activities   **15–25**
of both the ERC and the ERO in order that bodies having an interest
in European radiocommunications can express their views in the relevant
decision-making processes. This mechanism includes:

- An annual CEPT Radio Conference, organised by the ERO;
- co-ordination between interested parties and the ERO on specific
  projects;

- the possibility for international organisations to participate in much of the work of the ERC and its working groups;
- detailed spectrum investigations. One of the overall objectives of the ERO is the development of proposals for a European table of frequency allocations and utilisations. The detailed spectrum investigations analyse portions of the radio spectrum in depth to identify current use and future requirements. During this process, consultation with all interested parties is essential. The aim is to implement such a table by the year 2008.
- The publication of a regular newsletter by the ERO to keep all interested parties informed of the activities and decisions of the ERC.

The European Telecommunications Office

The European Telecommunications Office (ETO) was created by the European Committee on Telecommincations Regulatory Affairs (ECTRA) which is one of the three committees of the CEPT. A Memorandum of Understanding on the establishment of ETO came into force on April 30, 1994 signed by 21 countries representing more than 80 per cent of the weighted votes of the CEPT members.

ETO has two main functions concerning licensing and numbering.

On licensing, the objective of ETO is to propose and harmonise licensing conditions and procedures and to set up a one-stop-shopping procedure for the currently liberalised services. As regards numbering ETO's tasks are to form a centre of expertise and to support ECTRA in order to develop a European numbering strategy, advise NRAs in developing national numbering schemes, and to conduct studies and reports on numbering for the European Commission. ETO reports to two committees which direct its work, the ETO Administrative Council comprising the signatories of the MoU, and ECTRA itself. It is located in the same building as the European Radio Communications Office in Copenhagen.

2. INTERNATIONAL SATELLITE ORGANISATIONS

The International Telecommunications Satellite Organisation (INTELSAT)

15–26   The International Telecommunications Satellite Organisation or INTELSAT is the most important international provider of satellite communication services. It was originally established by an interim agreement in 1964 as a "not-for-profit cost-sharing co-operative" to provide high quality, reliable telecommunications by means of satellite to all nations of the world "on a global and non-discriminatory basis". It is now subscribed to by 110 member countries (parties). Parties can either directly enter the agreement, or designate one of their telecommunications entities to do so. For example, British Telecom is the signatory for the U.K. In most cases, however, the signatory is a government body.

At present, the INTELSAT system carries over half of the world's telephone calls, the bulk of international data circuits, and virtually all transoceanic and intercontinental television broadcasts.

The main objective of INTELSAT is to provide facilities for international communication services. The provision of INTELSAT's facilities for domestic public communication services is subject to the condition that it does not impair its ability to provide international public telecommunication services. The ethos behind INTELSAT was to be able to provide and to maintain a low-cost international system of communication by means of a pooling of international resources and through an equitable sharing of the costs and risks involved in the development of international satellite communications. Thus all members assume the risks and invest in INTELSAT in proportion to their actual use of the system for all services. Investment shares are adjusted annually based on the previous six months' usage.

Article V of the INTELSAT Agreement confirms the non-discriminatory basis upon which INTELSAT's services are to be provided. Indeed, all users, members and non-members alike, are to be charged the same for the same service, and rates are averaged based on total use of the entire global system.

One of the most important articles of the INTELSAT Agreement is Article XIV, which defines the rights and obligations of INTELSAT members, particularly as regards the question of them establishing or using separate systems to meet domestic or international requirements. Prior to the establishment, acquisition or utilisation of separate satellite facilities, INTELSAT members are required to consult with INTELSAT and to furnish all relevant information so as to ensure technical compatibility, to avoid significant economic harm, and so as not to prejudice the establishment of direct communications links. This article has in the past given INTELSAT a significant regulatory power in the field of international satellite services.   15–27

However, its scope and significance have been considerably reduced in recent years. It now applies only to public switched telecommunication services. Moreover, in 1985, the U.S. Congress passed legislation to the effect that if INTELSAT renders an unfavourable finding under Article XIV, the President may determine that it is nevertheless in the U.S. national interest to proceed with the separate system. U.S. policy has, indeed, tended to favour the stimulation of competition in the provision of international satellite services which, it is claimed, an international "co-operative" occupying a dominant position with both regulatory and operational powers, would tend to inhibit.

Structure

INTELSAT has a legal identity separate from the parties or signatories of which it is made up. Its organisational structure comprises: 15–28

(i) **The Assembly of Parties:** The principal organ of INTELSAT. It comprises all parties to the Agreement, and is concerned with the general policy and long-term objectives of INTELSAT.

(ii) **The Meeting of Signatories:** Composed of signatories to the INTELSAT Agreement, *i.e.* parties or their designated telecommunication entities. It deals with amendments to the INTELSAT Agreement and the Operating Agreement, future programmes and financial and other matters.

(iii) **The Board of Governors:** The Board's main responsibility is for the design, development, construction, establishment, operation and maintenance of INTELSAT facilities.

(iv) **The Director General:** The Chief Executive and legal representative of INTELSAT.

The Operating Agreement

15–29   The Operating Agreement is entered into by the parties or by other signatories, where a party has so designated a telecommunication entity. The Agreement sets out the financial commitments of each signatory. In general, these commitments are related to the signatory's utilisation of the INTELSAT facilities. This is expressed as a percentage of all utilisation of the INTELSAT facilities by all signatories.

Access to INTELSAT

15–30   It was announced in March 1994 in the U.K. that BT, the U.K. signatory to the Operating Agreement, and the Department of Trade and Industry would allow satellite service providers to have direct access to INTELSAT's satellite capacity. Previously, service providers had to buy capacity from the Signatory Affairs Office (SAO), the arms-length division of the signatory, BT.

However, non-signatory users will have to make an investment in INTELSAT in order to cover some of the contribution which has up to now been carried by BT. So while direct users will pay less for their actual transponder capacity than will users that go through the SAO, it is unclear whether the total cost will be any lower once the user investment has been taken into account.

The European Satellite Communications Organisation (EUTELSAT)

15–31   The Provisional European Satellite Telecommunications Organisation (EUTELSAT) was established in 1977 under the interim EUTELSAT Agreement to set up and operate a space segment for satellite communications and to make the necessary arrangements for this, particularly with the European Space Agency. The founder members of the provisional EUTELSAT were the 17 members of the Conference of European Postal and Telecommunications Administration (CEPT). A convention creating the definitive organisation was held in September 1985, resulting in a signed agreement among 26 European governments. By March 1993, membership had risen to 35 states.

The structure of EUTELSAT closely resembles that of the other international satellite organisations. The organisation is governed by a Convention, which is entered into by parties (sovereign states). The Convention sets out the constitution and aims of EUTELSAT. An Operating Agreement details the functioning of the organisation, for example defining utilisation charges. It is entered into by signatories, that is states or their designated telecommunications entities.

The administrative structure of EUTELSAT closely resembles that of INTELSAT. It is composed of an assembly of parties, a board of signatories, and an executive, headed by a Director General.

The main objective of EUTELSAT is the construction, establishment, operation and maintenance of the European space segment and the provision of the space segment required for international public telecommunications services in Europe. A guiding principle of EUTELSAT is "non-discrimination as between Signatories" (Article III (D)). The EUTELSAT Convention has a similar provision to Article XIV of the INTELSAT Agreement. Article XVI gives EUTELSAT the power effectively to decide whether potential competitors are likely to provide an unacceptable level of competition to the services provided by EUTELSAT, that is, to cause "any significant harm". If the assembly of parties decides that it would, coordination with that competitor can be refused.   15–32

As the technological and regulatory environment in the field of telecommunications generally has changed enormously since 1977 when EUTELSAT was provisionally established, many of the provisions of the EUTELSAT Convention, including particularly Article XVI, have been considered increasingly anachronistic in what is becoming a more competitive market.

Both the U.K. Department of Trade and Industry and the European Commission have focused attention on what they fear may be potentially anti-competitive practices which are effectively sanctioned by the EUTELSAT Convention. In particular, the Commission called for an amendment to the EUTELSAT Convention in order to avoid any distortion of competition and to allow full use and best allocation of the existing space segment which would give users direct access to space segment capacity and then give space segment providers the right to market space segment capacity directly to users.[1]   15–33

While access to INTELSAT has already been liberalised to a certain extent, the provision of direct access to EUTELSAT is still some distance away. EUTELSAT is, indeed, facing problems in implementing the necessary restructuring before direct access can become a realistic possibility. At present direct access is restricted to signatories, such as BT, whose usage

---

[1] See also, "The provision of and access to satellite capacity" (1994) E.C. communication; here the Commission argues that service providers should be afforded direct access to space segment capacity.

of capacity has been linked to their investment in the organisation. Therefore anyone else who wants direct access has to become an investor as well,
at least temporarily. Given the current high level of borrowing to which
EUTELSAT is committed, such investment is an unattractive prospect to
most operators.

The EUTELSAT assembly of parties has, however, set up a study group,
composed of parties and signatories, to investigate ways of facilitating
access for non-signatory entities. The study group is also looking into tariff
structures and "equitable arrangements for guaranteeing multiple access
to the EUTELSAT space segment". In the meantime, a "multiple access"
arrangement is currently in force between four EUTELSAT members, and
is open to all members who wish to join. The arrangement allows individual users to avoid having to deal directly with their local signatory
(which may indeed be their most immediate competitor), and to approach
the signatory in any of the other three countries to acquire space segment.
The parties to this arrangement are the U.K., the Netherlands, Germany
and France.

The International Maritime Satellite Organisation (INMARSAT)

15–34   The International Maritime Satellite Organisation (INMARSAT) was
established in July 1979. Its original purpose was "to make provision for
the space segment necessary for improving maritime communications, to
assist in improving distress and safety of life at sea, communications, efficiency and management of ships, maritime public correspondence services
and radio-determination capabilities". This purpose has since been
extended, in 1985, to include aeronautical satellite communications and,
in 1989, to include the provision of land mobile communications.

The constitution of INMARSAT very closely resembles the make-up of
both INTELSAT and EUTELSAT. It consists of the founding Convention
and the Operating Agreement. There are currently some 69 Member States
(parties) in INMARSAT. Each party must either sign the Operating Agreement itself or designate a competent entity, public or private, subject to
the jurisdiction of that party, to sign the Operating Agreement.

INMARSAT is financed by the contributions of the signatories to the
Operating Agreement. Each signatory has a financial interest in the organisation in proportion to its investment share which is determined in accordance with the Operating Agreement. INMARSAT's capital requirements
include the costs of acquiring space segment and, pending receipt of revenues, the funds required for operating, maintenance and administrative
costs. The investment shares of signatories are determined on the basis of
utilisation of the INMARSAT space segment.

15–35   The structure of INMARSAT also reflects those of INTELSAT and EUT-
ELSAT. The Assembly has responsibility for overall policy formation. It is
composed of all the parties. The Council, consisting of 23 representatives

of signatories, has a more executive role. Responsible to and under the direction of the Council are the Directorate and the Director General. The Director General is the Chief Executive and legal representative of INMARSAT.

The Operating Agreement contains a clause very similar to Article XIV of the INTELSAT Agreement and Article XVI of the EUTELSAT Agreement. If a party or any person within its jurisdiction intends to establish separate space segment facilities to meet any or all of the maritime purposes of the INMARSAT space segment, that party is obliged to notify INMARSAT in order to ensure technical and financial compatibility with the INMARSAT system (Article 8). Both the Council and Assembly will then express views on the notification in the form of a non-binding recommendation with respect to technical compatibility.

However, as with the similar provisions in the INTELSAT and EUTELSAT Agreements, concern has been mounting as to the tendency for powers such as these to restrict access to satellite services and thus to restrict competition. With a view to reaching a satisfactory solution of the situation, INMARSAT's Assembly has established an Intersessional Working Group (IWG) to review the current appropriateness of Article 8. The IWG has now concluded its report and has made a recommendation to the Assembly which provides that no system which falls within the scope of Article 8 and has been notified shall be deemed to cause significant economic harm. This recommendation is yet to be adopted by the Assembly.

In the early part of 1995 INMARSAT'S signatories approved the formation of a new and separate company (initially known as INMARSAT-P) to provide a global mobile telephone satellite service.

PART TWO

European Union

CHAPTER 16

European Union Telecommunications Policy[1]

By Peter Alexiadis, of Coudert Brothers, Brussels

SECTION I: REGULATORY FRAMEWORK

Historically, the European telecommunications sector has been charac-   **16–01**
terised by a strong public service monopoly tradition and the industrial
policy desires of Member States to create national champions. It is widely
acknowledged that these traditions have restricted the growth of diverse
and price-competitive telecommunication services being offered throughout
the European Union (E.U.). In light of its overall importance in the context
of the broader economic well-being of the E.U.—estimated to account for
over six per cent of total GDP by the end of the century—the regulatory
initiatives of the Commission in the telecommunications sector assume
major significance within the context of the E.U.'s internal market programme. These initiatives date back to the Single European Act 1985, with
the release of the Green Paper on Telecommunications in 1987[1a] providing
a blueprint for future Commission action in the sector.

The desire to create a free and open market for telecommunications has
meant that regulatory reform in the telecommunications sector at E.U. level
has been based on a balanced programme of liberalisation and harmonisation. The former policy goal is directed at injecting greater competitive
alternatives and consumer choice into an environment which was hitherto
dominated by national telecommunications operators (TOs). The latter
goal is directed at creating minimum equal conditions of competition in
the absence of which new market entrants would be placed at a competitive
disadvantage. By the early 1990s, Commission policy has also raised to

[1] I am grateful to Peter Sandler of Directorate-General XIII and to Dirk van Liedekerke of Coudert
Brothers (Brussels) whose comments on the draft were greatly appreciated.
[1a] COM(87)290.

even greater importance the systematic application of E.C. competition policy in the creation of a more competitive marketplace. Most recently, the Commission has sought to set its principal goals of liberalisation, harmonisation and competition policy in the context of universal service obligations, namely, the provision of basic telephony sources to all citizens of the E.U. at the lowest possible price.

16–02   The focus of Commission initiatives is directed at liberalising the market for the provision of services, ensuring that terminal equipment is sold without reference to national Member State boundaries, and providing new competitors with access to telecommunications infrastructure. Overriding these broad policy goals is the recognition that the TOs, because they operate at various functional levels of the market, must not be permitted to take advantage of their privileged status as judge and competitor, as the case may be. The recognition of this fact has meant that the Commission has accorded high priority to the separation of the regulatory and operational activities of national TOs in all of its policies.

Although not falling within the ambit of telecommunications *stricto sensu*, the regulatory initiatives in the related fields of broadcasting and satellites in effect constitute the audiovisual dimension to the Commission's work in the telecommunications sector.

1. THE SPECIAL ROLE OF ARTICLE 90

16–03   The legal form which E.C. Commission regulatory interventions have taken in the telecommunications sector reflects the fact that the principle objects of that regulation—the TOs—were (and, in many cases, still are) large public undertakings in respect of which specific provisions of the Treaty of Rome apply. Article 90(1) of the E.C. treaty recognises that Member States may grant special or exclusive rights to undertakings; the holding of a monopoly in the provision of telecommunications services is a classic instance of such special or exclusive rights. Nevertheless, Article 90(1) confirms that public undertakings and undertakings entrusted with such rights remain, in principle, subject to the provisions of the Treaty, especially those concerning competition rules (see Section II, "Competition Rules"). It explicitly confirms and that Member States shall neither enact nor maintain in force any measure regarding these undertakings which are contrary to the rules contained in the Treaty. This reinforces the obligations imposed on Member States by Article 5 of the Treaty not to hinder the achievement of the objectives outlined in the Treaty.

Article 90(2) of the E.C. Treaty, addressed to the undertakings themselves, provides that they are subject fully to the Treaty's rules, but only in so far as the application of those rules does not prevent them from carrying out their assigned tasks in the "general economic interest". This exception to liability under Article 90(2) has been narrowly construed by both the Court of Justice and the commission,[2] so that its utility to TOs as a shield to competition law actions is very limited. To the extent that a national court of a TO's Member State is called upon to determine the issue of how broadly the Article 90(2) exception is to be construed, however, the Article 90(2) exception does provide some insulation for TOs allegedly in breach of the competition rules.[3]

Finally, Article 90(3) enables the Commission to address appropriate directives or decisions[4] to Member States in order to ensure that they respect their Treaty obligations under Article 90(1).[5] Although the Commission has recently developed an informal practice of public consultation with regard to directives to be adopted under Article 90(3), the provision itself does not require the consultation of any other Community institution for the adoption of Community acts, thereby expediting the decision-making process and removing the uncertainties attendant in the E.U. legislative enactment process. In practice, however, the Commission does consult both the Member States and the European Parliament, with its most recent practice being to publish such proposed directives for consultation.

The Commission has used Article 90(3) as the legal basis for a number   **16–04** of its principal regulatory initiatives in the telecommunications sector, namely: the so-called Terminal Equipment Directive 1988[6]; the Services Directive 1990[7]; the extension of the operation of the Services and Terminal Equipment Directives to satellites in October 1994 under the Satellite Communications Directive[8]; the additional revisions to the Services Directive scheduled to occur by January 1, 1996 upon the adoption of the finalised draft Commission Directive on the Liberalisation of Cable Television Networks[9] and the proposed amendments to be introduced to the Services Directive by the draft Commission Directive on Mobile and Personal Communications (scheduled to be adopted by summer 1995, to take effect as of January 1, 1996).

By way of contrast, the Commission's other major regulatory initiative, the Open Network Provision (ONP) Framework Directive,[10] was taken

---

[2] *Ijsselcentrale* [1991] O.J. L28/2; *Magill TV Guide* [1989] O.J. L78/43; *ERT v. DEP* [1991] E.C.R. 2925; *Screensport/EBU* [1991] O.J. L63/32. See also *RTT/GB-Inno-BM* [1991] I E.C.R. 5941.
[3] *ERT v. DEP, supra; Corbeau* [1993]I E.C.R. 2533; *Almelo v. Ysselmy* [1994] I. E.C.R. 1477.
[4] Decisions were taken by the Commission concerning the respective Spanish and Dutch laws which reserved express courier services below certain weights and tariffs to their national operators; see *Dutch express delivery* [1990] O.J. L10/47 and *Spanish courier services* [1990] O.J. L23/19; see *The Netherlands/PTT Netherlands v. Commission* [1992]I E.C.R. 565.
[5] In exercising its decision-making powers under Art. 90(3), the Commission has a very broad discretion as to whether or not it should take any action. Consequently, the Court of First Instance has held that the failure of the Commission to act on a request for it to take action under Art. 90(3) is not a reviewable act under Art. 175 of the E.C. Treaty; *Ladbroke Racing Ltd. v. Commission* (1994), unreported.
[6] Directive 88/301; [1988] O.J. L131/73.
[7] Directive 90/388; [1990] O.J. L192/10.
[8] Directive 94/46; [1994] O.J. L268/15.
[9] On December 21, 1994, the Commission adopted a proposal for a draft Commission Directive amending Directive 90/388 regarding the abolition of the restrictions on the use of cable television networks for the provisions of telecommunications services. The Commission will take a formal decision after presenting the proposed directive both to the Member States and the European Parliament, and after having consulted with interested parties during the course of 1995. A draft Directive was published in [1995] O.J. C76/8.
[10] Directive 90/387; [1990] O.J. L192/1.

under the auspices of the harmonisation powers contained in Article 100A of the E.C. Treaty. Under the Article 100A procedure, which serves as the main legal basis for the implementation of the internal market programme, the Commission proposes legislative acts (in particular, directives) for adoption by the Council through the elaborate conciliation procedure with the Parliament as introduced by the Maastricht Treaty (Article 189b).

The difference in choice of legal procedure reflects the Commission's perceived differences in function for these respective Articles of the Treaty. The Commission perceives Article 90 to be the appropriate tool to create an open single market in telecommunications equipment and services by eliminating special or exclusive rights which are directly in conflict with basic Treaty objectives. According to the Commission, its directives on Terminal Equipment and Services (and revisions thereto) are justified under Article 90(3) of the E.C. Treaty because they are aimed at removing those very measures (the granting by Member States of exclusive rights to provide both equipment and services) which are inconsistent with several Treaty provisions relating to the free movement of goods and services and the abuse of a dominant position (Articles 30, 37, 59 and 86 of the Treaty).

The Limits of Article 90(3)

16–05   The use of the Article 90(3) procedure has generated much controversy, with a number of Member States having challenged the ability of the Commission to avail itself of the procedure with respect to both the Terminal Eqipment Directive[11] and the Services Directive.[12] According to those Member States, the Commission incorrectly adopted those Directives on the basis of Article 90(3), and should have taken the usual route of infringement proceedings against the Member States pursuant to Article 169 for their failure to fulfil an obligation under the Treaty. It was argued further by some Member States that a policy for the restructuring of the telecommunications sector—which is in effect what the two Directives were allegedly designed to accomplish—falls within the exclusive competence of the Council under Article 100A. The Member States who challenged the Commission's competence took the position that the Commission must accept the existence of national monopolies for the provision of terminal equipment or services, and should take action only in individual cases where it determines that a state monopolist has abused its dominant position, discriminated against imported terminal equipment or taken other actions contrary to Treaty provisions. In short, they argued that Article 90(3) should be used solely as a corrective mechanism, and not as a broad *fiat* for legislative action.

The Court of Justice, however, confirmed in both cases the power of the Commission under Article 90(3) to specify in a directive obligations incumbent upon Member States arising from Article 90(1). While the court

---

[11] *France v. Commission* [1991] E.C.R. 1223.
[12] *Spain v. Commission* [1992] I E.C.R. 5833.

agreed that Article 90(1) permits Member States to grant undertakings special or exclusive rights, it took the position that such rights are not necessarily compatible with the Treaty. When the existence of such rights cannot be separated from their illegal exercise, both may be eliminated by means of directives addressed to the Member States pursuant to the powers contained in Article 90(3). The court took the view that, as the exercise of exclusive rights in these areas could restrict intra-Community trade, it was therefore legally justified for the Commission to specify that such rights had to be abolished (subject only to the Terminal Equipment Directive's specific exceptions regarding the protection of the public network from damage and the need to ensure the inter-operability of terminal equipment).

For technical reasons, however, all references in the Terminal Equipment **16–06** and Services Directives to the abolition of "special rights" were annulled by the Court. The Commission had, according to the Court, erred in treating the expression "special or exclusive rights" as a composite phrase, and was under an obligation to differentiate between the two different types of rights. As regards such special rights, the court held that the Commission had failed to specify the *types* of special rights which were intended to be caught by the Directive and the reasons why such rights were considered to be contrary to any Treaty provisions.

It has taken until the adoption of the Satellite Communications Directive in October 1994 for a legal definition of "special rights" to come into existence.[13] The essence of that definition is that undertakings which are granted rights through any legislative, regulatory or administrative instrument on criteria other than those which are "objective, proportional and non-discriminatory" will be considered to enjoy "special rights".[14] The strict application of this definition will mean that, regardless of the number of licences or rights granted by a regulatory authority, unless such grants are made on completely transparent grounds, they will be considered to be "special" in nature and the regulatory obligations imposed under the Terminal Equipment Directive and Services Directive, and even the ONP Framework Directive will apply to those undertakings with full force.[15]

---

[13] According to the amended version of the Services Directive, "special rights" means rights that are granted by a Member State to a limited number of undertakings, through any legislative, regulatory or administrative instrument, which, within a given geographical area: limits to two or more the number of such undertakings, otherwise than according to objective, proportional and non-discriminatory criteria; or designates, otherwise than according to such criteria, several competing undertakings; or confers on any undertaking or undertakings, otherwise than according to such criteria, any legal or regulatory advantages which substantially affect the ability of any other undertaking to import, market, connect, bring into service and/or maintain telecommunication terminal equipment in the same geographical area under substantially equivalent conditions.

[14] See Art. 1(b) of the of the Satellite Communications Directive. This statutory formula is based on the criteria used by the Commission's Competition Directorate (DGIV) to determine whether or not a selective distribution system is "open" or "closed".

[15] Query whether the new definition of "special rights" will be operative in light of the pending judgment of the ECJ in the Art. 177 reference for a preliminary ruling by the English Divisional Court in the case of *R.v. The Secretary of State For Trade & Industry, ex-parte British Telecommunications plc* [1994] O.J. C380/3, in which the meaning of "special or exclusive rights" in the U.K. licensing context is at issue.

There are further jurisdictional limits on the extent to which the Commission can rely on its regulatory powers under Article 90(3). Thus, the Court of Justice made it clear that while the Commission may use Article 90(3) to specify general legal obligations on Member States already found in the E.C. Treaty, it *cannot* use those powers to micro-manage their commercial affairs. For example, the inclusion in the respective Terminal Equipment and Services Directives of an obligation on telecommunications authorities to terminate leasing or maintenance contracts on no more than one year's notice was held by the court to fall *outside* the powers of the Commission. The Court considered that Article 90 of the Treaty confers powers on the Commission only in relation to State measures and that anti-competitive conduct engaged in by undertakings on their own initiative (such as the anti-competitive aspects of the leasing or maintenance contracts) could be called into question only by individual decisions adopted under Articles 85 and 86 of the Treaty. Moreover, the decision of the Court suggests that the Commission should use Article 90(3) only when swift action is required or when the proposed legislation does not require considerable consultation and debate.

Article 87

16–07   To date, the Commission has not availed itself of the power to use Article 87 of the E.C. Treaty[16] as a means of clarifying the application of E.C. competition rules to the telecommunications sector through the adoption of Council Directives or Regulations. However, there is some indication to suggest that the scope of Article 87—which caters for subsidiarity issues, specific industrial sectors and policy clarifications—may, to a certain extent, well be suitable for the purpopses of further liberalising in the sector.

2. TERMINAL EQUIPMENT

16–08   The dual themes of liberalisation and harmonisation have been effected with respect to the telecommunications equipment market through the

---

[16] Art. 87 provides that:
"1. Within three years of the entry into force of this Treaty the Council shall, acting unanimously on a proposal from the Commission and after consulting the European Parliament adopt any appropriate regulations or directives to give effect to the principles set out in Art. 85 and 86.
If such provisions have not been adopted within the period mentioned, they shall be laid down by the Council, acting by a qualified majority on a proposal from the Commission and after consulting the European Parliament.
2. The regulations or directives referred to in para. 1 shall be designated in particular:
 (a) to ensure compliance with the prohibitions laid down in Art. 85(1) and in Art. 86 by making provision for fines and periodic penalty payments;
 (b) to lay down detailed rules for the application of Art. 85(3), taking into account the need to ensure effective supervision on the one hand, and to simplify administration to the greatest possible extent on the other;
 (c) to define, if need be, in the various branches of the economy, the scope of the provisions of Art. 85 and 86;

adoption of the Terminal Equipment Directive, and through the adoption of measures aimed at the mutual recognition of type approval procedures and the creation of uniform standards.

Competition in the Terminal Equipment Market

Until the adoption of the Commission's Terminal Equipment Directive in   **16–09** May 1988, there had been no Union-wide market for terminal equipment, with the supply of such equipment being the domain of domestic Member State producers, who in practice supplied their national TOs in each case. In some cases, a branch of the national TO was itself responsible for the production of such equipment. The pace of technological change meant, in certain cases, that some Member States relinquished the right of their TOs to have a monopoly with regard to all types of telecommunications equipment, preferring to maintain that monopoly only with respect to specific categories of terminal equipment such as telephone sets and PABX units. It was the complete inability of the internal market programme to break down these patterns of national markets in telecommunications equipment which necessitated the adoption of the Directive.

Under the Directive, it is clear that all exclusive rights for the importation, marketing, connection, bringing into service or maintenance of telecommunications terminal equipment had to be abolished. The rules applicable to special rights, by way of contrast, were in principle left unclassified after the European Court of Justice's judgment (see above), although the new definition of "special rights" introduced by the Satellite Communications Directive is designed to overcome this problem. The relevant type of equipment caught by the Terminal Equipment Directive is defined in general terms, so that new types of equipment can fall within its scope in the future.

The focal point of the Directive is the establishment of the principle of   **16–10** the free movement of terminal equipment between Member States, subject only to local type-approval procedures which are justified by "essential requirements" of the public network such as the maintenance of the security of the network, the achievement of interoperability and the protection of data.[17]

(d) to define the respective functions of the Commission and of the Court of Justice in applying the provisions laid down in this paragraph;

(e) to determine the relationship between national laws and the provisions contained in this section or adopted pursuant to this Article."

[17] Art. 3 of Directive 88/301 authorises Member States to submit terminal equipment to compliance control in order to check whether they satisfy certain essential requirements. The issue arose in *Rouffeteau & Badia* [1994] I. E.C.R. 3257 as to whether French legislation which provided that telecommunications terminals intended for connection to the public network could not be marketed in France unless they satisfied such requirements went beyond what is permitted by Art. 3. Under the French law, the advertising of telephones which did not satisfy the prescribed approval procedure was prohibited and subject to criminal penalties. Two French traders who were prosecuted under the legislation asserted that the provisions in question violated both Art. 30 of the E.C. Treaty and Directive 88/301, especially in light of their contention that the equipment was destined for export. The Court of Justice upheld the French law, holding that: "Neither [Article 30 of the EC Treaty nor Directive 88/301] precludes national rules

The abolition of all exclusive rights is supplemented by a series of provisions in the Directive aimed at the creation of transparent market conditions through the following:

- the publication of the criteria (wherever available) which must be complied with by technicians in connecting terminal equipment to the public network;
- rendering the network termination points physically accessible; and
- publishing all specifications for terminal equipment.

To ensure that the Commission can monitor the implementation of these obligations, the Directive contains provisions according to which Member States are required to provide:

- an Annual Report on the measures taken; and
- a list of all existing specifications and of all new specifications.

Most importantly, under the Community rules as specified by the Directive, TOs can, in principle, no longer have regulatory activities in addition to their operational activities, as Member States are under an obligation to separate the two.[18] Thus, regulatory matters such as the granting of type-approval and the drafting of technical specifications must, in principle, be entrusted to a body independent of the national TO.

Mutual Recognition of Type-approval Procedures

16–11 The opening up of the market for terminal equipment is also closely linked to the establishment of a regime whereby producers need not comply with costly and unnecessary national procedures for conformity assessment.

Effective as of November 1992, the Council's Mutual Recognition Directive[19] has prescribed a system for the full mutual recognition of approval

---

which prohibit traders, with penalties for infringement, from importing terminal equipment which has not been approved for release for consumption, possessing it with a view to sale, selling distributing or advertising it, even if the importer, holder or vendor has clearly stated that such equipment is intended solely for export, where there is no certainty that it will be re-exported, and is therefore not suitable for connection to the public network."

[18] Recent judgments of the ECJ have clarified the nature of the obligation contained in Art. 6 of Directive 88/301, e.g. (i) in the *Decoster* and *Taillandier* cases, the Court of Justice confirmed its former case law by holding that as long as a particular body was engaged in economic activities of a commercial or industrial nature, it was not necessary that it should have a legal personality separate from the State for the competition rules to apply to it; (ii) in *Decoster* and *Lagauche*, the Court took the view that a mere administrative separation of regulatory and operational functions was not sufficient to meet the criteria set forth in Art. 6; (iii) reaffirming its previous case law, the Court took the view in *Decoster* that, irrespective of the terms of Directive 88/301, Art. 3(f), 86 and 90(1) of the E.C. Treaty must be interpreted as forbidding a cumulation of regulatory and commercial functions in one organisation; (iv) in *Evrard* and *Lagauche*, the Court was prepared to take the view that the existence of an appeals procedure was sufficient to overcome the usual conclusion that Art. 30 of the E.C. Treaty would be infringed wherever a public undertaking was active in a market and was also acting in a regulatory capacity in determining the issue of access (*i.e.* the approval of equipment) to the market of products of actual or potential competitors. See *Procureur du Roi v. Lagauche & Ors, Evrard* [1993] I E.C.R 5267; *Ministere Public v. Decoster* [1993] I E.C.R. 5335; *Ministere Public v. Taillandier* [1993] I E.C.R. 5383.

[19] Directive 91/263; [1991] O.J. L 128/1. The Commission has commenced infringement actions under Art. 169 of the E.C. Treaty against the following countries for their respective failures to transpose into national law the terms of Directive 91/263: Belgium ([1994] O.J. C275/16); Ireland [1994] O.J. C288/3); Greece ([1994] O.J. C316/10); and Luxembourg ([1994] O.J. C254/10).

procedures for terminal equipment for the purposes of marketing such equipment and its connection to public networks. Thus, the successful testing for type approval in one Member State should mean in principle that a similar or equivalent test in another Member State prior to authorisation for connection to its public network is not necessary.

The mutual recognition procedure is activated where the terminal equipment in question satisfies a number of so-called "essential requirements" specified in Article 4 of the Directive. These requirements include:

- user safety;
- the safety of employees of public network operators;
- electromagnetic compatibility (wherever necessary);
- protection of the public network from harm;
- effective use of the radio frequency spectrum; and
- the interoperability of the equipment used in the network.

Where terminal equipment has been produced in conformity with harmonised E.C. standards, there is a presumption of compliance with the above requirements.

A producer can choose between utilising one of two alternative conformity procedures which will result in the designation of a CE label for the equipment, namely:

- the E.C. "declaration of conformity procedure", whereby certificates are issued by the relevant Member State authorities upon all the relevant tests having been performed; or
- the E.C. "type-examination procedure", whereby the producer implements a system of self-certification subject to a regime of surveillance at the E.C. level.[20]

The self-certification scheme is in keeping with the Commission's so-   **16–12** called "new global approach towards testing and certification", adopted formally in 1989 (see below). Under the new approach, there is a movement from testing and certification for the purposes of ensuring the technical specifications of the product to ensuring the level of safety required of a product; this latter goal can be achieved at a number of different stages in the development and production of a product.

The Mutual Recognition Directive constitutes the second phase of the Commission's overall mutual recognition programme, replacing a process which began in 1986 with the introduction of a Directive[21] providing for the mutual recognition of conformity tests carried out in other Member States by NRAs pursuant to common European technological specifications (NETs). The 1986 Directive was overhauled because it left open the possibility of administrative procedures for type-approval being repeated in other Member States. By way of contrast, the new Directive[22] has overcome

---

[20] These procedures were laid out in Council Decision 90/683 EEC regarding the modules for the various phases of the conformity assessment procedures which are intended to be used in the technical harmonisation directives [1990] O.J. L380/13. Most recently, see Directive 93/68; [1993] O.J. L220/1.
[21] Directive 86/361; [1986] O.J. L217/21.
[22] Directive 91/263; [1991] O.J. L128/1.

the requirement for multiple testing and even provides that approval in one Member State is sufficient for the equipment to be sold on the territory of another Member State. In so doing, the Mutual Recognition Directive has abandoned NETs, with the telecommunications-specific standards organisation ETSI being granted responsibility for the formulation of a harmonised European Technical Standard which will in turn lead to mandatory Common Technical Regulations (CTRs).[23] At the time of writing, there are indications that the Mutual Recognition Directive may be the subject of amendments, to be introduced by the end of 1995, which provide for an even more comprehensive self-certification regime which will be limited to the proscription of certain "essential requirements".

Standards-setting

16–13   European standardisation policy has been a major driver in the creation of an internal market. Until 1983, the Commission had sought to pursue European standardisation policies through the enactment of a large number of detailed directives in specific sectors. These directives under the "old approach" to standards-setting had all the hallmarks of over-regulation, seemed to be characterised by long delays in their adoption and were often abandoned prior to their adoption.

The adoption of the Technical Standards Directive in 1983[24] marked the onset of the "new approach" to standards-setting. Hereafter, the principle vehicle for harmonisation became common standards, rather than specific detailed directives. Attention turned to the harmonisation of "essential requirements" by the Acts adopted by the Community institutions, rather than exact technical uniformity, it being recognised that legislative intervention is only desirable in exceptional circumstances and that most Member State rules protecting safety and health pursued common objectives. The Directive brought much-needed transparency into the standardisation movement, requiring that Member States keep the Commission informed about new proposed standards, withdraw obsolete standards, consult closely with the Commission, ensure the mutual recognition of tests and the establishment of harmonised rules on certification.

In May 1985, a Council Resolution known as the "New Approach to Technical Harmonisation and Standardisation"[25] extended the policy underlying a standards-based approach to the assessment of conformity. The object of the Resolution was to set forth the key elements which were to be found in directives to be subsequently adopted in this area. Once again, the approach was minimalist, focusing on the need to take action only with respect to "essential requirements" and to ensure the free move-

---

[23] In November 1994, 3COM became the first company to obtain a CE mark from the Dutch National Regulatory Authorities for a network attached device under the "new approach" *notwithstanding* the absence of a CTR; this was achieved on the basis that 3COM could demonstrate that the product met *essential requirements* and *existing* harmonised telecommunications standards. To date, the authorisation has not been challenged by any other Member State regulatory authorities.

[24] Directive 83/189; [1983] O.J. L109/8. (as amended).

[25] [1985] O.J. C136/1.

ment of goods without verification for products that satisfied those essential requirements.

The Commission's new approach in the field of standards was further **16–14** completed in 1989, when the Commission introduced a Resolution on its so-called "Global approach to testing, certification and conformity assessment".[26] In an effort to ensure that Member States would recognise the conformity certificates issued by other Member States, the Council adopted a Resolution setting forth the various types of "modules" which could be used in the different phases of conformity assessment and the criteria under which they could operate in order to designate a CE mark for telecommunications equipment which satisfied the essential requirements. The adoption of Council Decision 90/683/EEC[27] (see above) laid out the two types of modules which can be used to designate the CE mark.

Of particular interest are the standards-setting procedures set forth in two directives which cover the major safety issues affecting telecommunications equipment, namely, the Low Voltage Directive 1973[28] and the Electromagnetic Compatibility Directive 1992.[29] The former Directive requires that Member States may not impede the free flow of electrical equipment (which is a broader category than "telecommunications" equipment) with a voltage rating of 50–1000V AC or 75–1500V DC if all essential requirements regarding the safety of such equipment when properly installed and used are satisfied; this Directive is in fact a precursor to the more recent "new approach" to harmonisation. The Electromagnetic Compatibility Directive applies to all electrical and electronic appliances (including telecommunications equipment) which may cause electromagnetic interference. The essential requirements protected under the Directive are that apparatus covered by it must be constructed so that electromagnetic disturbance generated does not exceed a level allowing radio and telecommunications equipment and other apparatus to operate as intended, and is not itself interfered with. A manufacturer may choose to follow harmonised standards or its own designs, as long as the essential requirements of the directive are met. In either case, the manufacturer must make a declaration, the accuracy of which must be subsequently checked by the authorities, and a CE mark must be attached.[30]

The role of ETSI

The establishment of a specific standards-setting body for the telecommuni- **16–15** cations industry became an imperative in light of the peculiar characteristics of that industry and the special regime for mutual recognition of terminal equipment.

[26] [1990] O.J. C10/1.
[27] [1990] O.J. L380/13.
[28] Directive 73/23; [1973] O.J. L77/29.
[29] Directive 89/336; [1989] O.J. L139/19 (as amended).
[30] Under the original Directive, an E.C. mark was prescribed; the CE mark has been introduced by Directive 93/68 (see *supra*).

Pursuant to the 1987 Green Paper's recommendations that a new European standardisation body be established outside the CEN/CENELEC framework (the usual standards-setting bodies in the E.U.), ETSI[31] was formed in 1988. The role of ETSI, in line with the "new approach", is to formulate standards in the fields of telecommunications, information technology and broadcasting.

Members of ETSI are drawn directly from national administrations, TOs, manufacturers, users and research bodies. This process constitutes a radical departure from previous policy, in so far as it allows for the direct participation of interested parties at European level rather than through their national delegations headed by their national standards body. At present, over 20 countries are represented within ETSI, although non-European companies and organisations are often invited to participate as observers at meetings of ETSI's Technical Assembly. The criterion for membership is that all members must be "established" in the geographic territory covered by CEPT,[32] which means that members may be drawn from undertakings of non-E.U. origin with a presence in the E.U. through a local subsidiary.

The official recognition by the Commission in its Green Paper on Standardisation[33] that standards-setting should not be used as a means of erecting barriers to entry is particularly important in the field of telecommunications, and poses a danger with respect to which ETSI is acutely aware. A particular manifestation of the danger of erecting barriers to entry is presented by the existence of intellectual property rights in connection with ETSI standards. The two main types of conflict which may occur with regard to such rights concern patents which relate to the standards themselves and copyright held in respect of the actual texts embodying those standards.

16–16   On November 23, 1994, ETSI's General Assembly apparently resolved satisfactorily the dispute regarding its common policy regarding potential conflicts among its members, which was based on the waiving of copyright in standards and the compulsory licensing of patents for an equitable remuneration.[34] Under the new procedure, ETSI's Director will ask patent holders who own patents which are essential to standards formulated by ETSI to grant irrevocable licences to make, sell, lease, repair, use or operate patented equipment or to use the patented methods. In so far as a patent owner refuses to license its technology, ETSI will be obliged to stop work on the proposed standard if no alternative becomes available.[35] Standards approved by ETSI have the status of voluntary standards. It is then up to the national governments and the Commission to transform them into mandatory standards. In some cases, these standards are adopted on an

---

[31] The European Telecommunications Standards Institute.
[32] The European Conference of Postal & Telecommunications Administrations.
[33] Green Paper on the development of European standardisation [1991] O.J. C20/1.
[34] The need for a new policy stemmed from the complaint under Art. 85 and 86 of the E.C. Treaty brought by the Computer and Business Equipment Manufacturers Association (CBEMA) against ETSI's supposedly compulsory licensing programme; *CBEMA v. ETSI*, August 30/31, 1993, Asence Europe.
[35] Members of ETSI who refuse to grant licences may be required to provide written explanations to justify their refusal.

interim basis wherever it is felt that a provisional solution is necessary. The procedures for the adoption of such standards are, in general, transparent, with weighted national voting taking place subsequent to public hearings being conducted. Standards priorities for ETSI at the time of writing relate to the development of ISDN standards and for digital cellular networks.

Although initially formed so as to operate completely independently of CEN and CENELEC, cooperation and coordination with those bodies continues to be necessary, particularly with respect to those grey areas which overlap into the telecommunications sector and in order to maintain a unified standardisation policy. ETSI also liaises closely with a number of international organisations, including the standardisation sector of the ITU.

3. Provision of Services

Aside from ad hoc Commission investigations in individual cases to correct **16–17** anti-competitive abuses, it was not until the adoption in June 1990 of the Services Directive that the Commission systematically sought to open up the market for the provision of telecommunications services throughout the E.U. The Services Directive was adopted in order to coincide with the adoption of the Council's ONP Directive. As discussed above, the Court of Justice held that the Commission was justified in adopting the Services Directive on the basis of its powers contained in Art. 90(3) of the E.C. Treaty, subject to the annulment of those sections of the Directive which related solely to undertakings enjoying "special" rights (which the Directive had failed to define).

The focal point of the Services Directive is the requirement that Member States abolish all exclusive and special rights for the provision of all telecommunications services other than for *public switched voice telephony*, which is the only form of "reserved service" available to the TOs until 1998 and, in some other cases, until 2003 (see discussion below on "Future of Reserved Services", paragraph 16–19).[36] According to the terms of Article 1 of the Directive, voice telephony is defined as:

> ". . . the commercial provision for the public of the direct transport and switching of speech in real-time between switched network termination points . . ."

On the basis of this definition, the reservation to TOs of public voice telephony services should arguably be limited to services including *all* of the following elements:

- commercial provision (for profit, ar at least covering costs);
- for the public (that is, not to a Closed User Group);
- direct transport and switching of speech in real-time (which would

---

[36] With the exception of the U.K., Sweden and Finland, all of which have liberalised the provision of telecommunications services in advance of the Commission's timetable.

exclude any store and forward applications for technically advanced voice facilities); and
- between public-switched network termination points (that is, not a service which is accessed via a dedicated leased line).

16–18   The reservation of public switched voice telephony, on the basis of these elements, is relatively narrow.[36a] In practice, given that the public voice telephony reservation is an exception to the general rule that telecommunication services should be open to competition, it should be up to the TO to prove to the national regulator that a service falls within the reserved category if there is any doubt as to its status.[37] The Commission will in turn have the ultimate say on whether a service is reserved or unreserved, subject to the right of appeal to the Court of Justice.[37a]

The narrowness with which voice telephony services can be construed aside, the opening up of competition in light of the adoption of the Services Directive does not cover the full spectrum of telecommunications services. For example, there are certain types of services which are *not* within the scope of the Directive, such as television services and radio-broadcasting (expressly falling outside the definition of "telecommunications services"), and telex, mobile radiotelephony and paging (expressly excluded under Article 1(2)).[38] The liberalisation of these other types of telecommunications services has to occur through recourse to national and E.C. competition rules[38a] and/or other Community rules such as those relating to free movement.

Similar to the situation which prevails under the Terminal Equipment Directive, there is a clear demarcation drawn in the Services Directive between operating and regulatory functions. As from July 1, 1991, Article 7 provides that the granting of operating licences, the control of type approval and mandatory specifications, the allocation of frequencies and the surveillance of usage conditions has had to be carried out by a body independent of the national TOs. This division of powers is important, especially in light of the fact that the provision of non-reserved services by third parties may be subject to licensing procedures at Member State level. Such licensing must be effected on the basis of objective, non-discriminatory and transparent criteria.

Given the complementarity of the Services Directive and the ONP Directive, there is also express provision in the Services Directive for the definition of conditions of access to the network for providers of services on leased

---

[36a] But nevertheless accounts for over 80 per cent of the turnover of TOs.
[37] Green Paper on Telecommunications 1987, para. 4.1.2 expressly states that the exclusive provision of service must be narrowly construed.
[37a] In this regard, refer to the Communication by the Commission to the European Parliament and the Council on the Status and Implementation of Directive 90/388 on Competition in the Market for Telecommunications Services, COM (95) 113 final.
[38] In addition, Member States were permitted until the end of 1992 to prohibit economic operators from providing "simple resale" services. The concept of what constitutes a simple resale (essentially the use of leased lines connecting at each end with the public switched network) is often problematic. There are varying interpretations by Member States as to the extent to which the addition of a value added service takes that service outside the notion of "simple resale".
[38a] For example the provision of GSM licences on a duopoly basis.

lines, the provision of packet and circuit-switched data services to the general public, and the publishing of technical interfaces necessary for use in the public network.

Future of Reserved Services

The Commission prepared a comprehensive report on the implementation   **16–19** of the Services Directive at Member State level.[38b] It is expected that the Commission will initiate a series of infringement proceedings against those Member States who are found to continue to maintain special or exclusive rights over non-reserved services or which have introduced licensing regimes which limit, hinder or distort competition.

The fact that voice telephony continues to be a "reserved service" is a reflection of the fact that it is the major source of income for national TOs, the financial viability of which would be problematic without revenues derived from such services, and would affect their ability to perform their services of "general economic interest" on the basis of Article 90(2). Nevertheless, as a result of the Commission's detailed review in 1992[39] on whether the reservation of the monopoly over public switched voice monopoly was justified, the 1993 Council Resolution[40] considered the liberalisation of all voice telephony services (whilst maintaining universal service) to be a major E.U. telecommunications policy goal. The principal legal message stemming from the Council Resolution was that monopoly is a disproportionate to guarantee universal service in the telecommunications sector. The review also confirmed that, following the political agreement of the Council of Ministers on June 16, 1993, the liberalisation of all public voice telephony services should be effected by January 1, 1988, subject to derogations for:

- Greece, Spain,[41] Portugal and Ireland up to the year 2003; and
- for Luxembourg, given the relatively small size of its network, up to the year 2000.

The full impact of such service liberalisation may be felt more strongly in light of the political commitment at the Council of Ministers meeting on November 17, 1994 to liberalise the provision of infrastructure within the same timeframe (see below).

This political commitment has been given further legislative impetus with the adoption of a series of proposed measures in the summer of 1995 which establish the legal foundations for a post-1998 liberalised telecommunications environment, and which increase significantly the scope of those services exposed to competition under the Services Directive, namely:

- the proposed Liberalisation of Cable Television Networks Directive (see para. 16–37);

[38b] Refer to COM (95) 113 final.
[39] "Communication on the 1992 Review of the situation in the telecommunications services sector", SEC(92) 1048 final, October 21, 1992.
[40] Council Resolution 93/C213/01; [1993] O.J. C213/1.
[41] Although Spain has expressed its political commitment to liberalise by January 1, 1998.

- the proposed Mobile and Personal Communications Directive (see para. 16–38);
- the proposed ONP Interconnection Directive (see para. 16–26); and
- the proposed "Full Competition" Directive (see para. 16–38).

4. INFRASTRUCTURE—THE ONP APPROACH

16–20   The legislative packages relating to equipment and services are not sufficient in and of themselves to create a truly single internal market for telecommunications. The potential would remain for TOs to abuse their privileged monopoly positions on their network facilities (infrastructure) and because competitors may require selected services from the holders of the infrastructure monopolies (*e.g.* itemised billing, calling-line identity, and so forth). The usual problems confronted by independent third parties in such circumstances arise with regard to the imposition of technical standards, and unfairness in tariffs.

The ONP Framework Directive

16–21   To cater for these problems, the Council adopted a Framework ONP Directive, the aim of which was to lay down minimum common principles regarding the general conditions for the provision of the network infrastructure (and services) by the TOs to users and competitive service providers, especially for the purposes of transfrontier service provision. The policy goal behind doing so was to provide the competitive environment necessary to promote free and fair competition among all operators and to stimulate the development of a wide range of non-reserved value added telecommunications services by third party operators, particularly at the pan-European level. In effect, the philosophy underlying ONP is the rendering of infrastructure providers as "common carriers" for certain purposes.[42]

Ultimately, this should help to establish integrated broadband European telecommunications services. Member States are not required to limit their exclusive and special rights on the provision of the network infrastructure, although the Framework ONP Directive does not treat the provision of infrastructure as a reserved activity; indeed, the future liberalisation of alternative infrastructure may result in a need to review the scope of the Directive in the future.

The pro-competitive goals of the Commission concerning telecommunications infrastructure are to be achieved by the three-pronged liberalisation package contained in the Framework ONP Directive, which provides for:

- the creation of uniform usage conditions for the infrastructure network;

---

[42] Although the notions of ONP and "essential facilities" are conceptually distinct, the latter notion is used in the Commission's Art. 86 administrative practice to satisfy similar policy goals.

- uniformity and transparency in the tariff principles used for the use of the network; and
- the creation of uniform technical interface standards, in so far as is possible.

At the heart of ONP is the principle that its harmonised conditions ought   **16–22** to be based on objective, transparent and fully publicised standards. As with other internal market programme initiatives, one of the ultimate aims of ONP is the establishment of a system of mutual recognition of licensing procedures, which would enable a private operator licensed in one Member State to provide services in another without the need to complete the equivalent procedures.

More specifically, as regards uniformity in usage and supply conditions, depending on the area of usage, there is, *inter alia*, provision for: maximum delivery periods; minimum standards for the quality of service; the establishment of minimum contractual periods; conditions for the resale of capacity; shared use; third party use; and conditions for interconnection with public and private networks. The guidelines adopted for common tariff principles provide that they should be cost-oriented, properly publicised, and applicable to all users on non-discriminatory terms.

With respect to technical interfaces in ONP, it is a general principle that existing interfaces should be adopted and, where their use is not suitable for new services, they should be developed pursuant to a standardisation programme which is already in existence or being developed by bodies such as ETSI.

Specific Applications of ONP

The implementation of ONP principles contained in the Framework Direct-   **16–23** ive is to be effected for specific telecommunications services through the introduction of a series of directives and recommendations in the following areas:

- leased lines;
- packet- and circuit-switched data services;
- Integrated Services Digital Network (ISDN);
- voice telephony service; and
- mobile services (as applicable).[43]

To date, three new legislative instruments have been adopted under the ONP framework.

First, priority has been given to the regulation of leased lines, with the

---

[43] In the field of mobile services, a study has been prepared for the Commission which contains recommendations for the adoption for an ONP Mobile Services Directive, broken down into four sections (technical standards, interconnection and access, usage and supply conditions, tariffing principles). The ONP Framework Directive also anticipated legislation with respect to new types of access to the network and access to the broad-band network (both of which are the subject of further study).

Council having adopted its ONP Leased Lines Directive in June 1992.[44] Leased lines are in fact telecommunications facilities which provide for transparent transmission capacity between network termination points, and which do not include the capability to effect on-demand switching from the public network itself. The liberalisation of leased lines is widely regarded as being critical to the development of value added services, with the Commission being aware that existing discrepancies in Member State regulation do not facilitate the growth of such services. New market entrants must have access to leased lines to be able to compete in the market for services provision.

16–24   The ONP Leased Lines Directive is expressed to apply on a mandatory basis to four types of leased lines, namely, ordinary quality and special quality voice bandwidth leased lines respectively, and 64kbit and 2Mbit nominal bit rate digital leased lines respectively. Moreover, the ONP conditions apply to national, intra-Community and international leased lines. ONP leased lines must be offered without discrimination to all potential users, adopting relevant international standards (where not available, to be developed by ETSI). In addition, so-called "one-stop shopping" ordering procedures will be developed for ONP leased lines. Users will be able to make enquiries for international lines at a single point of sale. Further, there should be publication of supply and usage conditions, which would include information on maximum provision times, the quality of service, conditions for the resale of capacity and so forth.

No usage restrictions can be applied for the provision of non-reserved services conducted on ONP leased lines and, for such services, an ONP leased line can be connected to other leased lines or to public switched networks in any configuration. Finally, tariff charges for ONP leased lines are to be cost-oriented, and may contain fees representing the rental charges for the relevant period, an initial connection charge and an access charge (the latter two figures representing the average costs incurred in completing those two functions).

Second, separate Council Recommendations have been adopted in 1992 regarding the harmonisation of ISDN access arrangements[45] and the harmonised provision of packet-switched data services.[46] Unfortunately, Recommendations are not binding on Member States, which means that mandatory requirements in relation to ISDN or packet-switched data are not available at the time of writing.[46a]

Third, a proposed ONP Voice Telephony Directive,[47] in respect of which

---

[44] Directive 92/44; [1992] O.J. L165/27. The Directive is the subject of an Art. 177 reference from the English Divisional Court, in the case of *R. v. The Secretary of State for Trade & Industry, ex parte: British Telecommunications plc* [1994] O.J. C380/3. The Commission is also bringing a series of actions against various Member States which the Commission claims have failed to implement properly their obligations under Directive 92/44, namely: Spain ([1994] O.J. C304/13); Ireland ([1994] O.J. C275/17); and Luxembourg ([1994] O.J. C254/10).

[45] Recommendation 92/383; [1992] O.J. L200/10.

[46] Recommendation 92/382; [1992] O.J. L200/1.

[46a] The use of a mandatory approach was rejected by the Commission for data (because of the competitive market at the time) and for ISDN (because of the limited existence of a market at the time).

[47] COM(92) 247 final, August 27, 1992; Common Position reconfirmed by Council on June 20, 1994.

a common position was reached on June 30, 1993, was a victim of the so-called new "Conciliation Procedure" introduced into Article 189 of the E.C. Treaty by the Treaty on European Union, resulting in the European Parliament rejecting the proposed Directive in July. However, because the Directive was rejected primarily on the basis of procedural disagreements between the Council and the Parliament rather than on matters of substance, the Commission has been able to reintroduce the proposed Directive in amended form onto the legislative agenda in the form of a joint Council and European Parliament Directive.[48]

The principal features of the proposed ONP Voice Telephony Directive are: **16–25**

- It seeks to define the parameters of guaranteed universal access to voice telephony for users.[48a]
- It establishes a framework for interconnection agreements for voice telephony, the main elements of which are: users and competitive service providers would have a right to equitable and non-discriminatory access to networks, with TOs having to meet legitimate requests for interconnection from other TOs; charges will be non-discriminatory; the use of European or international standards would be encouraged; and numbering plans will be controlled by the national regulatory authority.
- It sets out a number of principles for the charging of access charges for interconnection (in particular, in order to fairly share the burden of universal service obligations), specifying that they must be cost oriented, non-discriminatory, fully justified, based on regulatory obligations imposed by the relevant Member State and approved by the national regulatory authority.
- Miscellaneous matters are covered such as: publication of/access to information; targets for supply time and service quality; conditions for the termination of offerings; user contracts; access and usage conditions and essential requirements.

The Future of ONP

Basic ONP principles are likely to evolve to address three main areas: universal service, interconnection agreements and dispute resolution mechanisms between users and operators. Nevertheless, albeit necessary to stimulate new competition, ONP principles are becoming increasingly viewed as being excessively onerous for smaller competitors to comply with in markets characterised by a powerful incumbent TO (especially with respect to   **16–26**

---

[48] A draft proposal for a European Parliament and Council Directive on the Application of Open Network Provision (ONP) to Voice Telephony was adopted by the Commission on February 1, 1995. See [1995] O.J. C122/4 (see Common Position adopted on June 13, 1995).

[48a] The scope of voice telephony defines the concept of the Universal Service Obligation (USO) in Europe. Only services falling within the scope of voice telephony can be compensated out of a USO fund, if loss can be shown.

accounting and reporting obligations). Unfortunately, the definition of "TOs" to which present ONP obligations apply refers to the special or exclusive rights enjoyed by those organisations; as explained above, in the absence of transparent licensing conditions, any number of small licensees may satisfy this definition at present.

Ongoing liberalisation should bring with it the introduction of new series of infrastructure providers and service providers, which should in turn mean that it may be much more appropriate to establish other criteria such as market power or market dominance (*e.g.* 25 per cent of the market, control of a bottleneck), as the basis of the definition of TOs to which ONP obligations will apply.[48b] Indeed, it may be the case that stronger obligations will be imposed if significant market power is found to exist (*e.g.* cost accounting measures), with a general framework possibly applying to all providers of public telecommunications infrastructure or services (*e.g.* non-discrimination principles, use of dispute resolution mechanisms and so forth). At the time of writing, the Commission has shown definite signs that its thinking is already moving in this direction, and the Commission's Competition Directorate has commissioned a number of studies in order to formulate a series of working models for the appplication of ONP interconnection principles in a multi-vendor environment.

On July 19, 1995, the Commission adopted in draft form a Directive based on Article 100A which seeks to establish a harmonised framework for interconnection in telecommunications in the context of ONP (the ONP Interconnection draft Directive). In so doing, the proposal is also designed to ensure the provision of Universal Service and the interoperability of telecommunications services throughout the E.U. The ONP Interconnection draft Directive is directed towards removing current prohibitions on cross-border interconnection by setting out the fundamental rights and obligations of the economic actors in this area, which are to be given effect by the telecommunications National Regulatory Authorities (NRAs) of the respective Member States.

The most important features put forward under the proposed new regulatory framework for interconnection are:

- application of the principles of transparency, objectivity, and non-discrimination to guarantee equitable interconnection agreements, in particular between new entrants and the powerful incumbent TOs;[48c]
- priority given to commercial negotiations between interconnecting parties, while reserving some conditions to be set *a priori* by NRAs; and
- the prescription of clear responsibilities for NRAs, in accordance with

---

[48b] See on the general revision of the ONP legislation: the Commission's Communication to the Council and the European Parliament on the Present Status and Future Approach for Open Access to Telecommunications Networks and Services, COM(94) 513.

[48c] As regards the competition law ramifications of interconnection agreements, see Coudert Brothers Study, Final Report to the European Commission (June 1995) on "Competitive Aspects of Interconnection Agreements in the Telecommunications Sector."

the principle of subsidairity, including effective mechanisms for dispute resolution at the national and European level.[48d]

5. MOBILE COMMUNICATIONS

In April 1994, the Commission adopted a Green Paper on a Common **16–27** Approach to Mobile and Personal Communications in the European Union.[49] The adoption of the Green Paper was necessitated by the rapid expansion of mobile communications into the personal communications market and by the continuing high penetration of mobile communications.[50] These technological and commercial reasons prompted the need on the part of the Commission to develop a coherent policy framework for the sector.

The Green Paper has identified basic principles and action lines for further discussion, the aims of which were to:

- permit the development of an E.U.-wide market for mobile services, equipment and terminals;
- identify common principles for the provision of mobile infrastructure, the development of mobile networks and services, and the supply and operation of mobile terminals;
- promote the evolution of the mobile communications market into mass personal communications services, with particular emphasis on pan-European services; and
- facilitate and promote trans-European networks and services to ensure that the sector develops in a manner consistent with the public interest.

Five major policy changes to the present regulatory environment were **16–28** highlighted in the Green Paper as necessary to remove barriers to further development, namely:

(i) The abolition of remaining exclusive and special rights in the sector, subject to appropriate licensing conditions.

(ii) The removal of all restrictions on the provision of mobile services, both by independent service providers and by mobile network operators.

[48d] Other issues addressed in the ONP Interconnection draft Directive include:

- principles for interconnection charges and cost accounting systems;
- accounting separation and financial accounts;
- essential requirements (security of network operations, maintenance of network integrity, interoperability of services, protection of data);
- numbering (provision of numbers and numbering ranges for all public telecommunications services);
- technical standards;
- publication of and access to information.

[49] COM(94) 145 final.

[50] Calculated in the Green Paper itself as resulting in total user numbers ultimately exceeding 200 million in the E.U. (compared to a current total subscriber base of 135 million for traditional fixed telephony).

(iii) Full freedom for mobile network operators to develop their own networks, including the right to provide these themselves or use third party infrastructures to operate their mobile network, and the removal of restrictions on sharing infrastructure.

(iv) The provision of unrestricted combined services via the fixed and mobile networks, within the overall time schedule set by the Council Resolution of July 22, 1993 for the full liberalisation of public voice services via the fixed network.

 (v) The facilitation of pan-European operational and service provision. This should include the further development of mutual recognition of type approval of mobile terminal equipment and the coordination of licensing and award procedures in order to facilitate the development of trans-European networks.

The broad consultation process launched by the adoption of the Green Paper culminated in the adoption of a Commission Communication on November 23, 1994, which contains a comprehensive action programme including: (i) the full application of the competition rules; (ii) the development of a Code of Conduct for Service Providers; (iii) full access of service providers to the market; (iv) the promotion and availability of frequencies and numbers; and (v) the promotion of targeted programmes to support the market entry of emerging mobile technologies.[51]

6. UNIVERSAL SERVICE

16–29   To date, the Commission has pointed out that ONP measures serve as the basis for the definition of universal service and as an appropriate framework for interconnection.[52] The regulatory trade-off which the Commission will strive to achieve is the maintenance of universal service in a competitive environment which is not achieved through the imposition of high access/interconnection charges by the incumbent which could deter entry.

On November 15, 1993, the Commission adopted a Communication to the Council and the European Parliament on developing the concept of universal service in the new competitive environment.[53] According to the Commission, the benefits of liberalisation should, through lower costs and the introduction of new services, help the provision of universal service, particularly for services in the peripheral regions of the E.U., where improvements in the service will strenghten economic and social development. At the same time, regulatory safeguards will ensure that universal service is provided to all customers with special needs or in remote areas. For this reason, the consolidation of the current regulatory environment will require that the principle of public service address, *inter alia*, the following: (1) universality (access to all at an affordable price); (2) equality

---

[51] COM (94) 942.
[52] Council Resolution [1993] O.J. C213/1.
[53] COM(93) 345 final.

(access independent of geographic location); and (3) continuity (continuous provision at a definite quality).

A major aim of the Communication is to illustrate how the cost of uni-   **16–30** versal service can be covered in future. The cost should be met from a combination of:

- greater direct contributions from subscriber revenue and from rebalanced tariffs;
- access charges paid by new operators and service providers who are not subject to universal service obligations; and
- where appropriate, funding in the peripheral regions from the Community Support Framework.[53a]

The Council Resolution invited the Commission to draw up common access charge principles in consultation with National Regulatory Authorities, and to submit a report to the European Parliament and the Council by January 1, 1996. To date, the Commission's approach has been to focus on the need to provide universal service, rather than on the identity of those parties who should be obliged to provide such services. In doing so, the Commission has set targets which the Member States are to achieve, but it is up to the latter to determine the extent to which and the identity of operators from whom such contributions are to be exacted. The most recent elaboration of the Commission's thinking in this regard are reflected in Part II of the Infrastructure Green Paper (discussed below, paragraphs 16–32 to 16–33).

7. INFRASTRUCTURE LIBERALISATION

The 1993 Council Resolution prescribed that the Commission had to pub-   **16–31** lish by January 1, 1995 a Green Paper on the future policy for telecommunications infrastructure and cable television networks.

Infrastructure Green Paper

By the end of summer 1994, the Commission had sent to Member States   **16–32** a draft communication entitled "Lifting Constraints on Alternative Infrastructure for the Provision of Competitive Telecommunications Services".[54] The concept of alternative infrastructure embraces cable television operators, railways, water and power utility companies. All of these infrastructure providers are in a position to offer broadband services to selected sites fairly quickly over networks which are at present used for television and for internal communications, as the case may be.

[53a] See discussion of Infrastructure Green Paper below, where the role of access charges is significantly downplayed, with greater emphasis being placed on the provision of a Universal Service Fund.
[54] Draft Communication to the Council and the European Parliament, Brussels, July 22, 1994.

In its Communication the Commission concluded that service providers should have a "free choice of underlying infrastructure" with respect to those telecommunications services which have already been liberalised at E.U. level. The services in question are those already liberalised pursuant to the Services Directive, namely, private voice (Closed User Groups), data and value added services. Providers of public switched voice telephony services would also be able to exercise such a choice once those services were liberalised from January 1, 1998. The measures outlined in the communication above were in turn transformed by the first section of the Commission Green Paper on Infrastructure, adopted in October 1994.[55] Part I of the Green Paper follows the Commission's "Action Plan Towards an Information Society in Europe" and the report of the so-called Bangemann Group on "Europe and the Global Information Society" (see below), both of which identified infrastructure liberalisation as one of the main initiatives to be taken in order to open the way for the development of the network and applications on which the information society relies (remote accessing of databases, telebanking, distance learning, and so forth).

The Commission cites the policy imperative behind its liberalisation of infrastructure proposals as being the fact that, with the exception of the U.K., high-bandwidth services remain constrained by the limited availability (and expense) of necessary transmission infrastructure in many areas, especially for corporate communications. Accordingly, permitting providers of alternative infrastructure may increase the availability of high speed digital leased lines in the E.U. in the short term and satisfy the demand for high capacity infrastructure.[56]

16–33   The recommendations contained in Part I of the Green Paper are based on the guiding principle that, where the provision of telecommunications services is open to competition, there should be a free choice as to the underlying infrastructure over which such services are provided, subject to the establishment of the appropriate safeguards. On the basis of this principle, Part I of the Green Paper puts forward the position that immediate action is necessary and desirable, within the timetable for the full liberalisation of voice telephony (January 1, 1998) and subject to the necessary safeguards, to remove immediately restrictions on the use of own or third party infrastructure in the following areas:

- for the delivery of satellite communications services;
- for the provision of all terrestrial telecommunication services already liberalised (including the use of cable television infrastructure for this purpose);[57] and

---

[55] Communication to the Council and the European Parliament, Green Paper on the Liberalisation of Telecommunications Infrastructure and Cable Television Networks, Pt. I (Principle and Timetable), Brussels, October 25, 1994, com (94) 440.

[56] Studies undertaken by the Commission show that opening cable TV networks and alternative infrastructure networks would help to overcome the problems of high pricing levels and lack of suitable capacity, which result at least in part from current exclusive provision of infrastructure in most Member States.

[57] This concerns voice and data services for corporate networks and Closed User Groups, as well as all other telecommunications services, other than the provision of voice telephony services to the general public.

- to provide links, including microwave links, within the mobile network for the provision of mobile communications services.

Despite its commitment to the liberalisation of infrastructure, the Commission recognises that its approach must be conditioned on the appropriate conditions having been established so as to ensure that universal service is not undermined in a competitive environment. This will require that those providing Universal Source Obligations receive some form of an appropriate access charge or other contribution for allowing interconnection. Similarly, any approach to infrastructure needs to be consistent in regulatory terms with the Commission's approach in the Mobile Green Paper. At the same time, the Commission should be sensitive to the potential anti-competitive practices which alternative infrastructure monopolists may engage in; appropriate behavioural safeguards should therefore be introduced to ensure that cross-subsidisation, the bundling of services and the the provision of services on a discriminatory basis cannot occur.

On January 25, 1995, the Commission adopted Part II of its Infrastructure Green Paper,[58] which puts forward three major policy positions:

(i) Interconnection/interoperability

Emphasis is placed on the need to clarify the application of competition rules to interconnection agreements in order to remove current restrictions on the interconnection of all types of communications infrastructure, which includes all forms of public and private networks (terrestrial, mobile, cable TV, satellite). ONP principles should also be extended to the public telecommunications network, but only insofar as is permitted under the ONP Framework Directive. Basic ONP rules will be adapted to develop common principles for a multi-infrastructure provider/multi-service provider environment, and a proposed directive on ONP relating to interconnection will set forth the relevant principles which will govern interconnection charges for interconnection to the networks and services of TOs.

16–34

(ii) Licensing

A number of principles will need to be respected in the licensing process by Member States, who will be responsible for the implementation of licensing principles which are to be open, non-discriminatory and transparent. Restrictions on the number of licences will need to be justified in terms of these conditions derived from the Sources Directive, namely, trade regulations, essential requirements, such as public service obligations, frequency spectrum issues (on the grounds that numbers need be unified in advance), environmental and town planning considerations. The supply of telecommunication services distinct from the ownership of infrastructure should not be the subject of any individual authorisation procedures save for

16–35

---

[58] A Common Approach to the Provision of Infrastructure for Telecommunications in the European Union, Pt. II Brussels, January 25, 1995 com (94) 682.

licensing-by-category requirements, general authorisations or requirements for service providers to make statements concerning services to the regulatory bodies in each Member State where they intend to offer services. Providers of voice telephony services to the public will be obliged to contribute towards the funding of universal service, although the manner of their contributions has yet to be decided. Moreover, there will occur a harmonisation of licensing procedures at European level, with the introduction of a directive by the end of autumn 1995 which establishes a regulatory framework for licensing.

(iii) Universal service

16–36   A clear policy position is taken to the effect that the scope of universal service will not be extended in the short to medium term beyond voice telephony services and leased lines. In the longer term, the scope of universal service obligations may be extended in accordance with expanding consumer needs and technological progress (*e.g.* multimedia services). The cost of fulfilling universal service obligations should be funded on the basis of the net cost of supplying non-profitable public telephones, emergency services and other socially necessary services.

Finally, the methods by which universal service obligations are to be financed is a matter which is left to the Member States, acting in accordance with one of two policy options, namely: (1) an access charge regime; or (2) an independently managed universal service fund.[59]

A Communication on the follow-up to the Green Paper was sent to the Council and the European Parliament in May 1995.[59a] The Communication reports on the results of the consultation round following the issuing of the Green Paper and contains a calendar for the legislative proposals and action to be taken within the 1998 liberalisation programme. Most of the package of the legislative proposals by the Commission to the other Community institutions is to be submitted by January 1, 1996. Further to that Communication, the Commission on July 19, 1995 agreed on two fundamental legislative measures which represent the core of the Commission's post-1998 regulatory environment, namely a draft Article 90 Directive establishing "full competition' by January 1, 1998 (see below) and a proposal for a Directive under Article 100A which establishes a harmonised framework for interconnection in telecommunications in the context of ONP (see discussion at para. 16–36).

Liberalisation of Cable-TV Networks, Alternative Infrastructure, Mobile Infrastructure and Services

16–37   As an additional element of its policy to liberalise infrastructure, the Commission took the step on December 21, 1994 of adopting a proposal for

---

[59]   The Commission has a distinct preference for the adoption of a Universal Service Fund regime, into which operators would make a contribution expressed as a percentage of their turnover; this approach was supported by other Community propositions ending in consultation with the Commission. This approach constitutes a distinct shift in policy from the Commissions position outlined in November 1993.
[59a]   COM(95)158.

a draft Commission Directive on the Liberalisation of Cable Television Networks (see above), the policy aim of which is to allow multimedia telecommunication services to be transmitted on cable networks throughout the E.U. by *January 1, 1996.* The Commission will only adopt the draft Directive formally after having presented it to the Member States and to the Parliament, and after having consulted with other interested parties during the course of 1995.

The focal point of the draft Directive is its amendment of Article 4 of the existing Services Directive (see above) by obliging Member States to "withdraw all restrictions for the supply of transmission capacity on cable TV networks and allow operators to use the cable operators to deliver their services" and to ensure that restrictions on the direct interconnection of cable TV networks is abolished. The draft Directive would also introduce competition safeguards to prevent operators from using a dominant position in one market to impose predatory prices in another, and would require Member States to adopt measures allowing the monitoring of cross-subsidies between reserved and liberalised activities when a single operator provides both. However, these competitive safeguards are introduced primarily as "soft law", being contained in the recitals to the draft Directive rather than in its operative provisions. Since the draft Directive only concerns the provision of non-reserved services, it would not affect the right of Member States to maintain monopolies on the provision of voice telephony until 1998.

The Commission has emphasised that the effect of adopting the draft **16–38** Directive would be that liberalised access to cable infrastructure would reduce the cost, increase capacity, and encourage the use of new technology, thereby facilitating the greater use of multimedia services.[60] The Commission estimates that cable operators would be able to offer such multimedia services more cheaply than telecommunications operators, financed at least in part from the use of excess cable capacity for mobile communications. Moreover, cable-TV networks are particularly well suited for the carriage of TV signals because the broadband co-axial cables used have sufficient capacity to allow enhanced quality signals without the need for significant upgrading.[61]

The draft Directive is scheduled to be implemented into national law by January 1, 1996.

By the summer of 1995, the Commission had gone even further in its liberalisation programme with the adoption of a series of proposed direc-

---

[60] Commission press release IP/94/1262 of December 21, 1994. According to the Commission, many existing national regulations restrict the use of cable-TV networks by prohibiting operators from offering interactive or multimedia services. Many of these new services, such as home shopping, home banking and specialised on-line databases, require the accurate transmission of more digital information than most traditional telecommunications networks are able to carry.

[61] According to the studies submitted to the Commission, interactive services cannot be carried efficiently by most local national telecommunications networks because the most common connection to households used by these operators is twisted copper-pair cabling, able to carry data of up to 257 kbits/second reliably. While this is adequate for the transmission of voice telephony, requiring 64 kbits/second, the transmission of even standard-quality moving images demands cables which are able to carry at least 2,000 kbits/second. The broadband co-axial cables used by cable operators are more suitable for multimedia services because they can commonly carry up to 500,000 kbits/second reliably.

tives under the aegis of Article 90 which broaden fundamentally the scope of the Services Directive, namely:

- The Commission proposed on June 21, 1995 the adoption in draft form of a Directive on mobile and personal communications which extends expressly the terms of the Services Directive to such communications by removing all special and exclusive rights over mobile services and the removal of restrictions on the self-provision or the use of third-party infrastructure for mobile networks. The draft Directive also mandates the grant of DCS 1800 licences in every Member State in order to facilitate pan-European roaming.
- Most importantly, the Commission adopted in draft form on July 19, 1995 a "Full Competition" Directive which implements the political agreement reached by Member States to liberalise all telecommunications services (including public switched voice telephony) and telecommunications infrastructure by January 1, 1998, with transition periods for certain Member States. In addition, the draft Directive calls on Member States to take all necessary steps before 1998 in order to ensure that their markets are fully competitive by the agreed deadline. As a necessary extension of the draft Directive on the Liberalisation of Cable Television Networks, the draft Directive specifies that restrictions on the use of alternative infrastructure (such as electricity, gas and water) should be removed by January 1, 1996 for those services which have been already liberalised. It also specifies that licensing conditions and interconnection rules should be set down *by 1997*. The draft Directive lays out the general features and principles to be used for interconnection in the pro-competitive environment of post-1998, which is intended to complement the provision of the draft ONP Interconnection Directive (see above at para. 16–26). Finally, the draft Directive provides legislative foundation to the commitment that Universal Service must be safeguarded, but requires that this should not unnecessarily distort competition. Thus, it is proposed that fair schemes be established for the sharing of the net cost of Universal Service Obligations between the incumbent operator and competing public operators, but it also obliges the Member States to communicate such schemes to the Commission in order that they be vetted under E.C. competition rules.

8. TRANS-EUROPEAN NETWORKS

16–39   Under the general title of "trans-European networks", Article 129(b) of the E.C. Treaty, introduced by the Treaty of European Union, provides that:

"Within the framework of a system of open and competitive markets action by the Community shall aim at promoting the interconnection and inter-operability of national networks as well as access to such networks."

Article 129(c) goes on to provide that:

"The Community shall establish a series of guidelines covering the objectives, priorities and broad lines of measures envisaged in the sphere of trans-European networks. These guidelines shall identify projects of common interest. The Community's activities shall take into account the potential economic viability of the projects."

Clearly, the policy underlying trans-European networks has conflicting strains. On the one hand, there is a recognition that liberalisation and harmonisation programmes are necessary to break down national barriers in order to develop and establish a framework for genuine pan-European service provision. On the other hand, the policy seems to suggest that the Commission envisages that groups of operators may need to enter into cooperative joint ventures in order to create pan-European networks. Where the operators in question are "national champions", the pursuit of the latter policy must be interpreted carefully in light of the competition rules. It may often be difficult to draw the line between cooperation which unifies the E.U. and an agreement which results in the *de facto* cartelisation of the E.U. telecommunications market. Not surprisingly, it has been very difficult to achieve political consensus among the Member States as to the precise scope of the phrase "trans-European networks".

9. BANGEMANN "INFORMATION SOCIETY" HIGH LEVEL GROUP

Since February 1994, an "Information Society" High Level Group under the chairmanship of Commissioner Martin Bangemann, independent of the Commission itself, has been in operation. The task of the High Level Group is to point the way to the new opportunities presented by the information society, especially in the light of the White Paper on Growth, Competitiveness and Employment.[62] This will entail: **16–40**

- the further development of European networks and harmonisation of technical norms;
- removing the final obstacles to economic network utilisation;
- identifying new applications opportunities and, where necessary, providing targetted aid;
- adopting the framework of competition for the new realities of multimedia; and
- supporting social consensus on the use of modern information and communication technologies.

The realities of the convergence of telecommunications and media tech-

[62] COM(93) 700.

nologies, coupled with the adaptability of computers, means that the High Level Group will play an increasingly important role in the creation of a new regulatory structure for the future information society. This is especially so with regard to clarifying the role and function of trans-European networks in an information society. The release by the High Level Group in May 1994 of the report entitled "Europe and the Global Information Society" has served as an important point of reference for the Commission's policies on the liberalisation of infrastructure.[63] Indeed, the Group's calls for the rapid realisation of liberalisation has served as the political motor behind the Commission's initiatives relating to infrastructure. One of the side-effects of the Group's recommendations is the stimulation of the debate regarding the appointment of a "Euro-regulator" for telecommunications matters.

10. THE AUDIOVISUAL DIMENSION/BROADCASTING AND SATELLITES

16–41   The regulation of broadcasting in the E.U. embraces both the regulation of technical and transmission issues on the one hand, and the regulation of issues relating to content, in particular copyright, on the other. The initiatives in these sectors are numerous and diverse, reflecting in large measure some of the particular industrial policy concerns of the E.U.

Green Paper on Satellite Communications

16–42   In November 1990, the Commission adopted its Green Paper on Satellite Communications.[64] In its Green Paper, the Commission recognises that the current regulation of earth and space segment satellite communications in the Member States still reflects, on the whole, the situation throughout the 1960s–70s, where the only technically and economically feasible application of satellite communications was their use as an additional transmission path to carry international or long distance traffic for TOs. The rapid development of satellite broadcasting in the late 1980s demonstrates, however, that this limited role for satellite communications is no longer appropriate.

   In a nutshell, the Commission's policy and aims in this field are as follows:

   • A liberalisation of the earth segment, including the abolition of all exclusive or special rights in this area, whether relating to:

      (i) receive-only terminals (subject to type approval procedures when connected to the public switched network); and
      (ii) transmit/receive terminals, subject to licensing approval and licensing procedures where justified.

---

[63] Recommendations to the European Council, Brussels, May 26, 1994.
[64] COM(90) 490 final.

- Unrestricted access to space segment capacity (subject to licensing procedures to safeguard exclusive or special rights) on the basis of equitable, non-discriminatory and cost-oriented criteria. It is intended that current coordination procedures on economic harm by other providers of space segment capacity contained in the Conventions underlying INTELSAT, INMARSAT and EUTELSAT will be reviewed to avoid discrimination between commercial operators.
- Full commercial freedom throughout the E.U. for space segment providers, (e.g. EUTELSAT), including direct marketing of satellite capacity to service providers.
- To ensure interoperability throughout Europe, the introduction of harmonisation measures for the purpose of developing a uniform transmission standard.

Satellite Communications Directive[65]

As discussed above, the adoption by the Commission on October 13, 1994,   **16–43** of the Satellite Communications Directive was designed to amend the Terminal Equipment Directive and the Services Directive respectively as a means of liberalising the satellite communications equipment and services sectors, with the effect that private operators will be able to offer their products directly to consumers. In addition, the amendments serve to redefine the concept of "special rights" used in those Directives.

The central aim of the Satellite Communications is to abolish, wherever applicable, any special rights in this area which restrict progress by allowing private operators throughout the E.U. to offer satellite-based services directly in competition with TOs. However, the Commission is fully aware that it is necessary to avoid, for example, harmful interference between satellite telecommunications systems and other space-based or terrestrial services, and therefore the Directive allows Member States to maintain certain authorisation procedures in these areas (e.g. essential requirements).

As a result of these amendments, the Terminal Equipment Directive has been amended so as to provide that:

- "terminal equipment" also means satellite earth station equipment; and
- "special rights" are rights that are granted by a Member State to a limited number of undertakings concerning importation, marketing, connection, bringing into service and maintenance of telecommunications equipment.

The amendments to the Services Directive specify that the Member States   **16–44** have to withdraw all those measures which grant:

- exclusive rights for the supply of telecommunications services other than voice telephony for the public (which continues to be a reserved

---

[65] Directive 94/46; [1994] O.J. L268/15.

service until 1998),[66] and the provision of direct television broadcasting links; and

- special rights which limit to two or more the number of undertakings authorized to supply such telecommunications services, otherwise than in accordance with objective, proportional and non-discriminatory criteria; or
- special rights which designate, otherwise than according to such criteria, several competing undertakings to provide such telecommunications services.

Member States have to communicate the criteria on which authorisations are granted for the operation of transmitting earth stations.

Some Member States had already liberalised to a significant extent the activities covered by the Directive and had thus anticipated its adoption.[67]

The Satellite Communications Directive entered into force on November 8, 1994.

Satellite Earth Station Equipment

16–45   An open market for earth station equipment was created as a result of the adoption in October 1993 of Directive 93/97 on the approximation of legislation relating to the equipment used in ground stations for satellite communications.[68] Member States had until January 1, 1995 to implement fully the terms of the Directive, which not only provides for the free movement of such products, but also for harmonisation procedures for certification, testing, and so forth.

Mutual Recognition of Satellite Licences

16–46   In January 1994, the Commission adopted a proposal for a Directive concerning the mutual recognition of national authorisations for satellite telecommunications services.[69]

The proposed Directive aims to achieve the mutual recognition of national authorisations for satellite services on the basis of harmonised conditions, which is established in a procedure set out in the Directive. A national authorisation granted in accordance with such harmonised conditions will automatically allow the provision of the service concerned in all Member States. In order to facilitate their provision during a transition period before harmonised conditions are adopted, the Directive provides for an interim procedure based on regulatory "one-stop shopping". This will provide applicants intending to operate a service in more than one Member State the possibility of obtaining the necessary national licences at a single location and with a simplified procedure.

[66] Such as the satellite links used by the TOs for calls to other continents.
[67] France, Germany, the U.K., the Netherlands.
[68] [1993] O.J. L290/1.
[69] [1994] O.J. C36/2.

Among the satellite services which are expected to profit from the proposed framework are VSAT networks, satellite-based mobile data communications and fleet management systems, SNG (satellite news gathering) units and satellite-based personal communications services and networks. The proposal is a key element of the E.C.'s satellite communications policy, as set out in the Satellite Green Paper of 1990. The Commissions current approach to licensing in the satellite sector is, however, likely to be subsumed into a more global licensing approach which flows from the conclusions drawn in the Infrastructure Green Paper, and likely to be announced by the end of autumn 1995.

Access to Space Segment Capacity

On December 22, 1994, the Council adopted a Resolution on further devel- **16–46a** opment of the Community's satellite communications policy, especially with regard to the provision of, and access to, space segment capacity.[70] The Resolution emphasises that in an increasingly competitive environment, the availability of suitable, transparent and non-discriminatory, access arrangements to space segment resources is essential and that the effective management by the Member States of orbital and related frequency resources is closely connected with these arrangements.

The Resolution identifies as basic goals for the further development of the satellite communications policy:

- non-discriminatory access, for all providers and users of satellite services throughout the community, to space segment capacity;
- urgent adjustment of the intergovernmental satellite organisations such as INTELSAT, INMARSAT and in particular EUTELSAT in the light of the Community regulatory framework and the market requirements in accordance with the Treaty obligations;
- comparable and effective access to third-country markets, in parallel with the Community market liberalisation; and
- effective management of orbit and frequency resources within the framework of the ITU.

Standards Directive[71]

A Standards Directive on satellite television transmissions was originally **16–47** introduced on July 15, 1991 (amended proposal of December 6, 1991) for the purpose of filling the vacuum left by the expiry of Council Directive 86/529 at the end of 1991, which had imposed the obligation to use the MAC/Packet system on broadcasts transmitted from high powered broadcasting satellites but not low or medium powered telecommunications satellites, even if the latter were used for broadcasting.

[70] [1994] O.J. C379/5.
[71] Directive 93/38; [1992] O.J. L137/17.

The original proposal to succeed this Directive was to impose the D2 MAC transmission standard on all satellite broadcasts, whether transmitting via a broadcasting or telecommunications satellites, in order to develop HDTV using the European HD-MAC standard and using D2 MAC as an intermediate step. Aside from E.U. equipment manufacturers, this proposal prompted fierce industry opposition from all quarters. According to the compromise Directive adopted in 1992, only new broadcasters commencing services after the end of 1994 are obliged to use the D2 MAC standard, with existing broadcasters being entitled to simulcast in D2 MAC.

At the time of writing, the repeal of the Directive is under review, given that it expressly excludes digital technology, which is becoming increasingly important.

Digital Television

16–48 The Commission has agreed on two papers[72] relating to digital television services. The Commission takes the view that digital video will not only transform the internal structure of the broadcasting business, but will also accelerate the introduction of information highways and multimedia convergence between different businesses like broadcasting and telecommunications.

The Commission has concluded that the E.U. must act in order to:

- intensify and co-ordinate research and development;
- encourage standardisation;
- ensure free competition and consumer protection; and
- look for common elements with other global players, especially with the USA and Japan.

A proposal for a directive to be adopted by the Parliament and the Council on the use of standards for the transmission of television signals will be neutral on analogue or digital technology. Instead of imposing any specific transmission standard, it will concentrate on the 16:9 aspect ratio for wide-screen television services. In particular, cable operators must deliver 16:9 services to the consumer in the same 16:9 format. New TV receivers with screens over 42 cm must have a standardised interface socket so consumers can add the appropriate decoders as new systems become available. The initiative led the industry into formulating by September 1994 a voluntary Code of Conduct which permits satellite conditional access system providers the possibility of offering a closed "simulcrypt" system on the basis that other operators could utilise the same proprietary system.[72a]

A modified Proposal[72b] was put forward by the Commission in October 1994 and was agreed by the Council on December 22, 1994. Subsequently, however, the proposed Directive has been before European Parliament

---

[72] 1. COM(93) 577 final.
  2. Council Resolution [1994] O.J. C187/3.
[72a] In turn, the use of a common interface would allow the use of a single decoder with different access codes to the various pay-TV channels.
[72b] See [1994] O.J. C321/4.

which, during the course of its second reading, has tabled a number of amendments which in the view of many commentators may change fundamentally the scope and objectives of the proposed Directive. Under a proposed new Article 4c, the focus of the Directive may be shifting from a standards directive to one of open access for all programmers to all networks (in effect, a "must carry" obligation). To the extent that this amendment may come into effect in the final version of the Directive, it threatens to impose severe regulatory burdens on smaller content providers who may wish to compete on a pan-European level. In effect, an overly broad interpretation of such "must carry" obligations might transform cable TV operators into facilities providers which are deemed to be providing little more than a commodity product.

Television Without Frontiers Directive[73]

The so-called Television Without Frontiers Directive applies to all forms of television and was required to have been implemented into all Member State laws by October 3, 1991. Its purpose is to lay down minimum standards for television broadcasts in the E.U. whereby "it is necessary and sufficient that all broadcasts comply with the law of the Member State from which they emanate."   **16–49**

The principal provisions of the Directive—and those which have generated most controversy—relate to the freedom of reception, a quota for European works, a quota for independent productions, the regulation of advertising and sponsorship (both in terms of type and quantity), the protection of children and the creation of a right reply. A satellite broadcaster will be regulated in general in the country granting the frequency or satellite capacity, or in the country of uplink. Member States are, however, free to lay down more detailed or stricter rules in the areas covered by the Directive.

During the final quarter of 1994 and January 1995, fierce debate surrounded the issue of whether the terms of the Directive and the imposition of European "quotas" should, *inter alia*, be extended beyond broadcasting so as to cover other forms of communication (such as video-on-demand). At the time of writing, it would appear that strong opposition from a number of Member States has prevented the extension of the Directive into communications media other than television.[74]

Copyright for Cable and Satellite Directive[75]

The reality of cross-frontier broadcasting has meant that there has arisen the need to accompany audiovisual policy measures with effective protec-   **16–50**

---

[73] Directive 89/552 [1989] O.J. L298/23.
[74] Similarly, another important regulatory measure of high potential significance to the provision of multimedia services (in this case, teleshopping), an amended proposal for a Distance Selling Directive ([1993] O.J. C308/18), was not adopted as scheduled at the Council meeting of December 8, 1994. It is still hoped that the proposed Directive will be adopted in the first half of 1995.
[75] Directive 93/83 [1993] O.J. L248/15.

tion of copyright in all Member States. Accordingly, the Cable and Satellite Directive was adopted in 1993, which advocates the harmonisation of copyright in a number of key areas so that the holders of such rights may benefit fully from the European dimension of broadcasting. In summary form, these measures are:

- There should be no distinction in copyright terms between transmissions via broadcasting and telecommunication satellites; both should constitute a "communication to the public" by satellite of copyright works, with Member States being obliged to provide a right for authors to authorise or prohibit such communications to the public.
- The copyright law of *only one Member State* should apply to the broadcaster. It is specified that the relevant Member State should be that where a broadcasting organisation takes a single decision on the content and transmission of programme-carrying signals. This will not necessarily be the same place as the country of uplink, which poses a potential conflict not only with the provisions of the Television Without Frontiers Directive, but also with the proposal in the draft protocol to the Berne Convention; the latter advocates rights clearance not only in the originating country but also in the country or countries of reception in certain (albeit limited) circumstances.
- The primary liability for cable retransmission of copyright works will continue to fall on the cable operator, rather than on the originating broadcaster. The collective licensing (but not statutory or compulsory licensing) of cable retransmission rights is encouraged.

16–51   Disappointingly, the Directive fails to deal with the piracy of encrypted television signals on the basis that, according to the Commission's discussion paper which preceded the proposed Directive, the protection of encrypted signals "surpasses by far the framework of the measures proposed". In practice, however, issues involving encryption and signal piracy are entwined with copyright questions. A broadcaster's ability to encrypt and to limit the market of viewers can be central to its negotiations with rights holders. Encryption is also the principal mechanism by which a satellite broadcaster ensures that it controls reception and retransmission and that it is paid for its services. Sensitive to industry pressure on this point, the Commission is considering the issue, although not within the framework of the Directive. Interestingly, the Council of Europe has already issued its own (non-binding) proposal on this issue.[76]

11. PUBLIC PROCUREMENT

16–52   The fact that an estimated 70 to 90 per cent of national TO's procurement requirements are satisfied by national suppliers illustrates why one of the

---

[76] Recommendation to Member States of the Council on the Legal Protection of Encrypted Television Services, No. R (91) 14, adopted by the Committee of Ministers on September 27, 1991 at the 462nd meeting of the Ministers.

most important areas in which the E.C.'s public procurement liberalisation policies are to take effect is in the telecommunications sector.

Award Procedure

The original directives concerning the liberalisation of public procurement **16–53** specifically excluded from their scope the water, energy, transport and telecommunications sectors (the so-called excluded sectors). A separate slightly less liberal regime was adopted in September 1990 to remove this anomaly (the Utilities Directive[77] for these excluded sectors. The lesser degree of openness in this Utilities Directive testifies to the political sensitivity of regulation in these sectors and to the difficulties inherent in the Member States reaching a compromise thereon.

As with the original Supplies and Works Directive,[78] the Utilities Directive for the excluded sectors only applies to contracts exceeding certain monetary value thresholds. For telecommunications supply contracts, the threshold is ECU 600,000 (as compared to ECU 400,000 for the other excluded sectors). By way of contrast, the threshold for all works contracts, regardless of the sector, is ECU 5 million. As regards the assessment of the value of the contract, the extent to which use must be made of European standards and technical specifications, and the award criteria which must be utilised, the Utilities Directive contains rules similar to those contained in the Supplies and Works Directive for the non-excluded sectors. Comparable advertising requirements also apply, and there is an obligation to provide annual indicative notices forewarning bidders of intended procurement plans.

With respect to the provision of time limits, a distinction is drawn between open and restricted tenders. In open procedures, the limit for the receipt of tenders is 52 days from the date of despatch of the notice of publication (unless there has been an indicative notice, whereupon the limit is reduced to 36 days).

In the case of restricted tenders, the time limit for the receipt of applica- **16–54** tions to bid shall generally be five weeks from the date of despatch of the notice. The time limit in which bids must be received may be established by agreement, but in cases where such an agreement cannot be reached, the limit shall generally be three weeks from despatch of the invitations to bid.

Of interest to non-E.U. firms is the introduction of provisions in the Utilities Directive specifically targeting such firms. Where a non-E.U. firm bids for a contract, the Directive permits a tender to be rejected in one of two circumstances, namely:

   (i) where the proportion of products originating in non-E.U. countries exceeds 50 per cent of the total value of products in the tender; or

---

[77] Directive 90/531; [1990] O.J. L297/1.
[78] Directive 93/36; [1993] O.J. L199/1.

(ii) where bids by non-E.U. firms are not lower than bids by E.U. firms by more than three per cent.

These respective local content and price differential rules are expressed to apply only to tenders comprising products *originating* in those third countries with which the E.U. has not concluded some form of an equal-access agreement. The E.U. thus hopes to use the threat of exclusion from a liberalised E.U. market as a means of opening up foreign markets to E.U. enterprises. Despite their prima facie discriminatory nature, these dual provisions are generally considered to be much less restrictive towards non-E.U. firms than is the case under existing "buy America" legislation and its equivalents in other non-E.U. countries.

Remedies

16–55   The Utilities Directive for the excluded sectors is of little practical use unless effective remedies are available which permit its enforcement. Consequently, the Council adopted in 1992 a Remedies Directive[79] for the excluded sectors. According to the terms of the Remedies Directive, Member States must ensure that effective and rapid procedures are made available for the review of decisions taken within the scope of the Utilities Directive. In particular, the measures adopted must contain the following elements:

- effective interim measures in the event of an infringement in the award procedure for a contract; or
- the ability to set aside decisions taken unlawfully (which would include the improper tender procedure or the award procedure, or even on the basis of discriminatory technical or other specifications); or
- measures to correct infringements occurring in the award procedures or to prevent damage occurring; and
- awards of damages to parties harmed by the infringements.

16–56   Member States are under a duty to ensure that these measures are capable of being enforced effectively by non-judicial bodies which are subject to judicial review. Moreover, given the diversity in legal cultures and the possibility that formal laws may prove to be ineffective in practice, the Commission reserves the right to intervene where manifest infringements have occurred under a contract award procedure, by requiring that appropriate steps be taken to correct the infringement. The Member States are also required to conduct regular examinations of their procurement procedures.

The necessary implementing measures for both the Utilities and Remedies Directives were required to be in force under national law by January 1, 1993 at the latest, except in the cases of Spain (January 1, 1996) and Greece and Portugal (January 1, 1998).

[79] Directive 92/13; [1992] O.J. L76/14.

12. DATA PROTECTION

The Commission's attempts to create a pan-European data protection   **16–57**
regime based on two pieces of draft legislation, namely:

  (i) a proposed Framework Directive[80] concerning the protection of all
      forms of personal data; and
 (ii) a Proposed Telecoms Directive[81] which specifically addresses the
      protection of personal data in a telecommunications environment.

Framework Directive

Further to the political compromise reached at the Council of Ministers   **16–58**
meeting on December 8, 1994, a common position was reached on February 6, 1995 on the so-called Framework Directive for the Protection of
Personal Data. A common position on the proposed Directive was reached
by the Council of Ministers at the end of February 1995, following which it
was sent to the European Parliament for its second reading. Once adopted,
Member States will have a period of three years in which to transpose the
terms of the Directive into national law.
     In general terms, the Directive provides that:

  • the release and processing of personal data is based on the prior
    informed consent of the individual;
  • the individual has the right to obtain access to such information relat-
    ing to himself and also has the right to have erroneous information
    regarding himself corrected;
  • judicial remedies will be provided to protect an individual's rights;
  • the transfer of personal data to third countries can only occur where
    adequate data protection is afforded (with the establishment of a
    mechanism to ensure that this requirement is satisfied); and
  • national trade associations and national supervisory authorities may
    make additional provision for data protection in particular sectors,
    and are encouraged to participate in the drawing up of Community
    Codes of Conduct.

     In addition, the processing of personal information regarding racial or   **16–59**
ethnic origin, political opinions, religious convictions, trade union membership, health or sexual activity is prohibited under the terms of the Directive
unless express consent is provided by the individual in question or the use
of such information is necessary for medical reasons. Individuals also retain
a residual right to object to the handling of information which concerns

---

[80] Amended proposal for a Directive on the protection of individuals with regard to the processing and free
   movement of personal data [1992] O.J. C311/30.
[81] Amended proposal for a European Parliament Directive concerning the protection of personal data/privacy
   in the context of digital telecommunications networks (in particular integrated services digtial network
   (ISDN) and digital mobile networks) [1994] O.J. C200/4.

them on the basis of "significant and legitimate reasons related to their particular situation".

In arriving at a common position on the Framework Directive, a number of political compromises were made by the Member States which will alter the text of the draft Directive, with the following amendments being of particular importance:

- "Manual" data (data on paper) will be exempted from the operation of the Directive for 12 years from the date upon which the Directive enters into force.
- National authorities will be able to legislate on the possession and use of information by the press.
- Access to information which is valuable for research and development and statistical purposes can be restricted unless such a restriction violates individual rights.
- Data contained in images and sounds relevant to issues of public safety are to be excluded from the Framework Directive although the Directive itself envisages the need for subsequent revision so as to take into account technological advances.

Proposed Telecoms Directive

16–60 The proposed Telecoms Directive is designed to cater for the emergence of advanced digital public telecommunications networks within the E.U., which has given rise to particular requirements in the realm of personal data protection and privacy of the user (especially since the introduction of ISDN and digital mobile networks). Aside from the Commission's desire to create an internal market for telecommunications and the free flow of transborder communications, the Commission is also concerned to assure that the processing of personal data should not be used to provide TOs with an undue competitive advantage over other service providers.

The proposed Telecoms Directive is applicable to the processing of personal data by TOs in the provision of public telecommunications services in public digital telecommunications networks in the E.U. and is also applicable, where technically feasible, to the processing of personal data in connection with services provided via analogue networks. The basic institutional structure established under the proposed Framework Directive is also adopted under the proposed Telecoms Directive and is intended to ensure that discrepancies in enforcement as between Member States are kept to a minimum.

Some of the provisions which underpin the proposed Telecoms Directive are:

- subscribers are to be afforded encryption facilities upon a breach of network security having been detected;
- detailed provisions regulating the collection and storage of billing data;
- the possibility of avoiding calling-line identification;

- subscriber rights with respect to information contained in directories, communications surveillance, and unsolicited calls and telefax messages (notwithstanding the operation of the Proposed Directive on Distance Selling);
- mandatory requirements as regards technical features and standardisation to ensure compatible standards; and
- the availability of judicial remedies to protect an individual's rights.

With political agreement having been achieved with respect to the Proposed Framework Directive, more rapid progress can now be expected on the Proposed Telecoms Directive.

13. MISCELLANEOUS MATTERS

The types of E.U. legislative and policy matters which fall within the rubric **16–61** of "telecommunications" are many and varied. Space precludes any detailed assessment of these additional matters, although the table which follows outlines the more important E.U. initiatives in the more technical aspects of telecommunications regulation.

Chronological Landmarks

Council Recommendation of November 12, 1984 concerning the implementation of harmonisation in the field of telecommunications (84/549/EEC; [1984] O.J. L298/49).

Council Recommendation of November 12, 1984 concerning the first phase of opening up access to public telecommunications contracts (84/550/EEC; [1984] O.J. L298/51).

Council Decision of July 25, 1985 on a definition phase for a Community action in the field of telecommunications technologies—R&D programme in advanced communication technologies for Europe (RACE) (85/372/EEC; [1985] O.J. L210/24).

Council Resolution of June 9, 1986 on the use of videoconference and videophone techniques for intergovernmental applications (86/C 160/01; [1986] O.J. C160/01).

Council Directive of July 24, 1986 on the initial stage of the mutual recognition of type approval for telecommunications terminal equipment (86/361/EEC; [1986] O.J. L217/21).

Council Regulation of October 27, 1986 instituting a Community programme for the development of certain less-favoured regions of the Community by improving access to advanced telecommunications services (STAR programme) (3300/86/EEC; [1986] O.J. L305/1).

Council Directive of November 3, 1986 on the adoption of common technical specifications of the MAC/packet family of standards for direct satellite television broadcasting (86/529/EEC; [1986] O.J. L311/28).

Council Decision of December 22, 1986 on standardisation in the field of information technology and telecommunications (87/95/EEC; [1987] O.J. L36/31.

Council Recommendation of December 22, 1986 on the coordinated introduction of the Integrated Services Digital Network (ISDN) in the European Community (86/659/EEC; [1986] O.J. L382/36).

Council Recommendation of June 25, 1987 on the coordinated introduction of public pan-European cellular digital land-based mobile communications in the Community (87/371/EEC; [1987] O.J. L196/81).

Council Directive of June 25, 1987 on the frequency bands to be reserved for the coordinated introduction of publc pan-European cellular digital land-based mobile communications in the European Community (87/372/EEC; [1987] O.J. L196/85).

Council Decision of October 5, 1987 introducing a communications network Community programme on trade electronic data interchange systems (TEDIS) (87/499/EEC; [1987] O.J. L285/35) and Council Decision of April, 1989 amending decision (87/499/EEC; [1989] O.J. L97).

Council Decision of December 14, 1987 on a Community programme in the field of telecommunications technologies—research and development (R&D) in advanced communications technologies in Europe (RACE programme) (88/28/EEC; [1988] O.J. L16/35).

Commission Directive of May 16, 1988 on competition in the markets in telecommunications terminal equipment (88/301/EEC; [1988] O.J. L131/73).

Council Resolution of June 30, 1988 on the development of the common market for telecommunications services and equipment up to 1992 (88/C 257/01; [1988] O.J. C257/1).

Council Decision of April 5, 1989 amending Decision 87/499/EEC introducing a communications network Community programme on trade electronic data interchange systems (Tedis) (89/241/EEC; [1989] O.J. L97/46).

Council Resolution of April 27, 1989 concerning standardisation in the fields of information technology and telecommunications (89/C) 117/01; [1989] O.J. C117/1).

Council Decision of April 27, 1989 on high-definition television (89/337/EEC; [1989] O.J. L142/1).

Council Directive of May 3, 1989 on the approximation of the laws of the Member States relating to electromagnetic compatibility (89/336/EEC; [1989] L139/19).

Council Resolution of July 18, 1989 on the strengthening of the coordination for the introduction of the Integrated Service Digital Network (ISDN) in the European Community up to 1992 (89/C 196/04; [1989] O.J. C196/4).

Council Decision of December 7, 1989 on the common action to be taken by the Member States with respect to the adoption of a single world-wide high-definition television production standard by the Plenary Assembly of the International Radio Consultative Committee (CCIR) in 1990 (89/630/EEC; [1989] O.J. L363/30).

Council Resolution of January 22, 1990 concerning trans-European networks (90/C 27/05; [1990] O.J. C27/8).

Council Resolution of June 28, 1990 on the strengthening of the Europeanwide cooperation on radio frequencies, in particular with regard to services with a pan-European dimensions (90/C 166/02; [1990] O.J. C 166/4).

Council Directive of June 28, 1990 on the establishment of the internal market for telecommunications services through the implementation of open network provision (90/387/EEC; [1990] O.J. L192/1).

Commission Directive of June 28, 1990 on competition in the markets for telecommunications services (90/388/EEC; [1990] O.J. L192/10).

Commission Decision of July 30, 1990 setting up a Joint Committee on Telecommunications Services (90/450/EEC; [1990] O.J. L230/25).

Council Directive of September 17, 1990 on procurement proceures of entities operating in the water, energy, transport and telecommunications sectors (90/531/EEC; [1990] O.J. L297/1).

Council Recommendation of October 9, 1990 on the coordinated introduction of pan-European land-based public radio paging in the Community (90/543/EEC; [1990] O.J. L310/23).

Council Directive of October 9, 1990 on the frequency bands designated for the coordinated introduction of pan-European land-based public radio paging in the Community (90/544/EEC; [1990] O.J. L310/28).

Council Resolution of December 14, 1990 on the final stage of the coordinated introduction of pan-European land based public digital mobile cellular communications in the Community (GSM) (90/C/329/09; [1990] C329/25).

Judgment of the Court of March 19, 1991 in Case C-202/88: *French Republic v. Commission of the European Communities* (Competition in the markets in telecommunications terminal equipment) (91/C 96/04; [1991] O.J. 91/C 96/04).

Council Directive of April 29, 1991 on the approximation of the laws of the Member States concerning telecommunications terminal equipment, including the mutual recognition of their conformity (91/263/EEC; [1991] O.J. L128/1).

Council Directive of June 3, 1991 on the frequency band to be designated for the coordinated introduction of digital European cordless telecommunications (DECT) into the Community (91/287/EEC; [1991] O.J. L144/45).

Council Recommendation of June 3, 1991 on the coordinated introduction of digital European cordless telecommunications (DECT) into the Community (91/288/EEC; [1991] O.J. L144/47).

Council Decision of June 7, 1991 adopting a specific research and technological development programme in the field of communication technologies (1990 to 1994) (91/352/EEC; [1991] O.J. L192/8).

Council Decision of June 7, 1991 adopting a specific programme of research and technological development in the field of telematic systems in areas of general interest (1990 to 1994) (91/353/EEC; [1991] O.J. L192/18).

Council Decision of July 22, 1991 establishing the second phase of the Tedis programme (Trade electronic data interchange systems) (91/385/EEC; [1991] O.J. L208/66).

Council Decision of July 29, 1991 on the introduction of a single European emergency call number (91/396/EEC; O.J. L217/31, 06.08.91).

Guidelines on the Application of EEC Competition Rules in the Telecommunications Sector (92/C 233/02; [1991] O.J. C233/2).

Council Resolution of November 18, 1991 concerning electronics, information and communication technologies (91/C 325/02; [1991] O.J. C325/2).

Council Resolution of December 19, 1991 on the development of the common market for satellite communications services and equipment (92/C 8/01; [1992] O.J. C8/1).

Council Directive of February 25, 1992 coordinating the laws, regulations and administrative provisions relating to the application of community rules on the procurement procedures of entities operating in the water, energy, transport and telecommunications sectors (92/13/EEC; [1992] L76/14).

Council Decision of March 31, 1992 in the field of security of information systems (92/242/EEC; [1992] O.J. L123/19).

Council Directive of April 28, 1992 amending Directive 89/336/EEC on the approximation of the laws of the Member States relating to electromagnetic compatibility (92/31/EEC; [1992] O.J. L126/11).

Council Directive of May 11, 1992 on the adoption of standards for satellite broadcasting of television signals (92/38/EEC; [1992] L137/17).

Council Decision of May 11, 1992 on the introduction of a standard international telephone access code in the Community (92/264/EEC; [1992] O.J. L137/21).

Council Directive of June 5, 1992 on the application of open network provision to leased lines (92/44/EEC; [1992] O.J. L165/27).

Council Resolution of June 5, 1992 on the development of the integrated services digital network (ISDN) in the Community as a European-wide telecommunications infrastructure for 1993 and beyond (92/C; [1992] O.J. C158/1).

Council Recommendation of June 5, 1992 on the application of open network provision to public packet switched data services (92/382/EEC; [1992] O.J. L200/1).

Council Recommendation of June 5, 1992 on the application of open network provision to ISDN (92/383/EEC; [1992] L200/10).

Commission Decision of July 15, 1992 amending the lists of standards institutions annexed to Council Directive 83/189/EEC (92/400/EEC; [1992] O.J. L221/55).

Judgment of the Court of November 17, 1992 in Joined Cases C-271, C-281 and C-289/90: *Kingdom of Spain and Others v. Commission of the European Communities* (Competition in the markets for telecommunications services) (92/C 326/07; [1992] O.J. C326/8).

Council Resolution of November 19, 1992 on the implementation in the Community of the European Radiocommunications Committee Decisions (92/C 318/01; [1992] O.J. C318/1).

Council Resolution of November 19, 1992 on the promotion of Europe-wide cooperation on numbering of telecommunications services (92/C 318/02; [1992] C318/2).

European Parliament Resolution of April 20, 1993 on the Commission communication "Towards cost orientation and the adjustment of pricing structures—Telecommunications tariffs in the Community" (A3-0117/93; [1993] O.J. C150/37).

European Parliament Resolution of April 20 1993 on the Commission's 1992 review of the situation in the telecommunications services sector (A3-0113/93; [1993] O.J. C150/39).

Council Directive 93/38/EEC of June 14, 1993 coordinating the procurement procedures of entities of entities operating in the water, energy, transport and telecommunications.

Council Resolution of July 22, 1993 on the technology and standards in the field of advanced television services (93/C 209/01; [1993] O.J. C 209/1).

Council Decision of July 22, 1993 on an action plan for the introduction of advanced television services in Europe (93/424/EEC; [1993] O.J. L196/48).

Council Resolution of July 22, 1993 on the review of the situation in the telecommunications sector and the need for further development in that market (93/C 213/01; [1993] O.J. C213/1).

Telecommunications: open network provision (ONP) list of standards (third issue) (93/C 219/02; [1993] O.J. 93/C 219/02).

Telecommunications: open network provision for leased lines (93/C 277/04; [1993] O.J. C277/9).

Council Directive 93/97/EEC of October 29, 1993 supplementing Directive 91/263/EEC in respect of satellite earth station equipment (93/C97/EEC; [1993] O.J. L290/1).

Council Resolution of December 7, 1993 on the introduction of satellite personal communication services in the Community (93/C 339/01; [1993] O.J. C339/1).

Commission Decision of December 21, 1993 on a common technical regulation for the general attachment requirements for public pan-European cellular digital land-based mobile communications (94/11/EC; [1994] O.J. L8/20).

Commission Decision of December 21, 1993 on a common technical regulation for the telephony application requirements for public pan-European cellular digital land-based mobile communications (94/12/EC; [1994] O.J. L8/23).

Council Resolution of February 7, 1994 on universal service principles in the telecommunications sector (94/C 48/01; [1994] O.J. C48/1).

Commission Statement concerning Council resolution on universal service in the telecommunications sector (94/C 48/06; [1994] O.J. C48/8).

European Parliament Resolution of May 6, 1994 on the communication from the Commission accompanied by the proposal for a Council resolution on universal service principles in the telecommunications sector (A3-0317/94; [1994] O.J. C205/551).

Commission Decision of June 15 1994 on amendment of Annex II of Council Directive 92/44/EEC (94/439/EC; [1994] O.J. L181/40).

Council Resolution of June 27, 1994 on a framework for Community policy on digital video broadcasting (9/C 181/02; [1994] O.J. 183/3).

Commission Decision of July 18, 1994 on a common technical regulation for attachment requirements for terminal equipment interface for ONP 2,048 kbits/s digital unstructured leased line (94/470/EC; [1994] O.J. L194/87).

Commission Decision of July 18, 1994 on a common technical regulation for general terminal attachment requirements for Digital European Cordless Telecommunications (DECT) (94/471/EC; [1994] O.J. L.194/89).

Commission Decision of July 18 1994 on a common technical regulation for telephony application requirements for Digital European Cordless Telecommunications (DECT) (94/472/EC; [1994] O.J. L194/91).

Telecommunications: Open network provision (ONP) for leased lines—Concilation procedure (94/C 214/04; [1994] O.J. C214/4).

Commission Directive 94/46/EC of October 13, 1994 amending Directive 88/301/EEC and Directive 90/388/EEC in particular with regard to satellite communications ([1994] O.J. L268/15).

Commission Directive of November 18, 1994 on a common technical regulation for the pan-European integrated services digital network (ISDN) primary rate access (94/796/EC; [1994] O.J. L329/1).

Commission Decision of November 18, 1994 on a common technical regulation for the pan-European integrated services digital network (ISDN) basic access (94/797/EC; [1994] O.J. L329/14).

Council Resolution of December 22, 1994 on the principles and timetable for the liberalisation of telecommunications infrastructures (94/C 379/03; [1994] O.J. 379/4).

Council Resolution of December 22, 1994 on further development of the Community's satellite communications policy, especially with regard to the provision of, and access to, space segment capacity (94/C 379/04; [1994] O.J. C379/5).

Basic policy documents published by the Commission in this field:

Green Paper on the development of the Common Market for Telecommunications services and equipment (COM(87)290, 30.06.87).
Green Paper on a common approach in the field of satellite communications in the European Community (COM(90)490, 20.11.90).
1992 Review of the Situation in the telecommunications services sector (SEC(92) 1048) and Communication to the Council and European Parliament on the consultation on the review of the situation in the telecommunications sector (COM(93) 159 final).
Green Paper on a common approach in the field of mobile and personal communications in the European Union (COM(94)145, 27.04.94).
Green Paper on the liberalisation of Telecommunications Infrastructure and cable television networks: Part One (COM(94)440, 25.10.94).

SECTION II: COMPETITION RULES IN THE TELECOMMUNICATIONS SECTOR

16–62   Aside from the legislative powers contained in Articles 90 and 100A of the E.C. Treaty and in the new provisions on trans-European networks introduced by the Treaty on European Union (see Section I), the process of liberalisation in the telecommunications sector has been accomplished ad hoc in the implementation of the E.C. Treaty's provisions on the free movement of goods and services[82] and, primarily, through the enforcement of the competition rules contained in Articles 85 and 86 of the E.C. Treaty[83] through the bringing of infringement actions (often in combination with Article 90). By 1991, the Commission had developed a body of administrative practice in the application of competition policy to the telecommunications sector which it synthesised in the form of interpretative guidelines. Most recently, the Commission's principal focus of attention has been on the appraisal of strategic alliances in the telecommunications sector, primarily under Article 85 of the E.C. Treaty, but also under the so-called Merger Control Regulation.

The enforcement of the competition rules in the telecommunications sector has been singled out as a high priority by the E.C. Commission's Competition Directorate[84] because, as the formal legislative tools which harmonise and liberalise the sector take effect, the greater the perceived need of incumbent TOs to protect their positions through practices which run counter to competition policy goals. In particular, the convergence of technologies in the traditionally distinct telecommunications and broadcasting sectors and the growth of "multimedia" applications has given rise to a new range of potential competitive relationships and alliances which will require the creative application of E.C. competition rules.[85]

1. INFRINGEMENT ACTIONS

16–63   The adoption under Article 90(3) of the E.C. Treaty of the Terminal Equipment Directive 88/301 and the Services Directive 90/388, as amended by the Satellite Communications Directive 94/46 and the proposed draft Directives on the Liberalisation of Cable Television Networks mobile and

---

[82] In the absence of telecommunications-specific harmonisation measures, the principal means of opening up national markets was through reliance on the direct applicability of the E.C. Treaty provisions on the free movement of goods (Art. 30, 36 and 37) and on the freedom to provide services (Art. 52 and 59).

[83] Art. 85(1) prohibits anti-competitive agreements or concerted practices, but allows for a special exemption from this rule under Art. 85(3) where the agreement produces certain efficiencies and other pro-competitive effects. Art. 86 prohibits unilateral abusive behaviour by undertakings in a position of single (or, more recently, collective) market dominance.

[84] Commissioner Karel Van Miert, "Competition Policy in the Development of Telecommunications", Speech given in Brussels on May 17, 1994 (hereafter "Van Miert Speech").

[85] See discussion in Coudert Brothers, "Overview and Analysis of the Legal and Regulatory Barriers to the Take-off of Multimedia Applications in Prepartion for the Infrastructures Green Paper" (December 1994) *Competition Policy* S.C.4.

personal communicators and "Full Liberalisation" (refer back to Section I), are designed to eliminate anti-competitive practices supported by Member State laws. As such, they do not address the types of anti-competitive behaviour undertaken by TOs independently of national laws. Such anti-competitive behaviour is policed by the E.C. Commission pursuant to its enforcement powers under Articles 85 and 86 of the E.C. Treaty.

By the late 1980's, the Commission had commenced an active programme of competition policy enforcement which it had foreseen in its 1987 Green Paper,[86] belying the observation that:

> "in the early years of the Community, telecommunications was such an integral part of the national infrastructure, that there were, in fact, no cases in which there was sufficient Community interest for them to be considered by the Commission's services in the context of Community competition law".[87]

Ever since the so-called *Telespeed* case,[88] the Commission has taken an increasingly high profile interest in the risks of anti-competitive activities occurring in the telecommunications services and equipment sectors. Enforcement action pursuant to Article 85 has usually been effected with respect to the collective actions of TOs when acting in the context of their international commitments on tariffing principles such as ITU rules. By way of contrast, Article 86 actions have addressed a variety of anti-competitive abuses, especially the extension of monopoly power (or market dominance) into adjacent product or service markets.

Article 85

The following cases have been investigated by the Commission since the   16–64
early 1980's in the context of alleged violations of Article 85 of the EC
Treaty.

SWIFT v. CEPT[89]

SWIFT was a joint venture set up between European and American banks.   16–65
It offered customers value-added services using new technologies for inter-bank funds transfers. Access was gained to the international telecommunications network through the utilisation of leased lines for private use made available by the TOs (represented by CEPT—Conference of European Post and Telecommunications Administrations) and charged out at a flat monthly rate. The introduction of this new means of transmitting large quantities of data meant that leased lines were more economical if used on the basis of a flat rate calculation (as opposed to telexes, which were

---

[86] Green Paper on the Development of the Common Market for Telecommunications Services and Equipment, COM (87) 290 of June 30, 1987.
[87] Overbury & Ravaioli, "The Application of EEC Law to Telecommunications" (1989) Fordham Corporate Law Institute, 271 at 289.
[88] *Italy v. Commission* [1985] E.C.R. 873.
[89] Commission press release, IP/90/188 of March 6, 1990.

charged for by volume). Consequently, the TOs, fearing a loss in their telex revenues, agreed pursuant to the CEPT recommendation, to impose a 30 per cent surcharge where the capacity of such circuits was to be used for third party traffic or in the case of interconnection with the public telecommunications network.

In addition to an alleged infringement of Article 85, the TOs were suspected by the Commission of having abused their individual dominant positions. After initial investigation, the Commission concluded that the recommendation could constitute an illegal price agreement prohibited under Article 85(1) of the E.C. Treaty. According to the Commission, such coordination limited the commercial autonomy of the TOs to the detriment of users. Under pressure from the Commission, the tariff was abandoned by the CEPT in February 1990. However, the Commission indicated that it would be prepared to consider an exemption under Article 85(3) to a recommendation which purported to harmonise tariff principles without any price fixing agreement insofar as this would bring economic advantages, for example by making tariffs more cost-related and transparent.

The MDNS Case (Managed Data Network Services)[90]

16–66   This case concerned a project by 22 TOs to cooperate, with a view to providing one-stop services throughout the E.U. The Commission investigated that cooperation and found that it could only approve the project if guarantees were given by the participants so as to ensure that there would be no discrimination against the provision of comparable facilities and terms to competitive private service providers. The MDNS arrangements could have restricted competition by leading to the coordination of prices, joint specification of products and the joint purchase of hardware and/or software; ultimately, this could have resulted in a consolidation of the market power of the participants and amounted to an effective neutralisation of the benefits which would otherwise have flowed to consumers. The Commission acknowledged that this cooperation could bring economic benefits to telecommunications users and accelerate European standardisation, provided that certain undertakings given to the Commission were respected. The project was eventually abandoned and may well have proved practically unworkable in the face of the Commission's requirements for its approval.

The CCITT Recommendations Case[91]

16–67   The Commission took the view that certain recommendations and certain proposals for their revision, proposed at the meeting in May 1990 of study group III of the International Telegraph and Telephone Consultative Com-

---

[90] Commission press release, IP/89/948 of December 14, 1989.
[91] Guidelines on the Application of EEC Competition Rules in the Telecommunications Sector [1991] O.J. C233/2, paras. 142–144.

mittee (CCITT),[92] could be deemed to be agreements between, or decisions by, associations of undertakings which might be restrictive of competition between the TOs and between private telecommunications services suppliers. At the CCITT meeting, the Commission took the view that should the final revision of the recommendations be incompatible with the E.C. competition rules, the Commission could intervene pursuant to those rules to bring such an infringement to an end.

In July 1991, a revised recommendation was adopted by the CCITT under a written accelerated procedure. However, the Commission took the view that some of the provisions could also be interpreted as being contrary to E.C. competition rules. The Commission considered that the prohibition of least-cost routing was anti-competitive, and it therefore indicated to the TOs and the national regulatory authorities of the Member States that, in implementing the new recommendation, they should not prevent least-cost routing.

Article 86

Belgian RTT Case[93]

In September 1988, a private supplier of value-added telecommunications **16-68** services filed a complaint with the Commission, alleging that the Belgian TO, the *Régie des télégraphes et téléphones* (RTT) had abused its dominant position because it had refused to lease its international telecommunications circuits for the transmission of third-party data traffic. The private supplier was prevented from using international leased lines to carry its customers' data to its processing centre in a neighbouring country and to return the processed data to its customers over such leased lines. The Commission indicated to RTT that its refusal to grant leased lines to the complainant could amount to an abuse of its dominant position under Article 86. Subsequently, the RTT granted the international leased circuits to the complainant with no other usage restrictions other than the prohibition of the simple resale of capacity. The Commission, using the powers of enforcement granted under Article 86, requested that the RTT ensure that all its customers were entitled to the same rights as the complainant. Consequently, the RTT undertook that international leased circuits could be used to carry third-party data traffic without any restrictions or particular conditions apart from the requirement that the circuit should not be used for simple data transmission. The Commission emphasised that an undertaking in a dominant position in a market for telecommunications services may not impose any restrictions on the use of such services unless they are "necessary to the task of providing the service of general economic interest with which it has been entrusted".

---

[92] The CCIT is an ITU Consultative Committee.
[93] Commission press release IP/90/67 of January 29, 1990.

The International Courier Services Case[94]

16–69   In this case, the Commission considered that the German Bundespost's attempt to extend its monopoly over the delivery of letters to express delivery services was in violation of Article 86. The German Bundespost accepted the Commission's position, and the case was closed without the need for a formal decision.

Unlike the Commission's intervention in the *Cordless Telephone* investigation (see below), Article 86 was relied on by the Commission as the basis of its infringement action because it was the action of the Bundespost itself, rather than German law, which was responsible for the anti-competitive behaviour.[95]

The German Modems[96] and Cordless Telephone[97] Cases

16–70   The Commission instituted two procedures in the mid-1980s against the German Bundespost's attempts to extend its monopoly for the provision of terminal equipment into the adjacent markets for cordless telephones and modems. The actions could have been based on Article 86, but were brought principally under Article 37 of the E.C. Treaty relating to the free movement of goods.

In both cases, the German authorities sought to justify the extension of Bundespost's monopoly by taking the position that both types of equipment constituted an integral part of the telephone network which should be supplied by Bundespost. Although both proceedings were closed prior to any formal decision being taken, the Commission took the view that the extension of the monopoly over terminal equipment was clearly contrary to Article 37(1) because of the potential to exclude non-German suppliers from access to the local market.[98]

The Telespeed Case[99]

16–71   In the first case involving telecommunications brought before the European Court of Justice, Italy instituted proceedings for a declaration that a decision of the Commission, relating to a proceeding against BT (when BT was a public monopoly) under Article 86 of the E.C. Treaty, was void. In its decision, the Commission held that the schemes adopted by BT, under which private message-forwarding agencies in the U.K. were prohibited from retransmitting to destinations outside the U.K., constituted infringements of Article 86.

The Court upheld the decision of the Commission. In doing so, it held

---

[94] [1985] E.C. Bull. 1, para. 2.10.
[95] To the extent that German legislation might be said to have prompted the actions of Bundespost, an action brought under Art. 90(1), in conjunction with Art. 86, could have been brought.
[96] Refer to Sixteenth Report on Competition Policy (1986), paras. 294–297; *cf.* Seventeenth Report on Competition Policy (1987) paras. 282–284. Also discussed in Ungerer, Telecommunications in Europe "European Perspectives (Commission of the European Committees)" (1992), pp. 169–170.
[97] [1985] E.C. Bull. 3, para. 2.1.43.
[98] The extension of the monopoly was also clearly contrary to Art. 37(2) of the E.C. Treaty, constituting a "new measure" introduced by Germany.
[99] [1982] O.J. L360/36; on appeal, *Italy v. Commission* [1985] E.C.R. 873.

that the statutory monopoly enjoyed by BT did not extend to message-forwarding services. As to the jurisdiction of the Commission, the Court ruled that the behaviour of TOs such as BT can be challenged under Article 86 because they are "undertakings" within the meaning of that Article insofar as they are conducting "business activities". As a result, it was not necessary for the Commission to proceed under Article 90 or to institute infringement proceedings against the U.K. Government under Article 169.[1] The application of Article 86 against BT in the circumstances is widely interpreted as indicating that reserved monopoly rights will be interpreted narrowly by both the Commission and the Court, particularly in light of new technological developments which give rise to a variety of value-added services.[2]

The Commission is aware of the fact that no formal prohibition decisions have been taken against TOs since the action against BT in 1982. This, however, does not mean that abusive behaviour is not taking place. On the contrary, it is felt by the Commission that formal complaints are not forthcoming against TOs for fear of "reprisals" against the complainants;[3] these companies often request that the Commission take ex officio proceedings against the TOs.

Actions under Article 90

In addition to taking individual action under Article 85 and/or 86, the Commission can challenge the conduct of TOs with "special or exclusive rights" or the Member States themselves which enact or maintain in force measures which are contrary to the provisions of the E.C. Treaty, under Article 90 of the Treaty.[4] **16–72**

Aside from the series of directives adopted pursuant to the powers conferred under Article 90(3),[5] the Commission has initiated a number of actions against individual Member States by using a combination of Articles 85, 86 and 90 for their non-compliance with Treaty rules. A good example of such actions is the Commission's challenge to Germany, Spain, Italy and Belgium for the grant by these Member States of a monopoly over the supply of modems to the national TO (see above), which was brought under Articles 90(1) and 37 of the E.C. Treaty. Another recent example is the Commission's challenge to a number of Member States regarding the continuation of monopolies for the provision of GSM mobile services.[6] These infringement actions are usually settled informally, although a number of them have concluded in formal decisions.[7]

---

[1] In addition, Art. 222 of the E.C. Treaty could not be used to shield BT from the scope of Art. 86 to the extent that the attempted extension of the scope of its monopoly was made for reasons of an economic nature, rather than in pursuance of its public policy function recognised under Art 90(2) of the E.C. Treaty.
[2] See Ungerer, *Telecommunications in Europe, supra,* at p. 168.
[3] Van Miert speech, *supra.*
[4] Refer to detailed discussion of Art. 90 in section I.
[5] Refer to section I.
[6] Including Ireland and Italy.
[7] Refer to discussion in section I.

2. COMMISSION GUIDELINES

16–73   On September 6, 1991, the Commission issued Guidelines on the Application of EEC Competition Rules in the Telecommunications Sector,[8] which not only serve to emphasise the applicability of E.C. competition rules to the telecommunications sector, but also indicate those particular types of practices and relationships which expose TOs (referred to at the time of the Guidelines as national telecommunications organisations, or PTTs) to liability.

The Guidelines do not of themselves create legal rights, but they do serve to illustrate the more problematic issues which face TOs, equipment manufacturers and users of equipment and services in the context of the application of Articles 85 and 86 of the E.C. Treaty. In addition, the Guidelines explore the close relationship between competition rules, the rules applicable to state monopolies under Article 90 of the Treaty and the parcel of harmonisation measures intended to create an integrated Union-wide market for telecommunications and equipment (*e.g.* the Open Network Provision). The Guidelines do not therefore concern the activities of Member States; such activities are of course capable of supporting anti-competitive activity by TOs, but these actions are the subject of separate Treaty rules (Articles 3(g), 5 (2), 7 and 90 of the E.C. Treaty, as combined with Article 85 and/or 86).

Relevant Product/Service and Geographic Market(s)

16–74   The application of Articles 85 and 86 of the E.C. Treaty involves an assessment of the relevant product, as well as geographic, markets affected by the alleged anti-competitive conduct.

The Commission maintains in its Guidelines that telecommunications services can be divided into a number of relevant product or service markets. Examples of such markets cited in the Guidelines are terrestrial network (facilities) provision, voice telephony, data communication and satellites. The Commission acknowledges that some of these categories may indeed have relevant sub-markets; for example, because satellites can be deployed for various uses in order to provide different voice and data services, these sub-markets may be separate and distinct in certain circumstances.

In the absence of full liberalisation across most Member States of the E.U., there remain significant differences between them as to the terms upon which telecommunications services may be licensed. The various national regulatory conditions are so diverse that the Commission also takes the view that each national territory within the E.U. is likely to constitute a separate geographic market. As the effects of harmonisation directives on equipment/services gain greater momentum, however, the relevant

---

[8] See *ante* [1991] O. J. C233/2.

geographic market is expected to expand to cover the whole E.U. and, for certain services, is likely to embrace global markets.

In the meantime, where purely domestic agreements or practices are involved, Articles 85 or 86 may not apply at all if there is no actual or likely effect on Community trade.[9]

Article 86

Given that each TO in each Member State holds a dominant position with respect to the provision of most telecommunications services and equipment, Article 86 is of immediate concern to the activities of TOs. The *Telespeed* case (see above) confirmed that state undertakings operating as commercial providers of telecommunications services are as much subject to Article 86 as those operating in the private sector.

16–75

The Guidelines provide examples of typical "abuses" in the telecommunications sector, namely: refusal to provide access to a network; discrimination; cross-subsidisation of non-reserved activities[10]; the abuse of standards; refusal to supply interface information; and restrictions preventing the use of leased circuits for the provision of value added data communication services, at least where simple resale is not involved.[11] The Commission cites the Belgian *RTT* case as an example of usage restrictions which constitute an Article 86 abuse (see above).

The imposition of access charges or other forms of surcharge on the use of a TO's circuits may also be considered in certain circumstances to be discriminatory and abusive. Only where such charges are imposed to recover actual costs and are applied on an equal basis are they likely to be approved by the Commission; in any event, such charges would have to be imposed by TOs on the parts of their businesses which were also using such circuits to provide service to customers. However, it is difficult to apply such rules in practice in the absence of effective unbundling of service provision and transparency of accounts.

16–76

Other abusive behaviour mentioned in the Guidelines may include the discriminatory quality of service provided to different classes of user: for example, where a TO is only prepared to provide particular service to a competitor on terms less favourable than would apply to an ordinary user which is not offering competitive services. In addition, proprietary rights affecting standards can also have a significant effect on competition and, in the hands of a dominant undertaking, can be used to enhance its posi-

---

[9] *Alsatel v. Novasam SA* [1988] E.C.R. 5987.

[10] Two basic requirements are recognised as being necessary to prevent cross-subsidisation taking place: (1) the existence of a transparent accounting system (namely, one which demonstrates fully allocated costs as between reserved and non-reserved services); and (2) a distinct structural separation between the division of the TO providing telecommunication services and that supplying equipment.

[11] Simple resale is defined in the Services Directive as the "commercial provision on leased lines for the public of data transmission as a separate service, including any such switching, processing, data storage or protocol conversion as is necessary for transmission in real time to and from the public switched network".

tion. Relying on its settlement in the *IBM* Case,[12] the Commission cites the need in that case for IBM to provide advance interface information to its competitors as a useful precedent in those cases where a dominant undertaking refuses to provide necessary information or to grant appropriate licences to enable manufacturers of competing products to provide compatible and interconnectable equipment.

Article 85

16–77　The Commission accepts that TOs should be allowed to combine and cooperate to achieve the goals of creating pan-European networks and services of the highest quality and at the lowest price, but not at the expense of proper respect for and adherence to the principles contained in Article 85.[12a]

The Guidelines draw a distinction common in the industry between anticompetitive agreements with respect to the provision of terrestrial facilities and reserved services on the one hand (involving horizontal competitive restrictions between TOs), and agreements concerning the provision of non-reserved services and terminal equipment on the other (which may also involve independent third parties). In the former category, particular prominence is attached to agreements relating to price, other conditions for the provision of facilities, the choice of telecommunications routes, and the imposition of technical and quality standards regarding the services provided on the public network. With respect to agreements relating to the provision of non-reserved services and terminal equipment, the Commission is concerned that a number of different types of strategic agreements may restrict competition, namely agreements between TOs that each TO provide a particular service to a customer in an overall "package" of services (as this could lead to price fixing or market coordination); cooperative arrangements between TOs and other independent service providers (as this could have a foreclosing effect on other third parties); and agreements between service providers other than TOs on any of the traditional anticompetitive grounds canvassed under Article 85.

Notwithstanding the regulatory legitimacy of the existing reservations on public switched voice telephony and infrastructure provision (at least until 1998 in most cases), the Commission acknowledges that agreements concerning such matters could hinder competition between the TOs themselves and also downstream. The Commission cites "hub" competition, where different TOs vie for the business of major users who may centralise

---

[12] See [1984] E.C. Bull. 10, at para. 3.4.1, and refer to subsequent Commission press release IP/88/814 of December 15, 1988.

[12a] For example, although the Commission is well disposed towards technical cooperation arrangements, aimed at improving the quality and availability of leased lines, it has signalled its concerns that the Global European Network ("GEN") arrangement between BT and the TOs from France, Germany, Italy and Spain (and more recently, those of Belgium, The Netherlands, Switzerland and Portugal) should not result in a price rise to users who wish to obtain access to GEN for data transmission. In particular, the Commission remains concerned by the price at which competitors would have to acquire the resources necessary to implement a system competitive with GEN. See GEN Notice; [1995] O.J. C55/3; *cf.* Press Release IP/95/443 of May 14, 1995.

their telecommunications requirements in a particular Member State, as an example of this. International agreements between TOs on prices are also identified as being capable of restricting hub competition significantly. An example of such risks is found in the CEPT "Recommendation on Tariffs for International Leased Circuits" (see above).

The Commission also observes that agreements which impose technical **16–78** and quality standards can have restrictive effects on competition, particularly if they block network access to users and service providers in any way. Only where such requirements will also benefit users will there be an opportunity for the relevant agreement to be exempted by the Commission pursuant to its individual power of exemption under Article 85(3).

The Guidelines take the position that there is also significant scope for unlawful restraints on competition regarding liberalised services. In this regard, the Commission points out that agreements between TOs for the purpose of providing "one-stop" shopping facilities (whereby major customers obtain all their telecommunication services from one operator) may be potentially anti-competitive. This concern is illustrated in the Guidelines by the Commission's investigation in the *MDNS* case (see above).

In its assessment of the anti-competitive potential of agreements between TOs and service providers, the Commission relies on established principles distilled from case law and Commission administrative precedent. For example, if the parties are already actual or potential competitors, their cooperation in the form of a joint venture is liable to reduce such competition. Because of the privileged position of TOs, there exist risks that such a joint venture would unfairly favour the joint venture partners to the disadvantage of other competitive service providers. On the other hand, where considerable benefits could flow from the venture, these should in most cases be sufficient to warrant the grant of an individual exemption under Article 85(3) pursuant to a notification to the Commission. Recognisable benefits would include the rationalisation of production and distribution, technological improvements, greater efficiency and the enhanced competitiveness of European industry.

Finally, the Guidelines envisage that agreements simply between service providers (non-TOs) may also be capable of restricting competition, especially where the service providers enjoy a significant market strength and where the agreements concern quotas, price fixing and market or customer allocation. A typical example would be agreements with regard to the adoption of specific network architecture and other proprietary standards. The same problems are likely to arise with respect to agreements affecting the exploitation of intellectual property rights in standards, such as those adopted by ETSI. If such agreements were to have the effect of creating or enhancing a dominant market position and of restricting the use of other standards and the interoperability of equipment, the Commission is unlikely to grant an Article 85(3) exemption.[13]

---

[13] This view is confirmed in the Commission's discussion of intellectual property in its Green Paper on the development of European standardisation [1991] O.J. C20/1; *cf. CBEMA v. ETSI,* discussed above in Section I, "Regulatory Framework".

Satellites

16–79   Satellites is an area which is addressed separately by the Commission in the Guidelines subsequent to the release of a Green Paper on Satellite Communications.[14]

The Commission points out that by co-ordinating the activities of TOs in this sector, and particularly through the pooling of their sales of space segment capacity, they are capable of restricting competition between each other. However, more importantly, international arrangements of TOs for access to space segment capacity can prejudice competition by other satellite service providers. The Commission notes that since up-linking is gradually being liberalised throughout the E.U., co-ordinated arrangements between TOs as to the provision and use of up-links may give rise to infringements of Article 85.

Finally, the bundling of up-link and space segment capacity provision also causes concern as it is liable to limit competition to the detriment of consumers and to strengthen the already dominant position of the TOs involved.

Joint Ventures

16–80   Although paying due regard to the particularities of the industry, the Commission in its Guidelines applies its traditional analysis to joint ventures between TOs and those entered into between TOs and service providers, based on established principles which have arisen out of the application of Articles 85 and 86 over many years. There have been a number of notifications of such ventures since 1990 which have been granted an exemption or negative clearance by the Commission,[15] some of which are discussed separately below.

Pursuant to its powers of exemption contained in Article 85(3), the essential attributes of a joint venture or cooperation agreement which are likely to find favour with the Commission are those which aim to produce a high technology product, promote technical progress and facilitate the transfer of technology to the E.U. This positive attitude of the Commission is reinforced where, as is often the case, such ventures are designed to make the European telecommunications industry more competitive in global markets.

International Conventions: ITU and CCITT

16–81   Given the Commission's involvement in investigating the CEPT Tariff Recommendations and the "D" series of CCITT Recommendations Con-

---

[14] "Towards Europe-wide systems and services. Green Paper on a common approach in the field of satellite communications in the European Community, Communication from the Commission", COM(90) 490 final, Brussels, November 20, 1990.
[15] e.g., prior to the adoption of the Guidelines, refer to: *Alcatel/ANT* [1990] O.J. L32/19 (promotion of R&D in space electronic equipment—cooperation agreement); *Konsortium ECR 900*, [1990] O.J. L 228/31 (development of mobile telephone network—cooperation agreement); *GEC-Siemens/Plessey* [1990] O.J. C239/2 (defence industry electronics—joint takeover); *Eirpage* [1991] O.J. L306/22 (paging service operated throughout Ireland—joint venture).

cerning Private Circuits (see above), the Commission has also sought to provide further indications of its attitude to international conventions in the context of its interpretation of E.C. competition rules.

The Commission makes the point that its position in general is governed by Article 234 of the E.C. Treaty, which provides that rights and obligations concluded between one or more Member States on the one hand, and one or more third countries on the other before the Treaty entered into force are *not* affected by the provisions of the Treaty. However, according to the Commission, since ITU and World Administrative Telegraph and Telephone Conference (WATTC) regulations are terminated and renewed whenever they are revised or newly adopted, these will not be considered to pre-date the entry into force of the Treaty, which will therefore prevail in the event of a dispute. Even if this were not the case, the Treaty obliges Member States to take all necessary steps to eliminate any incompatibility between their international obligations and those to which they are subject under the Treaty.

The Commission confirms its view that CCITT recommendations require evaluation under E.C. competition rules. In the opinion of the Commission, such recommendations on international private circuits could amount to a collective horizontal agreement on prices and other terms of supply of such circuits. In practice, they may limit competition between TOs because they may lead to a coordination of sales policies. Similarly, such recommendations are also likely to restrict competition by service providers and other third parties insofar as the recommendations to restrict the use of private circuits continue in force. The Guidelines conclude with the Commission stating that it reserves the right to examine the compatibility of other CCITT recommendations with Article 85.

Conclusions

Since the Guidelines were issued in 1991, the telecommunications regulatory landscape has changed dramatically. The liberalisation of certain markets through the adoption of the Telecommunications Equipment and Services Directives, the effects of public procurement obligations on the holders of exclusive or special rights, the growth of competitive alliances for the provision of liberalised services, the increasing globalisation of telecommunications markets and the imminent liberalisation of alternative infrastructure, are factors which have radically altered many of the presuppositions on which the Guidelines are based. With the gradual erosion of special or exclusive rights for telecommunications infrastructure and services, the Guidelines are in need of significant revision.   16–82

3. COOPERATIVE JOINT VENTURES AND OTHER COOPERATIVE ALLIANCES

Since the early 1990s, the principal focus of the Commission's regulatory concerns under Article 85 of the E.C. Treaty is on the new series of   16–83 cooperative relationships developing among members of the sector as a result of the opportunities presented by the ongoing process of liberalisation. These new relationships not only cut across reserved and non-reserved services, but also combine equipment and service providers, TOs and satellite operators, and also those cooperative relationships designed to provide new service offerings.

Because of the rapid shifts in technology and customer preferences, the Commission is wary to ensure that markets do not develop in ways which will be restrictive of competition in the future. The Commission's traditional joint venture policy (as discussed above in the Guidelines) has therefore had to be reappraised in order to adapt to these different market and technological pressures. According to Commissioner Van Miert,[16] one of the Commission's main concerns in examining cooperative relationships involving TOs is the extent to which TOs could use their remaining monopoly power to eliminate competition in the market of the joint venture company; most of the assurances requested by the Commission have been designed with this concern in mind. Most importantly, the evolution of product and geographic markets over time also means that the Commission may be called upon to make periodic reassessments of joint ventures. Finally, Commissioner Van Miert takes the view that it may be necessary that, with respect to one and the same strategic alliance, certain activities should receive a *negative clearance*, whereas others may have to be prohibited altogether. In any event, despite the relative importance of the sector to the E.U.'s industrial development, the requirements for exemption under Article 85(3) "will by no means be waived because of the special features of the telecommunications sector".[17]

(i) Alliances for Advanced Services to Multinationals

16–84   A number of strategic alliances have recently been formed by national TOs to provide advanced telecommunications services on a one-stop shopping basis to multinational businesses. The services usually include data communications services, which are open to competition in the E.U., as well as voice services which can be provided to closed user groups such as multinational corporations.

In reviewing these joint ventures, the Commission has recognised that the market is new and rapidly developing, in part due to changes in technology, in part due to the current liberalisation of telecommunication services in the E.U. The Commission has stressed the advantages offered to users by one-stop shopping and one-stop billing facilities and the availability of "global" services of uniform quality. Generally speaking, it has sought to ensure that the benefits of the alliances outweigh any negative impact on competition by raising barriers to new market entry, by obtaining undertakings aimed at ensuring that:

- third parties will have non-discriminatory access to the telecommunication networks of the TOs which are parties to such alliances;

---

[16] Van Miert speech, *supra.*
[17] *ibid.*

- TOs will not be permitted to cross-subsidise the activities of their alliances from the exploitation of their public telecommunication infrastructures and the operation of reserved services; and
- TOs will not be permitted to bundle reserved and non-reserved services.

The Commission's policy is illustrated in the following joint ventures in respect of which it has granted individual exemptions under Article 85(3):

Infonet[18]

At the time of the Commission's review, Infonet was owned by five TOs   **16–85**
of E.U. origin (subsidiaries of France Telecom, Deutsche Telekom, Telefonica of Spain, RTT Belgium and PTT Telekom of the Netherlands) and various non-E.U. telecommunications operators (MCI Telecommunications, Telecom Australia, Singapore Telecom, Swedish Telecom, the Swiss PTT and Kokusai Denshin Denwa of Japan). Infonet offers global value-added network services, including data and voice communications services, on a one-stop shopping basis. Having obtained satisfactory undertakings against discrimination, bundling and cross-subsidisation, the Commission announced its intention to send a favourable comfort letter.

FNA[19]

FNA was established by six European and six non-European TOs to   **16–86**
develop a wide range of telecommunication services for customers active in the financial services sector (essentially banks and insurance companies). The joint venture will offer services especially adapted to the requirements of such companies, thereby combining voice, data and image services. It will offer its services worldwide through networks of the parents on a one-stop shopping and one-stop billing basis.

After receiving undertakings designed to prevent discrimination, bundling and cross-subsidisation, the Commission sent a favourable comfort letter. According to the Commission, centralised management and the optimisation of the existing national networks and international lines will increase quality and reduce costs, and the greater availability of bandwidth flexibility will benefit end users and service providers.

BT-MCI[20]

BT and MCI created a joint venture known as "Concert"[21] for the provi-   **16–87**
sion of enhanced and value-added global telecommunications services to

---

[18] Art 19(3) Notice; [1992] O.J. C7/3.
[19] Press release IP/93/988 of November 15, 1993.
[20] [1994] O.J. L223/36.
[21] BT, the former U.K. monopolist TO, supplies telephone exchange lines to homes and businesses, local, trunk and international telephone calls to and from the U.K. and other telecommunication services and equipment for customers' premises. MCI is a telecommunications common carrier in the USA which provides a broad range of U.S. and international voice and data communications services including long-distance telephone, record communications and electronic mail services to and from the U.S.

multinational or large regional companies. The parents contributed to the joint venture their existing non-correspondent international network facilities, including Syncordia, BT's existing outsourcing business. The emerging market to be addressed by the joint venture covers a wide range of existing global transborder services, including virtual network services, high-speed data services and outsourced global telecommunications solutions specially designed for individual customer requirements. This market is characterised by services and features (one-stop shopping, end-to-end and seamless service, and so forth) required by multinationals and other big international users, which are difficult to provide at present under the existing framework of cooperative relationships established by TOs.

The creation of this joint venture was held to infringe Article 85(1), as BT and MCI are at least potential competitors of the joint venture and of each other in respect of the global products to be offered by Concert, and are also actual competitors in the broader telecommunications market.[22] Moreover, Concert was considered to be a vehicle for the parents to pool their respective intellectual property rights and to cross-license each other and the joint venture on an exclusive basis.

The restrictions of competition resulting from the creation of the joint venture were nevertheless found to satisfy the conditions for exemption under Article 85(3) by improving telecommunications services and technical/economic progress in the E.U. to the benefit of consumers. Moreover, according to the Commission, there is significant third-party competition from AT&T's Worldsource and ATLAS (a joint venture of France Telecom and Deutsche Telekom),[23] other existing alliances such as Unisource and ISPS (see below), and future alliances which are likely to be formed. Moreover, the formation of the joint venture was found to be indispensable for the parent companies to enter the market successfully because it shortened the lead time for entry and substantially reduced the costs and risks.

16–88   In the application of its ancillary restraints doctrine, the Commission concluded that an obligation of the parent companies not to compete with the activities of their joint venture was ancillary to the joint venture's creation and successful operation, and thereby fell outside the scope of Article 85(1). It reached a similar conclusion regarding an obligation for the parent companies to obtain all of their requirements from the joint venture.

As an example of the application of the principle of subsidiarity, the Commission took the view that the regulatory constraints to which BT and MCI were subject in the U.K. under national licensing requirements ensured that no undue discrimination or cross-subsidisation could occur.

In addition, BT and MCI concluded agreements pursuant to which BT acquired a 20 per cent stake in MCI, which would make it the largest single shareholder in MCI, with proportionate board representation and investor

---

[22] Having so concluded, the Commission was nevertheless of the view that it was not possible to draw a more precise picture of the existing structure of this emerging market given that its principal feature was that it was constantly evolving.

[23] Notice to third parties on ATLAS joint venture [1994] O.J. C377/9. More recently, refer to the Phoenix Notice [1995] O.J. C184/11.

protection. The Commission concluded that this acquisition of a minority interest fell outside the scope of Article 85(1), because on the facts of this case the presence of BT's nominees on the MCI board and BT's possible access to MCI's confidential information was unlikely to give rise to the coordination of competitive behaviour. The basis for this conclusion was the fact that BT had undertaken in the acquisition agreement not to increase its shareholding for 10 years and not to seek to control or influence MCI. Further, both American corporate and antitrust laws would impede any access or misuse by BT of any piece of confidential MCI information.

By way of contrast, the *inability* of the ATLAS joint venture (as notified) to be able to address the global needs of multinational companies in competition with other strategic alliances such as BT/MCI's Concert joint venture, was identified by Commissioner Van Miert as being a key reason behind the Commission's long delay in granting regulatory approval to the ATLAS joint venture.[23a]

(ii) Alliance for Advanced Paging Services

The Commission's decision in *Eirpage*[24] illustrates clearly how broadly the notion of "trade between Member States" is to be interpreted under Article 85(1). Although the paging services to be run by the Eirpage joint venture (made up of the equipment manufacturer Motorola and Irish Telecom) were restricted to the territory of Ireland, they were regarded as having an effect on trade between Member States because the formation of the joint venture was said to: (i) have a *dissuasive* effect on market entry by other competitors due to the fact that Eirpage would be the only provider of those services; (ii) create the potential to *attract* other providers of paging services from competitors of various Member States; (iii) *stimulate* further imports or investments through the sales of paging equipment prompted by the joint venture; and (iv) inevitably form part of the Motorola group's broader E.C. and *worldwide* strategy.

16–89

It was only upon receipt of a number of significant contractual amendments that the Commission felt in a position to exempt the Eirpage joint venture under Article 85(3). First, to ensure that there would be no foreclosure of third party competitors, Irish Telecom and the relevant licensing authorities had to assure that potential competitors of Eirpage would be treated on an equal footing. Second, sales agents for paging services had to be free to promote their own complementary paging services (although not those of the joint venture's competitors) and had to be free to compete with Eirpage upon its termination.[25] Third, Eirpage was required to operate at arm's length from its parents, with Irish Telecom providing assurances that it would not cross-subsidise the joint venture from its other operations, nor grant it any preferential tariffs for the use of its facilities (*e.g.* leased lines). Fourth, assurances were given that Eirpage would cooperate with

---

[23a] See Press Release IP/95/524 of May 24, 1995; *cf.* Press Release IP/95/791 of July 18, 1995.
[24] [1991] O.J. L306/22.
[25] Originally, the agency contract provided for a non-solicitation of Eirpage customers obligation for three years after the termination of the contract.

all paging equipment manufacturers or dealers to ensure that their products could be used on the Eirpage system so that Motorola would not obtain any undue advantage at this level of the market. Finally, the original three year post-term ban on competition between the parents was removed, with the Commission confirming its traditional policy that the parents must be free to compete with one another immediately upon termination of the joint venture.

(iii) Alliance for the Resale of Mobile Telephony Air-time

16–90   The Commission announced in 1994 that it intended to take a favourable view towards the notification of a joint venture and cooperation arrangement in the field of mobile telephony services[26] concerning the operations of CMC, a French reseller of mobile telephony air-time (*i.e.* a "service provider"). CMC is jointly owned by Matra Communications (French manufacturer of telecommunication equipment), Cellcom (U.K. mobile telephony service provider), Talkline (a German mobile telephony service provider) and Norauto (a French retailer of automobile accessories).

According to the Notice, CMC, Cellcom and Talkline have signed a memorandum of understanding in which they agree to cooperate for the purpose of providing the pan-European distribution of mobile telephony services aimed at establishing a one-stop shopping and billing service facility to any of their respective customers. They also intend to procure jointly terminal equipment, establish regional stocks, apply common distribution methods and allow customers to obtain after-sales service for the equipment in any country in which another party to the MoU operates. A customer of one party will have access to SIM-cards (service subscription) and mobile phone services provided by any other party to the MoU. The parties will share know-how and software applications for subscriber handling, billing and customer service and intend to develop jointly a European SIM-card enabling cooperative services; to achieve this, they will adopt a joint training programme, recruit foreign language-speaking staff and exchange trading concepts and materials.

The Shareholders Agreement originally prohibited the parent companies from competing with CMC and restricted CMC to operating only in France as an air-time reseller. At the Commission's request, these provisions were deleted.

(iv) Alliance for Telecommunications Equipment

16–91   In 1992, the Commission expressed its intention to grant an individual Article 85(3) exemption for a joint venture in the field of telecommunications equipment involving STET, Italtel, AT&T and NTI.[27] The technical cooperation consists of exchanges of technical and commercial information, with the commercial cooperation consisting of exclusive cross-

---

[26] *CMC-Talkline*, Article 19(3) Notice; [1994] O.J. C221/9.
[27] *STET, Italtel, AT&T and AT&T–NSI*, Art. 19(3) Notice; [1992] O.J. C333/3.

distribution and purchasing agreements.[28] The products covered by the agreement are all high-technology products based on sophisticated electronic components and complex software programmes, namely telecommunications equipment for public and private switching systems; operating systems; public transmission systems; and certain private terminal equipment.

The Commission distinguished between the markets for public and private network equipment, because of their different demand structures, in order to assess the effects on competition of the notified agreements. The main users of public network equipment are public and private network operators, which have been granted exclusive or non-exclusive rights to operate the public network. Private network equipment is sold to a wide range of customers, from private and public enterprises to retailers, in a market that has already been liberalised as a result of the adoption of E.U. procurement rules.

The Commission initially objected to certain aspects of the notified agreements, namely, the territorial protection covering the distribution of public network products and the threat to competition posed by the vertical integration of the parties in their respective markets. During the administrative procedure, the parties agreed to abolish all territorial restrictions except where the costs involved in placing a licensed product on the market warranted protection from competitors for a limited period of time. As regards the issue of vertical integration, the Commission was concerned by the possibility of selling components and sub-assemblies at prices that did not reflect cost (in the case of AT & T) and the potential creation of barriers to entry on domestic markets (in the case of STET). AT & T provided sufficient evidence for the Commission to withdraw its objection, whereas STET demonstrated that the position of Italtel was not so strong as to prevent the entry of competitors and undertook to sell products only at prices more reflective of its costs.

(v) Alliance for Ground-to-Air Services

In May 1994, the Commission received a notification for negative clearance **16–92** or exemption for a joint venture agreement[29] between subsidiaries of BT and France Telecom for the development of telecommunications services operating on board aircraft and relayed by terrestrial means (*i.e.* a terrestrial flight telephone system). The Commission invited comments from interested third parties, but did not express its preliminary conclusions on its review of the joint venture.

---

[28] The technical cooperation makes provision for the joint development of products as well as the granting of exclusive licences and patents on a reciprocal basis. The commercial cooperation specifies that certain products will be purchased jointly from third parties, and that other products exchanged between the parties to the agreement will be sold at an internal price reflecting real cost. In so far as the distribution of the products is concerned, any market where any one of the parties to the agreement is already present is to be that party's exclusive territory. The technical and commercial cooperation between the parties which is managed by the technical and commercial management boards (one for public network products and one for private network products) will be accompanied by the acquisition by each party of a minority holding in the other's capital.

[29] *Jetphone* [1994] O.J. C134/5.

(vi) Consortium for Development of GSM Mobile Network[30]

16–93   The Commission granted negative clearance to a cooperation agreement between AEG, Alcatel, and Nokia relating to the formation of a consortium, ECR900, for the joint development, manufacture and distribution of a new pan-European digital cellular mobile telephone system. In developing the system on the basis of uniform standards, it was intended that it would be operable throughout Europe without regard to national and geographic frontiers. The national TOs and undertakings acting on their behalf were at the time the only potential buyers of the new system. Moreover, demand for all/part of the system was channelled through invitations to tender; such invitations involved orders for supply and installation, but not development orders. The consortium was to be used for the purpose of submitting such tenders in response to such invitations.

As regards the content of the cooperation agreement, provision was made for territorial restrictions on individual action by members of the consortium. Technical documentation was to be freely exchanged where several parties were involved in development activities, but was not to be exchanged where only one party was involved. Various time limits existed with regard to the free exploitation and licensing of jointly obtained information in the event of the agreement expiring. Members of the consortium in breach of the terms of the agreement were to be penalised by the loss of the right to use technical documentation.

16–94   The Commission's decision that the cooperation agreement did not have as its object or effect the restriction of competition within the meaning of Article 85(1) was reached after a three-pronged examination. First, as regards the development and manufacture of the system, the Commission concluded that this could not take place by the efforts of a company acting alone for a number of reasons, including the high individual cost of developing such a system; the tight deadlines for tender within which the system had to be developed (the installation and operation of the equipment being scheduled for the first quarter of 1991); the shortage of qualified personnel who could develop the system in that short time period; the limited amount of demand in this market; and the fact that even the invitations to tender anticipated collective bidding. Accordingly, no single member of the consortium was able to use its own production improved by individual development in order to achieve a competitive advantage over the other members. Secondly, the Commission adduced similar reasons for its finding that the joint distribution provided for by the agreement did not constitute a restriction on competition. Finally, the Commission held that the ban on a party from using the technical documentation because of its infringement of the agreement fell outside Article 85(1).

According to the Commission, if such a ban on the use of technical documents were disallowed, the infringing party would receive unjustified benefits which would lead to an undeserved competitive advantage *vis-à-vis* the

---

[30] *Konsortium ECR 900* [1990] O.J. L228/31.

other parties to the agreement. In the words of the Commission "[s]uch competition not based on performance is not protected by Article 85."

(vii) Satellites

The Commission's attitude towards strategic alliances in the field of satel- **16–95** lites and broadcasting is significantly more circumspect than is its attitude toward other joint ventures in the telecommunications sector. It is particularly concerned about the foreclosure effects of exclusive relationships which restrict access to space segment capacity.

For example, in June 1995 the Commission launched a series of investigations into a number of global mobile satellite systems (MSS). Following the notification of Inmarsat-P, a major MSS, the Commission requested that the rival consortia of Globalstar and Iridium provide the Commission with a comprehensive description of their systems from a technical, financial and commercial point of view. Despite the fact that MSS systems are global and prima facie competitive, the Commission is most concerned about the terms and conditions of the distribution policies chosen by the respective consortia and the nature of their links with cellular terrestrial networks, and also the issue of access by a competing MSS to infrastructure owned by a partner to one of the consortia being assessed.[30a]

Intrax[31]

Following the publication of its "favourable position" in an Article 19(3) **16–96** Notice, the Commission granted "clearance"[32] by means of a "comfort letter" to a joint venture (Intrax B.V.) between the Dutch TO and *Nederlands Omroepprodukite Bedrijf N.V.* (NOB), the main Dutch television facilities house, for the provision of a satellite news gathering service. According to the Commission, other news providers wishing to compete with Intrax on the Dutch market would not be faced by any major barriers to entry. These providers would not be foreclosed by the cooperation agreement between the Dutch TO and NOB because their agreement is nonexclusive: the Dutch TO remains free to provide "uplinking" services (the transmission of signals to satellites) to parties other than Intrax; NOB is free to provide technical facilities to parties other than Intrax; and both the Dutch TO and NOB are free to compete directly with Intrax in the provision of satellite news gathering services. Furthermore, the Commission took a favourable view of this cooperation agreement on the basis of a number of additional grounds, namely potential competitors of Intrax would be able to obtain uplinking services from the Dutch TO or other parties (as the market for such services in the Netherlands has been liberalised); competitors would be able to rent capacity through the Dutch TO from satellite-operating consortia such as EUTELSAT and INTELSAT

[30a] Press Release IP/95/549 of June 7, 1995.
[31] *Intrax*, Art. 19(3) Notice [1993] O.J. C117/3. See also Commission press release IP/93/907 of October 25, 1993.
[32] It is not clear from the Commission's press release whether "clearance" constitutes negative clearance.

on the same terms as Intrax (the Dutch TO, the Dutch signatory of the EUTELSAT Convention, undertook not to grant Intrax preferential terms); they would also be able to rent capacity on EUTELSAT satellites through the national signatories in other Member States or on independent satellites.

Astra[33]

16–97　In another case, the Commission took the highly unusual step of refusing clearance or exemption to a joint venture (BT Astra SA) notified to it by BT and *Société Européenne des Satellites* (SES), a Luxembourg-based private satellite operator.[34] The purpose of the joint venture was to offer broadcasters of U.K.-originated TV programmes a package service consisting of a facility—provided by BT—for beaming telecommunications signals to the Astra 1A satellite operated by SES (uplink) and a facility—provided by SES—for receiving the signals on that satellite before beaming them back for reception on the earth's surface (transponder capacity). Before the launch of Astra 1A, all satellites in Europe were owned exclusively by telecommunications operators.

In the *market for uplink services*, the Commission found that BT and SES were in direct competition with one another and that various restrictions in the agreement eliminated this competition. Although SES provided its uplink service in Luxembourg, the Commission concluded that it was technically feasible and commercially viable for U.K. programme providers to seek uplinking facilities in other Member States. The joint venture agreements obliged BT to consult with SES in setting the price for the uplinking service and prevented SES from offering commercially preferential terms to programme providers for the use of its facilities in Luxembourg. By bundling BT's uplinking service with the transponder capacity of SES, in other words, SES and other providers of such service were foreclosed from satisfying customer needs.

16–98　In the *market for transponder capacity*, the Commission also concluded that BT and SES were in direct competition with one another because of the participation of the former in international satellite-operating consortia such as EUTELSAT and INTELSAT and the latter's ownership of the Astra 1A satellite. Even though BT could only offer customers low-powered satellite capacity at the time of the joint venture, the Commission found that low-powered satellites (operated by the international consortia) and high-powered satellites (operated by SES) offered similar capabilities—with the exception of Direct to Home (DTH) reception—and therefore formed part of the same service market. In addition to BT's participation in international satellite-operating consortia, the Commission found that BT had the

---

[33] [1993] O.J. L 20/23.

[34] The Commission has struck down joint ventures as incompatible with Art. 85(1) on only two other occasions in the past: *Wano-Schwarzpulzer GmbH ("Black Powder")* [1978] O.J. L322/26; and *Screensport/EBU Members* [1991] O.J. L63/32. The *Screensport* decision, also in the field of telecommunications, indicates the depth of the Commission's concern with the effects of foreclosure on third parties where new or rapidly developing industries are affected.

financial and technical capabilities to enter independently the market for the operation of satellites. The joint venture agreements prevented SES from marketing transponder capacity to U.K. programme providers except through BT Astra and compelled SES to consult with BT in setting the price charged for the capacity to such providers. BT or SES independently from the joint venture were foreclosed from offering transponder capacity to customers of the services offered by the Astra 1A satellite.

Although the Commission conceded that the Astra 1A satellite supplemented consumer choice by providing privately-owned medium satellite capacity in competition with EUTELSAT and INTELSAT, it found that these advantages did not flow from the joint venture itself. Having found that the joint venture did not improve satellite services or impose restrictions which were strictly indispensable to its operation, the Commission refused to grant an Article 85(3) exemption. The case illustrates the Commission's serious concerns to prevent any foreclosure effects flowing from bundled agreements in new or rapidly developing fields in telecommunications, particularly if it believes that less restrictive options are available. The case is also notable insofar as the Commission has determined the legal effect which provisions which are void under Article 85(2) have on the joint venture agreement; rather than leaving this determination to national courts under local law, the Commission provided existing customers of the joint venture with the opportunity to renegotiate their contracts under radically changed commercial circumstances.[35]

Eurosport Mark III[36]

In an Article 19(3) Notice, the Commission took a favourable view toward the revised cooperation agreements involving Eurosport, a joint venture between a consortium of national television companies belonging to the European Broadcasting Union (EBU) and News International/Sky Channel. The purpose of the joint venture is to run a pan-European television channel devoted entirely to sports programmes broadcast by satellite and cable. The Commission had prohibited the first "generation" of these agreements in 1991 on the ground that they granted the joint venture privileged access to sports events and foreclosed other satellite and cable services.[37] Thereafter, Sky Channel withdrew from the joint subsidiary and the Eurosport consortium found a new partner in *Télévision Française 1*. The joint venture, in its new legal and operational form, was notified to the Commission under the name Eurosport Mark II. The name of the joint venture was changed once again to Eurosport Mark III to reflect the addition of new partners (Canal +, ESPN and *Générale d'Images*) and further changes to the structure and operation of the joint venture.

The Notice does not state whether the notified agreements contained com-

16–99

---

[35] BT filed an appeal to the Court of First Instance challenging the Commission's decision, which it subsequently decided to withdraw: action brought on March 17, 1993 by *British Telecommunications plc v. Commission* [1993] O.J. C113/4; proceedings removed from the register of the Court by Order of September 22; [1993] O.J. C277/8.
[36] *Eurosport Mark III*, Article 19(3) Notice [1993] O.J. C 76/8.
[37] See *Screensport/EBU Members* L63/32.

petitive restrictions which had to be eliminated; rather, it merely notes that the public service obligations of the consortium members will be carried over to the programming of the joint venture itself. There is no indication in the Notice as to why the new agreements do not give rise to the foreclosure concerns present in the prior versions of the notified agreements.

Aerospatiale/Alcatel Espace[38]

16–100 The Commission has announced its intention to take a favourable view towards a cooperation agreement between Aerospatiale and Alcatel Espace covering civilian and military telecommunications satellites, whose scope may be broadened in future to include other areas such as earth observation. The agreement provides for business cooperation, coupled with technical and industrial specialisation. It is aimed at rationalising the activities of the parties, but does not provide for the integration of their production lines. Consequently, each of the parties retains its own research and production facilities. The cooperation agreement forms part of a strategy of alliances pursued by the parties with regard to satellites, the aim of which is to establish a vertically integrated industrial facility of sufficient size to meet the requirements of a rapidly developing world market. The formation of this alliance must be seen in the framework of the general trend within Europe towards greater vertical integration in the satellite market, in particular the deregulation of telecommunications and the entry into force of the E.C. public procurement rules which must guarantee equality of opportunities to suppliers. In contrast, the U.S. market is dominated by a limited number of companies which are very large and vertically integrated. American companies have the benefit of a large, deregulated and unified domestic market and of government space programmes, and are very active outside the U.S., although not yet in Europe. However, with entry barriers in the E.U. falling, the American companies are expected to be increasingly active in Europe. The response of European companies to these developments is to increase in size and pursue vertical integration, through mergers and large-scale strategic alliances.

16–101   The key element in the cooperation between the two parties consists of a cooperation procedure based on a general principle of mutual provision of information and consultation on all measures relating to the areas covered by the agreement as well as on joint commercial action. In taking a favourable attitude towards the cooperation agreement, the Commission emphasised that the cooperation and joint commercial action envisaged left the parties a substantial margin for independent action.

International Private Satellite Partners (IPSP)[39]

16–102 In December 1994, the Commission granted negative clearance to notified agreements for the creation and operation of IPSP, company created under the form of a U.S. limited partnership, which had been formed to provide international business telecommunications services to businesses in Europe

---

[38] Art. 19(3) Notice [1994] O.J. C47/6.
[39] [1994] O.J. L354/75.

and North America through the use of its own satellite system. Services would be provided on a one-stop shop basis, and bulk transmission capacity would also be offered to third parties to the extent that the capacity of the satellites is not fully utilised by IPSP or its partners. The IPSP venture was created between nine partners, most of which are private companies active in the telecommunications and aerospace sectors, and will introduce a new competitor into the market for advanced telecommunications services.

Although the Commission did not expressly characterise IPSP as a joint venture, it applied its traditional test with regard to the creation and operation of joint ventures, namely, to verify whether or not the partners are actual or potential competitors. The Commission concluded that this was not the case, for a number of reasons: only the general partner had the necessary authorisations and licences to launch and operate the satellite; none of the partners held the necessary authorisations and licences to provide international telecommunications services in all the countries inside the footprint of the satellites; and none of the partners could reasonably be expected to make the required investment alone assuming the substantial risks associated with it. The Commission added that IPSP, as a new competitor, could be expected to increase the level of competition in a fast-growing segment of the overall telecommunications market which had until recently been reserved to companies holding exclusive rights. It is clear that the Commission considered that the "private venture", which included only two incumbent TOs (namely, STET and Kingston) would be a welcome new competitor on the market for enhanced value-added telecommunications services which is starting to be more and more dominated by big strategic alliances between TOs, and would also be a welcome alternative to the currently existing satellite facilities for the provision of space segment capacity (which up to now had mainly remained in the hands of three international satellite organisations, INTELSAT, EUTELSAT and INMARSAT).

In granting negative clearance to certain clauses in the agreements which **16–103** have otherwise often been seen as restrictions of competition, the Commission distinguished between three categories of clauses: (a) clauses which fell outside the scope of the E.C. EEA competition rules since they did not contain restrictions of competition; (b) provisions which were considered to be non-appreciable restrictions of competition; and (c) provisions which were considered to be ancillary restrictions, since they were directly related and necessary to IPSP and did not exceed what was required for the creation and operation of IPSP. Most importantly, the Commission considered that STET's *exclusive distribution right* for the services in Italy fell outside Article 85(1), on the ground that STET had in any event a legal monopoly for the provision of various services in some of the areas to be addressed by IPSP. STET's exclusive promotion rights for the services, which will replace its exclusive distribution rights once the Italian market has been deregulated, were considered to be non-appreciable restrictions of competition; this view was justified on the basis that IPSP was expected to have a market share below 5 per cent of the two markets affected by the operation.[40]

---

[40] It should be noted that the five per cent market share argument was used independently of any consideration with regard to turnover. Compare with Commission Notice 1986 on agreements of minor importance [1986] O.J. C231/2 (as amended).

Conclusions

16–104 The Commission's review of cooperative alliances in the telecommunications sector has by and large been characterised by the following trends in regulatory evaluation under Article 85 of the E.C. Treaty:

- a tendency to take a much more commercially realistic view of the notion of "actual or potential" competition when reviewing joint ventures for the provision of new products or services;
- a concern about the foreclosing effects of certain joint ventures, especially where those ventures enjoy a "bottleneck monopoly" (actual or *de facto*) over certain "essential facilities";
- a gradual broadening of the relevant product and geographic markets for telecommunication services;
- a desire to control potential anti-competitive effects through the imposition of behavioural undertakings on the parents and on the joint venture itself;
- a reluctance to characterise joint ventures for the provision of telecommunication services as being "concentrative" in nature, and thereby subject to the more liberal and speedy review procedure available for "concentrations" (see paragraphs 16–105 to 16–112 below); and
- a respect for the principle of subsidiarity, by permitting national regulatory authorities to provide the appropriate regulatory framework, wherever appropriate, with respect to certain types of potentially anti-competitive conduct.

4. MERGERS AND CONCENTRATIVE JOINT VENTURES

16–105 The so-called Merger Control Regulation[41] applies with equal force to "concentrations" in the telecommunications sector as in other sectors. Under the Merger Control Regulation, mandatory notification to the Commission's Merger Task Force is required for "concentrations" which exceed a "Community dimension".[42]

A concentration arises where two or more previously independent undertakings merge or where an undertaking acquires direct or indirect control of the whole or part of another undertaking. This definition embraces joint ventures which are concentrative in character, namely, those which perform the functions of a full function, autonomous entity without giving rise to the risk of coordination of the competitive behaviour of the parents or (until recently) between the parents and the joint venture itself.[43] Although the rules may

---

[41] Regulation 4064/89; [1989] O.J. L395/1.

[42] For a detailed review of the Regulation, reference should be made to the following texts: Jones & Gonzalez–Diaz, *The EEC Merger Regulation* (1992) (1995 ed. forthcoming); Cook & Kerse, *EEC Merger Control* (1991); Downes & Ellison, *The Legal Control of Mergers in the European Communities* (1991).

[43] Commission Notice of December 21, 1989 on the notion of a concentration under Regulation 4064/89 on the control of concentrations between undertakings; [1994] O.J. C385/5.

have become slightly more flexible, generally speaking, a significant body of administrative practice requires that the formation of a concentrative joint venture should result in the withdrawal of the parents completely from the relevant product/service market in which the joint venture is to operate. With the exception of joint ventures formed between equipment suppliers, the majority of joint ventures between telecommunication service providers are considered to be cooperative—and therefore reviewable under Article 85 of the E.C. Treaty—because of the risk of coordination between the parents, especially where the parents have their own infrastructure.[44]

A concentration must also have the required "Community dimension" to qualify for notification and review under the Merger Control Regulation which requires that:

- the combined aggregate worldwide turnover of all the undertakings concerned must exceed ECU 5 billion; and
- the aggregate Community-wide turnover of each of at least two of the undertakings concerned must exceed ECU 250 million;
- *unless* each of the undertakings concerned achieves more than two-thirds of its aggregate Community-wide turnover within the same Member State.

If the merger or concentration satisfies the Community dimension cri- **16–106** terion, notification to the Merger Task Force must occur before the merger or concentration is put into effect. The Commission examines, within a period of one month in simple cases and up to five months in the more difficult cases, whether the merger or concentrative joint venture is compatible with the common market. The Commission is under a legal obligation to prohibit those concentrations which create or strengthen a dominant position resulting in a significant restriction of competition. For example, the Commission is anxious to avoid the strengthening of positions of market dominance in the E.U. or on national markets as a result of vertical integration (a dominant TO may not be allowed to expand into the equipment market if it is likely to favour the use of its own equipment to the exclusion of other equipment suppliers on the market).

The Guidelines (see above) set forth the Commission's general policy position that restructuring within the European telecommunications industry is in general beneficial where it facilitates rationalisation and leads to the achievement of economies of scale. These conditions are said to be necessary if new technologies are to be developed (with the associated high R & D costs) and if Europe is to remain competitive in the world market.

The Commission's anti-competitive concerns are illustrated in the three cases decided to date which have entered into a "second phase" of investigation:

Alcatel/Telettra[45]

In January 1991 the Commission decided to carry out a second phase **16–107** investigation under the Merger Control Regulation to examine the pro-

---

[44] Refer to table, "Telecommunications Cases Decided Under Regulation 4064/89" at the end of this Chap.
[45] [1990] O.J. C315/13.

posed acquisition by Alcatel of a controlling interest in Fiat's telecommunications subsidiary, Telettra. The Merger Task Force's preliminary assessment was that the proposed merger would lead to high combined market shares for the new Alcatel/Telettra entity on the market for transmission equipment in. Spain (the transmission equipment market constitutes the second most important telecommunications sector, with a value of more than ECU 600 million in Spain in 1989).

The Commission approved the merger, subject to strict obligations imposed on Alcatel and firm assurances given by Telefonica, the Spanish TO. Sir Leon Brittan (Vice-President of the Commission and, at the time, responsible for competition policy) stated that the merger raised serious questions of competition policy because of the parties' high combined market share in Spain for transmission equipment (approximately 80 per cent). Although this would normally be unacceptable, he had obtained satisfactory commitments from Alcatel to agree to buy Telefonica's shares in Alcatel, thus opening up competition between suppliers of equipment to Telefonica. Telefonica also gave its assurance that it would pursue a diversified buying policy and agreed to clarify its technical approval procedures. It was also concluded that Telefonica's industrial presence in Spain would no longer be a decisive factor in the awarding of contracts. Because of the concern that links between a TO and its suppliers might distort competition by giving suppliers privileged market access, the Commission imposed strict legal obligations to ensure that these commitments were fully respected.

MSG Media Service[46]

16–108 The Commission refused authorisation to a concentration notified in June 1994 by the German undertakings Bertelsmann, Taurus Beteiligungs (Kirch group) and Deutsche Bundespost Telekom to create a joint venture, MSG Media Service GmbH (MSG), with the object of supplying administrative and technical services to digital pay-TV operators.

The Commission's view was that such a concentration was incompatible with the common market because it was likely to create or reinforce dominant positions in a number of markets: (i) the market for administrative and technical services supplied to pay-TV operators and operators of other TV services; (ii) the market for pay-TV; and (iii) the market for cable networks.

16–109 **Administrative and technical services market:** According to the Commission, it was expected that the initial monopoly of MSG would be durable and that the market would be sealed off. Entry on the market by a competitor to MSG would be very unlikely given the fact that Telekom owns most of the cable network in Germany and controls its development, and because Bertelsmann and Kirch already have their own incomparable programme resources.

16–110 **Pay-TV market:** The Commission concluded that the concentration notified

---

[46] [1994] O.J. C160/4.

would enable Bertelsmann and Kirch to hold a dominant position on this market because they were the only suppliers of pay-TV services in Germany. In addition, the activities of both companies in the field of film rights, free TV and media guaranteed them an extremely strong position on the relevant market, and competitors who would have to use MSG services to supply pay-TV could be subject to numerous forms of commercial or technical discrimination. Moreover, MSG would serve as the vehicle for information regarding the activities of competitors.

Cable networks market: The Commission considered that the concentra- **16–111** tion was likely to strengthen Deutsche Telekom's dominant position on the market because private cable operators could not easily compete with Telekom if the latter controlled MSG.

Undertakings proposed by the parties (*e.g.* the adoption by MSG of a conditional access technology which should allow independent programme providers to use the decoders) were rejected by the Commission on the basis that these were behavioural (rather than structural) undertakings which are difficult to monitor from a technical point of view.

STET/Siemens[47]

The Commission was notified in September 1994 of the proposed concen- **16–112** tration between Siemens Telecomunicazioni SpA, an Italian subsidiary of Siemens, and Italtel, an equipment manufacturer belonging to the STET group. STET is the holding company which also controls Telecom Italia. The Commission took the view that the joint venture raised serious doubts as to its compatibility with the common market because Siemens, through the joint venture, would share the pre-existing shareholder link between the Italian TO (Telecom Italia) and the equipment manufacturer, Italtel. In particular, it was concerned about the impact that the transaction could have on public procurement opportunities and on the opening up of national equipment markets to eu-wide competition.

In the second phase of its investigation the Commission found that, despite the joint venture's high market share, it would not lead to a dominant position on the market, and formally approved the concentration on February 17, 1995. Three reasons were cited by the Commission in support of its view: (i) the possible development of large markets because of technological developments; (ii) the fact that the effects of standardisations and public procurement legislation would have a larger impact on opening national markets; and (iii) progress towards liberalisation of services and infrastructure would lead increasingly to the creation of a worldwide market for public telecommunications equipment. As regards the shareholding link between the new joint venture and Telecom Italia, the Commission recognised that any benefits of any privileged treatment to the joint venture imposed on Telecom Italia by STET would be shared with Siemens. During the case, STET reassured the Commission that its interests in the new

[47] [1994] O.J. C264/4.

Siemens/Italtel joint venture would not conflict with its interests in Telecom Italia.

Telecommunications Cases Decided under Regulation 4064/89

Transactions	Notification	Decision
Alcatel/Telettra (Case IV/M.042)	C315/13 14.12.90	17.05.91 (L122/48 1991)
Sunrise (Case IV/M.176)	C312/23 03.12.91	13.01.92
SAAB Ericsson Space (Case IV/M.178)	C314/24 05.12.91	13.01.92
Ericsson Kolbe (Case IV/M.133)	C329/23 18.12.91	22.01.92
Ericsson/Ascom (Case IV/M.236)	C151/14 16.06.92	08.07.92
Matra/Northern Telecom (Case IV/M.249)	C181/22 17.07.92	10.08.92
Siemens/Philips Kabel (Case IV/M.238)	C227/9 03.09.92 (withdrawn)	
CEA Industrie/France Telecom/Finmeccanica/ SGS Thomson (Case IV/M.216)	C27/3 30.01.93	22.02.93
Ericsson/Hewlett-Packard (Case IV/M.292)	C48/9 19.02.93	12.03.93
JCSAT/SAJAC (Case IV/M.346)	C158/6 10.06.93	30.06.93
Alcatel/STC (Case IV/M.366)	C209/17 03.08.93 C225/2 20.08.93	13.09.93
British Telecom/MCI (Case IV/M.353)	C226/3 21.08.93	13.09.93
Mannesmann/RWE/Deutsche Bank (Case IV/ M.394)	C321/13 27.11.93	22.12.93
RWE/Mannesmann (Case IV/M.408)	C30/11 01.02.94	28.02.94
BS/BT (Case IV/M.425)	C68/4 05.03.94	28.03.94
MSG Media Service (Case IV/M.469)	C160/4 11.06.94	09.11.94 (L364/1 1994)
Matra Marconi Space/British Aerospace (Case IV/M.437)	C221/12 09.08.94	
Tractebel/Distrigaz II (Case IV/M.493)	C222/4 10.08.94	
Marconi/Finmeccanica (Case IV/M.496)	C223/7 11.08.94	05.09.94
Bertelsmann/News International/Vox (Case IV/ M.489)	C227/7 17.08.94	06.09.94 26.09.94
Rheinelektra/Cofira/Dekra (Case IV/M.485)	C247/3 03.09.94	17.02.95
Siemens/Italtel (Case IV/M.468)	C264/4 21.09.94	14.10.94 (O.J. L161/27) 12.07.95)
Matra Marconi Space/Statcomms (Case IV/ M.497)	C269/5 27.09.94	18.01.95
Vox II (Case IV/M.525)	C333/5 29.11.94	30.11.94
Cable & Wireless/Schlumberger (Case IV/ M.532)	C334/5 30.11.94	22.11.94
Nordic Satellite Distribution (Case IV/M.490)	C53/6 04.03.95	pending
Omnitel (Case IV/M.538)	C53/8 04.03.95	27.03.95 inapplicability
TBT Communitations/BT/TeleDanmark/Tel- enor (Case IV/M.570)	C76/3 28.03.95	24.04.95
Blockbuster/Burda (Case IV/M.579)	C.78/15 30.03.95	27.04.95
CTL/Disney/SuperRTL (Case IV/M.566)	C96/3 20.04.95	30.05.95

PART THREE

International

CHAPTER 17

Australia

By Peter G. Leonard, Gilbert & Tobin, Sydney

Although occupying a continent only marginally smaller than the contin-   17–01
ental United States, Australia has a population of only 17 million. A
common international perspective of Australia is that it is a predominantly
rural and resource based economy with a cultural and business orientation
towards western Europe and North America. However, Australia is highly
urbanised. The Australian economy is increasingly services based and
rapidly integrating into the East and South East Asian economy.

An important part of economic change in Australia has been Government
initiatives to increase the international competitiveness of Australian businesses and open up former Government monopolies to competition. Telecommunications was one of the first industry sectors targeted for reform.
Telecommunications business in Australia is now undergoing a period of
radical change as a result of telecommunications network and services competition, allocation of satellite pay television licences and plans for introduction of competitive broadband networks. Demand for communications
equipment has also continued to grow at rates at least comparable to other
industrialised economies, expanding opportunities for equipment suppliers
and service providers.

TELECOMMUNICATIONS POLICY

In June 1991 the Australian Parliament passed a package of legislation: 17–02

- establishing a new regulatory regime;
- facilitating a merger of the Government owned domestic monopoly
  carrier, Telecom, and the Government's international carrier, OTC,
  as the Australian and Overseas Telecommunications Corporation
  (AOTC), later renamed Telstra Corporation Limited; and
- allowing sale of the Government owned domestic satellite operator,

AUSSAT Pty Ltd, to Optus Communications Pty Ltd, with licences to compete with AOTC (now Telstra) across a full range of infrastructure and services. Optus Communications now provides long distance and international services through Optus Networks and public mobile services through Optus Mobile; and

• allowing issue of a third public mobile services licence.

Basic policy premises underlying the Government's 1991 telecommunications reforms were that:

• competition in provision of telecommunications services will best secure lower prices, higher quality and greater innovation;
• effective competition in the provision of telecommunications infrastructure would be achieved by providing the second carrier, Optus Communications, with the incentive to rapidly roll out a comprehensive alternative fixed (wireline) and satellite based long distance network, and the right to establish local access networks;
• this incentive would be best provided by allowing Optus a "window of opportunity" through initial access to Telstra facilities at cost-based interconnection and carriage rates, subject to contractual commitments by Optus to roll out its network before further network competition is introduced on or after June 30, 1997, and implementation of preselection arrangements which enable a customer to nominate either Optus or Telstra as the customer's preferred long distance carrier;
• competition in provision of public mobile services would be best achieved through initially licensing the second fixed network carrier as a public mobile carrier and subsequently issuing a third public mobile licence;
• the regulatory framework should foster competition among all service providers (whether carriers or not), limited only to the extent necessary to foster network development by the carriers and to allow the carriers to exploit economies of scale and scope. Services competition may include both switched and switchless resale; and
• regulation should be progressively reduced or eliminated in telecommunications markets as effective and sustainable competition is established in those markets.

Regulation specific to the telecommunications industry was seen by the Government as primarily a transitional measure, required to the extent that telecommunications carriers in nascent competitive markets may be able to adopt policies and practices that are in their own business interests but which are contrary to the interests of consumers or which would undermine development of effective and sustainable competition from other carriers and service providers.

17–03   Optus commenced provision of public mobile services as a reseller of capacity on Telstra's analogue network in July 1992 and as a long distance carrier in November 1992. A third public mobile telecommunications licence was issued to Vodafone Australia Pty Limited (formerly called Arena GSM Pty Limited) with effect from December 31, 1992. Both Optus Mobile and Vodafone were allowed initial access to Telstra services and facilities at cost based interconnection and carriage rates. All carriers are required to comply with network and industry development commitments made to the Government as a condition of issue of their respective carrier licences and must also comply with other conditions of those licences.

Accordingly, there are currently two general carriers (Telstra and Optus) and three mobile carriers (Telstra, Optus and Vodafone) licensed under the key statute, the Telecommunications Act 1991 ("Telecoms Act"). The complex substantive and definitional provisions of this Act, which set out the parameters of the carrier's exclusive rights, are summarised below. In essence, the general carriers have the exclusive right to install and maintain lines and cables (using any delivery technology) between two separate properties. These lines and cables are known as "reserved line links". The general carriers also have certain exclusive rights in relation to being the primary suppliers of satellite services, subject to an important qualification in relation to international private satellite earth station networks and services, and to provide public payphones. The general carriers' exclusive reservations do not extend to provision of switching functionality, or network infrastructure other than line links and facilities ancillary to line links such as poles and conduits. Accordingly, switched resellers may compete against carriers in provision of voice, data and video services, but subject to limited exceptions resellers must obtain line links from a general carrier. Similarly, public mobile carriers must obtain line links from general carriers.

Limits on the number of carrier licences issued by the Government carriers are given effect by formal agreements entered into by the Commonwealth of Australia with all licensees (other than the Government owned carrier, Telstra) pursuant to section 70 of the Telecoms Act 1991. When the Government announced the legislative reforms which were ultimately embodied in the 1991 legislation, the Government also announced that AUSTEL would periodically review the development of competition and that the Government would conduct a major review prior to ending the current limits on the number of general and mobile carriers in July 1997. In May 1994 the Minister for Communications and the Arts, Michael Lee, announced a review to examine what changes in policy, legislation and regulation are required to be introduced following expiry of the current limits on number of the licences on June 30, 1997, having regard to experience to date under the existing regulatory regime. The Government constituted a Telecommunications Advisory Panel, chaired by the Minister, to advise the Minister on issues which arise during the review. Panel members are drawn from the main industry and consumer groups. An issues paper was issued in October 1994. The Government's current timetable calls for a decision to be made by the Government in July 1995 with the aim of introducing legislative changes into the Parliament around December 1995. The Government's aim is to pass any necessary legislation to enable business planning by prospective new entrants to commence well prior to July 1997.

17–04   Matters addressed during the telecommunications policy review include:

- the structure of carrier licensing and services regulation provided for in the Telecommunications Act 1991 including whether there is a need for continued limitations on the number of licences in any category, the basis for considering any future licence applications, and the necessary rights, obligations, powers and immunities of existing and future licensees in each category;
- how future interconnection arrangements, including interconnection charges, will apply to ensure full interconnection and interoperability among competing networks and services on an efficient and competitive basis;
- the continued need for the competitive safeguards regime as provided for in the 1991 legislation;
- the continued need for and appropriate form of future prices regulation, including price caps, interconnection pricing and the structure of telecommunications charges in general;
- the social dimension of telecommunications policy, including the appropriate future scope of the universal service obligation, and how it may be best defined, provided and funded;
- consumer issues, including the continued need for and appropriate forms of consumer safeguards;
- the future role of AUSTEL as a specialist industry regulatory agency and the continued need for and appropriate forms of technical and economical regulation. Specifically, the Government will determine how AUSTEL should be restructured, or indeed whether it will continue to operate, in the light of a proposal to extend the competition regulation role of the Federal Government's general competition regulatory body, Trade Practices Commission (to be renamed the Australian Competition and Consumer Commission). This proposal follows recommendations of the Independent Committee of Inquiry into National Competition Policy, "Hilmer Report" (August 1993) as proposed to be implemented by the Competition Policy Reform Bill, which is expected to be enacted in the second half of 1995. The Government is likely to restructure the respective roles of the Trade Practices Commission, the Prices Surveillance Authority, and other specialist regulatory agencies including the Australian Broadcasting Authority (ABA) and the Spectrum Management Agency (SMA).

AUSTEL as the industry specialist regulator provided significant policy recommendations to the Government review. AUSTEL has also developed a technical and economic model of carrier and service provider interconnection as a proposed framework for network and services interworking in a multi-carrier and service-provider commercial environment: "An Interconnection Model for the Multi Service Deliverer Environment" (1995), AUSTEL.

The Government also instituted a number of related policy reviews.

17–05   The Broadband Services Expert Group, constituted of a range of industry members and chaired by Mr Brian Johns, ABA chairman, examined the technical, economic and commercial preconditions for the widespread delivery of broadband services to homes, businesses and schools in Australia. The Group released its final report "Networking Australia's Future" in December 1994.

The Communications Futures Project, undertaken by the Government's Bureau of Transport and Communications Economics, reviewed future economic, technical, commercial, regulatory and policy implications of emerging information and communications services and technologies. The project's terms of reference were wide ranging and examined likely developments in services and technologies; implications of these developments for market participants in those industries; and implications of these developments for policy and regulation over the coming decade. The project team reported in stages, with a final report in December 1994.

The Government is also obliged to conduct a review of possible expansion in the number of free-to-air commercial television broadcasting licences permitted in each service area and whether to amend Australian content requirements imposed on subscription television broadcasting (pay television) licences. These reviews are also to be conducted prior to July 1, 1997 (Broadcasting Services Act 1992, s. 215). The broadcasting and telecommunications policy reviews have taken place concurrently but remained separate.

This overview now considers the current (pre 1997) telecommunications legislation in greater detail.

1. TELECOMMUNICATIONS LEGISLATION

The Telecommunications Act 1991 is administered by the industry regu-   **17–06** lator, the Australian Telecommunications Authority (AUSTEL), under the broad policy direction of the Minister for Communications and the Federal Government Department of Communications and the Arts (DOCA). There is no relevant state or territory legislation governing telecommunications, other than provisions of generally applicable statutes such as planning and environmental laws, and the New South Wales Government Telecommunications Act 1991, which integrates the telecommunications networks of the Government of New South Wales and constitutes the New South Wales Government Telecommunications Authority to control the integrated network.

The regulatory package, of which the Telecoms Act 1991 is the key statute, came into effect in July 1991 and includes:

- the Telstra Corporation Act 1991, which allowed for the merger of Telecom and OTC to form Telstra; and which facilitates price capping, price control and other regulatory supervision of the business and corporate activities of Telstra by the Federal Government;
- the Telecommunications (Applications Fees) Act 1991, which allows

the Government to set fees payable to AUSTEL, for customer equipment permits and class licensed services;

- the Telecommunications (Carrier Licence Fees) Act 1991, which allows the Government to set fees for carrier licences which recover the cost of AUSTEL's supervision of carriers;
- the Telecommunications (Numbering Fees) Act 1991, which allows the Government to set appropriate fees for allocation of telephone numbers, including "premium numbers", under the AUSTEL—administered National Numbering Plan;
- the Telecommunications (Universal Service Levy) Act 1991, which facilitates allocation of the universal service levy (*i.e.* the cost of provision of otherwise uneconomic standard telephone and payphone service in remote and rural Australia) between carriers; and
- the Telecommunications (Transitional Provisions and Consequential Amendments) Act 1991, which facilitated the transition from the old regulatory regime underpinned by the Telecommunications Act 1989.

17–07   Many of these statutes have been amended on numerous occasions since initial enactment, for example, the Telecoms Act 1991 was amended 14 times between July 1991 and mid 1995. This legislation forms the super-structure for a complex web of regulations, Ministerial directives and determinations, carrier licences, and class licences applying to provision of tele-communications services by non-carriers.

Regulations are made pursuant to empowering provisions in Acts of Parliament. In general regulations expand upon the statutory provisions pursuant to which the regulations are made. The most important regulations are:

(i) the Telecommunications (General) Regulations 1991, which among other things sets out the types of earth stations that may be used in private international satellite services where satellite carriage capacity is not obtained from or through a "general carrier" licensed under the Telecoms Act 1991;

(ii) the Telecommunications (Exempt Activities) Regulations, which among other things exempt carriers from certain state laws which may inhibit establishment of telecommunications lines and other infrastructure.

Ministerial directives and determinations are also made pursuant to specific empowering provisions in Acts of Parliament. There are approximately twenty ministerial determinations and directions, the most important of which are:

(i) the Telecommunications (Eligible Combined Areas) Determination No. 1 of 1991, which allows aggregation of certain properties for the purpose of determining areas within which the general carriers' exclusive rights to establish telecommunications line capacity do not apply;

(ii) the Telecommunications (Radcom Facilities) Direction No. 1 of 1991, which among other things specifies restrictions that must be imposed by AUSTEL in class licences permitting non-carriers to provide telecommunications services, in order to ensure that the carrier's exclusive rights are not unduly eroded;

(iii) the Telecommunications (Authorised Facilities) Direction No. 1 of 1991, which directs AUSTEL as to how it should exercise its powers under the Act to allow non-carriers to install, maintain or use telecommunications lines and satellite based facilities that would otherwise fall within the scope of the carriers' exclusive rights.

AUSTEL has exercised its statutory authority to make a series of declarations and issue class licences which regulate activities of non-carrier providers of telecommunications services, referred to generically throughout the legislation and subordinate regulatory instruments as "service providers". Key class licences, considered further below, are: **17–08**

- the *Service Providers Class Licence*, which applies to any person, other than a carrier, providing a telecommunications service over or partially by means of a telecommunications network operated by a carrier;
- the *International Service Providers Class Licence*, which applies to any person, other than a carrier, providing telecommunications services to and from Australia over or partially by means of a telecommunications network operated by a carrier;
- the *Public Access Cordless Telecommunications Services Class Licence*, which applies to public access cordless services as defined in section 25 of the Telecoms Act 1991.

The Telecoms Act 1991 regulates provision of both wired and wireless services. For example, public switched telecommunications services provided over satellites or terrestrial cellular networks are regulated by the Telecoms Act 1991, which prescribes that only licensed carriers may initially provide these services (although resale by non-carriers is generally permitted).

2. Carriers and Service Providers—Rights to Provide Infrastructure and Services

Telstra and Optus are licensed as "general carriers" to supply a full range of domestic and international services using whatever technologies they choose. The reservation to the carriers is framed in terms of provision of fibre or wire infrastructure ("line links" which connect "distinct places" up to the "network termination points" within those distinct places, as those terms as defined in the Telecoms Act 1991). Each separate freehold or leasehold property is a "distinct place". Any person may establish network infrastructure within a "distinct place". This has allowed development of quite large private networks. For example, state governments have integrated their networks and outsourced network management and mainten-   **17–09** ance to the private sector operating under the federal legislation. BT Australasia manages the New South Wales Government network under contract from the New South Wales Government. Pacific Star, a joint venture of Telecom New Zealand and Bell Atlantic, operates the Queensland Government network under contract from that Government.

In addition, a number of complex rules (set by specific provisions of the 1991 Act and ministerial determinations and directives to AUSTEL) must be applied in determining whether a private fixed or radiocommunications (radcom) link can be established by non-carriers and interconnected with carrier networks. Relevant directives include:

- Telecommunications (Eligible Combined Areas) Determination No. 1 of 1991;
- Telecommunications (Authorised Facilities) Direction No. 1 of 1991;
- Telecommunications (Radcom Facilities) Direction No. 1 of 1991.

The Telecoms Act 1991 also provides that a person other than a general carrier must not supply a telecommunications service within Australia, or between a place within Australia and a place outside Australia, by the use of satellite-based facilities. This general rule is then qualified to allow resale (including a chain of resale) of satellite-based capacity, so long as one link in that chain is a service provided by a general carrier.

17–10   The complex rules governing the exclusive rights of carriers may be summarised as follows:

- Non-carriers (service providers) can operate radiocommunications networks which are not directly or indirectly interconnected with a carrier's network, essentially without restriction (other than the need to obtain radiocommunications licences, as issued under the Radiocommunications Act 1992).
- Service providers can install radiocommunications links to carry third party (including public) traffic to a network node within an area in which Telstra supplies local call service on an untimed charged (flat fee) basis—for example, most of the metropolitan areas of Sydney and Melbourne—but only for "own use" beyond those limits. "Own use" excludes service provider's customer traffic: Telecommunications (Radcoms Facilities) Direction No. 1. This restriction is intended to prevent wireless bypass of the carriers' long distance networks, but to allow establishment of alternative wireless local access networks.
- Service providers and end users can install cable links within private properties, or between private properties up to the distance of 500 metres where both private properties are occupied by the same principal user and the link is used exclusively for "own use": that is, the link is solely for carriage of communications of the authorisation holder and is not used for sharing or third party carriage, except carriage on behalf of a carrier. Specific authorisation is required from AUSTEL before a user can establish links under the "500 metre rule",

but authorisations are readily granted: refer to the Telecommunications (Authorised Facilities) Direction No. 1 of 1991.
- Where specified types of buildings occupies more than one property title, such as multi-storey residential apartments and shopping complexes, the aggregated property titles are treated as a single "distinct place" and accordingly persons other than carriers may install and maintain line links between the properties in the building: (Telecommunications (Eligible Combined Areas) Determination No. 1 of 1991).
- Mobile carriers cannot install cable links between their facilities (for example, base station to mobile switching centre, or between mobile switching centres located in different cities), other than where permitted in accordance with such exceptions to the general carriers' exclusive rights as apply generally to service providers.
- Any person licensed to provide subscription television or subscription **17–11** radio broadcasting services (that is, pay television or pay radio), or who is otherwise licensed as a broadcaster under the Broadcasting Services Act 1992 or operating under a class licence under that Act, may install and operate cable capacity between private properties for the purpose of providing that person's services either alone or jointly with other licensed broadcasters, but excess capacity on these networks may only be used for telecommunications services provided by or through a general carrier (Telecoms Act 1991, s. 99). As a result, private cable networks can be established by or on behalf of one or more broadcasters, but these networks may only be used for provision of a "telecommunications service" as defined in section 4 of the Telecoms Act—other than a licensed broadcasting service of the one or more broadcasters operating the capacity—provided by or through a general carrier.

The practical result of this restriction is to limit use of non-carrier cable networks other than for subscription or free radio or television services. As "video-on-demand" is specifically exempted from licensing under the Broadcasting Services Act 1992 but is a "telecommunications service" within the broad definition of that term under the Telecoms Act 1991, on-demand video, audio or text services may only be provided by a general carrier (at least prior to July 1997).
- A service provider may establish its own earth station and obtain directly international satellite capacity—whether INTELSAT, INMARSAT or privately owned satellite capacity—for establishment of international private networks (Telecoms Act, 1991 ss. 92 and 103), clause 6 of the Telecommunications (Radcom Facilities) Direction No. 1 of 1991 and regulation 5 of the Telecommunications (General) Regulations.

Private satellite networks which are not interconnected with a carrier network may be used to provide telecommunications services to any person operating a satellite earth station within the classes listed in regulation 5 of these regulations. This effectively allows establishment of earth stations to downlink international television services and operation of private very small aperture terminal (VSAT) networks. However, as Optus has been granted the exclusive right to provide satellite capacity for carriage of satellite-based subscription television broadcasting services (Pay TV) prior to July 1997 (Broadcasting Services Act 1992, s. 132(2)), Pay TV services may not be provided into Australia other than by the use of an Optus satellite (and at least prior to 1997) other than by a person allocated one of the three satellite subscription television broadcasting service licenses pursuant to section 93 of the Broadcasting Services Act 1992.

17–12   • Where a service is supplied using a satellite based facility which is not supplied by general carrier, and this service is directly or indirectly connected with a carrier network, the service must be supplied either on an "own use" or "single-ended interconnection" basis. Note that these restrictions do not apply unless non-carrier facilities are used in the course of transmission.

"Own use" refers to the carriage of communications belonging exclusively to the service provider and specifically precludes sharing the eligible service and third party carriage, including resale (except for carriage on behalf of a carrier).

"Single-ended interconnection" refers to a communication which is carried (first) on a public switched telecommunications network (wherever situated) and then (second) on a private radcom link or network being unable to be carried again (third) on a public switched telecommunications network, whether within or outside Australia, as a "single transaction". In other words, a "single-ended" interconnected service may either originate or terminate in a public switched network, but cannot both originate and terminate in public networks. "Single transaction" means a communication that is not delayed in the course of its transmission by 30 seconds or more. The effect of this is that unless a communication is delayed for 30 seconds or more in its transmission, the communication cannot traverse a public switched telecommunications network more than once in its transmission. The definition is designed to allow enhanced services and computer bureaux operations which involve termination of one transaction and separate access in less than 30 seconds: Telecommunications (Radcom Facilities) Direction No. 1 of 1991, cll. 2 and 6.

General carriers also have the exclusive right to install "public payphones", being payphones located in a place to which the public usually has access, but not including a vehicle, vessel, aircraft or other means of transport, or a place to which a person (other than a Government, a public authority or an officer or employee of a government or of a public authority) is entitled to deny the public entry: Telecoms Act 1991, s. 93 and definitions of "public payphone" and "public place" in section 5. The practical effect of these provisions is that non carriers may provide public payphones in shops or business premises, but may not provide payphones in areas controlled by public authorities such as railway stations.

3. OTHER POWERS AND PRIVILEGES OF CARRIERS AND SERVICE PROVIDERS

Subject to general pricing and tariffing rules, carriers are free to exploit the   17–13
economies of scale and scope available to them as a result of their control
of telecommunications facilities (Telecoms Act 1991, s. 94). This allows a
carrier to discriminate in favour of itself (through internal transfer pricing)
or another carrier, but only in relation to provision of "basic carriage services" over the carrier's network infrastructure. "Basic carriage services"
are essentially services which facilitate voice, data, or image (video or text)
carriage and connectivity without added value (Telecoms Act 1991, s. 174).
A carrier may not discriminate between itself and service providers in provision of any telecommunications service which is not a basic carriage service—these services generically referred to as "higher level services".

There are a number of other important differences between carriers and
service providers:

• Each carrier is entitled to technical interconnection with another carrier and equal treatment with internal operations of that other carrier
as to timeliness of access and quality of interconnection services and
facilities: Telecoms Act 1991, s. 137 and conditions applying to general and mobile carrier licences.

This right includes "exchange side" interconnection; priority provi-
sioning; access to network planning information; calling line identification (CLI) and other call information; directory listings; "joint"
operational liaison, such as in relation to network emergency and traffic measurement; non discriminatory service restoration and conges-
tion re-routing; co-location of facilities; and duct and tower sharing.
Interconnection potentially covers a wide range of services and facilities which a carrier does not offer to any other party, or even within
its own organisation other than in some bundled form.

By contrast, no technical level of interconnection is prescribed by
service providers, but interconnection must not be any worse than a
similarly situated customer receives: section 184 of the Telecoms Act
1991 prohibits discrimination by a carrier against service providers
for the reason, or reasons including the reason, that the customer is
a provider of services rather than an end user. Service providers are
entitled only to request interconnection at a level and at rates determined by each carrier and tariffed as a product offering from time to
time tariffed. However, AUSTEL has power to require a carrier dominant in the market for provision of a particular kind of basic carriage
service to "unbundle" that service. This power could be used to
require service provider interconnect to be provided by a carrier closer
to the level of carrier-carrier interconnect. AUSTEL's "unbundling"
power is subject to limitations which prevent AUSTEL from exercising
the power in a manner which erodes the dominant carrier's ability to
exploit economies of scale and scope, and which require AUSTEL to
take account of the Government's policy intent that carriers remain

the primary providers of telecommunications networks and services (Telecoms Act 1991, s. 181).

17–14   • Carriers have a statutory obligation to supply telecommunications services and facilities to other carriers under the Telecoms Act 1991, s. 137. This obligation is only qualified by an obligation of "reasonableness", which is to be determined having regard to the Government's policy objectives of promoting competition; encouraging the efficient use of and investment in telecommunications infrastructure; and removing obstacles to customers having equal access to the telecommunications services supplied by the various carriers (Telecoms Act, s. 136). The legislation in effect imposes a significant burden on a carrier to establish that it should *not* supply interconnection services or facilities to another carrier.

By contrast, a carrier is under a limited obligation to deal with a service provider by interconnecting its service (Telecoms Act 1991 s. 234). In particular, a carrier is relieved of its obligation to deal with the service provider where the service is included in another carrier's tariff: that is, there is no supply "bottleneck" in relation to the carrier's tariffed service.

• Provision of interconnection services by Telstra to Optus and Vodafone is subject to regulated cost based pricing under section 140 of the telecoms Act 1991 and the Telecommunications (Interconnection and Related Charging Principles) Determination No. 1 of 1991.

No pricing principles apply to service providers, other than that the price at which services are made available must not be unfairly discriminatory.

• A general carrier may enter into any private, state or Commonwealth land (other than special national land, such as the Parliament) and construct facilities without specific consent from the land occupier and on giving reasonable notice to the occupier (Telecoms Act 1991, ss. 128 and 129). Compensation is payable for loss or damage caused by entry or construction of facilities.

While other parties are permitted to install line links on land owned or occupied by others—for example, broadcasters may establish private cable capacity—they do not have similar privileges of access and therefore must negotiate easements or other contractual rights of access with individual land owners.

• General and mobile carriers are exempt from a wide range of state planning, environmental, land use, local government and tenancy laws, and subject in turn to a national code allowing for streamlined consultative and approval processes (Telecoms Act 1991 ss. 117, 118 and 119 and Telecommunications National Code).

Other permitted builders of networks are fully subject to state laws.

17–15   • General and mobile carriers have a specific statutory exemption from the general law relating to "fixtures". Pursuant to section 123 of the Telecoms Act 1991, a facility that is supplied, installed, maintained or operated by a general carrier remains the property of a carrier, whether or not it has become a fixture.

By contrast, other network builders may need to specifically negotiate with land owners in order to ensure that conduit or other facilities erected in, on or over land remains the property of the network builder.

- Carriers can create binding contracts with customers by filing with AUSTEL a tariff setting out terms and conditions applicable to the use of that service. A customer then using the service referred to in that tariff is bound by the tariffed terms and conditions unless the customer and the carrier otherwise agree (Telecoms Act 1991, s. 200.). This power facilitates provision of services accessible by customers under dial code access arrangements: that is, provision of services which are accessed through customers dialling a four digit prefix (14XX) before the called number. Calls dialled using this prefix are then routed to the network of the carrier or service provider which operates under that prefix.

  By contrast, although service providers may utilise dial code access (through a four digit code which overrides a customer's preselected nomination of preferred carrier), a service provider would need to contract directly with a customer which selects its service using the four digit override code.

- Each carrier has clearly established indirect access rights as against each other carrier. In particular, each carrier must provide another carrier with interim access code arrangements for their directly connected customers, within the time frame specified in licence conditions applying to the providing carrier. For example, the general carriers must provide each other with exchange conditioned preselection on an universal basis by 1997. Preselection is automatic routing of calls to the customers nominated carrier, except where the customer dials the four digit override code to access another carrier or service provider.

  Service providers have no express statutory right of "indirect access". However, through its role of managing the National Numbering Plan AUSTEL has provided for four digit code access (14XX) by service providers. This gives the service provider with a right to an access number. Provision of access capability within the carrier networks remains a tariffed service offering by each carrier. Exchange conditioned preselection is not currently available for service providers.

As the concomitant of their powers and priviledges, carriers have certain obligations which service providers do not. These include obligations: **17–16**

- to contribute to the universal service levy fund. The nominated "universal service carrier" is entitled to claim in relation to loss making provision of standard telephone service throughout Australia on an equitable basis. Each carrier is required to contribute a proportion of the total universal service cost. This proportion is calculated by reference to the contributing carrier's proportionate share of total timed

telecommunications traffic across all telecommunications networks in Australia (Telecoms Act 1991, Pt. 13);
- to contribute towards the cost of AUSTEL administering its function of regulating competition between the carriers, through payment of carrier licence fees calculated by reference to the respective carrier's share of timed traffic;
- to report to AUSTEL on a quarterly basis and to internally account in accordance with a chart of accounts and cost allocation manual determined by AUSTEL;
- to report other information periodically to AUSTEL as to operations, efficiency and performance;
- to comply with industry development and network development commitments to the Commonwealth of Australia;
- to provide emergency call services;
- to meet minimum maintenance obligations to customers, including free maintenance in defined circumstances;
- to make available untimed (flat fee) local calls as an option for residential customers (if the carrier elects to provide local service);
- to participate in the Telecommunications Industry Ombudsman scheme, which deals with customer complaints concerning carrier services and billing;
- to establish and comply with carrier determined customer codes of practice.

In addition, carriers are subject to more extensive powers of direction by AUSTEL in the exercise of AUSTEL's function of protecting consumers and promoting competition (Telecoms Act 1991, ss. 39 and 46).

4. Public Mobile Services and other Mobile Services

17–17   Separate public mobile carrier licences have been issued to Telstra, Optus Mobile and Vodafone Australia.

In essence, the mobile licences confer on mobile carriers the exclusive right to provide radio-based public telecommunications services with full roaming capability, the intention being that the reservation to the mobile carriers will apply to all services which provide users with fully mobile access to carrier networks, whether based on cell switching technology or not.

Sections 25 and 26 of the Telecoms Act 1991 draw a difficult distinction between public mobile telephone services (reserved to mobile carriers) and public access cordless telecommunications services (open to competition). In effect, four different layers of test, including two layers of exemptions, are used to determine whether a telecommunications service is a public mobile telephone service (PMTS) or is a public access cordless telephone service (PACTS). The two primary criterion are a technology and services characterisation respectively. PMTS are PACTS within the services characterisation, but differentiated from PACTS (and therefore excluded from the definition of PACTS) by use of cell switching technology.

In the absence of any regulation made under the Telecoms Act 1991 to contrary effect (of which there are currently none), a service can only be a PMTS, under section 26 of the 1991 Act, if the service is supplied using cell switching technology. The services characterisation of a PACTS is that the service:

- is offered to the public generally; and
- a person can use it while moving continuously between places; and
- customer equipment used for or in relation to the supply of the service is not in physical contact with any part of the telecommunications network by means of which the service is supplied; and
- a facility that used for or in relation to supply of this service is connected to a telecommunications network operated by a general carrier. (Telecoms Act 1991, s. 25.)

If a service which falls within the services characterisation of a PACTS   **17–18** uses cell switching technology, it is a PMTS and therefore within the exclusive rights of the public mobile carriers. However, the technology characterisation is not necessarily determinative of whether competition will be permitted; the Government has power to make regulations under the 1991 Act declaring a service to either be a PACTS or PMTS. This regulation making power was included to allow the Government to deal with new mobile technologies which may cause undue erosion of the public mobile carriers' exclusive rights, such as public mobile services provided through use of lower earth orbit (LEO) satellite mobile communications systems. It is expected that emerging technologies for the delivery of wireless personal communication services (PCS) that provide full roaming capability would be regulated as public mobile services (requiring a mobile carrier licence), rather than under the PACTS class licence.

The definitions specifically exclude a telecommunications service supplied by a telecommunications network with the principal function of supplying telecommunications services between equipment connected to it and which is connected to one or more general carrier's networks, provided that supply of telecommunications services to equipment connected to the general carrier's network is at most an ancillary function of the primary network and the network cannot be used for carriage of a communication between items of equipment connected to a general carrier's network as a single transaction (Telecoms Act 1991, s. 25(2)). For example, land mobile trunked radio service can be interconnected into the PSTN as long as communication between the network and the public switched telephone network is at most an ancillary function of the network.

Section 25(3) of the 1991 Act exempts the service if it is a "one way only, store and forward communication service; or a service that performs the same functions as such a service". This provision clearly excludes services such as packet radio and paging from regulation as a PACT service.

Optus Mobile commenced provision of services as an airtime reseller   **17–19** on Telstra's analogue AMPS MobileNet services, pending establishment of Optus digital GSM network. Telstra and Optus Mobile now provide both

AMPS and GSM services. A third licensed mobile operator, Vodafone Australia, competes in provision of GSM services utilising Vodafone's own (the third) GSM network. The mobile carriers continue to rollout new GSM service areas on a national basis. The Government has announced that Telstra's AMPS network must be progressively phased out between 1996 and 2000.

PACTS, including CT-2, CT-3, paging and trunked land mobile, are open to competition. Services open to competition may be supplied subject only to class licence conditions set and administered by AUSTEL. Numbers must be allocated by the service provider in accordance with the AUSTEL administered National Numbering Plan. Radiocommunications frequencies are allocated in accordance with a National Spectrum Plan and applicable Band Plans and licensed by the Spectrum Management Agency.

The PACTS class licence requires PACTS suppliers to register with AUSTEL and to pay a fee based on a graduated scale depending on the population centres being served. These fees are intended to recover AUSTEL's costs of administering the licence. The class licence also sets out the terms and conditions which must be complied with by PACTS suppliers, including a requirement that each operator's facilities must meet technical standards determined by AUSTEL. The AUSTEL *Guide to the PACTS Class Licence*, notes that "there must be no technical restraint on the use of customer equipment to access alternative PACT services using the same technology. It shall be a matter for suppliers to agree among themselves and commercial terms and conditions governing inter-network operations and roaming".

5. COMPETITIVE SERVICES AND RESALE

17–20 The Service Providers Class Licence, which applies to provision of services over carrier networks and interconnected private networks, does not generally require registration of these service providers. Rather, the class licence simply sets out the rules with which service providers must comply. These rules cover technical and operational matters, but also include important limitations relating to international services (to prevent misuse of market power by foreign carriers involved in resale) and services provided over radiocommunications facilities that are connected to a carrier's network and that have been privately installed.

Class licences may be amended by AUSTEL at its discretion from time to time, although AUSTEL is subject generally to ministerial directions which impact the content of the class licences in key areas.

Non-carriers, such as resellers, may establish their own switching capacity and provide services directly competitive with carriers' basic carriage services. The Telecoms Act 1991 eliminated former reservations to carriers of basic voice, data and video services. The only restrictions on resale are those imposed in relation to use of carriage capacity supplied by a person other than a carrier, or imposed under AUSTEL's International Service Providers Class Licence in order to prevent misuse of market power in provision of international services.

However, the distinction between "basic carriage services" (BCS) and enhanced or "higher level services" (HLS) remains relevant through the application of complex pricing, price discrimination and tariffing rules to carrier's supply and internal use of "higher level services". The requirements are intended to facilitate fair competition between carriers and service providers in provision of telecommunications services. These rules, administered by AUSTEL, require carriers to file tariffs setting out the price structure and other terms and conditions applying to supply of basic carriage services. In view of the entrenched market position of Telstra, particular rules apply to pricing and other terms of supply of services by Telstra in service markets for so long as it remains dominant in those markets. These rules in effect require Telstra to adhere to its tariffed pricing schedules (which may schedule volume discounts) so that it cannot, for example, offer discounts "off-tariff" to particular customers. The 1991 Act does not define the relevant markets or set rules for determining whether Telstra remains dominant, relying instead on case law under Australia's general competition statute, the Trade Practices Act 1984.

Although the Telecoms Act 1991 allows for preferential (cost based) interconnection and access charges and for technically equivalent interconnection as between carriers, non-carriers (service providers and end users) must commercially negotiate interconnection and access charges and other terms and conditions with the carriers. For example, Telstra's National Connect Service tariff allows access and egress service for voice and data service providers, and its Vision Stream tariff governs use of Telstra broadband carriage capacity by video service providers such as pay television operators. The National Connect tariff is capable of access through use of a 14XX prefix by third parties who use a service provider's services. The tariff was released following negotiations between Telstra and the first national switched service provider (non-carrier), AAP Telecommunications. As a tariffed service, the terms and conditions are fully transparent and available to all service providers on a non-discriminatory basis.

6. INTERNATIONAL SERVICES

Provision of international facilities is, in general, reserved to general carriers   **17–21**
in like manner to provision of domestic network, subject only to special rules permitting use of private satellite earth stations to access international capacity. By contrast, international services, including resale of capacity, is open to competition subject to:

(i) compliance by the carriers with a Code of Practice relating to their dealings with international operators; and
(ii) compliance by all (non-carrier) service providers with the International Service Providers Class Licence administered by AUSTEL.

International services provided through the use of international cable capacity or satellite capacity obtained from a general carrier do not require any particular regulatory approval or prior registration with AUSTEL, except as noted below.

An international service provided on a double ended interconnection basis may only be provided after enrolment as an international service provider with AUSTEL. "Double ended interconnection" means that a service is provided to the public through the use of a public switched telecommunications network at either end of an international link. The purpose of enrolment is to enable AUSTEL to detect possible misuse of market power or misuse of regulatory status by carriers at either the Australian or the distant end which may lead to a lessening of competition. AUSTEL is empowered to refuse or cancel enrolment where it concludes that provision of a double ended interconnection service would not be in the "public interest". AUSTEL has indicated that it will in general assume that an international service which increases competition is in the public interest (AUSTEL, *Guide to the International Service Providers Class Licence*, paragraphs 40 to 44.

Service providers providing enhanced (value added) voice or data services to and from Australia do not need to enrol with AUSTEL and there are no relevant use restrictions. In determining whether a service is a basic or value added service, AUSTEL applies the regulatory distinction which operated under the former legislation, the Telecommunications Act 1989.

Public switched international services with "single ended interconnection", private satellite services connected to a carrier's network, and services deriving from use of indefeasible rights of use (IIU) of international cables must enrol with AUSTEL. Suppliers taking advantage of indefeasible rights of use, and providers of private satellite services (that is, suppliers using satellite capacity obtained from a person other than a general carrier), may only provide international services on a "single ended interconnection basis" or for own use. "Single ended interconnection" refers to the inability of a communication being carried once on a public switched telecommunication network and on a private radcom link or network being carried again on the public switched telephone network as a single transaction. "Own use" refers to the carriage of communications belonging exclusively to the eligible service provider and specifically precludes sharing of the eligible service and third party carriage, including resale. The purpose of these restrictions is to preclude undue erosion of the general carriers' exclusive rights.

In general, carriers' and service providers' commercial arrangements with other international operators, including international accounting rate and other revenue sharing arrangements, will not be subject to specific regulation. However, where it considers that there is a possibility of anticompetitive conduct AUSTEL may investigate and impose appropriate restrictions, such as requirements to disclose traffic statistics or other information to AUSTEL on a commercial-in-confidence basis.

7. CARRIER INTERCONNECTION AND CUSTOMER ACCESS

Telstra, Optus Communications and Vodafone Australia have negotiated **17–22** arrangements for interconnection, lease of capacity and joint use of radio sites, ducts, poles and other infrastructure. The conditions of interconnection, access and use were determined by commercial negotiation within the framework of the Telecoms Act 1991, the Telecommunications (Interconnection and Related Charging Principles) Determination and carrier licence conditions.

Part 8 of the Telecoms Act 1991 creates carrier access rights of two kinds. The first is the basic right of any carrier to connect its facilities to the network of any other carrier, and to have its calls carried and completed over that network. This right is expressed in section 137(3)(b) to be exercisable "on such reasonable terms and conditions as the carriers agree on or, failing agreement, as AUSTEL determines". The obligation of one carrier to carry communications of the other carrier across the first carrier's network only applies so far as the use of the other carrier's network is "necessary or desirable" for the purpose of supplying telecommunications services which the requesting carrier is licensed to provide.

The second kind of right referred to in Part 8 of the 1991 Act is "supplementary access rights" created by a licence condition of a carrier's licence. Carrier licence conditions create rights include rights to customer billing, operator assistance and listing in published directories published by the other carrier, together with rights of access to, or use of, facilities or land on which facilities are located. Unlike the basic interconnection right which is created by the 1991 Act (albeit on such terms as the carriers agree or are arbitrated), supplementary access rights are only created if a condition in a carrier's licence confers those rights. Where the right is created through such licence conditions, it in turn forms the basis for terms negotiated in the access agreement.

The access agreement between the carriers must also be in accordance with pricing principles specified by the Minister (Telecoms Act 1991, s. 141). The relevant ministerial determination, the Telecommunications (Interconnection and Related Charging Principles) Determination, required Telstra to make available interconnection and access services and facilities at "initial charges" based on incremental cost, until certain trigger mechanisms operate to allow initial charges to be superseded by commercially negotiated charges. Although the Determination did not specify the incremental costing in methodology to be utilised by Telstra, government policy statements made at the time of introduction of the 1991 Act were to the effect that Telstra should be entitled to recover all of its actual additional costs in providing access to and for usage of its network, including allowance for any additional assets required to achieve interconnection and for the opportunity cost of capital. Interconnecting carriers have negotiated from the viewpoint that Telstra should only be entitled to recover its directly attributable incremental costs (DAIC), whereas Telstra has contended that it should be entitled to a commercial return on the use of its relevant direct assets. In practice, as reliable cost data is either not available within Telstra or has not been provided to interconnecting carriers, prices have generally been struck by reference to comparative data derived from other countries and estimates of world's best practice.

17–23   The specified trigger thresholds for renegotiation of initial charges varied according to service segment: for example, different trigger thresholds apply in relation to intra-ICCA (intercarrier charging areas, a geographical area within which the connecting carrier establishes its point of interconnection) carriage; inter-ICCA (gateway to gateway) carriage; international and mobile services. The trigger mechanisms primarily utilise measures of percentage market share (for example, where the interconnecting carriers share of international traffic on a particular international traffic stream reaches a five per cent share of traffic generated in Australia on that stream), or network rollout (for example, initial charges no longer apply from the date that the second carrier commits to acquire its own dedicated facilities to service a particular route).

The fact that a trigger threshold is reached does not automatically disentitle the interconnecting carrier to initial charges: instead, these thresholds signal the shift in the onus of proof from Telstra to the interconnecting carrier as to whether Telstra continues to have dominant market power sufficient to require initial charges to continue. Below these thresholds Telstra may argue a case to AUSTEL that it has lost market dominance and above the thresholds it is open to the interconnecting carrier to argue a case to AUSTEL that Telstra continues to have dominant market power. In practice, Optus and Telstra have been able to agree on a transition from most initial charges to subsequent, commercially negotiated, charges, without a formal determination by AUSTEL as to whether Telstra continues to remain dominant.

The interconnection arrangements between Telstra and Optus included agreement as to technical deployment and administrative procedures for implementation of exchange conditioned preselection. By late 1994 preselection had been implemented in most major metropolitan centres and was starting to extend to surrounding regions. Interim customer access measures apply until preselection is made available on an area by area basis. Prior to exchange conditioning and balloting for preselection, callers wishing to use the services of Optus instead of Telstra dial "1" plus the usual long distance or international code. As Telstra exchanges are modernised and conditioned for preselection, customers wishing to use Optus specify their carrier of choice. This selection takes place through a ballot conducted area by area, according to a timetable and procedures negotiated between the carriers and approved by AUSTEL. In default of any over-ride selection (which can be made by utilising a 14XX prefix), all calls from each customer preselecting either Optus or Telstra are automatically routed through that carrier's network where available. Customers, by overriding their default selection of general carrier (by dialling a four digit access code), can use the other general carrier (or any service provider accessible through use of an override code). This enables a customer to take advantage of special offers available from another carrier or a service provider.

8. Technical Regulation and Equipment Supply

AUSTEL has responsibility for technical regulation of telecommunication   17–24
services, equipment and cabling (Telecoms Act 1991, Pt. 12). AUSTEL undertakes a consultative process to determine standards relating to:

* network integrity and safety;
* interoperability;
* compliance with recognised international standards; and
* end-to-end quality of service.

An AUSTEL permit is required for sale and connection of customer equipment which is intended to be connected directly or indirectly with a carrier network (Telecoms Act 1991, ss. 253 and 255). The purposes of the permit process are to:

* ensure the safety of persons working on or using services supplies by means of the telecommunications network;
* ensure interoperability of customer equipment with the telecommunications network to which it is connected;
* prevent damage to the telecommunications network as a result of the actions of customer equipment; and
* ensure non-interference to the service provide to other users of the telecommunications network.

The essential elements of customer equipment regulation are the testing of a sample of equipment (type testing) to the applicable technical standards by a test house accredited by AUSTEL and maintenance of that equipment to "type" by the manufacturer/supplier. Technical standards are determined and promulgated by AUSTEL after consultation with industry and Standards Australia and having reference to international standards.

These are limited exceptions to this broad requirement to obtain equipment permits, the most significant exemption being for certain data terminal equipment which is installed upstream (on the customer side) of a safety extra low voltage interface (AUSTEL Data Terminal Equipment: Notice under section 267 (No. 1 of 1993). The equipment supplier is responsible for arranging evaluation of equipment against relevant technical standards by any of the test houses accredited jointly by AUSTEL and the National Association of Testing Authorities. AUSTEL has accredited approximately 50 test houses in Australia and overseas, including in the U.K., New Zealand, USA and Canada.

The critical document in support of a permit application is a test report which supports the compliance of the candidate equipment with the technical standard(s). In its practice AUSTEL exercises a discretion to accept test reports which indicate non-compliance against some of the clauses of relevant technical standards if, in AUSTEL's judgment, non-compliance does not compromise the objects of technical regulation. AUSTEL exercises its discretion in relation to many permit applications.

17–25   It is a usual condition of a permit that customer equipment be labelled with a unique permit number for all equipment of that type. AUSTEL maintains a register of permitted equipment. It is also a condition of the permit that a permit holder is responsible for ensuring the equipment remains true to type. Equipment not of the type for which a permit was issued is considered non-permitted.

AUSTEL has raised the possibility of changing from an exclusively type testing approvals regime to a scheme which accommodates declarations by accredited manufacturers linked to operation of an approved quality scheme, such as the ISO 9000 series of quality assurance measures (AUSTEL Technical Standards and Conformity Introstructure in Australia: Occasional Paper, Technical Regulation, 1994). Technical regulation will also be reviewed in early 1995 as part of the Government's 1997 policy review. However, as at September 1994 there was no specific proposal to move away from a type testing approvals regime.

An item of customer equipment is of the same type as another item if:

- the items were produced specifications that differ in no material respects;
- the respective ways in which the items were produced differ in no material respect; and
- the form and functions of one item differ in no material respect from the form and function of the other item (Telecoms Act 1991, s. 20).

AUSTEL generally takes the view that where a certain type of customer equipment is made available through two or more separate distribution channels under different names, separate permits must be obtained for each named product. For example, where equipment is sold under OEM arrangements, the different items must be subject to different equipment approvals, unless the respective distributors are able to demonstrate that the items are technically identical and agree to association of the items under a single technical approval.

Approval is in respect of a particular item or type of equipment and not a particular supplier. Accordingly, provided an item has been approved and carries the necessary AUSTEL approval plate, such equipment may be sold or connected to the network by a number of different suppliers.

There are currently no common Australian and New Zealand equipment approvals, although progress toward this objective has been set as a goal by the respective authorities. It is likely that common standards will emerge over the next few years, associated with general economic integration between Australia and New Zealand under the Closer Economic Relations (CER) Treaty between the two countries. AUSTEL has also declared that equipment approval is not required for temporary operation in Australia of a cellular mobile telephone pursuant to international roaming arrangements between an Australian public mobile carrier and or overseas carrier

(AUSTEL International Roamers-Cellular Mobile Telephones: Notice under section 267).

A licensing system also applies to persons installing and maintaining customer equipment (Telecoms Act 1991, Pt. 12, Division 7).

9. INDUSTRY DEVELOPMENT ARRANGEMENTS

Certain categories of customer premises equipment (CPE) (*i.e.* private branch or exchanges (PBX), key systems and small business systems, primary telephone instruments and cellular handsets) are currently also subject to Industry Development Arrangements (IDAs). These arrangements, which came into force in 1989, constituted a transitional scheme from the previous Australian preference arrangements administered by Telecom to intended full liberalisation, currently scheduled for June 30, 1996. **17–26**

The scheme is points based. Suppliers qualify for authorisation to market their CPE products in Australia and/or connect them to carrier networks where they earn sufficient points allocated on the basis of:

- the degree of Australian and New Zealand production;
- the amount of turnover expended on research and development on these product types; and
- the extent to which these product types are exported.

Each of these factors is accorded a score, with each factor then weighted equally and aggregated.

The IDA scheme is currently administered by AUSTEL. Suppliers of these product types are subject to annual audit to ensure their ongoing compliance with these arrangements. Under 1992 amendments to the IDAs, suppliers of customer equipment covered by the IDAs whose sales of telecommunications and computing hardware, software and services to Government exceed $40 million will be permitted to move from the IDAs scheme to the Partnerships for Development Program (PDP). Under the PDP scheme, companies enter into an agreement with the Commonwealth of Australia to meet specified industry development objectives which are negotiated on a case by case basis. Those selling between $10 million and $40 million can enter into the Fixed Term Arrangements (FTA) scheme. The FTA scheme was introduced in March 1991 to replace the Government's offset scheme. It applies to companies with between $10 million and $40 million in annual IT sales to Government and entry is also by assessment of the company's business plan against industry development criteria. In addition, existing PDP or FTA companies may apply for exemption from IDA arrangements. AUSTEL will continue to provide technical approval for connection to the network and administer the IDA scheme for companies that do not transfer to PDPs or FTAs.

Other industry development arrangements are reviewed by the Telecommunications Industry Development Authority (TIDA). Optus Communications, Vodafone Australia and Telstra have agreed to provide industry opportunities by developing local joint ventures, local research and development and to achieve export sales by acting in concert with local suppliers.

10. OTHER COMPETITION LAW

17–27   Although the Telecoms Act 1991, as administered by AUSTEL, has become the key statute regulating competition in the telecommunications sector, general competition law, as primarily set out in the Trade Practices Act 1974, continues to be of general relevance.

Section 236 of the Telecoms Act 1991 provides that the Act authorises doing or omitting to do an act that is necessary to comply with, or otherwise to give effect to, the condition of a carrier's licence, a direction or determination by the Minister or AUSTEL under the Act, or an interconnection agreement between the carriers that has been registered with AUSTEL. This exception will generally only be of relevance in relation to carriers.

Many actions or omissions by carriers and non carrier service providers will remain subject to the Trade Practices Act 1974. In addition, AUSTEL has comprehensive direction making powers in relation to carriers, and to a lesser extent service providers, and may exercise these powers to enforce relevant provisions of the Telecoms Act 1991. For example, exclusive site agreements with PACTS operators could contravene the 1991 Act, if the effect of these arrangements was to substantially lessen competition in the PACT service market and the exclusivity was obtained for the purpose of deterring or damaging competitors. Either AUSTEL or the Trade Practices Commission could take action to prevent such practices.

It is likely that the reach of general competition law into telecommunications regulation will be considerably extended when the Competition Policy Reform Bill 1995 is enacted. This Bill will extend the reach of the Trade Practices Act 1974, establish the Australian Competition and Consumer Commission and National Competition Council and set a framework for pricing and access arrangements which will be applied across deregulated industries.

Under the provisions of the Bill access may be mandated to a facility where "access to the service would promote competition in at least one market, other than the market for the service"; in other words, where access would promote effective competition in a downstream or upstream activity. The Bill's focus on access to services provided by means of this facility, rather than on the facility itself, appears aimed at unbundling services into a delivery and content component.

17–28   The Bill allows for access pricing principles to be established on a case by case basis by the National Competition Council and through negotiations between parties conducted in accordance with these principles and arbitrated if necessary by the Australian Competition and Consumer Commission. The Bill does not specify principles for determining an appropriate access price. The Government's explanatory paper accompanying the Bill noted that an appropriate price would depend upon current utilisations of the facility; planned future utilisation; the extent to which the capital costs of the facility have been recovered; and keeping incentives to maintain the facility, as well as encouraging further investment including technical innovation. The Government noted that it may be reasonable for an owner to recover a proportion of fixed costs attributable to the user, through a capital or "reservation" charge, as well as the directly attributable variable or operating charge for the quantity of service used. Accordingly, principles determined in relation to one type of service may vary significantly from those applicable to another service. It is possible that any access principles determined in relation to a particular telecommunications service may be significantly different to the pricing principles which apply under Part 8 of the Telecoms Act 1991 in relation to intercarrier interconnection and access.

As noted in paragraph 17–04 it is possible that the Government's review of telecommunications policy post 1997 will conclude that the Australian Competition and Consumer Commission is a suitable body to administer part or all of the competition functions currently undertaken by AUSTEL, or that industry specific competition regulation under the 1991 Act should be subsumed within the expanded general competition regime.

11. Radiocommunications

Access to radiocommunications spectrum is governed by the Radiocom-   **17–29** munications Act 1992. This Act also establishes the powers and functions of the Spectrum Management Agency (SMA). The major functions of the SMA include frequency planning, technical regulation, licensing and pricing of spectrum access, and enforcement of standards and conditions.

The 1992 Act allows for three licence systems to provide users with access to spectrum: a new spectrum licence system; a new class licence system; and the (traditional) apparatus licence system.

Under the apparatus licence system, generally a licence is granted in relation to a specified transmitter operating from a specified site at a particular frequency or for multiple channels within a particular frequency band. Receiver licences are also required in some cases. Licences are issued subject to any limitations on use and interference specified in the spectrum plan and any relevant band plans. Licences are generally renewable annually, for a fee fixed annually. However, the SMA is free to set apparatus licence terms and fees for up to five years (section 103 of the 1992 Act).

As at mid 1995, there were over 90 transmitter licence categories and 10 receiver licence categories. The SMA had concluded that the current categories are unnecessarily complex, work against efficiency and flexibility, and have led to inconsistencies (SMA, Inquiry into the Apparatus Licence System: Discussion Paper, December 1993). The SMA propose to simplify the apparatus system, either by reducing the number of categories, relating categories to industry accepted service groupings or ITU definitions, or eliminating categories altogether.

The SMA has also announced its intention to simplify the licence fee structure for apparatus licences. In general, a charge will be levied for the issue of a licence, calculated as sufficient to cover the costs of issuing and registering the licence and costs associated with the assignment of frequencies and other technical work. Annual licence fees will be calculated to comprise two components: one to cover ongoing SMA administration costs, and another as a tax for spectrum access, reflecting the scarcity value of the spectrum used. The tax component would vary according to whether or not a premium had been levied on initial allocation of the licence through price based allocation under section 106 of the Radiocommunications Act 1992. Where price based initial licence allocation had occurred—whether through auction, tender or negotiated price—the annual licence fee would be relatively low. For other users the level of the annual tax component will be determined using a formula that reflects the scarcity value of the spectrum used, taking account of geographic location; spectrum location or band; bandwidth denied to other users; and area of coverage denied to other users.

17–30   There is no provision in the Radiocommunications Act 1992 to enable apparatus licences to be transferred between parties. Licensees may use the licences themselves, or authorise third party users pursuant to section 114 of the 1992 Act, but may not assign or transfer their licences to another person or organisation. The usual course of action on transfer of a business and associated radiocommunications licences is to request the SMA to cancel the licence in the name of the prospective transferor and to reissue the licence in the name of the prospective transferee. Such requests are routinely dealt with, although it must be noted that determinations as to price based allocation of frequencies in particular bands may render this option commercially unattractive in certain frequency bands (if, for example, reallocation by the SMA can only follow price based allocation of those frequencies).

Holders of apparatus licences are not entitled to automatic renewal of licences at all or at any particular fee. Where the SMA refuses to renew an apparatus licence, or renews the licence on different conditions, the SMA must comply with the 1992 Act. In accordance with section 130(5) of the 1992 Act, the SMA must give the licensee a written notice to the licensee stating that it is renewing the licence on different conditions, together with a statement of reasons. Refusals to renew apparatus licences and changes to licence conditions on renewal are reviewable by the Administrative Appeals Tribunal.

Although there are limits on the purposes for which licences can be used, in general they do not include a requirement that the licences are in fact used. Accordingly, licences may at the discretion of the SMA be granted prior to a licensee actually commencing provision of service. However, continuing non-use of a licence will of course be a relevant consideration for the SMA in determining whether to renew it.

The second category of licensing is issue of class licences. A class licence 17–31 authorises any person to operate radiocommunications devices of specified kinds and/or for a specified purpose, provided the operation is in accordance with the conditions of the licence. Such licences are used where individual frequency assignments are not required. The SMA has issued class licences for a number of categories of radiocommunications devices, including public mobile telephone handsets; cordless telephone handsets (in the frequency bands 857MHz to 861MHz and in the band from 861MHz to 856MHz); cordless telephone services and low interference devices, such as garage door openers and radio transmitters used for control of models.

The third category of licensing is grant of spectrum licences. The current apparatus licensing or class licensing systems will continue to apply in each frequency band until the Minister for Communications and the Arts determines under section 36 of the 1992 Act to designate a band for spectrum licensing. A moratorium on the issue of further apparatus licences then comes into operation pending development by the SMA of a conversion plan and marketing plan for this spectrum (section 38 of the 1992 Act). The marketing plan may cover how spectrum will be apportioned between spectrum licensees, the procedures and timetable for the sale of spectrum licences, and reservation of frequencies for public and community services.

The SMA has not provided any guidance as to how spectrum access charges might be calculated for licences converted from apparatus licences to spectrum licences. The SMA has indicated that in determining the financial terms of conversions, consideration could be given to the basis on which any particular licence has been issued, for example, whether the licence was issued by a price-based process or by administrative allocation. However, as there has not been spectrum licensing to date there is no precedent supporting this analysis.

If any apparatus licence issued under the 1992 Act is converted to a spectrum licence, section 53(2) of the 1992 Act requires that the spectrum licence must, as far as practicable, authorise the operation of radiocommunications devices to the same extent as, or to a greater extent than, they are authorised under the apparatus licence being replaced.

Spectrum licences may be issued for any fixed term nominated by the SMA up to 10 years and then revert to the Commonwealth for reallocation through the market process. The spectrum licence is like a leasehold, rather than freehold, title. Spectrum licensees will be able at any time through the term to change the nature of use, amalgamate, subdivide, sublicence, trade and mortgage spectrum licences. General competition law, and in particular the Trade Practices Act 1974, will regulate acquisition and trade in spectrum licences. The SMA will be able to buy spectrum licences in the market place and resume spectrum for fair and just compensation "where this is essential to meet spectrum management objectives".

As noted above, as at the date of writing this review the SMA had not issued any spectrum licences. However, the SMA had allocated microwave multipoint distribution station (MDS) apparatus (transmitter) licences for a period of five years pursuant to an open outcry auction style process. This followed "closed envelope" tendering for telecommunications carrier licences and satellite subscription television broadcasting (Pay TV) licences. It is likely that the SMA will continue to use a variety of price based allocation systems and to issue apparatus licences for varying terms, depending upon the SMA's view of how best to secure a premium for licence allotment.

12. Broadcasting Services

17–32   The Telecommunications Act 1991 extends to any "communication", widely defined and so including a video or audio service, whether one-way or two-way (interactive). If the service delivers television programmes or radio programmes to the public, it is also regulated as a "broadcasting service" under the Broadcasting Services Act 1992. For example, a video service uplinked from a place outside Australia and downlinked in Australia (whether through fortuitous spillover—by the use of a satellite dish to access transmissions on international satellites—or as a service marketed locally in Australia) is an "international (telecommunications) service", and as a "telecommunications service" regulated by the Telecoms Act 1991. As noted earlier, this characterisation has important implications in defining exclusive rights of Australian telecommunications carriers. If the service is interconnected with a carrier network, the service provider must comply with the restrictions imposed under the AUSTEL International Service Providers Class Licence and Service Providers Class Licence AUSTEL does not regulate content of broadcasting services or the number of services offered to the public.

"Broadcasting services" as defined in the Broadcasting Services Act 1992 are services that deliver television programmes or radio programs to people having equipment appropriate for receiving those services, regardless of the delivery technology or mechanism used, but not including the following services:

- data and text services;
- services that make programs available on demand on a point-to-point basis, including dial-up services (that is, services made available for a customer at that customer's request). Contrast services which restart at regular intervals, which can be accessed at any time but do not restart at the customer's request—often referred to as "dial-in" services. So called "pay-per-view" may be either dial-up or dial-in, but the expression is often used in Australia to only refer to dial-in. In other words, video-on-demand or music-on-demand services (as distinct from dial-in services) are not "broadcasting services" regulated by the Broadcasting Services Act 1992. The regulatory status of other download based services, such as some proposed video games services, is less clear, but these are probably also outside the definition;

- services or classes of services that the Minister may by notice in the *Government Gazette* declare do not fall within the definition of "broadcasting service". There are no declarations to date.

The main significance of whether a telecommunications service falls **17–33** inside or outside the definition of "broadcasting service" lies in the broadcaster's cabling rights, as discussed above. In brief, a broadcaster may lay cable to carry its own service, or to provide delivery services to other providers of broadcasting services, but can only use that capacity to carry non-broadcasting services, such as telephony or a fully interactive service, (being a point-to-point video service) for or on behalf of a general carrier.

One of the key features of the Broadcasting Services Act 1992 is the division of broadcasting services into a graduated system of categories that attract different levels of regulation. The categories of service are:

- national broadcasters (the Australian Broadcasting Corporation and Special Broadcasting Service);
- commercial broadcasting (free to air radiated television networks and radio networks);
- community;
- subscription broadcasting (Pay TV);
- open narrowcasting (like free to air broadcasting, but with limited appeal or reach);
- subscription narrowcasting.

Each category is subject to a different level and type of regulation. For "narrowcasting services" (both open and subscription), service providers do not require an individual licence and instead are regulated by class licence conditions, which affect all providers within the category. An operator need only obtain access to a suitable means of transmission, such as a telecommunications network or a radcom licence issued under the Radiocommunications Act 1992, and in its operations comply with the class licence. For example, if a radio based service is interconnected with a carrier network, three forms of licence are required: compliance with the ABA (service) class licence, compliance with the AUSTEL service providers class licence; and issue (on an individual basis) of a radiocommunications licence.

Subscription television broadcasting services—Pay Television—are subject to a higher level of regulation. An individual licence must be obtained. However, with the exception of satellite Pay TV services, these licences can be readily obtained.

The most tightly controlled category remains free-to-air commercial tele- **17–34** vision licences, where the economic interests of the existing television broadcasters are protected by restrictions on the number of commercial services to be made available in any particular "service area", at least prior to July 1997.

The ABA may issue as many non-satellite Pay TV service licences as it sees fit; licences are issued on a channel by channel basis, without limit to the number issued to any applicants. The ABA will allocate these licences unless the Trade Practices Commission (which must be notified of the application) raises an objection on competition grounds or unless an applicant is found not to be a suitable licensee. "Suitability" review by the ABA is perfunctory. As a result, Pay TV licences issued by the ABA, with the exception of satellite Pay TV licences issued in the period prior to July 1997, have little or no inherent value.

The ABA does not maintain any public register of applications for non-satellite subscription broadcasting (Pay TV) licences and, as noted above, there is no individual licence required for narrowcast services. Nor does the ABA currently maintain a register of issued Pay TV licences.

The Broadcasting Act 1992 draws an important distinction between broadcasting and narrowcasting services. In brief, broadcasting services provide programmes "that, when considered in the context of the service being provided, appear to be intended to appeal to the general public": Broadcasting Services Act 1992, ss. 14 and 16. By contrast, a narrowcasting service is limited in reception by:

- being targeted to special interest groups;
- being intended only for limited locations, for example, sporting arenas or business premises;
- being provided during a limited period to cover special events;
- providing programmes of limited appeal, such as ethnic broadcasting;
- for some other reason, including any additional criteria as the ABA may determine (Broadcasting Services Act 1992, ss. 17 and 18.

17–35   The key points of difference between broadcasting and narrowcasting services are as follows:

- A narrowcast service may be provided without an individual licence or a specific permission from the ABA or any relevant body. By contrast, a Pay TV provider must obtain a service licence from the ABA, and issue of satellite Pay TV service licences is restricted until July 1, 1997.
- Subscription television licences (both Pay TV and narrowcast) are subject to a condition that subscription fees must be the predominant source of revenue for the service. It has been generally assumed that this means that at least 50 per cent of revenue must be derived from subscription fees rather than advertising.
- All Pay TV services, however delivered, are subject to a condition that "the licensee will not, before 1 July 1997, broadcast advertisements or sponsorship announcements". Narrowcast services are not so restricted.
- A Pay TV licence is subject to foreign ownership restrictions. A foreign person must not have an individual interest of more than 20 per cent and all foreign persons must not have interests in excess of 35 per cent in aggregate. Special rules apply to satellite Pay TV Licence A. No restrictions apply to narrowcast services.

The Australis Media Group acquired four channel satellite Pay TV Licence B and owns most microwave MDS licences in the major metropolitan centres. Australis will deliver Pay TV to urban areas using a hybrid satellite/MDS system and provide direct broadcast satellite services in rural areas where the population is too sparse to justify commercial use of MDS transmitters.

Continental Venture Capital Limited (an Australian company) and Century Communications Corp (a cellular phone operator and the fifteenth largest cable operator in the United States) formed Continental Century Pty Limited to provide satellite subscription television services as owner of the four channel satellite Pay TV Licence A. Continental Century and Australis will joint venture satellite network operations and subscriber management.

The Australian Broadcasting Corporation is entitled to take up satellite Pay TV Licence C, covering two channels, and may also joint venture with the other satellite licensees.

The 10 satellite Pay TV channels already allocated will be limit of direct broadcast satellite (DBS) delivery in Australia until July 1, 1997. Australis or affiliated companies purchased almost all of the MDS licences in the major metropolitan centres. It is likely that MDS will be the mode of provision of Pay TV services to most customers pending cable rollout.

There is also growing competition between Optus and Telstra to rollout **17–36** broadband cable network. Optus as a long distance carrier will use its local broadband access network, to be established in a joint venture with partners including Continental Cablevision of the United States of America, to bypass Telstra's local loop and reach the customer, and to leverage from provision of ancillary telecommunication services, including interactive services, which are more conveniently provided by cable than by satellite. Telstra, in joint venture with Rupert Murdoch's News Corporation, will overlay broadband on its existing copper pair customer access network as a defensive move in order to retain telephony business and to derive additional revenues from itself providing and from the common carriage of video services.

Telstra is reported to be ahead of schedule with its broadband network passing 25,000 homes "primarily in selected Sydney suburbs". Telstra has indicated that the network will support up to 60 channels and will pass 1.7 million homes by the end of 1995 (of a total of around 5.5 million homes) and 4 million homes within four years. The Optus/Continental Cable consortium, expects to roll out its Optus Vision network to serve 3.3 million homes within three years. Both Telstra and Optus have revised their rollout projections on a number of occassions and may further adjust these projections.

The Optus pay television satellite monopoly will end in 1997. It is likely that a number of alternative satellite carriers will in the interim position themselves for DBS Pay TV supply to and within Australia. For example, major satellite systems, including PanAmSat-2, APSTAR-2, PALAPA-C,

GCSat-3 and others, may all make available transponder capacity for satellite transmission that will reach major Australian markets.

A number of operators have announced their intention to establish their own cable capacity. For example, Rowcom has indicated that it is in advanced discussions with US cable TV operator Cox Cable and a number of Australian companies in relation to the establishment of a cable network in south-east Queensland at a cost of approximately $500 million. The network is intended to reach 700,000 homes on the Sunshine Coast (immediately north of Brisbane), Brisbane and the Gold Coast. It would begin construction in 1995 and aim to run past every home in the three regions within fives to six years.

The APA has also allocated cable licences to Paynet Telecommunications Pty Limited, a consortium of Australian and Canadian investors. It proposes to provide cable services in Cairns and Townsville, regions outside the low diameter dish segment of the Optus satellite footprint. Paynet has entered into an agreement with Optus whereby Optus will assist in installation of the Paynet network and will have the right to provide telephony services over this network.

As of mid 1995 there was only limited availability of narrowcast television services primarily by way of microwave MDS transmission. Aside from limited field trials conducted by Telstra, cable TV services were not available. Australis commenced provision of satellite and MDS Pay TV services in January/February 1995.

13. FOREIGN INVESTMENT RULES

17–37   Australia's foreign investment policy is administered by the Foreign Investment Review Board (FIRB), a division of the Australian Treasury. Its function is to review foreign investment proposals and to make recommendations to the Federal Treasurer. Foreign investment rules comprise both legislation (the Foreign Acquisitions and Takeovers Act 1975) and policy guidelines.

Types of investments which must be submitted to FIRB for approval before the investment is made include proposals to establish new businesses in the telecommunications sector, and other sectors of the economy, valued at greater than A$10 million. Such proposals will be approved unless judged to be contrary to the national interest. Investments which introduce expertise and technology, or acquisitions for which an equivalent local buyer is not readily available, are generally approved.

Specific foreign ownership restrictions apply to investments in free-to-air broadcasters and providers of Pay TV services. Apart from these requirements, no telecommunications industry-specific foreign investment rules are applied by the Australian Government.

Conclusion

The June 1991 telecommunications legislative package signified the begin-   **17–38**
ning of a rapid transformation of the Australian communications market.
Optus, Vodafone, and the operators of the State Government networks
have generated substantial demand for network equipment and software
and in turn stimulated the market for telecommunications services. A
switched reseller, AAP Telecommunications, competes with Telstra and
Optus in the long distance business telephony market. The range of services
open to competition has been considerably expanded. Telstra, Optus, and
Vodafone vigorously compete in provision of digital GSM public mobile
services. A number of international corporations have moved their Asian
telecommunications network control centres to Australia, and it is likely
that Australia will become a significant hubbing centre for the region.

The Australian regulatory scheme has a number of unusual features when
compared to schemes in force or proposed in other jurisdictions, for
example the interaction between the roles of the Minister and AUSTEL in
sitting economic and technical principles for interconnection and commercial negotiation contractual arrangements between the carriers; and use of
economic-based regulation to control carrier pricing of basic carriage services which service providers require to assemble and provide services competitive with those available from carriers. The complex BCS/HLS scheme
has been observed with particular interest in jurisdictions that have followed Australia in abandoning services-based regulation. As higher level
services (HLS) services increasingly depend on intelligent network functionality made available utilising the same facilities as are used to provide
underlying BCS, the distinction will become increasingly difficult to
manage.

Australia has now moved through two phases of telecommunications
regulatory liberalisation: the first through the 1989 legislation establishing
AUSTEL as an independent regulator and expanding the scope for services
based competition; and the second through the 1991 legislative package
introducing network based competition and permitting full resale in the
services market. The policy reviews now underway will determine the shape
and timing of the third phase of liberalisation, which will certainly include
an increase in the number of carrier licences issued after July 1, 1997. More
immediate changes will be driven by commercial decisions of carriers and
other prospective operators of cable networks, particularly as provision of
cable-based broadcasting services becomes intertwined with provision of
telecommunications services. Broadband network rollout, particularly by
Telstra and the Optus/Continental Cablevision consortium, will pose a substantial commercial challenge to prospective new carriers looking to enter
the Australian market in 1997. In the meantime, there will continue to be
active competition in provision of services between the carriers, switched
and switchless resellers and providers of niche telecommunications services.

CHAPTER 18

Canada

By Lorne P. Salzman, McCarthy Tétrault, Toronto

1. HISTORY OF DEVELOPMENT OF TELECOMMUNICATIONS IN CANADA

Industry Structure

The Canadian telecommunications industry today is characterised by a 18–01
wide and growing variety of players offering high quality service. This variety is the result of the opening of the market to competitive entry—a trend which started in the late 1970s, gathered steam during the 1980s and culminated in virtually complete liberalisation by the early 1990s.

Unlike many other countries, Canada never had a single telephone company. Over the years, a series of regional monopolies developed with a variety of ownership arrangements. Some were public companies with shares listed on stock exchanges; some were owned by various governments—federal, provincial and municipal; some were subsidiaries of foreign carriers; and one was for many years owned by a combination of the federal government and nine other telephone companies.

Canada currently has approximately 60 of what might still be termed the traditional wireline telephone companies. They range in size from small rural companies with a few hundred network accesses, to Bell Canada with some 9 million network accesses. The nine largest telephone companies, which collaborate with one another through an organisation called "Stentor", accounted for about 78 per cent of Canadian telecommunication service revenues of CAN$ 17.4 billion in 1993, with Bell Canada alone responsible for about 46 per cent of the industry total.[1]

The nine Stentor telephone companies serve territories that are adjacent 18–02
to each other and have for many years cooperated with each other to offer

---

[1] Industry Canada, *The Canadian Telecommunication Service Industries; An Overview*, November 1994, Pt. 1, p.16.

coast-to-coast long distance service. Originally, the focus of their cooperation was on traffic interchange and revenue settlement, and the development of microwave and, later, fibre optic infrastructure. Recently, however, this cooperation has expanded to include the development of new national services, as well as government relations and advocacy. Customer services are marketed under the Stentor banner although the services are actually provided by each local Stentor member in its own operating territory.

At present, all but two of the Stentor members are private companies that are owned, through holding companies, by investors. Saskatchewan Telecommunications and Manitoba Telephone System are owned by provincial governments.

Joining the nine telephone companies in Stentor is Telesat Canada, the sole Canadian domestic telecommunications satellite operator. Telesat offers a variety of satellite-based services in the 6/4 GHz and 14/12 GHz bands, notably to broadcasters. Commencing in 1995 a related company, TMI Communications, which will provide mobile voice and data communications services to remote areas using a dedicated satellite system.

With the opening up of long distance competition, several dozen competitors now offer service. A few are facilities-based carriers which operate their own transmission infrastructure. The majority, however, are resellers. That is, they obtain transmission capacity from other facilities-based carriers, to which they add multiplexing and/or switching functionality. Of the long distance competitors, Unitel Communications has the most extensive network facilities. It is owned by Canadian Pacific, a railway based conglomerate (48 per cent); Rogers Communications, which has extensive cable television, broadcasting and cellular telephone interests (32 per cent); and AT&T (20 per cent). Other larger competitors include Sprint Canada, Fonorola and ACC Long Distance (which operates purely as a reseller).

18–03   Canada has one overseas telecommunication carrier, Teleglobe Canada, which handles traffic to and from all countries other than the USA. Teleglobe Canada has interests in various underseas cables, and is a participant in INTELSAT and INMARSAT. Canada-USA traffic is not handled by Teleglobe Canada but is exchanged directly between long distance companies in each country.

Canada has a well developed wireless industry. Two cellular operators complete in most areas. One is a national licensee, Rogers Cantel. The other is the local wireline telephone company or an affiliate. The telephone companies' cellular operators are linked through a consortium called "Mobility Canada". In late 1992 four national licenses were awarded for public cordless telephone service at 944 MHz: one to Rogers Cantel, one to Mobility Canada and two to independents—Telezone and Microcell 1–2–1. As of mid 1995 only Telezone has begun to provide service in Toronto, Montreal and Vancouver. In addition, a variety of companies offer paging and radio common carrier services (such as dispatch). The federal government has announced its intention to licence up to six providers of personal communications services, or "PCS", in the 1.9 Ghz band.

The licences are expected to be awarded late in 1995, with service start-up likely in 1997.

Canadian policy-makers and regulators have long pursued the goal of universality in telephone service—a goal that has been achieved even in remote Arctic communities. Virtually all Canadian homes (over 98 per cent) and businesses have access to good quality telephone service.

Canada also has a very high degree of cable television penetration, with about 95 per cent of households passed by cable service, and about 79 per cent of households as subscribers. Cable television is furnished on a mono-poly basis, although the telephone companies are strongly advocating that they be allowed to furnish certain types of video services that will compete with the cable television companies. Several of the latter offer point-to-point transmission capacity using their cable television infrastructure, although none has yet sought to offer switched voice telephone service using cable infrastructure. In a 1995 report to the federal government,[2] the Canadian Radio-television and Telecommunications Commission (CRTC) recommended that competition in the supply of broadcasting services by cable or other technologies be permitted. The telephone companies would not however be permitted to enter this market until the barriers to entry in the local telephony market have been removed, which is expected to occur late in 1996.

Institutional Structure

Canada is a federal state in which certain powers are constitutionally alloc-   **18–04** ated to the federal parliament and others to the legislatures of the 10 prov-inces. As a result of a series of court decisions, it has now been firmly established that the federal level has exclusive jurisdiction over virtually all aspects of telecommunications, radiocommunication and broadcasting.

Within the federal government, the Department of Industry, also known as "Industry Canada", has primary responsibility for telecommunications matters (as well as a number of unrelated other mandates). The Spectrum, Information Technology and Telecommunications Industries Sector (SITT) of Industry Canada is headed by an Assistant Deputy Minister, and encom-passes four branches:

(i) The Telecommunications Policy Branch is responsible for policy issues relating to telecommunications including those concerning radio spectrum, radio and satellite services, telecommunications legislation, and the exercise of government powers as permitted under the Telecommunications Act 1993[3] is also responsible for international institutional arrangements, such as the ITU and INTELSAT.

(ii) The Spectrum Engineering Branch manages the process of develop-

---

[2] CRTC, *Competition and Culture on Canada's Information Highway: Managing the Realities of Transition,* May 11, 1991.
[3] S.C. 1993, c.38, as amended.

ing technical standards for radio equipment and equipment connected to telecommunications networks.
(iii) The Radiocommunication and Broadcasting Regulation Branch
manages radio spectrum allocation, including the granting of radio
licences.
(iv) The Information Technologies Industry Branch is responsible for
fostering Canada's information technology industries, and develops
policies and programmes to facilitate new media industries and
infrastructure in Canada.

18–05   The second pillar of the institutional structure relating to telecommunications is the Canadian Radio-television and Telecommunications Commission. The CRTC is an independent tribunal empowered by the Telecommunications Act to review rates, tariffs and interconnection
arrangements of telephone companies and other regulated carriers, plus a
number of related telecommunications matters.
       As its name suggests, the CRTC also has regulatory jurisdiction over
radio and television broadcasting by reason of its authority under the
Broadcasting Act 1991.[4] Included is jurisdiction over cable television—
referred to as "distribution undertakings" in that statute. Radio, television,
(over-the-air and cable-delivered) and cable television undertakings must
all be licensed by the CRTC, and must abide by regulations established by
the CRTC.
       The CRTC has 13 full-time members and six part-time members. Each
member is appointed by the federal government for a five year term, which
can be renewed. It is headed by a Chairman, and two Vice-Chairmen—
one each for broadcasting and telecommunications. Only the full-time
members are involved in decision making relating to telecommunications
matters. The CRTC is supported by a full-time staff of about 450, of which
about 80 deal with telecommunications matters.

18–06   Canada also has a Competition Act 1986[5] which deals with criminal
matters such as price-fixing conspiracies, and as well with what are called
"reviewable practices". The latter includes conduct, such as tied-selling that
is usually benign but can, in certain circumstances, result in a substantial
lessening of competition. Conduct that is reviewable can be challenged
before the Competition Tribunal, a specialised tribunal which deals solely
with such matters. It has the power to prohibit the continuation of practices
that are found to be anti-competitive and in some instances can order other
relief to restore competition. Only the Competition Bureau, the department
of the federal government that enforces the Competition Act, can initiate
a proceeding before the Competition Tribunal. A private party cannot do
so. One of the categories of reviewable practice is abuse of dominant position—conduct where one or more dominant marketplace participants use
monopoly power to exclude competition. It has been suggested that this

---

[4] S.C. 1991, c.11, as amended.
[5] R.S.C. 1985, c.19 (2nd Supp.), as amended.

is an ideal mechanism to bring about increased competition in the many telecommunications markets that are still dominated by incumbent telephone companies. To date, however, abuse of dominant position has not been used in the telecommunications field, and virtually all initiatives to increase competition have taken place by way of proceedings before the CRTC.

2. EXISTING LEGISLATIVE AND REGULATORY ENVIRONMENT

The Telecommunications Act

The Telecommunications Act is the primary statute relating to telecommu-   **18–07**
nications matters. It came into force in 1993, replacing a variety of statutes, the most important of which had been the Railway Act 1908.[6] The Telecommunications Act 1993 is divided into five main parts.

Part 1 includes a statement of Canadian telecommunications policy, and a mechanism for the federal government to make further broad policy determinations which then become binding on the CRTC. It also provides for the ability of the federal government to review and, if considered necessary, vary, rescind or refer back decisions made by the CRTC. The CRTC is also given the power to exempt carriers from regulation.

Part 2 establishes Canadian ownership and control rules for facilities-based carriers and continues a licensing regime for underseas cables landing on Canadian shores.

Part 3 sets out the rules relating to rates and tariffs of Canadian carriers. The CRTC is given the explicit power to forbear from regulating in certain circumstances. Other powers granted to the CRTC include the power to order a Canadian carrier to provide service or to interconnect its facilities with those of another carrier, the power to compel reporting of information and the power to direct the construction or modification of facilities.

Part 4 sets out the administrative and procedural framework for CRTC activities, while Part 5 describes the CRTC's investigative and enforcement powers.

Telecommunications Policy

The Telecommunications Act contains a statement of Canadian telecommu-   **18–08**
nications policy that the CRTC is required to apply in discharging its duties. Before the coming into force of the new Telecommunications Act, no single statement of telecommunications policy guided the CRTC. Rather, policy was made by the CRTC itself by means of its decisions, augmented by the occasional pronouncements of the federal government.

Section 7 of the Telecommunications Act 1993 provides as follows:

[6] R.S.C. 1985, c.R-3, as amended.

"7. It is hereby affirmed that telecommunications performs an essential role in the maintenance of Canada's identity and sovereignty and that the Canadian telecommunications policy has as its objectives

(a) to facilitate the orderly development throughout Canada of a telecommunications system that serves to safeguard, enrich and strengthen the social and economic fabric of Canada and its regions;

(b) to render reliable and affordable telecommunications services of high quality accessible to Canadians in both urban and rural areas in all regions of Canada;

(c) to enhance the efficiency and competitiveness, at the national and international levels, of Canadian telecommunications;

(d) to promote the ownership and control of Canadian carriers by Canadians:

(e) to promote the use of Canadian transmission facilities for telecommunications within Canada and between Canada and points outside Canada;

(f) to foster increased reliance on market forces for the provision of telecommunications services and to ensure that regulation, where required, is efficient and effective;

(g) to stimulate research and development in Canada in the field of telecommunications and to encourage innovation in the provision of telecommunications services;

(h) to respond to the economic and social requirements of users of telecommunications services; and

(i) to contribute to the protection of the privacy of persons."

18–09   As can be seen, the policy statement has something for every constituency. Each individual policy objective is expressed in broad language, using words of flexible application, such as "to promote" or "to facilitate". Because all of the policy objectives, are accorded equal status, conflicts can easily arise. For example, the CRTC is told in (b) that high quality service should be affordable everywhere in Canada, and in (h) that it should "respond" to the economic requirements of users. Yet despite these interventionist objectives, in (f) the CRTC must foster increased reliance on market forces. When asked the question: "Should local rates be raised to cost-based levels?", the CRTC will have to decide which policy objective should take priority. Ultimately, the CRTC would come to an answer, but based primarily on its own best assessment of the public interest at the time.

Recognising that further pronouncements on telecommunications policy may be necessary, the Telecommunications Act permits the federal government to direct the CRTC as to broad policy matters relating to the policy objectives in the Act and these directions are binding on the CRTC. To date, no such directions have been issued.

The existence of a policy direction power by the federal government recognises that telecommunications issues can sometimes be of sufficient importance that they call for action at the political, not regulatory, level. There is nothing new in this statement. Canada has long permitted telecommunications decisions of the CRTC to be appealed to the federal government for review and possible variance—a power continued in the Telecommunications Act 1993. But whereas the appeal route resulted in policy after the CRTC had first made its own ruling on an issue (that is, the CRTC created a policy which the government disagreed with), the new policy direction power contemplates a more proactive role for the government. It remains to be seen if the new power will be utilised very often. If experience in the transportation field is any guide, where the federal government has had a somewhat similar policy direction power for a number of years, the power will be rarely exercised.

Eligibility to Operate as a Common Carrier

The 1993 Act imposes no licensing requirement on anyone who wishes   **18–10** to establish a telecommunications common carrier. Licences are, however, required for certain types of activities that carriers might want to undertake. Most significantly, a licence under section 17 of the Telecommunications Act 1993 is required to establish an undersea cable landing on Canadian shores; a licence under the Radiocommunication Act 1989[7] is required to operate most radio transmitting or (non-broadcasting) receiving devices; and a licence under the Broadcasting Act 1991 is required to establish and run a cable television system.

The 1993 Act also imposes, for the first time, Canadian ownership and control requirements on certain telecommunication common carriers that operate in Canada (referred to as "Canadian carriers"). The term "telecommunications common carrier" is defined to mean a person who owns or operates a transmission facility used by that person or another person to provide telecommunications services to the public for a fee. A "transmission facility" means any system for transmission between network termination points, such as fibre optic cable or microwave links, but excludes certain exempt apparatus such as switching or multiplexing equipment. Thus resellers that just perform their own switching but lease transmission facilities from other carriers would not be operating transmission facilities, as defined, and therefore are not subject to Canadian ownership and control requirements.

Once it has been established that a service provider is a telecommunications common carrier, there are three Canadian ownership and control requirements that must be met:

(i) individual Canadians must occupy 80 per cent or more of the positions on the board of directors of the company (The Act contemplates that only corporations will be eligible to operate as Canadian carriers.);

---

[7] R.S.C. 1985, c.R-2, as amended.

(ii) Canadians must own 80 per cent or more of the issued voting shares; and

(iii) the company must be controlled in fact by Canadians.

18–11   Regulations have been issued to further define what are "Canadians" for the purposes of these requirements and how to monitor and enforce compliance. A different standard of "Canadian-ness" has been establishing for those companies holding shares in telecommunications common carriers than for the Canadian carriers themselves. A holding company will be considered Canadian if two-thirds of its voting shares are held by Canadians and it is controlled in fact by Canadians.

The control in fact test is not discussed in the Act or the regulations. And because the 1993 Act is quite recent, the CRTC has not had an opportunity to issue any guidance on it. Still, the test is not unknown to Canadian law. In the airline field, a Canadian control in fact test has been in place for many years. In one recent decision the National Transportation Agency of Canada commented as follows (Decision No. 297-A-1993):

"In reviewing the Canadian ownership status of an air carrier, the Agency considers various factors in making a control in fact determination. There is no one standard definition of control in fact but generally, it can be viewed as the ongoing power or ability, whether exercised or not, to determine or decide the strategic decision-making activities of an enterprise. It also can be viewed as the ability to manage and run the day-to-day operations of an enterprise. Minority shareholders and their designated directors normally have the ability to influence a company as do others such as bankers and employees. The influence, which can be exercised either positively or negatively by way of veto rights, needs to be dominant or determining, however, for it to translate into control in fact."

As a last point on the Canadian ownership and control requirements, certain telecommunications common carriers were already foreign-controlled prior to when these requirements came into effect, notably BC TEL—the telephone company that operates throughout most of the province of British Columbia. These companies have been permitted to retain their eligibility to operate although with restrictions on expanding their territories of operations.

Special Licensing Situations

18–12   As mentioned above, the Telecommunications Act 1993 does not have a licensing regime for Canadian carriers. In certain circumstances, however, statutory provisions which require licensing of particular telecommunications facilities have been used by the federal government to limit entry into the following fields: Canada-overseas communications, satellite-based communications and radio common carriers such as cellular telephone operators.

In the case of Canada-overseas communications (but not Canada-U.S.

communications), the government has designated Teleglobe Canada as the monopoly facilities-based operator. This is maintained through the licensing of overseas cable landings pursuant to the 1993 Act and the licensing pursuant to the Radiocommunication Act 1989 of earth stations aimed at INTELSAT and other international satellites. The monopoly runs until 1997, although a forthcoming government review will consider if it should be extended.

Similarly, the federal government has designed Telesat Canada as the sole domestic communications satellite operator, employing the licensing provisions of the Radiocommunication Act 1989 as the entry control mechanism. Domestic satellites (referred to in the regulations under that Act as "space stations") must be licensed, as must most all transmitting earth stations.

3. REGULATION OF SERVICES

Basic Rules Applicable to Services

The Telecommunications Act accords the CRTC a great deal of power to regulate all aspects of the activities of Canadian carriers under its jurisdiction. All rates for services, and their terms and conditions, are submitted to the CRTC for approval before they come into force, unless the CRTC has determined that it should either forbear from regulating a particular service or exempt a carrier from regulation. The applicant for a new service or seeking a change in rate will also supply material in support of its application.   **18–13**

Minor filings will generally be reviewed and dealt with by the CRTC without public process. An application by a wireline telephone company which will result in significant rate changes, or which impacts competitors, will lead to the CRTC issuing a public notice inviting the public to make written submissions as input to the CRTC's decision-making process. Full public hearings will be held in cases where important regulatory principles are at stake, such as occurred with the opening up of long distance competition, or where a telephone company seeks to improve its revenues by increasing all of its rates.

The Telecommunications Act 1993 requires that all rates charged by a Canadian carrier for telecommunications services be just and reasonable. A Canadian carrier is required not to unjustly discriminate against or bestow an undue or unreasonable preference or disadvantage on any person, including itself, in relation to a telecommunications service or its rates. The CRTC determines compliance with the foregoing. It is under the heading of preventing unjust discrimination that the CRTC has made many important decisions opening up competition in what were previously monopoly activities.

With the passing of the 1993 Act, the CRTC has been given an important new power, to foster competition, sometimes referred to as the "in-and-   **18–14** out" power. Where the CRTC determines that a service provided by a carrier is competitive, it can order that carrier to discontinue the service if to do so would be an effective and practical means of fostering just and reasonable rates and no unjust discrimination (the "out" power). Conversely, where a service is provided by an affiliate of a Canadian carrier, but the market lacks competition, the CRTC can order the carrier itself to provide the service (the "in" power).

Local telephone service is charged at a flat monthly rate in virtually all parts of Canada, although recently the CRTC has approved in principle Bell Canada's application to implement a type of measured service for business users starting in 1997. Traditionally, local telephone rates were set on a value of service basis, with little regard for costs. Customers in smaller centres with a limited number of flat rate calling destinations paid less than those in big cities with hundreds of thousands of numbers to call. Business rates were several times residential rates and long distance rates were high. The net effect of these policies was for subsidies to flow from business, urban and long distance services to residential, rural and local services. The consequence was highly affordable local service, especially for residential consumers. This, in turn, achieved a very high degree of household penetration. Although there have been periodic calls for movement towards cost-based pricing, progress has been slow.

18–15   With the opening up of many telecommunications services to competitive entry, the CRTC has permitted regulated carriers a great deal of pricing flexibility, particularly for business services and long distance services. Discounts are permitted for high volumes, for long term commitments and other incentive schemes, provided that they are offered to all. The CRTC does not, at present, permit regulated carriers to enter into customer specific arrangements except in limited circumstances (for example, the construction of customer-specific network facilities).

In assessing if a carrier's rates are in the aggregate reasonable, the CRTC employs a rate base, rate of return form of regulation. When a telephone company believes that it requires more revenue generally, it will ask the Commission for a general review of its rates. On occasion, the CRTC itself has initiated such reviews. The CRTC calculates a carrier's revenue requirement for a test period, typically one year, and sets the rates at levels that will yield the required revenue. The revenue requirement is determined by totalling all of the carrier's expenses, including salaries, depreciation, taxes, etc., plus a return on equity at a level that is comparable to other companies of comparable risk.

18–16   In late 1994, the CRTC issued its regulatory framework decision. In it the CRTC outlined its approach to regulating Canadian carriers in an increasingly competitive environment. The decision encompasses several intertwined initiatives. First, the rate bases of the telephone companies will be split (in an accounting, and not structural, sense) into utility and competitive segments. The utility segment will comprise mainly monopoly local and access services. Secondly, a carrier access tariff will be implemented which is a regulated charge by the utility segment to recover network access costs of long distance companies, as well as a contribution towards keeping down local rates to consumers. The carrier access tariff will be included as an imputed cost in the cost justification that the telephone companies must submit when they apply for approval of rates for their competitive services. Thirdly, the utility segment will move to a form of price regulation to take effect in 1998, and which will replace the current earnings regulation approach. Fourthly, restrictions on local competition will be removed. Several follow-on proceedings have been initiated by the CRTC to implement the various components of this new regulatory framework.

When a telephone company proposes a new service, the CRTC requires that the carrier provide an economic evaluation study of the service over a multi-year period. The purpose of the study is to demonstrate to the Commission that the service covers its costs and will yield a return. The Commission applies an incremental cost approach, and has established detailed rules for calculating the various study components. As the regulatory framework decision is implemented, this approach will be maintained for the utility segment. For competitive services, however, a more simplified procedure will be adopted.

Competitive Services

In Canada today, most aspects of telecommunications business are open to competition. What follows is a subject by subject description of the key rules that apply. Some parts of Canada, notably the province of Saskatchewan whose government-owned carrier has been granted an exemption from federal regulation until at least 1998, have not progressed as far as the rest of the country.   **18–17**

(i) Terminal Equipment

This area is fully competitive, with both telecommunications carriers and independent suppliers participating in the sale, leasing and maintenance of terminal equipment. The CRTC has decided that the market is sufficiently competitive to justify forbearance from regulation of Canadian carriers in respect of their terminal equipment activities. Network addressing equipment must comply with certain standards as explained below.   **18–18**

(ii) Enhanced Services

This is another area that is fully competitive. A regulated carrier that offers enhanced services must do so pursuant to tariffs approved by the CRTC, although it has the option of avoiding regulation by offering the enhanced services through an affiliate.   **18–19**

(iii) Long Distance

Competitive entry is permitted both on a resale or facilities-based basis. Equal access has been implemented as of mid-1994, thereby permitting   **18–20** customers to pre-subscribe to their carrier of choice for all long distance traffic. New entrants pay contribution charges to the incumbent telephone companies to subsidise local calling rates. At present the only permitted provider of facilities-based, Canada-Overseas telecommunications is Teleglobe Canada.

In order to promote the use of Canadian facilities, all long distance providers are required not to route Canada-Canada or Canada-Overseas basic traffic through the USA.

(iv) Pay Telephones

18–21 Competitive supply of pay telephones is presently not permitted. The CRTC has, however, stated that pay telephones directly connected to an alternative long distance supplier will be allowed once that supplier establishes an operator services tariff approved by the CRTC.

(v) Local Services

18–22 Non-switched local services can be supplied by competitors, and are offered by cable television companies and other suppliers in a limited number of areas. Competition in the provisions of facilities-based switched local services has been endorsed by the CRTC and proceedings are underway to unbundle telephone company bottleneck services in order to open up the local services marketplace.

(vi) Satellite Services

18–23 By federal government policy, only Telesat Canada is licensed to offer domestic satellite service. A limited amount of competition does take place, however, by means of the resale of satellite services. Also VSAT user in the USA is permitted to establish a small number of VSAT terminals in Canada to access a USA satellite for transborder and Canadian domestic use.

(vii) Wireless Services

18–24 Cellular service is a duopoly in all parts of Canada. Paging and radio common carriage services are competitive, with a variety of suppliers in most communities. Except in the largest centres, new entrants are able to obtain licences without difficulty. Public cordless telephone service has been licensed to four suppliers. The CRTC has recently decided to forbear from regulation most aspects of cellular and public cordless telephone service, and to exempt paging and other radio common carrier services from regulation.

Where there are many potential applicants for the available RF spectrum assignments, Industry Canada uses a comparative selection process to determine the successful licensees. The best known of these was the 1983 award of cellular telephone licences. Applicants are invited to submit applications which are then evaluated by government personnel to determine which will be successful. Evaluation criteria are publicised at the time that Industry Canada makes the public announcement inviting applications. Auctions of spectrum do not occur.

4. REGULATION OF BROADCASTING

The Canadian broadcasting industry is highly competitive. Each market is **18–25** served by a number of AM and FM stations and over-the-air television stations. Cable television delivers local and distant television signals as well as Pay-TV and cable-delivered specialty channels. Competing cable television services have not, however, been permitted. A satellite-delivered direct-to-home service has been announced for start-up in 1995.

The Canadian Broadcasting Corporation, which is owned by the federal government, offers radio and television services in English and French languages in virtually all markets. Some provinces operate their own educational television services. Otherwise all broadcasting and cable television companies are privately owned.

Broadcasting is regulated by the CRTC exercising its powers under the federal Broadcasting Act 1991. The 1991 Act establishes a broadcasting policy for Canada, which strongly promotes Canadian programming objectives. The CRTC has made policy pronouncements and issued regulations on a great number of topics, such as quotas for Canadian programming and advertising limits. The federal cabinet is empowered to issue policy directions to the CRTC, and several have been so issued, including one which restricts licences to Canadians and Canadian-controlled corporations (as defined).

The CRTC has a broad mandate to regulate all aspects of broadcasting. **18–26** It is given the power to issue and revoke licences, and to establish licence conditions. The federal cabinet may set aside, or refer a CRTC licensing decision back to the CRTC for reconsideration—a power which is rarely exercised.

The CRTC has deregulated two classes of service that make use of excess capacity in television and FM radio signals, namely the vertical blanking interval (VBI) of television signals and the subsidiary communications multiplex operation (SCMO) channel of FM stations. Broadcasters are therefore free to use or market these services with minimal restrictions. Line 21 of the VBI is reserved, however, for captions for the hearing impaired.

A cable television company that offers telecommunications services will be regulated by the CRTC under the Telecommunications Act 1993, in addition to its regulation under the Broadcasting Act 1991.

5. REGULATION OF APPARATUS

As mentioned earlier, the supply of terminal equipment for connection to **18–27** the telephone networks is fully competitive in all parts of Canada. The tariffs of the telephone companies include specific language authorising such connection.

Most of the large telephone companies offer terminal equipment on a lease or purchase basis. The CRTC has recently decided to forbear from regulating the telephone companies in the sale, lease or maintenance of terminal equipment.

All terminal equipment that is connected to the public switched telephone network must be type-certified for compliance with the appropriate technical standards. Certification is performed by Industry Canada, although testing is carried out by a number of accredited testing laboratories and manufacturers.

6. TECHNICAL STANDARDS

18–28   The development of technical standards for telecommunication equipment has typically been the product of committees of interested representatives from government (Industry Canada) and industry (mainly manufacturers and carriers). Once a standard is agreed upon, Industry Canada will seek public input by way of an announcement in the *Canada Gazette*, an official publication of the federal government. When the standard is finalised, it will be published by Industry Canada, and then brought into force. In the case of equipment connecting to the telephone network, the telephone company tariffs will stipulate adherence to the applicable standard. In the case of radio standards, Industry Canada is able to enforce compliance through its control of licensing under the Radiocommunications Act 1989.

There are two standards setting groups of particular note. The Terminal Attachment Program Advisory Committee (TAPAC) has been in existence since the 1970s, and is responsible for setting standards for the attachment of terminal equipment to the telephone company networks. Typically, the Canadian standards have been similar to the applicable USA standards, although with some variations. The Radio Advisory Board of Canada (RABC) has been very involved in developing radio standards, including most recently the standard for public cordless telephone service.

18–29   A number of other organisations have an interest in standards. The Telecommunications Standards Advisory Council of Canada (TSACC) provides policy input to governments in the area of standards. Its membership includes representatives of various standards setting bodies and industry. The Consultative Committee on Communications (CCT) was established in 1993 with the mandate of trying to harmonise telecommunications standards among the signatories to the North American Free Trade Agreement, namely Canada, the USA and Mexico. It comprises representatives of industry and government from all three countries.

The Canadian Standards Association (CSA) has established standards in a wide variety of fields. Best known are its standards for connection to the electric power grid. CSA has established a Steering Committee on Telecommunications (SCOT), comprising membership from industry, users and some governments. Its telecommunications standards are voluntary.

With the opening up of facilities-based long distance competition in 1992, the CRTC recognised that the interconnection of networks would raise a number of technical issues. The solution was to require the establishment of joint technical committees between competitors and the telephone companies in order to work out issues of technical compatibility and the like. As well, the Canadian Interconnect Liaison Committee (CILC) has operated as an industry-wide forum for trying to resolve problems related to implementing long distance competition.

The new Telecommunications Act gives the CRTC the explicit power to set technical standards for telecommunications facilities operated by Canadian carriers and for connection to them. To date, this new power has not been used. In the past, however, the CRTC has occasionally had cause to consider standards-related issues when opening up new markets to competition, such as terminal equipment. In that case, the CRTC required the telephone companies to amend their tariffs to allow attachment of equipment complying with stated standards.

CHAPTER 19

China

By Owen Nee, Coudert Brothers, Hong Kong

1. BEFORE TELECOMMUNICATIONS REFORM

During the Han dynasty, China boasted the most efficient communications   **19–01**
system in the world. Imperial couriers travelled the country across roads
of uniform width on scheduled trips delivering the emperor's directions to
his mandarins and the common correspondence of domestic commerce.
Sometime during the Ching Dynasty, China lost its lead in efficient organisation of the post, as well as the technological edge over the Western world.
Nevertheless, parts of China, such as Shanghai, before the Second World
War were technological wonders of telecommunications. New York Stock
Exchange quotations were telegraphed daily to Shanghai and appeared in
the local newspapers. By telegraph and radio, Shanghai traded on most of
the world's principal stock, bullion and commodities markets.
 During the 30-year period from 1949 to 1979, China's communications
system made progress in some areas and marched backwards in others.
Among the areas of improvement, domestic postal services improved in
China and the price of the mail decreased as a percentage of income. China
is one of the few major countries where overnight post intra-city is still an
expected service. National mail is both comparatively efficient and
affordable and remains the basic system for national communications.
Another area of post 1949 improvement was the telegraph service; while
previously the telegraph was limited to the principal cities, the Communist
Government expanded the service to cover all county and district centers.

 During the Nixon visit to China in 1972, international telecommunica-   **19–02**
tions took a giant step forward when the President personally authorised
the lifting of the Export Administration Act's prohibition against high technology exports to China, so that ITT could install a satellite ground station
permitting the live transmission of television pictures of the President's historic visit to China.

The type of communication that advanced the least during this 30-year period was the telephone. Prior to 1949 all major cities in China had telephone companies that operated an intra-city service with limited ability to make long-distance telephone calls to other major cities.

Thirty years after the communist take-over, the situation remained much the same, except now the local telephone companies were part of the post and telecommunications bureau formed by the local government's nationalisation of the former telephone company and assumption of control of the Nationalists' postal system. If one wished to place a telephone call to another part of China, it was necessary to go to the local Post and Telecommunications office, book a call with a teller, and wait until one was advised that the telephone booth was free and the call was through. There was no pretence made that the calls were not monitored; in fact, callers were instructed to speak in Putonghua, the common dialect, so as to make the job of the monitors more straight-forward. If one choose to use a different dialect, a rather stern voice from nowhere would enter the line and instruct that telephone calls in China only worked if the common language was used. In the posh hotels of the late seventies, such as the Peking Hotel in Beijing, a whole floor was reserved for the monitors who were capable of understanding most of the principal languages of the world. A guest wishing to make a call would register the destination, telephone number, and the name of the person to be called with the service desk on the floor, which would then call the guest's room when the call went through. Again, there was no pretence made that the calls were not monitored.

2. REFORM DURING THE 1980s

Forces Encouraging Telecommunications Reform

19–03   When the Deng Xiaoping reform era took hold in the late 1970s, a number of forces encouraged the reform of the telecommunications system. Probably one of the first was the agreement of many large cities to have new, Western-style hotels to encourage the tourist trade. Such hotels required in some case more than a thousand telephone lines, so that each room would have its own telephone and access to overseas dialling facilities; something that the existing telephone system was completely incapable of handling. In Beijing, the first modern telephone switching system installed was paid for by an assessment against a number of foreign investment projects that would use telephone lines from the new interchange.

Another factor requiring improved telecommunications facilities was the need for Chinese enterprises to communicate with the rest of the world. It made little sense to be the lowest cost producer of cheap, export goods, if a factory could not communicate with its overseas buyers. The unwillingness of Chinese factories or commercial enterprises to return a telephone call, answer by facsimile, or send a telex was legendary; but to a large extent this reluctance to respond to enquiries was due to a difficulty in obtaining outside lines for transmissions of commercial messages. Chinese industry therefore put pressure to bear on the government to improve communications.

In the early 1980s, when there was roughly one telephone for every 250 people in China, it was far easier to call overseas than it was to call across town. Foreign businesses in China would appoint a daily caller, who on behalf of all other personnel in the office, would constantly call business numbers around town until one was through and then would find the individual who wanted to speak with that successfully called company.

The emphasis on the expansion of international telecommunications was comparatively easier than expanding domestic communications, since both required the expenditure of foreign exchange and only the international business generated foreign exchange earnings.[1] This expansion of international telecommunications was concentrated in the principal business cities of China, which also happened to be the main tourist destinations, and included Beijing, Shanghai, Guangzhou and Dalian. During 1982, for example, the Chinese Ministry of Post and Telecommunications ("MPT") added 1,000 long distance telephone circuits in the whole of the country, but almost all of these circuits were concentrated in the principal coastal cities, as were the additional 170,000 telephones added in that year.[2]   **19–04**

However, not all development was concentrated in the international telecommunications area. In April 1982, the Chinese Ministry of Post and Telecommunications issued its first formal request to lease transponders for domestic communications on the INTELSAT satellite. The lease provided nationwide broadcasting for two colour television stations through one of the transponders and 200 channels for intercity telephone communications through the remaining one-quarter of the transponder leased.[3]

Factors Discouraging Telecommunications Reform

The principal factors discouraging the further and faster development of the Chinese telecommunications market were policy and economic concerns. The policy concern related to the ability of the public security organs to monitor traffic through a vastly expanded telecommunications network. The economic consideration was one of financing the expansion.   **19–05**

Because of the central MPT monopoly over telecommunications, during the 1970s and early 1980s all financing of telecommunications improvements was through the central government budget. This substantially limited the amount of money available from the central government, since the MPT needed to compete with all other national ministries for a portion of the national budget.

Rapid Reform in Telecommunications

In the mid-1980s, changes in Chinese policy toward telecommunications allowed the rapid modernisation of China's telephone system, which was   **19–06**

---

[1] C. Brown, "Telecommunications" (July/August 1982), *China Business Review* 43–44.
[2] *ibid*, p. 43.
[3] *ibid*, p. 44.

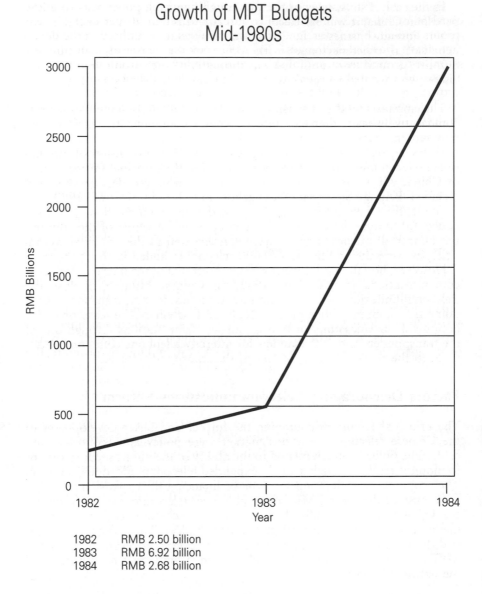

Growth of MPT Budgets
Mid-1980s

Year	
1982	RMB 2.50 billion
1983	RMB 6.92 billion
1984	RMB 2.68 billion

then estimated to lag over 20 years behind any other industrialised country. In part, after monitoring foreign commercial traffic for some years, the Chinese discovered that there was little of value to be gained in tapping into the foreign commercial traffic of visiting businessmen. Probably more important was the introduction of the so-called "diversified financing policy" which permitted broader experimentation with methods of financing new equipment.

One of the first changes of the diversified financing policy was to allow local post and telecommunications bureaus to retain 90 per cent of their profits and all of their foreign exchange earnings to be utilised in the development of the local exchanges.[4] The MPT took on the role as a coordinator of industry expansion, establishing compatibility specifications for equipment and exchanges imported, as well as adding the necessary political vetting of possible suppliers. By not returning telecommunications' profits to the central budget, the three year expansion in telecommunications budgets, including both the central and local MPTs budgets, was dramatic.

Along with the increased retention of profits and retained foreign exchange, the MPT transferred responsibility for financing network development to the local post and telecommunications bureaux (P&Ts). As a result, central financing decreased from 60 per cent of total telecommunications investment in 1983 to less than 9 per cent in 1989.[5]   **19–07**

Another aspect of the diversified financing plan was the use of international loans in order to finance the importation of foreign telecommunications equipment. During the 1980s, China borrowed in excess of US$1 billion in order to develop its telephone services. Loans were provided by Australia, Austria, Belgium, Canada, France, Italy, Germany, the Netherlands, Japan, Norway, Spain, Sweden and Switzerland. Loans from Japan made up almost 30 per cent of the total. Tianjin, Shanghai and Guangzhou added several hundred thousand local telephone lines with Japanese loans of approximately US$200 million. The Swedish Government provided an interest free loan of US$150 million to promote the supply of Ericsson exchanges of more 600,000 lines and long-distance exchanges of approximately 10,000 lines. Part of the Canadian Government's mixed loan of US$100 million was used to develop telecommunications in a number of medium-size and small cities in Hebei, Jiangxi and Shaanxi provinces (including more than 400,000 urban lines and more than 17,000 long-distance lines) and a 2,000 kilometer digital microwave telecommunications system covering Jilin, Liaoning, Guangdong and Xinjiang. Nearly 200,000 urban lines were added in Beijing and two digital microwave trunk lines from Wuhan to Chongqing and from Xian to Chongqing were financed by the French government's US$100 million mixed loan.[6]

By the end of 1989, the estimated total debt load of China's telecommunications industry was approximately RMB5 billion ($900 million) divided roughly three to two between funds expended on general construction (including plant and central office installation) and on system improvements. Foreign loans from governments and banks, according to the MPT, accounted for 63 per cent of general construction loans and 55 per cent of system improvement debts.[7] A list of the principal soft loans for China's telecommunications projects in the 1980 appears at the end of this chapter.   **19–08**

---

[4] *ibid.*, p. 43.
[5] L. Sun, "Funding Telecom Expansion" (March/April 1993), *China Business Review* 31.
[6] "China Develops Telecommunications with the Help of Foreign Loans" (September 23, 1991), *China Economic News* 3–4.
[7] L. Sun, "Funding Telecom Expansion" (March/April 1993), *China Business Review* 32.

Localities with the Highest Telecom Debt
(1989)

Municipality/Province	Debt Ratio*
Tianjin	51.8%
Shanghai	42.9%
Guangdong	22.3%
Beijing	17.6%
National Average	18.9%

* Debt as a proportion of total annual investment.[8]

The third part of the diversified financing plan introduced in the mid-1980s was that the local P&Ts were given more flexibility to raise their charges for domestic and long-distance calls, as well as charges for the installation of a telephone or lines. The percentage increase in the various charges differs region to region and city to city in part depending on the amount of debt incurred in order to finance the expansion of the telecommunications network.

19–09   Because of the relationship between debt incurred and local telephone charges, the MPT has issued guidelines on the permissible levels of debt for the telecommunications industry. The guidelines stipulate that major cities such as Beijing, Shanghai, Tianjin and Guangzhou, which have a large population base of potentially wealthy local subscribers, may have debt up to 40 per cent of the total investment in the local telephone network. Other coastal cities and provinces, where the demand is strong, but not as strong as the central cities, should keep their debt ratio to below 30 per cent; for remote provinces and cities with poor populations and uncertain revenues, the debt ratio to total telecommunications investment should not exceed 10 per cent.[9]

Results of Reform

19–10   The diversified financing plan of the mid-1980s was very successful. Digital telephone lines in China increased from 273,000 in 1986 to 1,475,000 in 1990 and the total number of telephones in major Chinese localities increased from 4.1 million in 1986 to 8.0 million in 1990.[10] At a conference on the telecommunications industry in 1993, Mr. Yang Taifang, the Minister of Posts and Telecommunications stated that in the 14 years of reform since 1979, the national MPT and the local P&Ts had invested in excess of RMB 50.1 billion in the improvement of the country's telecommunications system. This investment had caused city and long-distance telephone lines to increase by 6.7 and 11.2 times respectively since 1985, which in turn

[8] *ibid.* Original source MPT.
[9] *ibid.*, p. 32.
[10] S. Gorham and A.M. Chadran, "Telecom Races Ahead" (March/April 1993), *China Business Review* 19–20.

has increased the penetration of telephones from 0.38 per cent to 1.63 per cent by 1992.[11] From 1986 to 1990, national telecom traffic grew at an annual average rate of 24.4 per cent and telephone main lines at 17 per cent. This growth reached its zenith in 1993 when telecom traffic grew by 68.5 per cent, a total of 10.8 million office exchange lines were installed, and 5.86 million new telephone subscribers were added. By the end of 1993, telephone main lines totalled 17.33 million, equivalent to a penetration of around 1.46 per cent and the number of telephone sets stood at 25.44 million, a penetration of 2.15 per cent while in the major cities the penetration rate was almost nine per cent.[12]

In addition to quantitative improvements in the system, during the same period the local telephone networks substantially converted to automatic connection with 99 per cent of local calls automatically connected and over 90 per cent of long-distance calls. Digitalised exchanges now make up 84 per cent of all exchanges, and long-distance digitalisation has now reached 98 per cent. There are currently 36,000 kilometers of fiber optic cable connecting China's major cities, a network which should be completed in 1995. By the end of 1993, public telecom enterprises had registered 638,000 cellular mobile telephone users, which constituted a five-fold increase for the latter as compared with the figures for 1992.[13]

3. PROGRESS TO THE YEAR 2000

While China's progress in modernising its telecommunications system is **19–11** impressive, the road ahead is a long one since China has established very difficult goals to be achieved by the year 2000. Among China's goals are the following:

- A 16-fold increase in the major communications capacity of the telecommunications system as compared with 1980.
- A national automatic telephone exchange network adopting SPC exchange and fiber optic transmission technologies with the objective of increasing office exchange capacity to 100 million lines, toll circuits to 1.4 million and the telephone set penetration to five to six per cent. In fact, current estimates are that the latter figures on penetration are conservative and that nationwide the penetration figure will be more like seven per cent and in the large urban centers approximately 40 per cent.
- Constructing a digital data transmission network, public packet switching network and ISDN; development of communications information services as well as other value added services.[14]

In order to achieve these very ambitious goals, China will need to have an efficient regulatory framework for its telecommunications industry and find additional sources of finance for its planned expansion.

[11] (May 3, 1993) *China Economic Review* 16.
[12] Zhu Guofeng, Vice Minister MPT, "Telecom development—the Chinese Way" (March 1994), speech at Buenos Aires World Telecommunications Development Conference.
[13] *ibid.*
[14] *ibid.*

Organisation of Telecom Industry

19–12   The basic three level approach which has existed in China since the mid-1980s will continue through the end of the century. At the top of the pyramid is the Ministry of Post and Telecommunications ("MPT"), which sets important policies such as interconnection and technological development policies. The MPT also is in charge of ground station connections to satellite transponders of voice and data communications. The next level of organisation is the provincial post and telecommunications bureaux, which include bureaux in all provinces and the cities of Beijing, Shanghai and Tianjin. The P&Ts are responsible for developing the operating networks within the provinces and this responsibility includes arranging for the investment funds necessary for the development of the network. The lowest level of administration are the county and district offices of the P&Ts, which handle service administration, maintenance, installation of new telephones, and billing and collections.

In April 1994, China's National People's Congress approved a restructuring of the MPT to divide its functions in two: on the one hand, the MPT will continue its govermental administrative function of regulating the 30 or so semi-autonomous P&Ts both in regard to equipment standards, financing methods, and policy matters; on the other hand, a separate national long distance telephone company will be created which will buy and sell services to the local P&Ts.[15]

19–13   One aspect of the MPT's mission is to control the projects in which foreign investors are permitted to invest. The MPT's Post & Telecommunications Industry Corporation (the PTIC) reviews all proposals for foreign investment in the telecommunications sector, which are forwarded to it by the local P&Ts. The PTIC also plays a role in directing the P&Ts in their purchases of equipment to established PRC joint venture enterprises, rather than permitting the P&Ts to import from abroad. Notwithstanding the declared role of the PTIC as an industry gatekeeper, by ceding both the planning and financial responsibility for the development of the provincial networks to the provincial P&Ts, the MPT has created a situation where its guidance as to policy and technical matters may become less relevant to local officials intent on expanding their own telecommunications system locally.

Financing Methods

19–14   As noted above, concessionary loans from European and Japanese banks played a significant role in the expansion of China's telecommunications system from 1985 to 1990. However, in part due to pressure from the United States to reduce concessionary lending, countries such as Germany, Sweden, Canada, France, Norway and Australia have all agreed to reduce

[15] (May 26, 1994), *Telecom Markets* 247/5.

concessionary lending to projects which are commercially viable. More-over, the MPT itself is likely to insist that concessionary loans be made available to the least economically viable provinces, rather than the coastal areas which can afford to repay normal commercial borrowings.

Replacing the absence of concessionary funding, local P&Ts in the coastal areas have begun to fund their own expansions with internally gen-erated monies. Because of the new, higher toll charges, local P&Ts have found that capital investments in digital switching equipment can be repaid within two years of the original investment.[16] For the period from 1991 to 1995, RMB 75 billion has been budgeted for the enhancement of local telecommunications facilities as part of the state planning process. During the three years 1991 to 1993, the investments of the P&Ts have more than doubled each year.

These large investments have been made possible by the increased rev-   **19–15**
enues of the system. In 1993, total revenue generated by the state-owned enterprises engaged in post and telecommunications operations amounted to RMB 39.2 billion, representing a growth of approximately 56 per cent over the previous year. Long distance telephone revenue was RMB 19.5 billion alone.[17]

4. EXISTING LEGISLATIVE AND REGULATORY ENVIRONMENT

Telecommunications Policy Statement

China's system of regulating its telecommunications industry like other   **19–16**
areas of industrial endeavor in China, is by Western standards, bizarre. While China has a number of laws relating to telecommunications, cable television, satellites and satellite receptors, China's principal law on tele-communications is not a law at all, but an opinion. On August 3, 1993, the State Council, China's cabinet level organisation of the executive branch of government, issued The State Council's Notice of Approval and Transmis-sion of the Opinions of the Ministry of Posts and Telecommunications on Further Strengthening the Management of the Telecommunications Mar-kets[18]—this document set forth China's national telecommunications policy. The MPT Opinion was sent to all provinces, ministries and major municipalities in China with the admonition that the addressees "please earnestly implement it."

The MPT Opinion mentions three reasons for the necessity of the revised national policy. First, in order to maintain the normal order in the com-munications system—that is, to ensure that each of the local systems func-tion together in one comprehensive system. Secondly, to ensure the security

---

[16.] See S. Gorham and A.M. Chadran, "Telecom Races Ahead" (March/April 1993), *China Business Review* 19
[17] *Prospectus: Shanghai Posts & Telecommunications Equipment Co. Ltd* (September 30, 1994) pp 18–19.
[18] *Guo Fa* [1993] No. 55 (herein referred to as the MPT Opinion). The MPT Opinion itself is dated June 30, 1993.

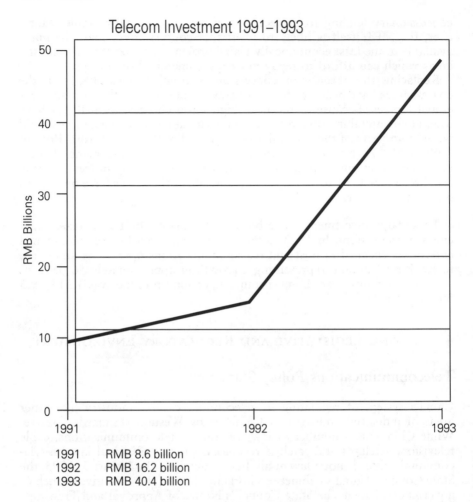

Telecom Investment 1991–1993

1991	RMB 8.6 billion
1992	RMB 16.2 billion
1993	RMB 40.4 billion

of the communications of the state and to provide quality communications services. It is somewhat strange that these two objectives should be linked together, but it is the natural outcome of an extensive fiber optic underground system, which China is currently building, that both state communications are more protected from spying and the consumer receives the benefit of a quality communications system. The third reason for the policy is to establish a "fair competitive environment".[19]

19–17   The first policy reform of the MPT Opinion was the institution of a formal telecommunications licensing policy. The local P&Ts are authorised to license telecommunications providers that operate solely within the local P&T's province; any operator that will cover an area broader than a single province or overlap between two provinces, must be licensed by the

---

[19] MPT Opinion, p. 2.

national MPT. The second policy statement reinforces the commitment of the local P&Ts and operators to a nationally integrated telecommunications system. All local telecommunications operators must abide by relevant national policies, accept management supervision from the national MPT, only charge fees determined in accordance with national policies on fees and charges, and not obstruct or hinder in anyway the full functioning of the national telecommunications system.[20] The MPT is given the responsibility for creating "a fair competition environment" for all P&Ts and local operators, including making arrangements for the compensatory sharing of equipment and facilities to provide an efficient national system.[21]

The most controversial portion of the MPT Opinion, however, is Article V, which restricts the role of foreign investment in the telecommunications sector:

"Foreign enterprises shall not conduct or participate in telecommunication operations within the territory of China. For all the public communication networks, wire or radio communications of the specific networks, the operation or participation in such business by all kinds of foreign institutions, enterprises, individuals and wholly foreign owned enterprises, Sino-foreign joint ventures and Sino-foreign cooperative enterprises established within our country are not allowed. Any means of introducing foreign equity investment in the operations is not allowed either."[22]

Although opinions differ, in general, most commentators believe that China has managed to enforce this ban on foreign participation in the operation of its telecommunications system. To date only a few companies have been able to breach this rigid exclusionary regime: **19–18**

Foreign Ventures in China's Network Services[23]

Company	Project
SCM/Brooks Telecommunications (U.S.)	Broadband network trial in Guangzhou with Galaxy New Technology.
Cable & Wireless/H.K. Telecom	Joint venture undersea cable agreement in place; revenue sharing to finance cellular telephones under discussion in Guangdong.
Champion Technology (Hong Kong)	Cellular network in Chengdu.
Star Paging (Hong Kong)	Cellular network investment in Guangdong.
Technology Resources (Malaysia)	Cellular network project with China Poly Industries Group.
Master Call (Thailand)	Announcement of $500 million cellular, local loop, optical fiber deal announced August 1994—no further developments.
MTC Electronic Technologies (Canada)	Cellular network project in Hubei and Shanghai on hold.

[20] MPT Opinion, Art III.
[21] MPT Opinion, Art IV.
[22] MPT Opinion, Art. V.
[23] (October 24, 1994), *Communications International*, 18.

Principal Laws and Regulations

19–19   Although the MPT Opinion establishes the most controversial portion of national telecommunications policy, there are a number of other regulations affecting telecommunications in China. The system used for the establishment of telecommunications rates is set forth in a Notice of the Ministry of Posts and Telecommunications Promulgating the Rules of Public Telecommunications Services issued on May 21, 1987, which replaced earlier similar directives, and sets forth the organisation system for the MPT and local P&Ts, the methods for establishing telephone charges and service rates, and the obligation of the users to pay for the service.

When dealing with such essential services like the telephone, which were in existence prior to the reform period, there tend to be few regulations, other than directives from the MPT itself. In areas of new technology, however, there is considerably more regulation.

(i) Satellite Reception

19–20   China regulates the reception of programming from satellites by regulating the manufacture, sale, and ownership of satellite receptors. The Regulations for the Management of Ground Satellite Television Broadcasting Receptors provide that the "production, importation, sale, installation and use of satellite receptors" are all subject to state licensing.[24] Only the State Council's electronic industry administrative bureau can authorise the production of satellite receivers.[25] The sale of receivers can only be conducted by enterprises that have been licensed by the local P&Ts.[26] Any enterprise or individual that wishes to import foreign made satellite receivers needs a licence from the State Council Ministry of Radio, Film and Television; anyone importing components needs a licence from the electronics ministry; and individual importation of satellite receptors by hand or through the post is prohibited.[27]

While the basic licensing system is set forth in the State Council Decree, the Ministry of Radio, Film and Television was given the power to issue detailed implementing regulations for the management of ground satellite television broadcasting receptors. These regulations[28] make clear that only certain units may apply for a permit to have a satellite receptor: units involved in education, scientific research, news, finance or economic affairs; hotels and guest houses for foreigners; offices and apartments exclusively used by foreigners. The detailed regulations go on at much length as to the licences required and the punishments in store for anyone obtaining a satellite receptor without receiving the appropriate permissions. In order to make the situation even more difficult, on January 14, 1994, the State

---

[24] State Council Decree No. 129, Art.3.
[25] State Council Decree No. 129, Art.4.
[26] State Council Decree No. 129, Art.5.
[27] State Council Decree No. 129, Art.6.
[28] Detailed Implementation Regulations for the Management of Ground Satellite Television Broadcasting Receptors (February 3, 1994).

Administration of Industry and Commerce, the body which supervises the registration of business entities in China, announced that units engaged in the sale of satellite dishes must have registered capital of at least RMB 1 million and be capable of providing after sales service support.[29]

For all the strictures contained in the Decree and Detailed Regulations, China's attempts to limit the distribution of satellite dishes has proved quite impossible. According to a market survey of the penetration of Star TV in Asia, the satellite dish recipients increased in China from January 1993 to October 1993 by 536 per cent[30]. One estimate of the total number of satellite dishes in China in early 1994 was 600,000 receptors.[31]

(ii) Cable Television

Cable television systems have been in existence in China for many years.   **19–21**
Originally organised in urban areas, cable wiring was installed in buildings in order to assure adequate television reception. Although these separate systems were not linked together, in the early 1980s the State Council approved a proposal to require all new buildings to be equipped with Master Access Television (MATV) to provide better reception and eliminate the overabundance of aerial antennas. In time, these systems began to be patched together in small cable systems—most of which played local, State approved television programming, but other uncensored programming began to appear.[32]

Therefore in the early 1990s, China adopted the Singapore model of cable television regulation, which permits the establishment and licensing of cable systems while prohibiting direct reception of foreign broadcasts. The cable systems are preferred since they permit easy censorship of content. In 1990, the State Council adopted the Provisional Regulations Governing Cable Television[33] which put the Ministry of Radio, Film and Television in charge of licensing authorised cable systems. The regulations also set guidelines for the construction and financing of cable systems. Since that time the Ministry has introduced several additional regulations to strengthen their control of the cable business.[34] The Ministry has stated that it approved over 661 cable licences in 1993 and now every major city in China has an approved and licensed cable system.

Most urban networks are 300 Mhz cable networks that carry between   **19–22**
12 and 13 channels. The typical installation cost in Beijing is approximately 200 to 250 RMB and monthly fees of about 10 to 15 RMB are charged to subscribers. These networks rely on the television stations and other

---

[29] New China News Agency, January 14, 1994: State Administration for Industry and Commerce formulates five provisions concerning ground satellite receptors.

[30] (March 7, 1994), *Business China: Media* 3.

[31] S. J. Schenfeld, "Cable Ready" (September/October 1994), *China Business Review* 25.

[32] *ibid.* pp. 24–25.

[33] *Guo Fa* [1990] No. 94, November 2, 1990.

[34] Rules for the Implementation of the Provisional Regulations Governing the Supervision of Cable Television, Decree No. 5 of the Ministry of Radio, Film and Television, April, 20, 1991; Rules Governing Cable Television, Decree No. 12 of the Ministry of Radio, Film and Television, February 3, 1994.

domestic programmers for their programming. With the low fees charged, the networks are not in a position to produce their own programming or to purchase more appealing programs. The three largest urban cable systems are in Beijing, Tianjin, and Shanghai. The Beijing Cable Network has approximately 300,000 subscribers,[35] the Tianjin system is utilising a microwave system to join together a number of local residential systems, and the Shanghai Cable TV Network, which is China's largest, has almost 1.1 million subscribers.[36] It is highly likely that these large urban networks with many subscribers and not enough programming will seek out cooperative arrangements with foreign program suppliers so as to fulfil their large requirements for air-time programming.

(iii) Control of Radio Frequencies

19–23 As with cable networks, radio stations are subject to a licensing system, which is administered at the national level by the State Radio Management Organisation.[37] The system is similar to cable networks in the sense that if a radio station will only have a broadcast range within one province, then the station may be licensed by the provincial radio management bureau; however, if the broadcast range covers more than one province or crosses a provincial border, then the national level State Radio Management Organisation must issue the licence.[38]

Another important function of the State Radio Management Organisation is the assignment of radio frequencies, including those frequencies used for cellular telephone communications.[39] As one would expect, cellular radio frequencies are required to fall within a certain range and, from within that range, the State Radio Management Organisation or its provincial affiliate will assign the frequency to the approved cellular operator.

The regulations provide that "no unit or individual is allowed to assign, lease or lease in disguise the frequencies designated to it without approval".[40] Certain early cellular telephone investments were done by forming a joint venture either with the local P&T, the local military, or the local police, each of whom have radio frequencies assigned to them for their use. The foreign company would pay all costs related to the establishment of the cellular business and the only contribution from the local Chinese partner was the use of the assigned radio frequency.

(iv) Equipment Supply and Investment

19–24 In April 1994, China's MPT announced a list of five preferred suppliers of switching and transmission equipment: Alcatel, American Telephone & Telegraph, NEC, Northern Telecom and Siemens. Although other suppliers

---

[35] S. J. Schenfeld, "Cable Ready" (September/October 1994), *China Business Review* 27.
[36] C. X. Yuan, "The Development and Prospects of Shanghai's Telecommunications" (January 24, 1995), IIR into Shanghai Conference.
[37] Regulations of the People's Republic of China on the Management of Radio Operations, Decree No. 128 of the State Council, September 11, 1993.
[38] *ibid.*, Art 13(1).
[39] *ibid.*, Art 23.
[40] *ibid.*

are not excluded, the preferential designation is certainly helpful in dealing with each of the local P&Ts. Almost immediately after the MPT's announcement, two of the preferred suppliers, AT&T and Northern Telecom, announced plans to create joint ventures in China for the manufacture of switching and other telecommunications equipment.[41]

It is likely that such equipment supply joint ventures will be the most likely method for foreign participation in the Chinese telecommunications market. The MPT Opinion still prohibits foreign participation in the operation of a telecommunications network, although apparently—in part due to a shortage of capital—this is now being interpreted as permitting lease arrangements whereby the rental paid takes into account the profitability of the system as a whole.

5. DEVELOPMENT OF COMPETITION

At the same time that National People's Congress authorised the restructuring of the MPT into a separate regulatory body and a commercial enterprise, the Congress also ratified the formation of two separate enterprises which are licensed to compete with the MPT and the local P&Ts in certain telecommunications areas. Approval was granted for the formation of Lian Tong, an enterprise authorised to offer voice services, and Ji Tong, an enterprise authorised to provide value added services. The two new companies are not intended as direct competition for the MPT, but to cooperate in the development of a vast market.[42]

19–25

Lian Tong

Lian Tong, which is sometimes referred to in English as the China United Telecommunications Corporation, is a joint venture between the Ministry of Electronic Industries, the Ministry of Railroads, and the Ministry of Energy, each of which originally contributed one-third to the company's registered capital. These three ministries have existing private networks and the intention is to offer existing spare capacity for voice services on these networks, once joined together, to other enterprises and commercial users. Both the railways and electric power grids in particular have additional capacity to connect commercial users to their existing networks. As of April 1994, Lian Tong signed an agreement with NYNEX to advise on the technical parameters for such a second network.

19–26

The company has received additional investment from 13 large domestic companies including the China International Trust and Investment Corporation, Everbright International Trust and Investment Corporation, the China Huaneng Group, China Resources Group, and China Merchants

---

[41] (May 26, 1994), *Telecom Markets* 247/4.
[42] (May 26, 1994), *Telecom Markets* 247/5.

Holdings. Each of the 13 investors has invested 80 million yuan for a total of approximately US$120.25 million in equity.[43]

Ji Tong

19-27   Ji Tong was established at approximately the same time as Lian Tong and currently has in excess of 30 different shareholders, including some of the same investors as in Lian Tong. The proposed business of Ji Tong are value added services in the areas of information services, mobile telephone services, cable TV, and the manufacture of telecommunications equipment.

Ji Tong has signed an agreement with BellSouth to form the Beijing Ji Tong-BellSouth Communications & Information Engineering Company, which will provide network planning, design and engineering services for the value added network. Although still in the initial stages of planning, the proposed network should be nationwide.

Another proposed Ji Tong project is its cooperation with the International Business Machines Corporation for three "golden projects". Included within these planned value added networks are three separate services: Golden Bridge—a public information network for data exchange between universities, research institutes and commercial users; Golden Customs—a network for messages and electronic data information with respect to automated customs clearance and foreign trade; and Golden Card—a validation of credit card transactions network.

In the mobile telecommunications field, Ji Tong intends to start in the paging, trunk mobile radio and CT2 business areas, as these particular areas are of less concern to the MPT. Ultimately, Ji Tong would like to become involved in satellite mobile communications.

[43] (July 30, 1994), *China Daily*, p. 1.

Foreign Soft Loans to China Telecom Projects 1984–1989

Country	Project	Value $Million	Year	Purpose
Belgium	Shanghai Bell Telephone Equipment Factory	$10 Million	1984	Belgian government coinvestment in Bell Telephone Co. switching equipment joint venture (all money concessional).
Sweden	Shanghai Telex Machine Factory	$2.2 Million	1984	Loan for technology transfer by Phillips Sweden and factory upgrading.
Belgium	Shanghai Bell Telephone Plant	$11 Million	1985	Equipment and parts purchase by Shanghai Bell Telephone equipment joint venture.
France	Beijing Telephone Network, Phase 1	$66 Million	1985	Contract with Alcatel for 100,000 lines in 14 exchanges, plus trunk circuits, optical fiber cable and terminals.
Japan	Shanghai, Guangzhou and Tianjin Telephone Networks	$135 Million	1985	OECF commitment for 150,000 lines worth of equipment (actual contracts may not have come out to such a high value). Loans have been effectively tied.
Sweden	Liaoning Telephone Network, Phase 1	$21.8 Million	1985	Contracts with Ericsson for switching systems and digital microwave communications in 4 cities; covers a total of 68,000 subscriber lines.
Italy	Chongqing Telecommunications Equipment Plant	$15 Million	1986	To support transfer of Italtel pulse code modulation (transmission) technology
Sweden	Liaoning Telephone System, Phase II	$14.3 Million	1986	
Canada	Jiangxi Telephone System	$6 Million	1987	Contract to Northern Telecommunications for switching systems.
Norway	Guangdong Power Network Microwave Communications	$4.7 Million	1987	Concessional portion in form of grant; contract to A/S-NERA.
Norway	Beijing Marine Satellite Earth Station	$8.2 Million	1987	

Country	Project	Value $Million	Year	Purpose
Sweden	Tianjin No. 2 Cable Plant Upgrading	$2.7 Million	1987	To support technology transfer for polyolefin-sheathed telephone cable by Kabmatic Company.
France	Beijing Telephone Network, Phase II	$103 Million	1988	Contract with Alcatel and Cables de Lyon for 155,000 lines in 10 exchanges plus 12 satellite exchanges, trunk circuits, optical fiber cable and plastic cable.
Netherlands	Yangtze Optical Fiber and Cable Company	$22.7 Million	1988	Mixed credit for Dutch part of investment in Philips Optical Fiber and Cable joint venture in Wuhan.
Netherlands	Suzhou No. 1 Wire Communications Factory	$7 Million	1988	For technology transfer of Philips digital switchboard technology and factory upgrading.
Norway	Guangdong Power Network Microwave Communications	$1.9 Million	1988	New contract for A/S NERA (followed similar 1987 contract).
Spain	Hubei Telephone Network	$20 million	1988	Contract to Accatel Spain to cover 43,000 lines, 18,000 trunk lines, and cable and tgransmission equipment for several cities.
Sweden	Liaoning Telephone Network, Phase III	$13.5 Million	1988	Loan promised for Ericsson Equipment for 50,000 lines switchying and 300 long distance lines for Dalian, Liaoyang, and Benxi in 1989.
Sweden	Guangdong Telephone Network, Phase I	$33.4 Million	1988	Covers contract with Ericsson for switching system for 131,000 local lines in several cities and digital microwave transmission
Germany	Shandong Switching, Phase II	$10.0 Million	—	Germans appraised project 1988 for future loan, to be offered after 1989.

Country	Project	Value $Million	Year	Purpose
Canada	Shanghai Television Tower	$40 Million	1989	Chinese media announced in 1989 that facility would be bought from Canada using mixed credits.
Canada	Telephone Systems in Jiangxi, Shandong, Hebei, and Shaanxi Provinces	$50 Million	1989	To fincance all or part of Northern Telecom sale of 300,000 lines worth of switching equipment in 1988.
Germany	Shandong Telephone System, Phase I	$20 Million	1989	Loan promised in 1987–88 for 25,000 lines of switching equipment in various cities and microwave transmission equipment
Spain	Zheijiang Telephone System	$15 Million	1989	Contract with Alcatel Spain for a reported 50,000 lines.
Sweden	Congqing Telephone Network	$12.1 Million	1989	Announced that China would import 40,000 lines of switching systems from Ericsson in 1989.

Source: (November/December 1989), *China Business Review* 26–27.

CHAPTER 20

Denmark[1]

By Karen Larsen, Berning Schlüter Hald, Copenhagen

1. HISTORY OF DEVELOPMENT OF TELECOMMUNICATIONS IN DENMARK

The Beginning

Visions are today frequently overtaken by technological developments 20–01 within the telecommunications sector. The ability to adapt visions and developments to one another and to reality is vital. However, this was apparently not the governing policy in the State-owned Danish Telegraph in the beginning of the 1870s when they tested the newly invented telephone sets. The outcome of the testing was that the telephone was interesting, but it would probably never out-compete the telegraphic media!

The inhabitants in Copenhagen thought differently, and privately owned enterprises installed the first telephone sets around 1880. A well-known Danish industrialist, Mr C F Tietgen, established the Copenhagen Telephone Company in 1882. Similar companies were soon established regionally in other parts of Denmark.

The private companies expanded their activities in a rapidly growing market. The state-owned Telegraph—now wiser—competed with the private companies and lobbied behind the lines with a view to the establishing of a state monopoly. Cables were put side by side by the competing parties, lots of small companies emerged, merged and sometimes disappeared.

For 14 years the Danish Government did not succeed in getting a bill passed instituting a State monopoly. Finally, in 1897, the Telegraph and Telephones Act was passed.

[1] It must be emphasised that the leglislation in Denmark is currently undergoing fundamental changes with a view to reaching the ultimate level of liberalisation no later than January 1, 1998. The contents of this Chapter was drafted prior to the finalisation, April 6, 1995 of the third Supplement to the Political Agreement on the Telecommunications Policy, and subsequently updated on June 15, 1995 at which time not all legislation implementing this third Supplement was issued.

Establishing of the State Monopoly

20–02 The Telegraph and Telephones Act 1897 has been the foundation for the carrying out of public telecommunications activities in Denmark. Basically, it is still in force. The guiding principle is that the State has an exclusive right to all such activities, but that the Minister may grant concessions to others to establish and operate specified telecommunications services. 100 years later—by year-end 1997—the monopoly will disappear in accordance with decisions made by the Member States of the European Union.

In 1897, 11 regional companies were operating in Denmark, and each was awarded a concession to operate in their area. The State had to take care of the regions which were not profitable. In addition to such operations, the State remained responsible for connections between regions and for the international operations.

The telephone concessions awarded to the privately owned regional companies were time-barred, and almost every renewal of the concession has involved the establishing of commissions to prepare, implement or at least evaluate the structure, primarily as to whether or not the telephone service should be operated solely by the Government.

In the late 1930s the Danish Government succeeded in becoming majority shareholder in the two major telephone companies. However, all the regional companies including the two major companies continued to act rather independently. Several attempts were made to unify the structure, but without success. The parties could not agree on changes to the structure.

Institutional Aspects

20–03 The concessionaries have always been subject to governmental control. The institution in which such regulatory powers have been vested has, however, changed over the years.

Originally, post and telephony (and telegraphy) belonged under the Ministry of Public Works, later the Ministry of Traffic and in the 1980s transferred to a newly established Ministry for Communications.

All the time the General-Directorate of Post and Telegraph held many of the regulatory and supervisory powers within the Ministry. Several other Committees or Councils have emerged and disappeared as supervisory bodies.

Following the establishment in 1990 of Telestyrelsen (the National Telecom Agency Denmark—NTAD) a separation between regulatory and operating functions within the telecommunications sector was seen. In addition, a separation between the post and the telecommunications sector was instituted, although NTAD was an agency under the General-Directorate of Post and Telegraph.

In October 1994, following the general election in September, a reorganisation of Ministries and their agencies were carried through. The reorganisation comprised the final separation of the post and telecommunications. Telecommunications are now under the Ministry of Research, together with all information technology.

Influence of the European Union/EEC

Denmark became a member of the European Communities in 1972. As described elsewhere, the E.C. (now E.U.) has for some years now been very active within the telecommunications field, putting together a policy relating to telecommunications, primarily with a view to liberalising the activities, and issuing directives to be implemented in the Member States. Thus, for many years the Danish telecommunications sector has had a very busy period adapting to that deregulation, liberalisation, and abandonment of ancient monopolies which were necessitated by the multitudes of initiatives launched by the Commission within this field.   **20–04**

In 1988 the Terminal Equipment Directive (88/301) had been agreed upon. In 1990 the Services Directive (90/388) and the ONP Framework Directive (90/387) were agreed upon and issued as the very essential and fundamental steps on the road of liberalisation. Further, several Green Papers had been issued, recommendations agreed, and the development accelerated. Reference is made to Chapter 16.

In particular the Services Directive meant that competition had to be instituted within the telecommunications sector except for telex, voice telephony, radio-based mobile services, satellite services, use of the network, distribution of broadcasting signals and infrastructure.

Political and Structural Background to the Present Legislation and Structure

Originally the Government was in charge of the operation of all telecommunications internationally, maritime services, telegraph, distribution of broadcasting signals, satellite earth stations, inter-regional cables, land mobile services, data transmission, etc. The concession companies held telephone concessions, regionally divided.   **20–05**

The major concessions could be terminated by giving one year's notice to redeem the shares privately owned at a price of 125/100 (in which case the concession elapsed). The concessions were to be renewed as of March 1, 1992, and accordingly action had to be taken at least one year ahead.

On June 22, 1990, a political agreement was reached between the Danish Government (at that time being a Conservative Liberal Government) and the Social Democrats on an amended telecommunications structure in Denmark in light of the completely changed conditions for the telecommunications sector in the 1990s.

The concessionaries and the state-controlled operations were made wholly owned subsidiaries of a holding company, Tele Danmark A/S, established for this purpose. Fifty one per cent of the shares in Tele Danmark A/S were agreed to be owned by the Danish State.

20–06   The exclusive rights of the regional telephone companies in the relevant regions as well as all state operated services were thus repealed and replaced with a nation-wide concession to the new parent company. It was agreed that the Government's right to redeem the private shares should continue as a right to redeem shares in the parent company instead. However, it took several years to implement the new structure due to this right of redemption being opposed by several shareholders.

It was decided to introduce B-shares not subject to redemption, but with voting rights representing one-tenth of the A-shares. This division made it possible to quote the company on the stock exchange. However, until further notice the company remains under Government control, the State owning 51 per cent.

In the 1990 agreement it was agreed to establish a governmental agency, Telestyrelsen (the National Telecom Agency Denmark, NTAD) to control and regulate the telecommunications sector. By establishing the NTAD, a clearer distinction was made between the regulatory and operational functions in accordance with E.C. policy (as stated in Directive 90/338, the Services Directive).

It was included in the political agreement that data communications services would be liberalised as of January 1, 1993, and that legislation making it possible to issue a licence to a competing operator covering the GSM mobile telephony services was to be expected shortly.

The political agreement is valid until March 1, 1997. It was also a part of the agreement that liberalisation required by the E.C. could be implemented, but that no further liberalisation should take effect until after January 1, 1994.

20–07   In early 1993, the Director General of the Postal & Telegraph Services finalised a paper on the future structure and liberalisation pace within telecommunications. Being ahead of the demands from the E.C. at that time, the paper was positive towards liberalisation of simple resale (telephony), satellite services, etc. Parallel with these Danish deliberations, the E.C. moved forward.

The first political supplement to the 1990 agreement was entered into on June 25, 1993 but dealt primarily with the capital structure of Tele Danmark A/S. Also the introduction on the New York Stock Exchange, which actually happened in 1994, was agreed.

The second political supplement entered into on February 8, 1994 had more substance, agreeing on the liberalisation as of January 1, 1994. Voice telephony was liberalised, although only to be provided through the Public Telecommunications Network. Tele Danmark A/S still operates the numbering scheme, and all lines must be leased from them.

Infrastructure is not expected to be liberalised ahead of E.C. requirements, viz. prior to January 1, 1998.

Following very lengthy negotiations, a third political supplement was **20–07a** agreed upon on April 6, 1995. As expected, infrastructure will not be liberalised ahead of E.C. requirements, although some restrictions are already lifted as of July 1, 1995. The agreement operates with a two-stage strategy for the implementation of the future Danish telecommunications policy. The legislation necessary for realising Stage 1 takes effect on July 1, 1995 by the lifting of Tele Danmark's exclusive rights in a number of fields, containing the following elements—

(i) the exclusive rights held by Tele Danmark A/S of transmitting radio and television broadcasting will be lifted to allow unrestricted programming transmission. The restrictions on ownership access to community antenna systems will be lifted so that such systems may be privately owned in the future. The hybrid network which until now is used exclusively for conveying radio and television shall be available for all telecommunications services.

   Accordingly, all restrictions on the use of leased lines will thereby be lifted, allowing for interested service providers to combine transmissions of programming with other services, for example, data, telephony, and interactive services (such as video-on-demand).

(ii) Liberalisation of local broadband networks, meaning that the exclusive right held by Tele Danmark A/S of establishing and owning broadband networks in local areas, *viz.* within the boundaries of a municipality, will be lifted. The same delimitation of a local area is valid in the context of community antenna systems.

   This opens a possibility for *e.g.* Danish State Railways, and public utility companies, etc., to offer the lines (within a municipality) to third parties. Requirements re-accounting to prevent cross-subsidising will be established.

   These new broadband services will be regulated according to existing telecommunications legislation.

(iii) Establishment of own (corporate) networks to be facilitated by way of an Executive Order following discussions with Tele Danmark A/S.

(iv) Lowest tariffs for broadband connections in order to increase use of new services.

(v) Improved conditions for mobile operators, meaning that the Interconnect Agreement between Tele Danmark A/S and the mobile operators be renegotiated to achieve better conditions and lower charges for the mobile operators.

(vi) Political constraints on Tele Danmark's organisation will no longer apply. According to the first political Supplement, Tele Danmark was obliged to let the original companies exist until March 1997. Immediately following the agreement on the third Supplement, Tele Danmark A/S merged all subsidiaries into the parent company taking effect as of January 1, 1995. Thus, KTAS, Jydsk Telefon, Fyns Telefon, Tele Sønderjylland and Telecom do no longer exist as separate companies, but as divisions of Tele Danmark A/S.

The Agreement states that additional liberalisation until January 1, 1998, apart from that set out by the third Supplement, should be made the subject of discussion between the parties.

2. EXISTING LEGISLATIVE AND REGULATORY ENVIRONMENT

Principal Legislative Instruments

20–08 The most basic and still valid legislation is the Telegraphs and Telephones Act passed in 1897. According to this Act, the Danish State has the exclusive right to establish and operate telegraphs and telephones.

The 1897 Act allows the minister for a period of up to 20 years to grant concessions on the establishment and operating of telegraphs and telephones. In cases where the facility connects Danish territory with foreign countries, concession can be granted for 25 years.

The monopoly of the State has always excluded certain installations for user's own purposes, including facilities established by municipalities and municipal institutions like the fire squads, electricity and water supply plants, etc. to the extent such facilities are operated exclusively as an internal facility for internal services.

Also, the railways have the right to establish and operate facilities exclusively for their own internal use; and finally the monopoly excludes facilities inside the borders of an estate or between estates belonging to the same owner for the owner's internal use, provided that the facility does not have to cross the sea.

It is part of the third political agreement, *cf.* above, to facilitate the establishment of own networks, *e.g.* by lifting the requirement of ownership to an estate, also allowing lessees to establish and use facilities for internal purposes, and to make the joint use of an internal network more widely accessible. Additionally, it allows the establishment of radio-based own networks across the territorial sea (under licence obtained for this purpose under the Radio Communications Act 1992).

Although still valid, fundamental and applicable, the 1897 Act is no longer to be considered the basic legislative instrument.

20–09 In 1990, the Telecommunications Act was enacted,[2] implementing the first political agreement on the structure of the telecommunications sector in Denmark, and as a consequence of E.C. policy to be implemented. The new structure instituted the required separation between the regulatory and operational functions.

The Telecommunications Act 1990 comprises all public radio and wire-based telecommunications services. It was amended in 1994 in accordance with the amendment to the political agreement dealing with further liberalisation and as of July 1, 1995 implementing the third Supplement.

[2] Now Consolidated Act 1994 (no. 474) to regulate certain aspects of the telecommunications sector and later amended by Act (no. 974) of June 14, 1995. (originally Act (no. 743) of November 14, 1990).

The 1990 Act divides the telecommunications services into two groups: the reserved area and the competition area of which the reserved area is further divided into two:

- Exclusive rights, which means there is only one provider of services in Denmark;
- Special rights, *i.e.* a limited circle of providers have been granted permission to offer specific service(s).

The 1990 Act states that the Ministry shall grant concession of the   **20–10** installation and operation of the transmission routes and exchanges linked to the following public radio and wire-based telecommunications services and their provision:

   (i) Telephony services.
  (ii) Text and data communications services.
 (iii) Permanent leased lines.
  (iv) Mobile communications and satellite services.

In accordance with these provisions, a concession has been granted to Tele Danmark A/S covering the exclusive rights area.[3] The contents of the concession, issued in pursuance of section 2 of the 1990 Act on Telecommunications, is also a basic legislative instrument.

The Minister is entitled to transfer areas from exclusive rights to special rights or to liberalise them totally. The concessionary shall tolerate the decisions made. Normally, at least one year's notice of such limitation of the concession will be given.

Special rights are introduced by way of legislation. For the time being   **20–11** the special rights comprise the GSM Mobile Telephone System,[4] in which area a second operator has been licensed, and the ERMES paging system[5] in which area both GSM operators are now licensed, but no third operator has yet been identified.

The installation and operation of services not subject to any concession or special rights are, thus, in principle liberalised. However, according to section 3 of the 1990 Act the Minister may lay down directions for the installation and operation of the services not subject to any concession or special rights, including a right to introduce notification or licensing arrangements and determine the conditions hereof. The directions will secure that the so-called essential requirements are fulfilled by the providers.

NTAD issues a public "List of Services, Transmission Routes and Exchange Equipment excepted from Tele Danmark A/S' Exclusive Rights".

Under the concession and thereby being an exclusive right falls the infra-

---

[3] (Now) Executive Order (no. 157) issued on March 10, 1994 on concession to Tele Danmark A/S (replacing Executive Order (no. 852) of December 11, 1990).
[4] Public Mobile Communications Act 1990 (no. 744) (the GSM Act) as amended by Act (no. 372) of May 18, 1994, now published as Consolidated Act 1994, (no. 473) and later amended by Act (1974) of June 14, 1995.
[5] Public Paging Act 1994 (no. 372).

structure, *viz.* the establishment and installation of all necessary cables, lines, etc. Any third party will have right to access on fair and equal terms in accordance with the subscription conditions approved by the National Telecom Agency, but ownership is exclusive. In accordance with E.C./E.U. rules, access shall be given on equitable, non-discriminatory and cost-oriented basis.

As a consequence, even subscription terms are forming part of the basic legislative environment.

20–12   The right to access on fair and equal terms appears from the Telecommunications Act 1990, the concession issued to Tele Danmark A/S and is also underlined in the guidelines issued by the NTAD and applied to the concessionary stipulating for instance a prohibition against cross-subsidising between the exclusive area, the special rights area and the liberalised areas. Guidelines are also issued regarding the accounting system, etc.

Alongside the 1990 Act, one of the basic legislative instruments is the Radiocommunications Act 1992[6] which states that the establishment and operating of facilities for radiocommunications in Danish territory can only take place in accordance with a licence issued by the Minister.

All applicable E.C./E.U. rules are implemented in Danish law, including the Directives on Tele Terminal Equipment, Services, etc. in so far as the date stipulated for latest implementation is reached.

An Act passed by the Danish Parliament will normally provide that details be given in executives orders and issued and signed by the Minister in charge of the regulated area. Such executive orders are often of even more interest than the Acts when evaluating the legislative environment.

20–13   Instructions to adminstrative bodies can be given by way of circulars, not binding the citizens but binding the civil servants/administrative bodies. A governmental agency will normally be established by a single provision in an Act, overall provisions laid down in an executive order and detailed provisions, for instance rules of procedure, etc., laid down in a circular which is also issued by the Ministry in charge.

No provisions of the Danish Constitution has any specific bearing on telecommunications.

Denmark is a signatory to various international conventions, for instance regarding satellite co-operation. In addition Denmark is also a contracting party to various conventions regarding the intellectual proprietary rights relevant, for instance in connection with broadcasting/cable distribution (Berne Convention, Rome Convention, The European Agreement on the Protection of Television Broadcasts).

Agencies of Regulation

20–14 The regulatory body responsible for supervising concessions and special rights within the telecommunications services is the National Telecom

---

[6] Radio Communications Act 1992 (no. 297).

Agency Denmark, established in accordance with the political agreement of June 22, 1990 and the Telecommunications Act 1990. The updated executive order regulating the activities of the NTAD was issued on July 26, 1994 (Executive Order No. 708). On the same date a circular describing the tasks to be undertaken by the NTAD was further issued.

The NTAD is a governmental agency under the Ministry of Research. The tasks to be performed by the NTAD practically cover all areas within telecommunications, whether reserved or liberalised, whether regulated by the Telecommunications Act 1990, the Radiocommunications Act 1992, the Telecommunications Terminal Equipment Act 1992 or any other Act relating to telecommunications.

The NTAD is divided into a number of sections, indicating the responsibilities vested in NTAD, including the following sections: **20–15**

- Security in telecommunications;
- administration, including all relations to the international co-operation organisations;
- broadcasting, including the so-called hybrid network and Community Antenna Systems and satellites;
- specification and approval (within radio-communications, telecommunications and EMC (Electro Magnetic Disturbances));
- all telecommunications technical matters, including updating within research and development;
- frequency planning, including participation in international co-operative work;
- licensing, including for instance land mobile radio terminals and radio terminals to be installed in ships and aeroplanes as well as licences to radio amateurs;
- supervision of operators and providers of services within the telecommunications area;
- coordination of the participation of the NTAD in E.C./E.U. co-operative work; and
- inspection.

Decisions made by the NTAD can either be appealed to the Ministry or to a special Tele Council in cases where decisions fall within the scope specifically made the final administrative responsibility of the Tele Council.   **20–16**

The executive order and adhering circular describing the tasks and the powers vested in the NTAD contain a lengthy list of acts and executive orders instituting the basis for the powers.

The concession issued to Tele Danmark A/S as well as the licences issued within the special rights area also contain a long list of tasks to be undertaken by the NTAD, *cf.* below (paragraphs 20–21 *et seq.*).

Competition Control

The competition areas concerning fair competition, pricing, etc., are not under the control of the NTAD. Competition control outside the conces-   **20–17** sion and special rights area are matters dealt with in the same way as regards any other enterprise. This means that the E.C. rules on competition as well as the national Competition Act 1993 are the applicable provisions.

The Danish Competition Act 1993 was adopted with effect as of January 1, 1990.[7] The Act is based on a principle of transparency in the market. It seeks to create a competition and efficiency oriented development of any business sector. Unlike the E.C. competition rules which are valid alongside the Competition Act 1993, the 1990 Act is not based on a principle of prohibition, but provides for intervention in agreements and marketing conduct that could have harmful effect.

A significant means of achieving the desired market transparency is a requirement that agreements and resolutions which result in or which could result in a dominant influence on a market sector must be notified to the Competition Council. Non-notified agreements are void. Notification implies that the agreement will be accessible to the public—which also means that competitors may obtain a copy.

The contents of the agreement is studied in order to judge whether it should be notified or not. The measures which the authorities can implement against anti-competitive activities are covered in sections 11 to 14 of the 1990 Act. Entities included and governed by concession legislation are excluded from these sections due to the fact that the concession legislation will include special remedies as well as the institution of a special body to supervise transactions and actions taken under the concession.

20–18   Basically, the applicability of sections 11 to 14 in the Danish Act on Competition 1990 means that if the Competition Council on the basis of a complaint or by its own initiative finds anti-competitive aspects of a certain character, the Council may call for negotiations, or alternatively request the anti-competitive actions to cease or declare an agreement, decision or term fully or partly void.

Similarly, in accordance with section 13, if a price policy is considered anti-competitive the Council may decide the guidelines for prices and terms and even fix such prices for a certain period up to one year.

The possibilities of sanctions within areas covered by the concession are even more efficient, quicker and much broader than according to the Competition Act 1990. The NTAD can order the concessionary to change its prices and terms with immediate effect, although in practice the parties will always consult in advance. Although the same sanctions and remedies in principle are included in the Competition Act 1990, normally the Competition Council will initially have to apply milder sanctions.

Within the telecommunications sector the competition rules have been applied for instance relating to sole distributor agreements between the concessionary and a major manufacturer and supplier of PABC systems, based on the rather interesting delineation of the market. In this actual case, a distinction was made between large and small PABC systems in order to establish the market share and evaluate the possibility of substitution.

[7] Competition Act 1993 (no. 114).

As mentioned above, Articles 85 and 86 of the Treaty apply alongside the Competition Act 1990. To the extent that trade between Member States can be affected, these rules have to be adhered to. The importance of the rules will probably increase as the liberalisation within the telecommunications sector in the E.U. is extended.

General Rules Governing Telecommunications Installations Services and Apparatus

The general rules today are based on legislation implementing the applic-   20–19
able E.U. directives, *viz.* in particular the Terminal Equipment Directive (88/301), the Service Directive (90/388) and The Open Network Provision Directive (90/387). Until these Directives were implemented in Danish law, the regulations were first and foremost found in the terms and conditions applicable to subscribers, issued by the concessioned telephone companies and approved by the regulatory body. According to the subscription terms, only equipment supplied by the telephone companies could lawfully be connected to the public telecommunications network. This meant for instance that a subscriber could buy a telephone appliance abroad, but it was illegal to connect such apparatus. Two of the major concessionaries offered different design in telephone sets to their subscribers, but in case a subscriber wanted the design of the other concession company, he had to move to that region.

In 1986, the Telephone Appliances Act was passed liberalising the possibilities. However, it hardly ever entered into force and the Act implementing the Terminal Equipment Directive was passed instead.

The exclusive rights still include the right of connection to the public networks. Inter-connect agreements are only of relevance within the special rights area.[8] Liberalised services still have to be provided through the public network and connected under terms stipulated by the concessionary as approved by the NTAD. Regarding apparatus, however, any appliance which is type-approved can be connected to the public network.

The remaining barrier is the telecommunications infrastructure. Any public offering of any telecommunications service has as a prerequisite that the infrastructure belonging to Tele Danmark A/S is used.

Licensing Regimes including Main Licence Categories

It goes without saying that a concession is necessary within the exclusive   20–20
area, and that a licence is necessary within the special rights services.

Regarding most services within the competition areas, guidelines on the exercising of rights to offer services are laid down in executive orders. However, far from all areas require a licence.

The reason for not having an extensive licensing regime is that one of the conditions for the legal provision of a liberalised service as stipulated

---

[8] Inter-connect agreements will, after July 1, 1995, be of relevance in additional areas, following the liberalisation on establishing and owning broadband networks in local areas.

in the executive orders, is that the service shall be connected to the public network—thus, a subscription arrangement must be entered into with the concessionary, and the details will be regulated in the subscription agreement. If necessary, the NTAD as regulatory and supervisory body is vested with an express authority to demand that the concessionary disconnects the provider's access to the telecommunications infrastructure.

Within broadcasting, the regime is quite different. A licence issued by at least one authority is always a prerequisite. Reference is made to paragraphs 20–46 *et seq.*

The Radiocommunications Act 1992 normally provides that a licence must be applied for and obtained as a prerequisite to use radiocommunications equipment (in addition to the necessary approval relating to the equipment). Within the 1992 Act and adhering executive orders, both individual and general licences can be issued.

3. REGULATION OF SERVICES

Analysis of Rights to Provide Telecommunications Services

20–21   Telecommunications services are divided into the reserved area (exclusive rights and special rights) and the competition area.

The Telecommunications Act 1990 comprises all telecom services. The exclusive rights are provided according to concession to Tele Danmark A/S and the conditions laid down therein in accordance with the provisions of the 1990 Act.

Special rights are introduced by way of law. For the time being the special rights comprise the GSM Mobile Telephone System and the ERMES system.

As regards the competition area, this is either totally unregulated or regulations have been issued by the NTAD in accordance with section 3 of the Telecommunications Act 1990 entitling the Minister to issue executive orders covering the competition areas. According to this section, the Minister is entitled to establish rules for the installation and operation of the services, transmission routes and exchanges not subject to any concession or any special right. The NTAD issues a list of the exceptions.

Section 3e of the 1990 Act empowers the Minister to issue regulations covering the contents of telecommunications services which provide sound or picture, further entitling the Minister to decide that such services shall be subject to licence or notification requirements, linking these provisions to the provisions of the Act on Radio and Television.

Section 3f of the 1990 Act empowers the Minister to regulate the offering of use to third parties of infrastructure by public utilities having established networks for their internal use, as a consequence of the third Supplement.

The Exclusive Area

20–22   The areas covered by the exclusive rights above, *cf.* paragraphs 20–07 *et seq.* on the Principal Legislative Instruments, are telephony services, text and data communications services, permanent leased lines, mobile communications and satellite services.

As stated above, infrastructure is the most important of the remaining parts of the monopoly (the exclusive right), although other areas still remain monopolistic for the time being.

(i) Infrastructure

The right to establish infrastructure is dependant upon a concession issued 20–23 by the Government, and the provision of liberalised services is dependant upon access to the monopolised infrastructure, *cf.* below.

However, as excluded in section 3 of the 1897 Act, infrastructure can be established and operated for purely internal purposes, specifically stated with regard to the railways and the municipalities and their institutions (waterworks, electricity providers, etc.) and additionally by private legal or physical entities within the borders of their own estate.

There are no hindrances to asking a third party to operate the facility ("facility management"), but only, for purely internal purposes.

The regulations and restrictions only cover services handled for a third party. In addition to facility management, intra-group management is excepted. Generally, although not clearly stated in any Act, "one single legal person" will be interpreted to cover a group consisting of limited liability or private companies qualifying as a "concern" as defined in the Danish Presentation of Accounts Act or controlled by the same persons.

An executive order[9] has been issued liberalising networks for telecommunications purposes (voice, text or data) located within the boundaries of a property and intended for connection to and communication via the public network. This will be revised and, in accordance with the third political Supplement, the possibilities of extending, the rights to establish infrastructure for internal purposes will be examined.

As mentioned, Tele Danmark A/S has been granted the exclusive right 20–24 to lease lines in and out of Denmark. Leased lines are defined as any transmission route, permanently established between two or more monopoly termination points. (A monopoly termination point being defined as the interface in the telecommunications network where telecommunications terminal equipment or internal networks are connected to the public telecommunications network.)

A leased line may consist of cables, radio relay systems or satellite connections. The leased line may be based on analogue or digital technology. All means of providing telecommunications services were until recently covered by the exclusive rights. However, satellite networks are now liberalised. Further, due to the liberalisation in accordance with the 3rd Supplement on local broadband networks, the public utilities and others can now offer third parties to lease their networks. Alternative operators preparing for the future total national liberalisation have already initiated negotiations and/or entered into agreements on a local basis.

[9] Executive Order no. 109 of March 10, 1993 on Internal Networks.

The monopoly on infrastructure is expected to be totally abolished no later than January 1, 1998, in accordance with E.U. policy. According to the political agreements in force, liberalisation ahead of E.U. demands requires unanimity.

(ii) Conditions to the Concession

20–25   The conditions, which may be stipulated as prerequisites to hold the concession, are listed in the Act on Telecommunications 1990 and specified in the executive order. The important conditions are, *inter alia*, to accept the control and supervision exercised by the NTAD, including,

- Approval of the concessionary company's accounting rules and financial reporting, etc., and approval of changes in or deviations from these rules;
- approval of prices and price principles and amendments hereto as regards new and existing services, as well as demanding changes of current prices and principles;
- approval of terms of subscription and amendments hereto covering the concessionary's network and services within the exclusive area;
- approval of terms and amendments hereto as regards connection to the public tele network for operators offering services comprised by special rights (interconnect agreements);
- approval of terms and amendments hereto as regards connection to the public tele network for suppliers of services comprised by section 3 in the Act on Telecommunications (liberalised services), including the right to demand current terms and conditions changed;
- collection of information as regards co-communication agreements with tele operators in other countries as well as the right to demand in harmony with E.C./E.U. legislation that co-communication agreements are changed;
- the right to demand that the concessionary observes regulations laid down by the E.C./E.U.;
- the fixing of specific regulations as regards the concessionary observing appropriate decisions made within the international co-operation in the telecommunications area;
- the fixing of further regulations as regards the participation of the concessionary in international co-operation organisations including specification tasks; and
- the right to demand information on any matter, and the right to issue orders as regards violation of the Concession ascertained. The NTAD can fix penalties to ensure that the concessionary complies with orders made in accordance with the Concession.

20–26   Violation of the conditions entitles the Minister to withdraw the concession. The concession further stipulates that the Minister is entitled to transfer exclusive areas to become a special right or to liberalise the service.

The concessionary pays an annual royalty to the Government of four per cent of the net income on operations within the exclusive rights.

The control exercised by NTAD comprises the issue of accounting regulations. Furthermore, NTAD has issued general guidelines of July 8, 1993 for the relationship between the telephone companies' activities in the liberalised area and in the reserved area. The strict separation of funds and accounts relating to the reserved area, and of funds and accounts relating to the competition area is stipulated.

Special Rights

Mobile Telephony based on the GSM network is a special right,[10] and can be provided according to a licence issued by the Government. The provision of Mobile Telephony based on NMT (the nordic system) is still an exclusive right under the concession issued to Tele Danmark A/S. However, the frequencies used for NMT 900 (applicable also in Switzerland and the Netherlands) are reserved for the GSM network and are consequently subject for reallocation, if necessary.   **20–27**

Two GSM Mobile Telephone operators are licensed, Tele Danmark A/S (via its subsidiary Tele Danmark Mobil A/S) and Dansk Mobil Telefon I/S (called Sonofon—a consortium composed by Danish GN Store Nord, Swedish Nordic Tel and American Bell South). The second operatorship was issued as the result of public competitive tendering. The licence terms are fairly similar to the concession terms.

The competition between the two operators is rather tough. The NTAD felt that it was necessary to regulate by way of an executive order the rules governing marketing subsidies to distributors when selling the apparatus and subscriptions, and the only penalty fixed by the NTAD relating to a violation of the licence conditions was imposed upon Tele Denmark A/S in this respect.

Special guidelines are issued by NTAD dealing with "Certain competitive aspects of public mobile telecommunications services. (Guidelines for the mobile area)"—based on and extending the general guidelines of July 8, 1993 for the relationship between the telephone companies' activities in the liberalised area and in the reserved area.

The guidelines are issued to ensure that the two GSM operators are able to compete on equal terms in relation to the exclusive area concessioned to Tele Danmark A/S, and that the GSM operators' activities in the special rights area do not result in distortion of competition in the liberalised area, and that subscribers are given as good and low-cost services as possible.   **20–28**

A strict prohibition on cross-subsidising is included, accounts shall be divided into sections, and the accounts of both operators are supervised by the NTAD.[11]

ERMES (the European Radio Messaging System) which is a land-based public radio system was transferred to be a special right in 1994. The infrastructure established by the operators of the GSM network can be used as the basis for operating ERMES. Thus, both GSM operators will get a

---

[10] cf. Public Mobile Communications Act, Promulgation Order (no. 708) of July 26, 1994 and as amended by Act (no. 474) 1995.

[11] It is part of the 3rd Supplement that the Interconnect Agreement shall be renegotiated, prices lowered, and possibly introducing the right for the operators to make use of each other's masts.

licence. However, no third operator has been appointed, even though the public tendering has been carried through no-one applied for the licence, and in fact only Tele Danmark A/S operates ERMES.

Liberalised Services

20–29   The list issued by the NTAD of services, transmission routes and exchange equipment excepted from Tele Danmark A/S' exclusive rights comprises the following liberalised services:

Voice telephony services, data communications, pay-phones (not set up in streets and squares to which the public have direct access), value-added services, internal networks, satellite services, video links (licence issued to Radio Denmark, TV2 and local TV stations for TV purposes), installations for user's own purpose, CATV community antenna systems, and special radio related applications.

The starting point for a regulation of a liberalised area will be the non-profit safeguarding of the interest of the general public which may limit the access to the public telecom network or the public telecom services, the so-called essential requirements.

The procedures to be stipulated can for instance allow for the following considerations—

(i) The satisfaction of the needs of specific user groups, such as disabled persons.

(ii) Information to the users of public telecommunications services about the terms and conditions governing connection to the service, including information about tariffs, quality and times of delivery of the service.

(iii) Safeguarding of essential requirements, including the reliability of service and integrity of the public telecommunications network, operational compatibility of the services as well as data protection.

(iv) Safeguarding of issues of importance to the public, including alarms and emergencies.

(v) Safeguarding of universal telecommunications services of importance to the public.

(vi) The secrecy of telecommunications.

(vii) Implementation of E.C. rules.

(viii) Limitation of the supply of services so that the services are offered only to specific countries.

20–30   Regarding most areas, guidelines on the exercise of the rights to offer service within the competition areas are laid down in executive orders.

Providers must compete against the facility-based services offered and provided by the concessionary Tele Danmark A/S. In order not to distort competition between the concessionary and the other enterprises, it is a fundamental requirement that non-discriminatory access is possible on equal and fair terms which is explicitly stated in the executive order on the Concession. Such requirements will also apply to public utilities and bodies now able to offer lines to third parties.

According to section 5 in the Concession, Tele Denmark A/S must propose terms and conditions applicable for connection to the public network regarding all liberalised services to be approved by the NTAD.

"Interconnection" in the narrow sense is in principle still only of relevance within the mobile telecommunications services because all other services require a subscription agreement with the concessionary, Tele Denmark A/S, as owner of the infrastructure. Naturally, interconnect agreements have been entered into between the two GSM operators and approved by the NTAD, but apart from this field no "Standard Interconnect Agreements" are issued.

Leased lines are made available by the concessionary, Tele Danmark A/   20–31
S, in accordance with the "Terms of Subscription on Leased Lines", a contract that stipulates the mutual rights and obligations of the customer and Tele Danmark A/S. The terms of subscription are approved by the National Telecom Agency.

All terms, conditions, prices, charging principles, etc., are made public and are available from either the concessionary or from NTAD.

An accounting system is issued by the NTAD and applied by the concessionary in order to make sure that the owner of the facility is not favoured in the competition. All activities within the exclusive area must be accounted for separately, and cross-subsidising between the exclusive area, the special rights area and the liberalised areas is prohibited. Separate accounting is also required with respect to the special rights area. Departments and/or subsidiaries within Tele Danmark A/S providing services within the liberalised areas must buy access to the facilities on the same terms and conditions as any other provider.

(i) Data Communication Services

The provision of data communication services was liberalised as of January   20–32
1, 1993, based on the Executive Order on Data Communication Services Provided through the Public Telecommunications Network and was, thus, one of the first services to be liberalised.[12]

According to the executive order, it is allowed to provide on a commercial basis data communication services, *viz.* communication of data through various network termination points (whether by means of wire connections, radio waves, optical means or other electromagnetic means). This liberalisation does not include telex services, mobile radiocommunications, paging, satellite communication and radio/television broadcasting.

The provider shall upon request inform anybody of the conditions, including quality, tariffs and times of delivery applicable to the service in question. Furthermore, the provider shall upon request define the interface specifications, as well as any public standards attached thereto, which are used for the service provided. Any condition that terminal equipment must fulfil certain requirements shall be publicly available, objective and non-discriminatory.

There is no demand that a licence to provide the service must be obtained.

---

[12] *cf.* Executive Order no. 986 of December 11, 1992 implementing Directive 90/388.

However, the NTAD manages and allocates numbers, series and addresses. Allocation takes place for a maximum period of five years at a time.

It is possible to apply for an exemption in whole or in part from having to comply with the requirements of the order.

(ii) Enhanced Services

20–33   Value-added services (VAS)/enhanced services are understood as services provided through use of the public telecommunications network and involving an added value as compared to the bearer service or telecommunications service used. Value-added services are not subject to regulation except for VAS provided by way of services in the so-called 900 series.

The 900 series services are those which are made available via the public telephone network where the telecom enterprises also carry out the registration of the use of the service and the invoicing and collection of payment.[13] The services are provided via a special series of telephone numbers beginning with the digits "900".

The establishing and operation of the service require prior notification to "The Service 900 Board" (specifically established for this purpose) as well as the provider having entered into an agreement with the telephone company. The terms and conditions, including tariffing principles for the services provided by the concessionary, are thus established in this agreement, subject to approval from NTAD.

20–34   The notification to the Board must be made for each category of service to be provided. Some categories can only be provided in case the service can be barred. In addition to information on the provider, the notification must include, *inter alia*, a short description of the substance of the service, and whether the service is provided exclusively by means of an answering machine or other automatic equipment.

The service must be provided in a way that the user is not required to call more than one 900-number. Various other conditions apply, for example that the provider shall ensure that each caller is informed about the price for using the service during the first 20 seconds of the call. The regulations are rather detailed, and services consisting of sex, violence, gambling and competition cannot be offered.

The exclusive right concessioned to Tele Danmark A/S of the telephony service must not be infringed. Thus, conference calls and call diversions cannot be offered, whereas "voice response" and "voice mail" services are considered as value-added services. They do not provide real-time transmission of speech as contained in the Concession.

(iii) Voice telephony

20–35   Taking effect on November 15, 1994, voice telephony was liberalised.[14] Consequently, it is now possible as a liberalised service to provide, for example conference calls.

---

[13] *cf.* Executive Order (no. 931) of November 23, 1992.
[14] *cf.* Executive Order (no. 905) on Public Voice Telephony Services Provided through the Public Telecommunications Network, November 2, 1994.

If the provision of a public voice telephony service includes a 900-Service (a regulated VAS), the Executive Order on the Service shall apply in parallel with the Executive Order on Public Voice Telephony Services. This is, of course, also valid regarding a combination of voice and data transmission. In this event, the Executive Order on Data Communication Services Provided shall apply in parallel with the Executive Order on Public Voice Telephony Services. Thus, all applicable provisions must be complied with.

Voice telephony, like any other liberalised service, can only be provided through lines leased from Tele Danmark A/S. Subscription is available to anybody on equal terms.

However, Tele Denmark A/S has lately obtained approval to grant rebates to major customers which are intended to make it possible for resellers to compete with the concessionary.

The liberalisation of voice telephony makes it possible to provide for instance services including real-time transmission of voice together with other services. The system is not an interconnect system, but a resale system. The service providers are not given the status of operators, and, thus, they are not given access to, *e.g.* the public numbering scheme, etc.

The provider of the voice telephony service must ensure that the all users   **20–36** by way of the service provided are given the possibility to,

- be called by other users of the service;
- be called by customers of the public telephony service and be called by customers of the public mobile networks;
- use the text telephone service and the emergency call number provided by Tele Danmark A/S for this service;
- make calls free of charge to the emergency service (no. 112). The provider shall ensure that the alarm call is directed to the relevant alarm control board or as agreed with the competent emergency authorities and Tele Danmark A/S.

The provider shall upon request inform anybody of the conditions, including quality, tariffs and delivery times that apply for the service of the provider concerned.

Furthermore, the provider shall upon request give out details of the interface specifications and, if applicable, any attaching public standard used in the service provided. Any limitation in access to the public switched telephone network shall be announced.

Finally, any specific requirements of terminal equipment to be used in obtaining the service must be accessible to the public, objective and non-discriminatory.

Resale is defined as traffic connected to the public switched network at   **20–37** only one end of the leased line(s) used in the service (single-end resale)— either at the end of the calling party or at the end of the called party.

Simple resale is defined as traffic accumulated by the service provider in the public switched telephone network handled through leased lines; the traffic is then distributed by the service provider through the public switched telephone network to the called party—double-end resale (connected to the public switched telephone network at both ends).

The provision of international simple resale (PSTN at both ends of the leased line) requires a licence from the NTAD. It is a condition for obtaining such licence that the country where the leased line terminates is considered to have a regulation corresponding to that established in Denmark. Only Sweden and the U.K. have been designated by the NTAD in the first instance.

The provision of international resale should be notified to NTAD in advance but requires no licence.

Only type-approved equipment may be connected, and the service provider is the customer towards Tele Danmark A/S and consequently the responsible contracting party, whether or not a violation of the terms were performed by the customer of the Service Provider. The Provider is also liable to pay for any consumption notwithstanding payment or non-payment from his customer.

The condition that only type-approved equipment may be connected is valid with regard to all services to be provided through the public network. Equipment is dealt with in accordance with the applicable E.C./E.U. directives implemented in Danish law, *cf.* section 5.

(iv) Relation to Exclusive and Special Rights

20–38   The executive orders regulating the liberalised services are naturally not as such applicable to the public telephony services covered by concessions. However, both operators may provide liberalised public voice telephony services in accordance with the provisions of the Executive Order on Equal Terms with other providers.

Satellite Services (Non Broadcasting)

20–39   The Telecommunications Act 1990 comprises fixed as well as mobile services, whether terrestrial or space oriented (by satellite). Thus, all ways of providing telecommunications services—whether terrestrial or by satellite fixed or by mobile link—are covered, and the exclusive rights concessioned to Tele Danmark A/S covered until recently all infrastructure. A leased line may consist of cables, radio relay systems or satellite connections, whether based on analogue or digital technology.

For the time being Tele Danmark A/S remains the signatory to the three major international satellite organisations—INTELSAT, INMARSAT and EUTELSAT.

Frequency planning is the responsibility of NTAD.

Denmark has no specific satellite legislation nor specific legislation dealing with telecom services via satellite. However, as stated above, satellites are covered by the Telecommunications Act 1990, and the regulations in the Radiocommunications Act 1992 are applicable also to such traffic.

20–40   As of May 15, 1995 an Executive Order on Satellite Services entered into force, implementing E.C. Directive 94/46, liberalising satellite networks and services.

A satellite network is defined as a configuration consisting of two or more satellite earth stations communicating via satellite. As a minimum, the network consists of one radiocommunications up-link from a satellite earth station to a satellite and a radiocommunications down-link from a satellite to a satellite earth station. Earth stations intended to be used as an integrated part of the public telecommunications network operated by Tele Danmark A/S are not covered by the Executive Order.

A setellite service is defined as telecommunication (first instance, voice, text, video, or data) conveyed through a satellite network.

The establishing and operating of satellite services is subject to prior **20–41** notification to NTAD, and further subject to the obtaining of the necessary licence under the Radio Communications Act 1992. Prior to making the network operational, an agreement about applicability of space segment capacity must have been entered into.

The access to the space segment and allocation of frequencies is administered by NTAD. The notification must inform, *inter alia*, whether this service is intended to be connected to the public switched telephony network, and, if this is the case, indicate the the telephone number and identify the fixed circuits rented, if any. In case the service is offered to third parties, any specific requirements of equipment to be used in obtaining the service must be accessible to the public, objective and non-discriminatory.

Radiocommunications Generally

Radiocommunications comprises any conveyance, transmission or recep- **20–42** tion of signals, text, pictures, sound or data or other information of any kind by use of radio waves. Ordinary optic signals are not covered by the definition as laid down in the Radiocommunications Act 1992.

The establishing and operation of terminals for radiocommunications on Danish territory requires a licence, *cf.* the Radiocommunications Act 1992.[15] The 1992 Act is applicable on transmitters as well as receivers. A terminal is regarded "established" when ready to transmit or receive and is connected or can immediately be connected to power supply and antenna. Further, the 1992 Act and the executive orders issued in pursuance here-of stipulate provisions covering the equipment and the approval hereof.

The radio frequency allocation and coordination is the responsibility of the NTAD with the objective, *inter alia*, to provide for a sufficient frequency spectrum for existing and new services.

The availability and allocation of radio frequencies is an essential element for the establishment of the single (E.U.) market for radiocommunications equipment and radiocommunications based services. Recent technological developments have led to a rapidly growing demand for frequency allocation, in particular for mobile radiocommunications, satellite communications and broadcasting, and frequencies may be reserved according to for instance E.C./E.U. decisions to develop common services.

[15] Radiocommunications Act 1992 (no. 297). Several Executive Orders are issued in pursuance of the Act.

Thus, demand has made the efficient use of frequencies even more important. When judging an application for a licence, the NTAD will evaluate the possibility of providing the service by other means, for instance by private subscribed lines. In this case, a licence will not be granted.

20–43   The 1992 Act contains a provision entitling the Minister to except categories of radio terminals from the licensing regime. The 1992 Act contains specific exceptions covering certain categories, such as terminals used for national defense purpose, and terminals operating to receive only on frequency bands below 30 MHz or on frequencies allocated to amateur-radio.

Excepted also are terminals which can only receive radio and broadcasting signals whether via telecommunications satellite or via direct broadcasting satellite (DBS). This distinction between telecommunications satellites and DBS's are sometimes of relevance under Danish law within the telecommunications field. Accordingly, in the 1992 Act a special exception was added covering the terminals receiving broadcasting signals via DBS. Such terminals can be established and operated without obtaining a licence provided always that the equipment is legal. The 1992 Act contains provisions on radiocommunications equipment, and the E.C. Directive on Tele Terminal Equipment comprising radiocommunications equipment to be connected to the public network is implemented in Danish law.

Licensing according to the 1992 Act is required alongside other requirements to be fulfilled according to other legislation, for instance the Radio and Television Broadcasting Act 1992.

As mentioned above, the licensing authority will ensure that the use of the radio terminal secures an efficient use of the radio frequency spectrum and that the use of the radio terminal has due consideration to hindering unauthorised receipt of radiocommunications not destined to the public. These considerations are mentioned in the Act, but it is not an exhaustive listing. Services covered by the exclusive rights concessioned to Tele Danmark A/S or the special rights will not be licensed under the Radiocommunicaitons Act 1992 to other parties.

20–44   A licence can stipulate several conditions, *inter alia*, that the user shall possess certain qualifications (for instance by passing specific exams), technical requirements securing use of only certain allocated frequencies, type-approval of the equipment, quality assurance systems, time-barring of the licence, and the right to revoke the licence, whether due to the actual licensee or due to national considerations or the like.

A licence can be granted individually in accordance with an application received, or the NTAD can issue general licences covering specific terminals, granted normally in connection with the type-approval. The NTAD is responsible also for type-approvals.

Several executive orders have been issued in connection with the 1992 Act, vesting in the Minister the right to issue regulations on the details regarding the administration of the 1992 Act and the issue of licences.

Important regulations are found in the Executive Order on the Establishing and Operating of Certain Radiocommunication Installations.[16]

A range of non-public radio services in closed, landmobile radio networks can be established and operated solely according to a licence issued by the NTAD with reference to the 1992 Act. Such services must not be connected to the public network. Typically many will share the same frequency, this, *inter alia*, normally means that secrecy is not obtainable.

In special cases permission may be given for installation and operation of radio lines for telecommunications traffic (with connection to the public telecommunications network), for instance in cases where the need for communications traffic cannot be considered with other means of communication forms and where connection to the public network is of vital importance.

The future closed landmobile net TETRA—operating on a digital basis—   **20–45** is expected to be established from the beginning as a trunked network, characterised by a mutual use of many radio channels/telephone cables by a large number of users.

Regarding the Telepoints—*viz.* the special form of mobile radiocommunications in which transmission takes place only from the mobile to the telephone network and requiring that the base station is within a rather short distance—those are covered by the exclusive rights (the concession to Tele Danmark A/S) whereas the cordless PABC's are covered by the Radiocommunications Act 1992 and the Terminal Equipment Act 1992, thus requiring licence and type-approval.

Many developments are in fact based on cordless wiring, communicating by radio instead of cable, and many frequencies are already reserved for E.U. purposes. The limited number of frequencies available may result in the implementation of further liberalisation within the sector (including satellites) by way of enacting special rights. No final Danish decision on this question has yet been made.

4. REGULATION OF BROADCASTING

Basic Legislative Rules

The broadcasting activities are basically regulated in the Radio and Televi-   **20–46** sion Broadcasting Act 1991[17] and reflects that the broadcasting environment in Denmark has changed dramatically during the past 10 to 20 years.

Originally, only Radio Denmark could by received by the public. Since 1974 an increasing number of Danes have been able to watch television broadcast from neighbouring countries. In 1983, the first steps in introdu-

---

[16] Executive Order (no. 738), August 13, 1994 as later amended (most recently in February 1995).
[17] A revised Act was passed on December 18, 1992 and is now amended into Promulgation Order (no. 578) of June 24, 1994.

cing local television programmes were initiated. Today, about 50 local television stations and 50 very small local television stations operate as well as an increasing number of local radio stations.

Satellite channels were introduced in 1985, and it is foreseen that it will soon be possible to receive almost any programme distributed from all over the world.

A second Danish public service channel—TV2—competing with Radio Denmark was introduced in 1988 together with a number of regional channels.

The governing legal principle is that broadcasting in Denmark requires a licence for which anybody can apply. The licences issued are, however, subject to certain conditions, *cf.* below.

Terrestrial and Other Delivery

20–47   Since 1984, the end-users have received broadcasting signals in four different ways—

First, the two Danish national channels and some channels from the neighbouring countries can be received with an individual antenna. No restrictions will apply.

Secondly, it is possible to receive the Danish channels, neighbouring channels and satellite television through privately owned parabolic reflectors (one-way VSAT). Except for type-approval of the parabolic reflector, no restrictions are applicable for such private reception of signals.

Thirdly, it is possible for the individual household to receive the Danish channels, neighbouring channels and satellite television from private cable operators, *cf.* below.

Finally, it is possible to receive the above mentioned channels through the hybrid network.

20–48   The hybrid network is a broadband network, established by Tele Danmark A/S under the concession which until July 1, 1995 comprised the exclusive right to convey radio and television programmes through the publicly owned network. It was established in order to provide transmission capacity to distribute foreign television programmes to community antenna systems and other cable systems. The liberalisation according to the third Supplement means that the use of the hybrid network will no longer be reserved for this specific purpose.

The liberalisation means that others will be allowed to make use of the hybrid network, including as leased lines, in order to convey programmes to for instance community antenna systems as well as to make use of the network separately or additionally to convey all telecommunications services (data, telephony, television, etc.).

In addition, the establishing of broadband networks within the boundaries of a municipality is at the same time liberalised totally. This means that companies other than Tele Danmark A/S can own broadband networks to be used either by the owner or a third party renting the capacity for all telecommunications services. Others will be allowed to make use of the network as leased lines in order to convey programmes and at the same time to separately or additionally use the network to convey all telecommunications services.

The hybrid network consists of earth satellite (receive) stations and cables. From "termination points" the cables are conveyed to local main distribution points which actually are the monopoly termination points. Connection to cables owned by companies other than Tele Danmark A/S can take place in these main distribution points, or Tele Danmark A/S will distribute directly to the end-user.

The private cable operators are allowed to receive and transmit satellite television. As regards down-link and reception of signals, the operators can choose between being linked to the hybrid network or receiving the signals directly from the satellite. As regards distribution to the end-users, a licence issued by the NTAD in accordance with the provisions of the Executive Order concerning Community Antenna Systems[18] is required.

According to the Executive Order, a private cable operator must have a licence to establish the programme distribution system transmitting to the households connected.   **20–49**

A licence can be granted to owners of or tenants in houses connected to the programme distribution system. Furthermore, a licence can be granted to local authorities and regional telecommunications enterprises.

A licence does not include an exclusive right to establish programme distribution systems in the geographical area. Normally, the licence is granted only for transmission of signals within a local area. It is possible to insert a condition in the licence stating that the telecommunications enterprise at a later stage may buy parts of the programme distribution system if this is necessary for extending the public telecommunications network.

The licence from NTAD does not include the construction permit which must be obtained from other authorities, nor does it include copyright issues.

The NTAD can withdraw the licence to use the programme distribution system. Furthermore, the licence to establish the programme distribution system becomes invalid if the works do not commenced 12 months after the licence has been issued.   **20–50**

If a minor programme distribution system is established by an owner of one of the houses and less than 25 households are connected to the system a licence to establish the programme distribution system is not necessary. If a programme distribution system is further developed or extended a new licence is required.

---

[18] Executive Order (no. 72) on Community Antenna Systems, January 27, 1994 (issued in pursuance of Chapter 2 of the Act on Radio and Television Broadcasting 1992). The Executive Order will be amended in order to implement the third political Supplement abolishing the restrictions on ownership of these sytems, and as a consequence, abolishing the monopoly held by Tele Danmark A/S of conveying radio and television programmes. The necessary changes in the Radio and Television Broadcasting Act 1992 are implemented effective as of July 1, 1995.

It is a condition when choosing the programmes that the end-user (the connected households) is involved in the decisions concerning the programmes. This influence will be strengthened with the introduction of the third political supplement.

The NTAD has an obligation to ensure that the provisions of the Executive Order concerning Community Antenna Systems are not violated.

Regulatory Agencies and Bodies

20–51   Four different public bodies are involved in regulating and supervising the broadcasting and programming: The Satellite and Cable Council, local committees under the Town Council, and the Radio and Television Advertising Council. In addition, the NTAD is vested with regulatory powers.

(i) The Satellite and Cable Council

20–52   The purpose of the Satellite and Cable Council is to grant licences to companies applying to broadcast in an area larger than one local area. Furthermore, it is the task of the Satellite and Cable Council to ensure that the licensee abides to the provisions of the Radio and Television Broadcasting Act 1992, and the Council must withdraw the licence to broadcast in case the conditions are violated.

The Satellite and Cable Council is the body prosecuting any violation of the Radio and Television Broadcasting Act 1992.

According to the amended provisions of the Telecommunications Act 1990 following the liberalisation of the hybrid network, the Satellite and Cable Council is vested with the final administrative authority in all matters concerning telecommunications services offering sound or picture.

(ii) Town Council Committees

20–53   The second regulatory body is appointed by the town council of each local area. The purpose of these committees is to grant and withdraw licences and to ensure that the regulations of the Radio and Television Broadcasting Act 1992 are not violated. These committees have authority only with regard to companies broadcasting to one local area.

(iii) The National Telecom Agency

20–54   With regard to the cable operators, the regulatory agency is the NTAD. The NTAD has authority to grant licences to cable operators, to establish programme distribution systems and to use the programme distribution systems. The NTAD may withdraw the licence to use the programme distribution system and the NTAD must ensure that the regulations of the Executive Order concerning Community Antenna Systems are not violated.

(iv) The Radio and Television Advertising Council

20–55   The task of The Radio and Television Advertising Council is to consider issues concerning the contents of commercials. It is the obligation of the

respective cable operators to ensure that the Radio and Television Broadcasting Act 1992 and the regulations on commercials and sponsored programmes are not violated.

Regulation of Transmission v. Content

As stated above, a licence is required to broadcast programmes. With   **20–56**
regard to transmission routes and means of distribution, reference is made
to the remarks above (terrestrial and other delivery). In addition, the contents of the programmes is regulated.

Except for the two Danish national channels and the regional channels
the broadcasting companies are divided into two groups—one consisting
of companies wishing to broadcast to a larger geographical area than just
one local area; the other group consisting of broadcasting companies wanting to broadcast to one local area only.

A company wanting to broadcast to an area larger than just one local
area must obtain a licence from the Satellite and Cable Council. The
application for a licence must contain information such as name, address,
ownership, management, financial status, etc. and a definition of the geographical area covered by the broadcasting.

The applicant must also state whether the programmes are broadcast by
satellite and the application must be accompanied by a list stating from
which countries the up-link will take place. Further, the application must
include information concerning whether the programmes will be encrypted
and the conditions for getting a decoder.

Regarding the programmes, the applicant must include information   **20–57**
about the categories of programmes that are intended to be broadcast, the
intended air time for the programmes, whether the programmes include
Teletext, whether the target is specific groups, for instance children, and
at what time of the day the company intends to broadcast. Furthermore,
the application must contain information on the languages to be used in
the programmes including the language used in the sub-titles. Finally, it
must be stated at which date the company wishes to commence broadcasting. On the basis of the application the Satellite and Cable Council will
decide whether a licence can be granted.

A broadcasting company can loose its licence if the Satellite and Cable
Council so decides, or if the licence has not been used for one year. The
Satellite and Cable Council can withdraw the licence, finally or temporarily.

With regard to broadcasting companies wishing to broadcast to one local
area only the licence must be given by the committee appointed by the
local town council.

The companies eligible to obtain a licence must fulfil the following
conditions:

* The majority of the board members of the broadcasting company must
  be domiciled in the local area.
* Broadcasting must be the company's only object.

- No commercial company except for newspapers may have determining influence on the broadcasting company.
- No persons must be involved in the board or be responsible for programming with regard to more than one channel in a local area.

20–58   Broadcasting of programmes by cable can take place without a licence when no more than 25 households are connected.

The licence is only given for a limited period up to a maximum of five years. The licence can be renewed after the expiry of the period. The broadcasting company must broadcast independently and not broadcast programmes that are shown by other broadcasting companies at the same time or programmes broadcast by other companies according to a permanent and close cooperation with this broadcasting company.

The Radio and Television Broadcasting Act 1992 regulates the programming. Thus, more than 50 per cent of the viewing hours, excluding news, sports, competitions, commercials and Teletext, must be covered by European programmes. Further, the licensee must aim at 10 per cent of the viewing hours excluding news, sports, competitions, commercials and Teletext, or 10 per cent of the budget being related to European programmes from producers which are not broadcasting companies. A suitable number of the programmes must be of recent date.

Finally, the broadcasting company must ensure that programmes which may seriously harm the physical, mental and moral development of minors, are not broadcast. Programmes that in any way incite hatred of race, sex, religion, or nationality are not allowed.

20–59   Under The Radio and Television Broadcasting Act 1992 it is the obligation of the licensee to ensure that the regulations on commercials and sponsored programmes are not violated. These regulations are rather strict, including for instance prohibition against marketing liqueur and cigarettes.

Naturally, the broadcasting company is responsible to observe any Danish legislation applicable, and any broadcasting company must have an editor responsible under the press laws.[19] The Press Council must be informed of the identity of the editor.

The author of the text or the picture, the person who makes a statement, the responsible editor and/or the broadcasting company can be held liable for the contents of the programme. Under the Media Responsibility Act, the advertiser is liable in conjunction with the responsible editor regarding the contents of the commercials. Rectifications can be demanded.

Copyright

20–60   The broadcasting company and/or the cable operator must ensure that the Danish Copyright Act is not violated and must obtain the necessary consents from the right-owners and pay the fees due, if any.

---

[19] In particular the Media Responsibility Act 1991 (no. 348) and amended by Act (no. 106), December 23, 1992.

In addition to the Berne Convention, Denmark is party to the Rome Convention and the European Agreement of the Protection of Television Broadcasts. Thus, all right-owners are protected in Denmark, whether producers, authors, artists, composers, broadcasting companies, etc.

In Denmark, a collecting society—CopyDan, Cable Television Division—has been established for many years, taking care of the collection and distribution of the fees to the all right-owners concerning retransmission via cable. The liability to pay these fees falls on the cable operators. In addition to CopyDan, other collecting societies take care of fees covering music rights, etc. whether towards radio stations, broadcasting companies or other users of such rights.

The Cable and Satellite Directive adopted by the EEC in 1993 will effect the existing legal system in Denmark. The essential impact of this Directive will be that Denmark must leave the compulsory licence regime instituted in 1985 regarding cable retransmission. Today, the right-owner cannot forbid the transmission of his work (terrestrially and via direct broadcast satellite). The cable operator can distribute without permission and will be liable to pay to the right-owner "an equitable remuneration", *cf.* the Berne Convention, Article 11(2). In case of dispute, the remuneration is fixed by the Compulsory Licence Council (an administrative body). Regarding retransmission of programmes transmitted via telecommunications satellites, the right to authorise or prohibit such communication prior to its distribution was upheld at the time of implementing the compulsory licensing system. The Danish system can and will be upheld until 1998 at which time it will be replaced by a system abiding to the Directive.

20–61

It is legal to provide encrypted programmes for which a decoder is necessary, and piracy is becoming an increasing problem. The broadcasting and operating companies try to protect the encryption by practical means, lately for instance by introducing smart cards for use in the decoders allowing for regular changes in encryption codes. Thus, illegal cards will have to be replaced every time the code changes, and the rightful providers hope that this impracticality will limit the increasing use of illegal cards.

5. REGULATION OF APPARATUS

Regulation of Apparatus Supply and Installation

The regulation of tele terminal equipment is an area which has undergone a process of liberalisation since the late 1970s. Today, the area is totally liberalised and equipment can be freely traded and installed provided the equipment fulfils certain technical demands. Apparatus is not comprised in the concession issued to Tele Danmark A/S, and a large number of suppliers of equipment operate in Denmark.

20–62

The E.C. Tele Terminal Equipment Directive was implemented by the

Telecommunications Terminal Equipment Act 1992 (no. 230). An executive order stipulating details have been issued as well.[20]

The provisions of the executive order cover terminal equipment as well as any other equipment which can be connected to the public network and is intended to be connected either directly to a termination point in the public network or to be able to function together with the public network by way of direct or indirect connection to a network termination point for the purpose of transmitting or receiving information.

The connection can be made by way of cable, radio, optic system or any other electro-magnetical system.

20–63   An amendment to the 1992 Act has now been proposed (and passed) in order to implement the E.C. Directive regarding satellite equipment (93/97) to take effect no later than May 1, 1995.[21]

The Directive covers satellite earth station equipment (equipment used to transmit, transmit/receive or receive only radiocommunications signals distributed via satellite or other space based systems). Such equipment may be telecommunications terminal equipment as well, and it is proposed to implement the Directive in Danish law by extending the Telecommunications Terminal Equipment Act 1992 accordingly. Details stipulating authorised testing laboratories, marking, documentation regarding fulfilment of the essential requirements, internal manufacturing quality control, etc., will be issued. The regulations are expected to be very similar to those issued covering telecommunications terminal equipment.

It is only allowed to market approved equipment which is correctly marked. "To market" includes the sale, distribution and taking use of the equipment. The procedures are laid down in accordance with the E.C./E.U. rules, including the procedures based on either "Declaration on Conformity" or "Type-examination".

"Other equipment", *viz.* equipment which can be connected to a termination point without being intended for this purpose, can be marketed provided always that the equipment is clearly marked to this effect.

Terminal equipment which at the same time is radio equipment may require further licences with respect to establishing operation or use in accordance with the Radiocommunications Act 1992.

20–64   Radio equipment is as a general rule covered by the special 1992 Act and the executive orders issued in pursuance hereof.

A special executive order is issued concerning the establishing and operation of certain radio equipment related to the land mobile and other terrestrial radiocommunications services. A radio terminal can only be marketed, sold or distributed in case such equipment is approved and marked correctly except for equipment explicitly excluded from the approval procedure.

In addition to the requirement that the radio equipment is approved,

---

[20] The Executive Order in force is no. 737 of August 13, 1994.
[21] Directive 93/97 amending Directive 91/263.

radio equipment can only be sold in case the necessary additional licence, if any, is obtained by the buyer from the NTAD to establish and operate the radio terminal.

Type Approvals and Testing Procedures

The National Telecom Agency is the authority responsible for issuing regu-   20–65
lations and for the approval and licensing of equipment. The NTAD represents Denmark in ETSI.

In case a CTR (Common Technical Regulation) has been agreed, approval is applied for on the basis of the CTR. Otherwise, testing reports and documentation must be forwarded for examination, with reference to relevant national or international standards, if any.

Various standards and testing procedures are referred to and must be abided by. The regulations implement the E.U. directives. A list of approved laboratories will be provided by the NTA upon request.

6. THE FUTURE

Liberalisation has accelerated during the past years, and the concessionary   20–66
and other interested parties are preparing for a totally liberalised future. Recently, Tele Danmark A/S has announced several co-operations and alliances, and the company is expanding its activities to many new areas and many new countries, within or outside the concession. Several competitors are already established in Denmark.

The Danish company GN Store Nord (Great Nordic) recently applied for a licence to become second operator of more or less all telecommunications services, including real-time transmission of speech and including the right to establish its own infrastructure. The granting of such a licence could only happen provided the legislation was changed accordingly, at least making telephony a special right. Also Swedish Telia was taking a lead in expressing specific interest in becoming an operator on the Danish market, suggesting to use the infrastructure established by the railways and has recently launched an application to this effect. However, it seems rather clear that a second licence will not be granted without abiding to the public procurement rules because it will be considered contrary to E.U. rules.

The railways, the public utilities, and the cable operators are acting in order to make the most of the recent liberalisation of network ownership within the boundaries of the municipality. Alliances, acquisitions, etc., are being carried through.

Stage two of the third political Supplement entered into on April 6, 1995 will see the establishment of an entirely new statutory basis following the main principles on which agreement is reached at European level, creating the basis for efficient and fair competition after the introduction of full liberalisation on January 1, 1998. The parties have agreed that the principles of stage two should be fixed as soon as possible, *i.e.* before the autumn of 1995. The preconditions of advanced liberalisation are:

(a) the necessary clarification of future E.U. regulation in the area has been achieved; and

(b) the parties to the Agreement accept it and reach agreement on the general regulation of telecommunications after the introduction of full liberalisation;

the entire process being assumed to be completed on January 1, 1998.

The implementation pace of the individual initiatives during stage two will be decided as further deliberations are carried on at political level. So far, the Ministry of Research has issued a Draft paper named "Danish Telecommunications Policy, Stage 2. Real Competition—the Road to the World's Best and Cheapest Telecommunications".

CHAPTER 20A

France

By Philippe Shin, Coudert Frères, Paris

1. HISTORY OF THE DEVELOPMENT OF TELECOMMUNICATIONS IN FRANCE

History

Postal and telecommunications services have always been considered to be 20A–01 a state monopoly, since King Louis XI declared the post to be a state monopoly in the fifteenth century. In 1851, telegraphy was made a state monopoly and the first Ministry of Posts and Telegraphs was created in 1879. In 1889 telephony was declared a state monopoly under the responsibility of the Ministry of Posts, Telegraphs and Telephones ("MPT").

In 1942, postal service and telecommunications were reorganised into two separate departments within the PTT: the Direction Générale des Postes ("DGP") and the Direction Générale des Télécommunications ("DGT"). The DGT started using the popular name "France Telecom" in 1988.

Until the mid-1970s, France was an under-developed nation as far as telecommunications were concerned. The government initiated a plan to catch up with the other industrialised countries, which so far has proved successful. The French network is now developed enough to offer services that did not exist a few years back; the success of the Minitel epitomises the relevance of the policies pursued by the government and the maturity of the market.

Background

The growth of telecommunication media has produced a multiplication of 20A–02 offers of services by private sector enterprises, in particular in the field of radiocommunications. The competition within these new markets created

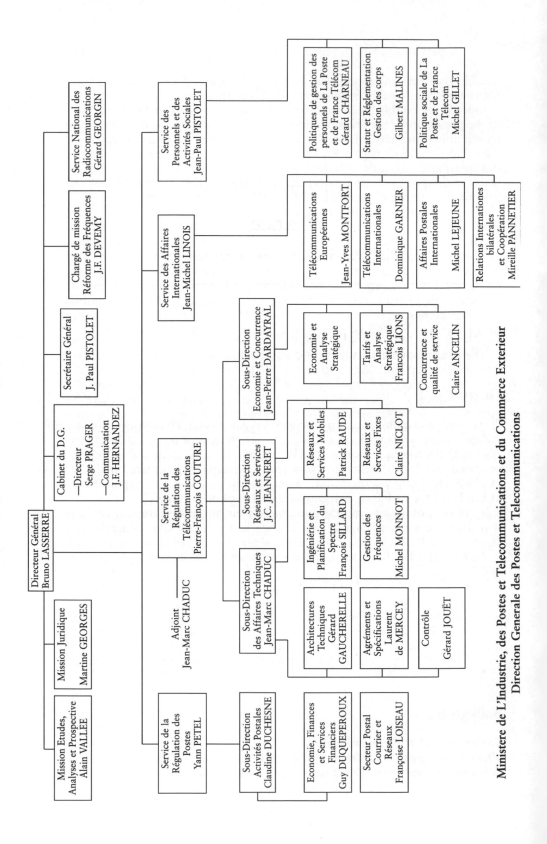

Ministere de L'Industrie, des Postes et Telecommunications et du Commerce Exterieur
Direction Generale des Postes et Telecommunications

outside the state monopoly has had to be organised. The existing legal framework was clearly insufficient for this purpose.

Indeed, the establishment and operation of the general network, as well as the provision by the State of telecommunications services, were governed by administrative law unsuited to a market-oriented style of management. Rules applicable to networks, services and terminal equipment were not distinguishable, even though the requirements, both technical and economical, for the provision of services differed considerably from those for the provision of terminal equipment or for the provision of networks.

The government started by granting licences to operate what were deemed to be "public services" (*i.e.* services to be naturally and only provided by the State) to private companies, whose capital nevertheless belonged to the State. The authorisations were granted in the sectors of services of data transmission and specialised international links. Then came the possibility to compete in the marketing of terminal equipment, including telephone instruments, and of services transmitting data over the public switched network (Télétel system). The MPT then authorised competition in the sectors of one-way paging and radiotelephony services.

These developments also revealed that the intermingling of the regulatory 20A-03 and operational functions within the MPT was no longer adapted to the needs of a rapidly developing environment. The private sector competitor could not totally rely on the fairness of the competition with a public operator who was simultaneously deciding what rules to apply. The *Direction de la Réglementation Générale* ("DRG") was established within the Ministry of Posts and Telecommunications in 1989, about the same time that the *Conseil Supérieur de l'Audiovisuel*, the supervisory authority for television and radio broadcasting, was constituted, both with the initial mission to participate in the preparation of the reform law of July 2, 1990.

In December 1993, the DRG was merged with the *Direction Générale de la Poste* and became the *Direction Générale des Postes et Télécommunications* ("DGPT"). The new DGPT assumes all the duties of the former DRG.

The Public Operator: France Telecom

The new status of *La Poste* (the public operator of the postal service) and 20A-04 of France Telecom as separate, autonomous legal entities organised under public law, governed by their respective boards of directors but under the "tutelage" of the Ministry of Posts and Telecommunications, was established by Law No. 90-568 of July 2, 1990. The legal charter ("statuts") of France Telecom was completed by Decree No. 90-1112 of December 12, 1990, and its operating charter ("cahier des charges") was laid down in the Annex to Decree No. 90-1213 of December 29, 1990.

France Telecom is the only fixed network public telecommunications operator. The primary purpose of France Telecom is to ensure all telecommunications services in domestic and international communications.

To this end, pursuant to Article 1 of its operating charter, its duty is to provide the following services: voice telephony between fixed points; telex;

basic terminal equipment rental and maintenance; public phones; telephone directory and subscriber information service; leased lines and packet switched or circuit switched data transmission services; one public radiotelephone service; any services it has provided on the public radioelectric network prior to December 31, 1990; any general interest services mandated by special interministerial rulings. It must also design, establish, develop, operate and maintain the public network necessary for the performance of those services, and ensure the interconnection of that network with the different foreign or international telecommunications networks, as well as with domestic networks it does not operate (*e.g.* mobile telephone networks).

20A–05   In addition, France Telecom may, within the framework of the Posts and Telecommunications Code and the rules of competition, provide any other telecommunications services; establish, operate and market independent networks; establish networks for the distribution of radio and television programs and share in their operation through equity participation; market and maintain any type of terminal equipment; offer any complementary service which enhances the use of telephones; offer engineering consulting services abroad; and, in general, engage in any activity in France or abroad which is directly or indirectly related to its statutory purpose.

Finally, France Telecom is also obliged upon request to lend support to various government needs and projects, such as special communications needs for defence, public security and safety, research and development projects, national and international standards efforts, or the development and application of technical requirements of terminal equipment type approvals (including laboratory testing if requested).

As a general rule, France Telecom prefers to limit its own direct activities to the operation of the public network and the traditional telecommunications services and to engage in other activities such as equipment manufacturing and marketing, value added services, etc., through a large and complex network of subsidiaries, equity participations and joint ventures.

Any legal action relating to licences, authorisation, denials, rulings, decisions or other acts of the MPT is to be brought before administrative courts. Litigation involving France Telecom with a user, subscriber, supplier, subcontractor, joint venture partner, etc., is to be pursued before regular courts of law.

2. Existing Legislative and Regulatory Environment

The Posts & Telecommunications Code—The Law of December 29, 1990

20A–06 The Posts & Telecommunications ("P&T") Code, in its sections relating to telecommunications, has been considerably modified by the Law no. 90-1170 of December 29, 1990, also known as the "LRT" (*loi sur la réglementation des télécommunications*). The draft law was prepared by the

DRG and broadly discussed with all interested parties of the French tele-communications sector.

The P&T Code and the LRT are the two main sources of rules for tele-communication issues in France. The P&T Code is divided into several parts, "legal" (*i.e.* whose content is considered law) and "regulatory" (*i.e.* whose content has not the value of a law voted by Parliament but that of a decree taken by the government).

Agencies of regulation—Organisation of the DGPT

The Minister of Posts and Telecommunications ("MPT") has the overall   20A–07 responsibility in the French Government for the areas of post and tele-communications. (The current name of the Ministry, as of May 1995, is "Ministry of Information Technology and Post".)

The MPT's responsibility includes enforcement of the applicable laws and regulations, the development and proposal of new policies and legisla-tion, representation of France in international telecommunications forums and activities, and exercising government control (tutelage, "tutelle" in French) over the public operators, La Poste and France Telecom.

The regulatory function of the Ministry is performed through a division called *Direction Générale des Postes et Télécommunications* (General Dir-ectorate for Post and Telecommunications, "DGPT"). The DGPT has a staff of 220, about 70 per cent of whom are at managerial level ("cadres"). Many of them have technical training and come from France Telecom. The National Service for Radiocommunications (SNR) which is attached to the DGPT, has about 230 employees.

First established in 1989 as the DRG (*Direction de la Réglementation*   20A–08 *générale*) to be instrumental in preparing the reform legislation of 1990, the DGPT was redefined in Decree No. 93-1272 of December 1, 1993; its task is to draft and implement the policy of the Government in the field of telecommunications. The DGPT's responsibilities include, in particular:

- ensuring compliance with legal rules;
- determining and adapting the economic, legal and technical frame-work for post and telecommunications activities;
- ensuring the conditions of a fair competition among the various actors in the post and telecommunications fields. The DGPT processes the prior declarations and requests for authorisation and approval addressed to the Minister of PTT;
- proposing any measure in favour of the development of new services and, more generally, of the competitiveness of services offered in the post and telecommunications sectors.

The main supervisory and advisory bodies assisting in the regulation of telecommunications are:

20A–09 *"Commission Supérieure du Service Public des Postes et Télécommunications"* (**Public Service Commission**). Established by Article 35 of Law No.90-568 of July 2, 1990 (the Reorganisation Law), it provides the "tutelary" control and supervision over the public operators, La Poste and France Telecom. Its functions are (1) to review the conditions under which La Poste and France Telecom perform their missions, (2) to advise the MPT in the preparation and modification of the operating charter (*cahier des charges*) of the public operators (this consultation is obligatory and the Commission's opinion is published), (3) to oversee, together with the MPT, the public operators' compliance with their operating charters and (4) to watch over the balanced development of the postal and telecommunications sectors, in particular by giving opinions about proposed national legislative changes as well as E.U. directives.

20A–10 *"Conseil National des Postes et Télécommunications"* (**National P&T Council**) **or CNPT.** Established by Article 37 of Law No.90-568 of July 2, 1990, it advises the MPT on (1) the role of posts and telecommunications in the economic and social life of the nation, (2) general principles to be applied to the legal regulation of these sectors and (3) the development and coordination of the activities of the operators.

20A–11 *"Commission Consultative sur les Réseaux et Services de Radiocommunications"* (**Advisory Commission on Radiocommunications Networks and Services**) (**CCR**) **and the** *"Commission Consultative sur les Services-Supports et Autres Services de Télécommunications"* (**Advisory Commission on Bearer Services and other Telecommunications Services**) (**CCT**). These are the two advisory commissions installed within the Ministry of Post and Telecommunications pursuant to Article L. 34-6 of the P&T Code as amended by the LRT and Decree No.91-664 of July 10, 1991. Each commission consists of an equal number of representatives of service providers, service users and "qualified personalities" (5-5-6 in the CCR and 5-6-5 in the CCT). The Chairman is appointed by the MPT. The purpose of these two commissions is to provide specialised technical advice to the MPT (in practice, the DGPT) in performing its regulatory functions.

In particular, each commission must be consulted on the following subjects, and their advice is transmitted to the Public Service Commission:

For the CCR: (1) authorisation procedure and technical operating (licence) conditions for independent radioelectric networks and services, (2) technical specifications and requirements applicable to such networks and services, (3) conditions for interconnecting independent radioelectric networks to the public network and (4) installations that can be established without authorisation and conditions for their operation.

For the CCT: (1) draft decrees determining procedures for authorising bearer services and for handling value added services (one of the most important functions of the CCT is to help the DGPT distinguish between genuine value added services and simple resale), (2) technical conditions for the performance of such services and (3) technical requirements to be applied to value added services using the public switched network or leased lines.

Both advisory commissions may be consulted on other technical matters, and they can intervene on their own initiative in subject matters within their respective technical domains.

Each advisory commission also serves as an internal review board to consider complaints relating to authorizations, licences or applications falling into their respective scopes of competence.

"Service National des Radiocommunications" (SNR) (National Radio- 20A–12 communications Service. Established by Decree No. 90-1138 of December 21, 1990 it is the field operating arm of the DGPT to carry out day-to-day tasks, throughout the French territory, of managing the radioelectric independent networks and stations and the utilisation of the frequency ranges allocated to the MPT, as well as controlling that the DGPT regulations are enforced.

Jurisdiction of other Bodies

While the MPT is competent with regard to postal service and telecommu- 20A–13 nications, the *Conseil Supérieur de l'Audiovisuel* (Supreme Audiovisual Council) ("CSA") is competent to regulate radio and television broadcasting.

Unlike the MPT, the CSA is not part of the government but is an independent administrative authority. It was established by Law No.89-25 of January 17, 1989. It consists of nine members, three of whom, including the chairman, are appointed by the President of the Republic, three by the President of the National Assembly and three by the President of the Senate, for a single term of six years, with rotation of one-third of the members every two years. It operates currently with a staff of about 220.

The line between jurisdictions of the MPT and the CSA is not entirely clear where they overlap in the area of networks and/or services using radioelectric frequencies. As a general principle, the MPT is competent for telecommunications services, and the CSA for audiovisual services; but where a telecommunication service utilises a radioelectric frequency that has been allocated to the CSA (see below), the licence granted by the MPT for the service is subject to the applicant's having also obtained an authorisation to use the frequency from the CSA. The reciprocality of this rule is not equally clear.

Radio frequencies are ultimately controlled by the *Service Technique et Juridique de l'Information* (STJI) attached to the Prime Minister's Office, which allocates them among the CSA, the MPT, the Ministry of Defence and other such primary users. The DGPT in turn allocates its frequencies among different operators using radio-electric transmission, and the CSA allocates its frequencies among the different radio and TV broadcasters within its scope of authority.

The rule for cable TV (CATV) is clearer, being expressly dealt with in 20A–14

both the P&T Code (Article L.34-4) and the Law on freedom of communication (Article 34-2) as amended by the LRT:

- A municipality or a group of municipalities is competent to establish
  or authorise the establishment of CATV networks within its territorial
  jurisdiction.
- The CSA is competent to authorise the operation of CATV networks
  including the use of such networks to deliver telecommunications services directly related to an audiovisual service.
- The delivery over cable of telecommunications services not directly
  related to an audiovisual service requires authorisation of the MPT.

Licensing Regime

20A–15 The only telecommunications services that are regulated by the government
are those provided to the public, *i.e.* those that are commercially operated
and destined to the public in general. Other services, including voice services, are entirely free; they require neither a licence nor a even a declaration to the DGPT. In other words, any enterprise or closed user group may,
without having to undergo any regulatory processing, provide its own
voice, data or picture services, from any type of infrastructure (switched
network, leased lines, or independent network).

In the matter of authorisations or licences (the French regulatory texts
use only the word "authorization", but in English it is sometimes more
convenient to use the term "licence") required under various provisions of
the P&T Code, some of the more basic rules are the following:

- Cases are handled by the DGPT, with the assistance of the advisory
  commissions described above; decisions are signed by the MPT.
- Authorisations are personal and cannot be transferred unless the
  authorisation expressly provides for such possibility.
- Authorisations under Code Article L.33-1 (radioelectric networks),
  L.34-2 (bearer or "support" services), L.34-3 (services using radioelectric transmission), L.34-4 (telecommunication services over CATV),
  and L.34-5 (telecommunication services using leased lines above a certain capacity) are published in the "Journal Officiel" together with
  the operating charter, if any.
- Authorisations for independent networks are only to be notified to
  the applicant; however, in practice the MPT has published the more
  important independent network authorisations such as the "3RP"
  licences (independent shared radio networks covering various regions
  of France).
- For a radioelectric network or service utilising frequencies allocated
  to the CSA, the applicant must obtain the CSA's permission for such
  utilisation, as well as the authorisation of the MPT under Code Article
  L.33-1 and/or L.34-3.
- The denial of an authorisation must be in writing and motivated.
- If the licence holder fails to comply with the obligations imposed by

the applicable legal and regulatory provisions and by the licence conditions, if any, the MPT may, after proper notice, either suspend the licence for up to one month, reduce the licence duration to one year, or revoke the licence.

- A radioelectric network licence under Code Article L.33-1 may be revoked without notice if the capital structure of the licensee company changes in violation of Article L.33-1/II (*i.e.* the 20 per cent limit on foreign ownership of the capital).
- Licences for radioelectric networks and/or services open to the public (*e.g.* SFR's mobile phone system, or the 3RD mobile data project) are granted in the course of open public tenders (*appel à candidatures*).
- Authorisations for independent networks and value added services are handled on a case by case basis upon individual applications; however, in the case of the local shared radio networks (3RP), the MPT chose to proceed by public tender.
- In the case of public tenders, the submission deadline is usually two months, and the award must be made within two months thereafter.
- There is no formal time limit on other licence applications, but independent network authorisations so far have been issued (or refused) within five or six weeks on average.
- There is also an application fee or a licence fee that varies according to the subject.
- Except for operation of the public switched network, public voice telephone service between fixed points, telex, public phones, and bearer or "support" services (basic data transport), France Telecom is subject to the same authorisation requirements as other private competitors.

Scope of Foreign Participation Permitted

There is no private, domestic or foreign, participation in France Telecom, **20A–16** which is a public entity organised under administrative law (*etablissement public à caractère industrie et commercial*). However, the move towards a transformation into a private entity, although resisted by its personnel, is inevitable; it can be confidently assumed that in the near future France Telecom will become a private company and that eventually it will be privatised.

There is no limitation on foreign participation in any private telecommunications operator or service provider, other than Article L.33-1.II of the P&T Code. Article L.33-1.II of the P&T Code provides that no licence may be granted for the establishment and operation of a radioelectric network for the purpose of providing a telecommunication service to the public, to any company in which more than 20 per cent of the capital or voting rights are held directly or indirectly by a person or persons of foreign nationality. Moreover, a foreign national may not acquire directly or indirectly an interest in a company which already has such a licence, if such acquisition would result in bringing such foreign participation in the company above the 20 per cent threshold. Violation of this rule may result in revocation of the licence.

A person of "foreign nationality" for the purpose of Article L.33-1.II is defined as a foreign individual or a company the majority of whose capital is *not* held directly or indirectly by an individual or legal entity of French nationality. However, the limitation does not apply to nationals (individuals or legal entities) of a Member State of the E.U.

The foregoing rule is parallel to the one in Article 40 of the Law No. 86-1067 of September 30, 1986 which provides that no foreign person may make an acquisition that brings foreign ownership or voting rights over 20 per cent in a company that holds a licence for radio or television broadcasting service using earth to earth radio waves in the French language. The definition of "foreign person" in that case includes an association whose top managers (*dirigeants*) are foreign nationals, and there is no exception regarding E.U. nationals.

3. REGULATION OF SERVICES

Analysis of Rights to Provide Services.

20A-17   1.   The only service monopoly reserved by law under Code Article L.34-1 to the public operator is the provision to the public of voice telephony service between fixed points, telex (*i.e.* teletype) service and public telephones. The definition of voice telephony service as the "commercial exploitation of the direct transmission of voice in real time between users connected to terminal points of a telecommunication network" in Code Article L.32/7° is similar to the definition used in the E.C. Directive of June 28, 1990 (90/388).

2.   Telecommunication services over a radioelectric network may be provided by persons other than the public operator under a licence from the MPT according to Code Article L.34-3. This includes, in particular, mobile or radiotelephony and mobile data transmission services. Candidates are selected by way of a public call for tender (*appel à candidatures*). The two currently existing mobile telephone networks are those of France Telecom (*Itineris*) and the SFR, a privately owned company. A third licence has been attributed to Bouygues Telecom, a consortium headed by Bouygues (construction, civil engineering, television) which counts among its partners Mercury and the U.S. RBOC Southwestern Bell. The third mobile phone network is to use the DCS 1800 standard, a byproduct of the GSM standard.

3.   Bearer services, (defined in Code Article L.32/9° as the simple transport of data without any processing of such data beyond what is necessary to its transmission and routing) are part of France Telecom's public service duties. They may also be provided by private providers, subject to a licence and an operating Charter granted by the MPT under Code Article L.34-2.

4.   The possibility of providing basic data transport over a cable radio television network is expressly mentioned in Code Article L.34-4. That

Article stipulates that the provision over a cable RTV network of telecommunication services, whose object is not directly related to an audiovisual broadcasting service, requires authorisation from the MPT upon recommendation of the municipality involved and concludes by stating in its last subparagraph that if that service is a bearer service, *i.e.* basic data transport as defined above, the licensing conditions of Article L.34-2 shall apply.

5.   All other telecommunication services not covered in the preceding **20A–18** Code articles (which leaves, by elimination, those customarily referred to as value added services, although the Code no longer uses that term) can be provided to the public by anyone under the catch-all provision of Code Article L.34-5. If such a service uses radio waves, a licence of the MPT is required according to Code Article L.34-3.

Value added services which do not use radio waves are open, but if they use line capacity leased from the public operator (which the public operator must provide together with the connection to the public network on reasonable and non-discriminatory terms), then, pursuant to Code Article L.34-5, prior notification of, or authorisation from, the MPT may be required depending on the total leased line capacity needed.

The purpose of this requirement according to Code Article L.34-5 is to allow the MPT to verify (a) whether the service involved is really a value added service as opposed to basic data transport (in which case a licence would be required according to Code Article L.34-2), and (b) whether the essential technical requirements are met. These "essential requirements" are defined in Code Article L. 32/12° as relating to user safety, network protection, proper frequency utilisation, interoperability and data security, in line with the definition used in the E.C. Services Directive.

6.   To facilitate the application of the appropriate approved criteria, Code Article L.34-6 provides for the establishment of two advisory commissions, one to assist the MPT in the area of radioelectric networks and services, and the other to assist in distinguishing between basic data transport service subject to Code Article L.34-2 and value added services subject to Code Article L.34-5.

7.   Finally, it should be noted that, pursuant to Article 5 of the operating charter of France Telecom promulgated by Decree No.90-1213 of December 29, 1990, the public operator itself is subject to the licensing and notification requirements of the applicable Code provisions described above, on the basis of fair and equal competition with other private operators, for establishing independent networks under Code Article L.33-2, and for providing any public telecommunication services under Code Article L.34-3 (services using radio waves), L.34-4 (services over a cable RTV network) and L.34-5 (value added services). It needs no licence for providing services still covered by the state monopoly, *i.e.* voice telephony between fixed points and telex pursuant to Code Article L.34-1 and basic data transport pursuant to Code Article L.34-2.

P&T Code	Art. L.34-1	Art. L.34-2	Art. L.34-3	Art. L.34-4	Art. L.34-5
Type of service	Telephone between fixed points and telex.	Bearer services.	Radio-electric services.	Services on cable networks.	Other services.
Regime of authorisation	Monopoly of the public operator.	Regime of authorisation and specifications.	Regime of authorisation.	Regime of authorisation except for services related to broadcasting (teledistribution, . . . governed by the 1986 law).	Freedom of establishment, provided essential requirements are fulfilled.
Conditions of application	Are excluded from the Monopoly: — vocal services on independent networks; and — voice transmission either coming from the PSTN or going to the PSTN (but not both).	Decree of December 30, 1992 relating to bearer services.	For the use of frequencies granted by the MPT, the authorisation is given both for the creation or modification of the network (Art. L.33-1) and for the creation of the service (Art. L.34-3). For the use of frequencies granted by another authority, the authorisation is granted pursuant to the respect of specifications following the requirements of Art. L.33-1.	— the authorisation is granted to municipalities or a group of municipalities; — if the service is a bearer-service, the authorisation is granted pursuant to the same conditions as those set forth in Art. L.34-2; Art. L.34-2 only concerns services open to the public; the authorisation for services for a closed group of users (independent networks) is granted pursuant to Art. L.33-2.	Decree No. 92-286 of March 27, 1992 Class I services: — prior declaration leased lines with a capacity less than five Megabits/s Class II services: — prior authorisation leased lines with a capacity greater than five Megabits/s.

Summary Charts

The P&T Code distinguishes between networks (the physical **20A–19** infrastructure) and services that can be provided on a network. The authorisation regimes for services (Articles L.34-1 to L.34-5) depend on the type of network being used to carry the services.

The typology of services follows.

For all services described in this question, we have indicated the corres- **20A–20** ponding authorisation regime according to the type of carrier network. Six different types of networks have been considered:

Network	Content	Authorisation regime
PSTN	Public Switched Telecommunications Network	Art. L.33-1
ISDN	Integrated Services Digital Network	Art. L. 3-1
PPSDN	Public Packet Switched Digital Network	Art. L.33-1
Leased	Leased lines	Art. L.34-2
Radio	Radionetwork	Art. L.33-1
Cable	Cable network	Law dated September 30, 1986[1]

Some services cannot be provided on some networks: corresponding boxes have been left blank. Some services are not usually provided on some networks or could only be provided following technical adaptations: these are shown as grey boxes.

1. Switched voice

2. Low speed (64/b)

a. Teletex-file transfer

PSTN	ISDN	PPSDN	Leased	Radio	Cable
L.34-5		L.34-5	L.34-5	L.34-3	L.34-4

b. Telewriting

PSTN	ISDN	PPSDN	Leased	Radio	Cable
L.34-5	L.34-5	L.34-5	L.34-5	L.34-5	L.34-4

[1] Law No. 86-1067 of September 30, 1986 relating to freedom of communication, modified by law No. 89-25 of January 17, 1989, and by law No. 90-1170 of December 29, 1990; the initial purpose of cable networks is the distribution of television and broadcasting services; however the 1990 law has enabled the provision of telecommunications services on cable networks (Art. L. 34-4).

c. Facsimile

PSTN	ISDN	PPSDN	Leased	Radio	Cable
L.34-5	L.34-5	L.34-5	L.34-5	L.34-5	L.34-4

d. Image: data base access/processing

PSTN	ISDN	PPSDN	Leased	Radio	Cable
L.34-5	L.34-5	L.34-5	L.34-5		L.34-4

e. Data communications: packet/circuit

	PSTN	ISDN	PPSDN	Leased	Radio	Cable
Packet			L.34-2	L.34-2		
Circuit	L.34-2	L.34-2		L.34-2	L.34-3	L.34-4

f. Teletext

PSTN	ISDN	PPSDN	Leased	Radio	Cable
				L.34-3	L.34-4

3. High speed
a. TV Conference

PSTN	ISDN	PPSDN	Leased	Radio	Cable
	L.34-1		L.34-1		L.34-4

b. TV Telephone

PSTN	ISDN	PPSDN	Leased	Radio	Cable
	L.34-1		L.34-1		L.34-4

c. Videotex

PSTN	ISDN	PPSDN	Leased	Radio	Cable

d. Facsimile

PSTN	ISDN	PPSDN	Leased	Radio	Cable
	L.34-5	L.34-5			L.34-4

e. High speed digital communication

PSTN	ISDN	PPSDN	Leased	Radio	Cable
	L.34-2	L.34-2	L.34-2		L.34-4

f. Videoconference

PSTN	ISDN	PPSDN	Leased	Radio	Cable
	L.34-1		L.34-1	L.34-3	L.34-4

g. Audioconference

PSTN	ISDN	PPSDN	Leased	Radio	Cable
	L.34-1		L.34-1	L.34-3	L.34-4

4. Mobile

a. Paging (domestic/international)

	PSTN	ISDN	PPSDN	Leased	Radio	Cable
Domestic					L.34-3	
International					L.34-3	

b. Communications on sea/land/air

	PSTN	ISDN	PPSDN	Leased	Radio	Cable
Sea					L.34-3	
Land					L.34-3	
Air					L.34-3	

c. Third party service providers (mobile and dispatcher)

PSTN	ISDN	PPSDN	Leased	Radio	Cable
				L.34-3	

d. Mobile data

PSTN	ISDN	PPSDN	Leased	Radio	Cable
				L.34-3	

5. Other
a. ISDN

PSTN	ISDN	PPSDN	Leased	Radio	Cable
	L.34-2				

b. EDI

PSTN	ISDN	PPSDN	Leased	Radio	Cable
L.34-5	L.34-5	L.34-5	L.34-5	L.34-5	L.34-4

EDI services can also be provided on a telex network under the regime of Article L.34-5.

c. Remote accessed image database

PSTN	ISDN	PPSDN	Leased	Radio	Cable
	L.34-5		L.34-5		L.34-4

d. Processing service

PSTN	ISDN	PPSDN	Leased	Radio	Cable
L.34-5	L.34-5	L.34-5	L.34-5	L.34-5	L.34-4

e. Voice mail

PSTN	ISDN	PPSDN	Leased	Radio	Cable
L.34-5	L.34-5		L.34-5	L.34-5	L.34-4

f. Message handling

PSTN	ISDN	PPSDN	Leased	Radio	Cable
L.34-5	L.34-5	L.34-5	L.34-5	L.34-5	L.34-4

g. *Electronic Mail*

PSTN	ISDN	PPSDN	Leased	Radio	Cable
L.34-5	L.34-5	L.34-5	L.34-5	L.34-5	L.34-4

h. *Enhanced facsimile*

PSTN	ISDN	PPSDN	Leased	Radio	Cable
L.34-5	L.34-5	L.34-5	L.34-5	L.34-5	L.34-4

i. *Resale/use of leased lines*

PSTN	ISDN	PPSDN	Leased	Radio	Cable
				L.34-3	L.34-4

Use and Ownership of Receive-only/Receive-transmit Satellite Earth Terminals

In accordance with the Green Paper on Satellite Communications,[2] the 20A–21
MPT has now authorised a number of satellite communication networks.
These include independent VSAT networks, SNG, and mobile networks
open to the public.

(i) Very Small Aperture Terminals (VSAT)

a. *Introduction*

The earth segment of a VSAT network consists of main stations (*station* 20A–22
maîtresse) ("hubs") and dependent stations. The main station handles
transmission with the satellite(s) and serves the dependent stations.

VSAT networks are generally set up by value-added service providers
using space segments leased from space segment operators. Although
potential users (large industrial or service groups) have shown a real interest in satellite networks, there are so far very few networks in operations.
On the other hand, reporters seem more and more willing to use satellite
services for news gathering and broadcasting purposes.

b. *The authorisation*

Scope and nature: An authorisation for the entire network including space 20A–23

---

[2] Towards Europe-wide systems and services—Green Paper on a common approach in the field of satellite
communications in the E.C.; Communication from the Commission; COM (90) 490 final, November
20, 1990.

segment and earth stations is granted by the MPT; therefore it is not necessary to request an authorisation for each new dependent station of the network. Such networks not open to the public are licensed as independent networks under Article L.33-2 of the P&T Code. The licence conditions cover the following subjects:

- the creation and operation of the network;
- the operation of the services for which the authorisation is requested;
- the installation of the dependent stations at the users' premises.

The authorisation is based on a technical approval granted by the satellite operator for each of the main stations of the network.

Dependent stations must be approved in accordance with ETSI standards. The approval covers the radiofrequency aspects and, if need be, interfaces with public networks.

20A–24 **Closed user groups and interconnections:** The operator has to notify the DGPT within three months of the identification number of the closed groups of users. The authorisation states whether interconnection of the network to the public network and/or other networks is authorised and specifies the conditions for such interconnection, if any.

20A–25 **Frequencies:** The following frequencies are available:

- uplink transmissions (*liaisons montantes*): 14.00–14.25 GHz
- downlink transmissions (*liaisons descendantes*): 12.50–12.75 Ghz.

Downlink communications are also available on the 10.7–11.7 GHz band; however, there is no guarantee that there will not be any radioelectric disturbances.

20A–26 **General conditions:** The authorisation is personal and cannot be transferred to another party. In addition, the authorisation can be withdrawn by the DGPT in case of "substantial" change in the shareholding of the operator which has not been previously approved by the DGPT.

20A–27 **Duration of the authorisation:** The authorisation is granted to the operator for a period of 10 years and can be renewed upon request made two years prior to the end of this period. Otherwise, the DGPT will notify the operator of the conditions of renewal one year prior to the expiration of the 10 year period.

20A–28 **Fees:** Under Decree of February 3, 1993, the operator of an independent radioelectric network utilizing satellite capacity, whether on a private use or shared use basis, must pay to the State a yearly fee for management expenses.

A network is classified in Category 1 if the number of broadcasting stations of the network installed on French European territory and overseas *départements* is not greater than five and if the location of these stations is known as from the filing of the request for authorisation.

A network is classified in Category 2 if the number of stations is greater than five. Only an annual ex-post declaration is required.

		One-way Network	Two-way Network
Category 1	Management fees	FF 3,000 per year	FF 3,000 per year and FF 500 per year and per station
Category 2	Management fees	FF 10,000 per year	FF 10,000 per year and FF 500 per year and per station

(ii) Satellite News Gathering (SNG)

Temporary usage of space capacity for news gathering and broadcasting **20A–29** of pictures and sounds is known as SNG (*stations terriennes pour liaisons vidéo temporaires*). Reporting stations are autonomous, unlike dependent stations in a VSAT network: they have access to the space sector without control of any other station. They are subject to the same regulatory regime as main stations in a VSAT network. Licences are granted by the MPT under Code Article L.33-2 for a period of five years subject to the same provisions as those applied to VSAT's.

Use of the space capacity must be notified 24 hours in advance to the DGPT. Any space segment can be used provided that a specific authorisation has been granted beforehand. The request for authorisation must indicate the name of an operating manager with power to coordinate and, if necessary, interrupt the broadcasting.

(iii) Mobile satellite networks

a. Mobile satellite networks open to the public

Those are licensed as radioelectric networks open to the public under Art- **20A–30** icles L.33-1 and L.34-3 of the P&T Code, and are connected to the public network.

b. Mobile satellite networks not open to the public

Under the 1990 law, these networks are considered as independent net- **20A–31** works and are licensed as such under Article L.33-2 of the P&T Code

The Decree of February 3, 1993 provides that the yearly management fees owed by the operator of a mobile satellite network amount to FF 100,000.

4. PTT INTERCONNECT AGREEMENTS

Regulatory Authority Involvement

(i) Introduction

As a matter of definition, "public network," according to Code Article **20A–32**

L.32-14°, means all of the telecommunications networks established or operated by the public operator to serve the needs of the general public.

As a general rule, the public operator, France Telecom, must allow access to the public network on an objective, transparent, and nondiscriminatory basis to other duly authorised network operators and service providers. This rule is stipulated in Article 11 of France Telecom's operating charter and is in conformity with the E.C. Open Network Provision Framework Directive of June 28, 1990 (90/387).

Article 11 of said operating charter further provides that France Telecom may ask for access charges taking into account: (1) tariff and geographic service constraints which are not recovered by the regular standard charges for the use of the network, and (2) the cost of any modification to the public network installations made necessary by the interconnection. Such access charges are to be specified in a contract negotiated between France Telecom and the operator requesting interconnection which is then submitted to the MPT for review. In case of disagreement, the MPT may set the access charges on the basis of real costs.

To ensure equal competition between France Telecom and other actual or potential operators, in all cases where France Telecom is licensed for an activity not covered by its monopoly, France Telecom must consider itself to be another operator—even if France Telecom happens to be the only licensee in a given case. If there is another operator licensed for the same kind of network or service, France Telecom must apply to its own licensed network the same terms and charges as those negotiated with the other operator(s). In either case, the terms for connecting France Telecom's licensed network to the public network are set forth in a quasi-contractual "frame document" which is subject to review by the MPT, and separate accounting must be established for the licensed network to prevent cross-subsidisation.

The MPT's involvement with interconnect agreements is, first, to authorise the interconnection where such authorisation is required (see further below); secondly, to lay down the basic conditions for such interconnection in the licence authorising the network and/or service; and finally, to review the financial terms in the interconnection contract or "frame document" and act as an umpire in case of disagreement. The degree of such involvement varies greatly with the kind of network and/or service at issue.

(ii) Networks Open to the Public

20A–33 As an exception to France Telecom's statutory monopoly on networks open to the public under Code Article L.33-1, the MPT may license other operators to establish radioelectric networks open to the public. Such radioelectric networks may be terrestrial or satellite based.

It is generally assumed that networks open to the public are to be connected to the public network. Thus the MPT licences authorising an operator to establish and operate such a network usually do not contain an express authorisation to interconnect but go straight to the terms and conditions to be included in the interconnection contract or frame document.

The licences involving terrestrial voice services, *i.e.* SFR's and France Telecom's GSM licences, Bouygues Telecom, DCS 1800 licence and France Telecom's POINTEL licence, specify many conditions in great length and detail, leaving little more than the actual figures of the interconnection contract. Such contracts are usually required to be submitted to the MPT within three months of the date of the licence, beyond which time the MPT would step in and "'arbitrate" the terms between parties.

Two other such licences, one involving a satellite paging system and the other a satellite mobile telecommunication system for airplanes, both operated by France Telecom, merely refer to the "frame document" to deal with interconnection terms and conditions, requiring such document to be submitted to the MPT within four months in the first case and two months in the second. The only condition in this regard specified by the MPT in the licence is the maintenance of separate accounting by France Telecom (with separate billing added in the second case.)

(iii) Independent Networks

In contrast to networks open to the public dealt with under Code Article **20A–34** L33-1, independent networks licensed under Code Article L.33-2 are closed, *i.e.* they are limited to a single user ("private use") or to one or more "closed groups" of users ("shared use"). Independent networks are infrastructures that are privately established for purposes of internal communication. Users may, however, resort to a third party to establish, and then operate these infrastructures. This is the case of radioelectric networks with shared frequencies (*réseaux radioélectriques à ressources partagées* or 3RP) or of VSAT networks. Thus, as France Telecom has no statutory monopoly in the area of independent networks, fixed networks as well as radioelectric terrestrial and satellite networks can be licensed to private operators under Code Article L.33-2, as long as they are not open to the public.

The notions "open to the public" and "closed group" are not defined in the law, and they have proven to be quite elastic in the MPT's application so far.

As independent networks by definition are intended to provide communication internal to a single user or between the members of a closed group, their interconnection to the public network can only be authorised exceptionally by the MPT on a case by case basis (Code Article L.33-2). This rule, however, has been applied in the same liberal spirit as the distinction between "closed group" and "public."

Thus, in the case of independent networks, the MPT has a significant task in deciding whether to allow the interconnection of the proposed network to the public network. In accordance with the second paragraph of Code Article L.33-2, the MPT must also specify in the licence the conditions for such interconnection so as not to permit the exchange of communications between persons other than those for whose use the independent network has been reserved.

In some such licences, there is a mention that the technical and financial terms of the interconnection are to be covered in a contract to be negotiated between France Telecom and the licensed operator, a copy of which must be filed with the MPT. In many other cases there is no mention of any interconnection terms or charges. We understand this to be the case because there is no need to negotiate special access charges in many such situations. The independent network may consist of leased circuits, which carry their own tariffs including regular standard interconnection charges, and there may not be any need to modify any public network facilities or to incur any other extraordinary cost on the part of France Telecom. Moreover, the practice of requiring the connection of each connection point of an independent network to the closest termination point of the public network within the same tariff zone would also tend to reduce the justification for additional access charges.

Interconnection Between Domestic Leased Circuits and Public Switched Networks

20A–35 For France, the question needs to be reformulated. The important distinction is not between leased and owned circuits but between networks open to the public and independent or private networks, regardless as to whether they contain privately owned or leased lines and facilities.

A network open to the public today is limited to radioelectric networks licensed under Code Article L.33-1 (*e.g.* the mobile phone network operated by the *Société Française du Radiotéléphone*—SFR). Such a network can be connected to the public switched network under conditions negotiated with France Telecom, approved by the DGPT and specified in the licence operating charter.

Independent networks, which may include "private networks" (including VSAT networks) intended for the internal use of a single user or "shared networks" for the use of a closed user group are subject to authorisation pursuant to code Article L.33-2. (Some purely internal or low capacity independent networks require only a declaration to the DGPT under Article L.33-3.)

As a general principle, such a network may be connected to the public switched network (or to a network open to the public like SFR's) only exceptionally, upon special approval of the DGPT, and on the condition that such connection does not enable communication between persons other than the members of the "group" for which it has been authorised. The conditions for such interconnection, if authorised, are specified in the individual authorisation.

The same rules apply to interconnection between international leased circuits and public switched networks.

Interconnection of Private Networks

20A–36 The correct relevant term in the French regulatory context is "independent" networks. Such networks can be connected together only if the persons thus permitted to communicate with one another qualify as a closed group for purposes of the DGPT authorisation under Code Article L.33-2. The notion of closed user group (*groupe fermé d'utilisateurs*, GFU) is not defined by the Code. The notion is at the borderline between services provided to the public (*i.e.* regulated or reserved services) and services not provided to the public (*i.e.* services that do not require any authorisation nor consultation of the DGPT), as well as between independent networks and networks open to the public. The notion of GFU has been clarified by the DGPT on a case by case basis; a GFU is any group of users bound by a common economic interest pre-existing the formation of the GFU and sufficiently stable to be precisely identified. These criteria refer to the relationships between a company and its customers, its licensees or its subsidiaries, or between companies involved in the same sectors.

Networks Open to the Public	Radioelecric networks.	Art. 33 LRT: Monopoly of the public operator over all telecommunications networks.
		Art. 33-1 LRT: Possibility of authorisation by MPT; specifications charter.
Independent Networks for Private or Shared Use	• Internal networks. • Independent networks other than radioelectrical, whose terminal points are separated by less than 300 m. • Low capacity radioelectical installations.	These three kinds of networks can be freely established.
		Other independent networks must be authorised by the MPT.

5. REGULATION OF BROADCASTING

General Principles

Broadcasting matters are governed by the Law on Freedom of Communication of September 30, 1986 ("LFC"), as modified by subsequent laws. Broadcasting is a generic term and comprises radioelectrical as well as cable broadcasting. Article 1 of the LFC provides for the general principle of freedom of telecommunication, including audiovisual communication, under the supervision of the CSA. 20A–37

The LFC has set forth in its Article 2 a fundamental difference between telecommunication interpreted as "any transmission, broadcasting or reception of signs, signals, documents, images, sounds or information of any type whether by wire, optics, radioelectricity or any other electromagnetic systems", while audiovisual communication is defined as "any disposal to the public or categories of the public, by means of a telecommunications process, of signs, signals, documents, images, sounds or messages of any type which do not have a private correspondence aspect".

Pursuant to Article 21 of the LFC, the Government decides which of the available frequencies are attributed to government entities and which are to be attributed by the CSA. The use of frequencies is generally considered as a matter of public interest and, therefore, must be authorised by the CSA (Article 22 of the LFC).

Pursuant to Article 26 of the LFC, national programme companies (listed in Article 44 of the LFC) are allowed to broadcast their programmes via the frequencies used by them at the date of publication of the LFC, *i.e.* October 1, 1990. These national programme companies are engaged in the following activities:

- nationwide radio broadcasting;
- nationwide television broadcasting
- regional television broadcasting;
- overseas radio and television broadcasting (for French overseas territories); and
- international radio and television broadcasting.

Specific rules

(i) For Operators other than National Programme Companies (Article 44 of the LFC)

20A–38 Authorisations to use frequencies for the purpose of radio or television broadcasting via radioelectric wave networks or satellite are delivered only after conclusion of an agreement between the CSA and the interested company (Article 28 of the LFC).

This agreement must determine:

- the duration and general characteristics of the programmes,
- the time envisaged for broadcasting of original French programmes,
- the percentage of turnover spent for the acquisition of French programmes,
- the time reserved for cultural and educational programmes, and
- the maximum time reserved for commercial advertisements.

These authorisations for broadcasting are granted for a maximum period of 10 years for television services and five years for radio services.

(ii) Specific Restrictions Concerning Interest in other Broadcasting Companies (Articles 39 to 41 of the LFC).

Pursuant to Article 39(1) of the LFC, it is prohibited for a person or company to hold, directly or indirectly, more than 49 per cent of the capital or voting rights in another company, which has been authorised to deliver only TV services via radioelectric wave network.   20A–39

Pursuant to Article 39(II) of the LFC, the interest held by a company in another one authorised to supply radio and TV services via satellites must not exceed 50 per cent. If this interest is higher than 33.3 per cent, another interest in a third broadcasting company must be lower than 33.3 per cent. In the case of two different interests exceeding five per cent each, an interest in a fourth broadcasting company is also limited to 5 per cent.

Any direct or indirect acquisition by foreign persons or companies, of an interest in the capital of a company which is authorised for radio and TV services in French language, broadcast via radioelectric wave networks, is limited to 20 per cent (Article 40 of the LFC). For this purpose, are considered as foreign all persons without French nationality, as well as all companies in whose share capital French nationals are not a majority.

(iii) Specific Rules for Cable Networks

The installation of networks broadcasting by cable radio sound broadcasting and television services is subject to two separate administrative authorisations: an authorisation for setting up a network and an authorisation for exploiting the network. In exceptional cases, a notification may be required.   20A–40

Establishing a cable network

The authorisation for establishing a cable network is within the jurisdiction of municipalities or groupings of municipalities (Article 34 *et seq.* of the LFC). Law no 92-653 of July 13, 1993 relating to the installation of broadcasting networks by cable of radio sound broadcasting and television services moreover specifies a regime for setting up networks. In particular, cities are required to respect the "aesthetic quality of the premises" where the cable networks will be installed.   20A–41

Operating a cable network

The authorisation for operating the network is within the jurisdiction of the CSA upon the proposal of the municipalities even if the possibility of exploiting the cable network has been extended to municipality licensees (*régies communales*) by Article 34 of the LRT. The authorisations to operate can only be issued to a single company or municipality or inter-municipality licensees. The authorisation specify their duration as well as the number and the nature of the services to be broadcast.   20A–42

(iv) Restrictions on Owners of Broadcasting Interests

20A–43 The LFC has set forth numerous provisions relating to the managers of the structure and the transparency of audiovisual companies, holders of an authorisation (Article 35 *et seq.*):

a. "it is forbidden to lend one's name, in any manner whatsoever, to any person who is a candidate for the issuance of an authorization relating to an audiovisual communication service" (Article 35);

b. "any legal or natural person whose interest in the capital or the voting rights at General Meetings of a company which is the holder of an authorization relating to an audiovisual communication service becomes greater than or equal to 20%" must inform the CSA within a period of one month as of when these thresholds are crossed (Article 38);

c. "a person holding one or more authorizations, each relating to the operation of a network broadcasting by cable radio sound broadcasts and television services, may not become the holder of a new authorization relating to a service of a similar type if the effect of this authorization is to increase the number of inhabitants of the areas served by the networks which it will be authorized to operate to above eight million"

d. Restrictions on the national level

"In order to prevent prejudice being caused to the concept of plurality on the national level, no authorization whatsoever, relating to a radio sound broadcast or television service by terrestrial hertzian wave or for the operation of a network broadcasting by cable radio sound broadcasts and television services, may be issued to a person who holds more than two of the following positions:

1° The holder of one or more authorizations relating to television services broadcasted by terrestrial hertzian waves allowing for the servicing of areas, whose population numbers four million inhabitants;

2° The holder of one or more authorizations relating to radio sound broadcasting services allowing for the servicing of areas, whose population numbers thirty million inhabitants;

3° The holder of one or more authorizations relating to the exploitation of networks broadcasting by cable radio sound broadcasting and television services allowing for the servicing of areas, whose population numbers more than six million inhabitants;

4° The editor or controller of one or more daily printed publications of political and general information, representing more than 20% of the total distribution, on the national territory, of daily publications of the same type, evaluated over the past twelve months preceding the date on which the request for authorization was made.

However, an authorization may be issued to a person who fails to fulfill the provision of this Article, subject to the condition that they comply with these provisions within a period set by the 'Conseil supérieur de l'Audiovisuel' and which can not exceed a period of six months" (Article 41-1);

e. Restrictions on the regional and local level

"In order to prevent prejudice being caused to the concept of plurality on the regional and local level, no authorization whatsoever relating to a radio sound broadcasting or television service, other than national, by terrestrial hertzian wave or for the operation of a network broadcasting by cable radio sound broadcasting and television services, can be issued for a geographic area determined by a person who holds more than two of the following positions:

1° The holder of one or more authorizations relating to television services, having a national aspect or not, broadcasted by terrestrial hertzian waves in the area concerned;

2° The holder of one or more authorizations relating to radio sound broadcasting services, having a national aspect or not, whose potential overall audience, in the area concerned, is greater than 10% of the overall potential audiences, in the same area, for all the services, public or authorized, of the same type;

3° The holder of one or more authorizations relating to the exploitation of networks broadcasting by cable, inside this area, radio sound broadcasting and television services;

4° The editor or controller of one or more daily printed publications of political and general information, having a national aspect or not, distributed in this area.

However, an authorization can be issued to a person who fails to fulfill the provision of this Article, subject to the condition that they comply with these provisions according to the conditions stipulated in the last paragraph of Article 41-1 above" (Article 41-2).

6. Regulation of Apparatus

Terminal v. Network Equipment

The manufacture, importation, sale and use of terminal equipment are **20A–44** open, subject to type approvals according to Code Article L.34-9, along the lines of the E.C. Terminal Equipment Directive of May 16, 1988 (88/301). There are no formal certification or type approval requirements other than those relating to terminal equipment.

Terminal equipment is defined under Code Article L.32-10 as "any equipment designed to be connected directly or indirectly to a termination point of a network for the purpose of transmitting, processing or receiving information". A telecommunication network is defined under Article L.32-2 as "any installation or group of installations that ensures either the transmission or the transmission and routing of telecommunication signals as well as the exchange of associated control and operational instructions between network termination points."

Accordingly, a distinction can be made between network equipment which forms an integral part of a network (*e.g.* telephone centre exchanges, switches, controllers, towers, cables, etc.) and terminal equipment which is connected to a network termination point, but is not a part of the network (*e.g.* telephone handsets, telecopiers, answering machine, telex terminals, PBXs, etc.).

Formal type approval requirements under existing French regulations apply to terminal equipment only. However, the "directly or indirectly" provision in the definition of terminal equipment indicates that equipment, within an independent network and connected to the public network—such as private exchanges—is considered as "terminal equipment" from the viewpoint of the public network.

Type Approval of Terminal Equipment

20A–45   Code Article L.34-9 sets forth the principle that the market for telecommunications terminal equipment is open to free competition. The liberalisation of the terminal equipment market in France dates back to 1986 and has been going on in line with the applicable E.C. policy developments.

Article L.34-9 specifies that any terminal equipment designed for connection to a public network, as well as any radioelectric terminal equipment regardless of whether it is to be connected to a public network, is subject to the prior approval of the MPT.

Article L.34-9 also provides that such type approval must be limited to the purpose of ensuring that the terminal equipment in question complies with the "essential requirements" defined in Article L.32-12, *i.e.* user safety, network or frequency spectrum protection, interoperability and data protection.

The Approval Process

20A–46   Code Articles R. 20-1 to 20-30 provide for the specifications and conditions of obtaining type approvals and regulate the qualification and certification of technicians authorised to install and maintain such equipment.

Approval is required only for terminal equipment specifically designed for direct and indirect connection with a network open to the public and for radioelectric equipment. Terminal equipment which may be connected to a network open to the public, but whose primary purpose is different, only requires that a prior declaration be made to the DGPT.

In accordance with the Mutual Recognition Directive, Articles R.20-17 to 20-21 provide for the recognition, without any formalities required, of the certification delivered by an institution officially authorised to deliver such certification. In addition, the results of tests carried out by laboratories officially designated by other Member States can be validly used in order to obtain a certification in France.

The applicant may choose between two approval procedures set forth by the Code:

(i) Type Approval Procedure: *"examen de type"* or *"CE de type"*

A request for approval is made to the DGPT. The request must include the **20A-47** following information:

- identity of the applicant who must be a French applicant;
- purpose of the equipment;
- technical specifications of the equipment;
- information for users;
- results of tests made by designated laboratories;
- declaration that no similar request for approval has been made to another institution; and
- representative samples of the equipment.

Certain terminal equipment (for instance, CB radioelectric equipment) are subject to a simplified process which only requires the submission of a prior declaration from the manufacturer or its representative.

Tests are carried out by DGPT-approved laboratories: *Laboratoire d'Essais d'Agréments* for tests in telecommunications field, and *Laboratoire Central des Industries Electriques* for test on electromagnetic compatibility and, if need be, on electric safety.

Following the tests in laboratories, the DGPT either examines samples (in case of a declaration to market only products conforming with the type tested in laboratories), or audits the chain of production on a systematic basis in the case where the declaration was made together with a request for an "insurance quality production" (E.C. standard EN-29002).

(ii) Alternative Procedure: Insurance Total Quality

Certification of the design and manufacturing process is made with regard **20A-48** to E.C. standard EN-29001. The whole industrial process is certified, *i.e.* studies, conception and manufacturing.

For both procedures, a formal declaration of conformity (production quality insurance) by the manufacturer or its representative to whom the type approval has been granted and a conformity stamp ("AGREE") on each product are required. Any modification on the design of a terminal equipment which has already been certified must be certified anew. The certification process, which usually lasts between three and seven months, will be much shorter and will be limited to the modification itself.

Marketing of Terminal Equipment

Violations of the above provisions can result in criminal sanctions, suspen- **20A-49** sion, revocation of the type approval, and even confiscation of the equipment. Code Article L.40 provides that control of the marketing of unapproved terminal is performed by authorised DGPT and SNR agents. Advertising of unapproved terminal equipment is a misdemeanour. The utilisation without authorisation of a frequency or a radioelectric station, or utilisation beyond the authorised conditions, is also prohibited.

7. TECHNICAL STANDARDS

Open Network Provision

20A–50 The French position on the voice telephony ONP Draft Directive has been determined by the DGPT as follows:

1. the opening of non standardised access points entails:

   • the possibility of refusing, for technical reasons pertaining to the integrity of the network, the repercussion on the user of the total cost implied by a new access point, and
   • the obligation to resort to standardization to determine special access at the European level.

2. Guarantees are required as to the application of the principle of subsidiarity, a greater flexibility in the fixation of prices and accounting, and regulatory powers of the ONP committee.

3. The ONP-Leased Lines Directive has become under French law the Decree of July 28, 1993. This Decree provides for the transparency and publicity of offers of leased lines, prices set by the public operator closer to costs, and the offer of a minimum set of compulsory leased lines.

CHAPTER 21

Germany

By Klaus-Jürgen Kraatz, Kraatz & Kraatz, Kronberg

1. HISTORY OF THE DEVELOPMENT OF TELECOMMUNICATIONS IN GERMANY

Industry structure

Germany did not have a private telecommunications operator prior to   21–01
1990, but there was a long established telecommunications industry providing telecommunications equipment to companies (*e.g.* private branch
exchanges), and in particular to Deutsche Bundespost for many decades
(*e.g.* Siemens or Standard Elektrik Lorenz AG (SEL), which is nowadays a
subsidiary of Alcatel, have been major suppliers for many years).

When Germany was unified in 1990 some 1922 Siemens equipment was
still working properly in former East Germany. There was even one special
network with such old equipment capable of transmitting facsimiles! As
foreign investment in Germany is not restricted, there have been other
major companies from abroad selling equipment in Germany for many
years although in smaller quantities due to the monopolistic structure of
the market.

Institutional Structure

The major change occurred on July 1, 1989. At that time the so-called   21–02
Postreform I entered into force. It basically aimed at:

i) reorganising the German PTT (Deutsche Bundespost—DBP);
ii) creating a new regulatory framework regarding telecommunications
services and terminal equipment.

DBP was originally a dependent public entity under the authority of the Ministry for Post and Telecommunication (*Bundesministerium für Post und Telekommunikation*—BMPT) employing exclusively public servants and public employees. DBP was divided up in 1989 into three separate public agencies *Deutsche Bundespost Postdienst* (postal services), *Deutsche Bundespost Postbank* (postal banking services) and *Deutsche Bundespost Telekom* (telecommunication services) each having their board of directors similar to private stock companies. This became possible under the Postal Constitution Act 1989 (Postverfassungsgesetz)[1] which defined the separation of the regulatory and the commercial activities of DBP.

Since then regulatory powers are remaining with the Federal Minister of Post and Telecommunication (*Bundesminister für Post und Telekommunikation*). The commercial activities are now exercised by the three public enterprises, Deutsche Bundespost Postdienst, Deutsche Bundespost Postbank and Deutsche Bundespost Telekom.

On January 1, 1995 Postreform II was enacted. It deals primarily with the reorganisation, incorporation and transformation of the three DBP entities into stock corporations under German private law. Thus DBP Telekom AG was created having an initial stock capital of DM 10 billion. Its shares are at present owned by a Government holding company, the Federal Office for Post and Telecommunication Deutsche Bundespost (*Bundesanstalt für Post und Telekommunikation Deutsche Bundespost*). It is intended to sell shares to private owners within the next two years.

Background to Legislation

21-03   The German Constitution of 1949 (*Grundgesetz*) provided until July 8, 1994 in Articles 73, 87f and 143a[2] that all PTT services have to be rendered by a Federal public entity. This was due to the fact that the competencies of the Federation for postal matters versus the Federal States (*Bundeslaender*) had to be secured. There are estimates that DBP Telekom's turnover would decrease by 10 per cent, the share capital by 15 per cent and the profits before taxes would reach only 40 per cent of those possible, had DBP Telekom remained unable to change its legal status to a private stock company. This comparison is based on actual figures and projections up to the year 2000.

Finally, the political parties in the German parliament (*Bundestag*) and the representatives of the Federal States in the second chamber (*Bundesrat*) reached an agreement on Postreform II in July 1994, which as a major objective will lead to the privatisation of all three public enterprises, *i.e.* DBP Postdienst, DBP Postbank and DBP Telekom. Their transformation into private stock companies occurred on January 1, 1995. The required amendment to the German Constitution by a two-thirds majority in the *Bundestag* and the consent of the *Bundesrat* on July 8, 1994 have opened

---

[1] *Gesetz zur Neustrukturierung des Post- und Fernmeldewesens und der Deutschen Bundespost (Poststrukturgesetz)*, June 8, 1989—BGBI I (= Federal Gazette) pp. 1026 *et seq.*, Art. 1.
[2] May 23, 1949—BGBI I pp. 1 *et seq.*

the way to the biggest reform in this public sector since the Dukes of Thurn and Taxis held the privilege of postal services by horse driven carriages in the 16th century.

In the future the offer of telecommunications services will be a private 21–04 entrepreneurial activity. The Federal Republic of Germany will keep the public activities and competencies, *i.e.* mainly the regulatory function in the sector of post and telecommunication. The new law is called the Law on the New Situation of Post.[3]

Article 1 of the law contains the Law on the Establishment of a Federal Agency for Post and Telecommunications[4] in order to enable the Federal Republic of Germany to exercise its special rights in the stock companies to be established. The Federal Agency will be supervised by the BMPT. It will be possible to sell up to 49 per cent of the shares in DBP Telekom once it has been privatised.

Article 3 contains the Law on the Transformation of the Enterprises of DBP into stock companies.[5]

Article 5 and 6 deal with modifications to the Telecommunications 21–05 Installations Act 1994 (*Fernmeldeanlagengesetz*—FAG)[6] and to the Law on Post (*Gesetz über das Postwesen*).[7] These modifications will be necessary due to the new structure under the regime of stock companies (AG) of DBP Telekom and DBP Bundespost. They will not change the present scope of the voice or network monopolies, etc. Both articles are limited in time up to December 31, 1997 due to the then occurring liberalisation of the telecommunications market in the European Union.

Article 7 deals with the Law on the Regulation of Telecommunication and Postal Services,[8] which is in many parts similar to the Postal Constitution Act as to regulatory matters.

2. EXISTING LEGISLATIVE AND REGULATORY ENVIRONMENT

Principal Legislative Instruments

Article 87, paragraph 1 of the German Constitution provided until recently 21–06 that Deutsche Bundespost is a federal administration having its own subentities. This explains why DBP and its public entities DBP Bundespost, DBP Postbank and DBP Telekom were public enterprises until December 31, 1994, which are by this structure under German law subject to a number of restrictions, in particular when raising funds from the public.

---

[3] *Gesetz zur Neuordnung des Postwesens und der Telekommunikation (Postneuordnungsgesetz)*, September 14, 1994, BGBI I pp. 2325 *et seq.*
[4] BGBI I (1994) pp. 232 *et seq.*
[5] *ibid.*, pp. 2339 *et seq.*
[6] *ibid.*, pp. 2363 *et seq.*
[7] *ibid.*, pp. 2368 *et seq.*
[8] *ibid.*, pp. 2371 *et seq.*

As far as telecommunications services are concerned the Postal Constitution Act distinguished three categories of services to be provided by DBP Telekom.

(i) monopoly services
(ii) mandatory services
(iii) optional services

21-07   The new Law on the Regulation of Telecommunication and Postal Services does only refer to mandatory services in section 8.[9]

The Telecommunications Installations Act 1989[10] describes in detail the monopoly services and their distinction from other services. Furthermore, there have been specifications made by the BMPT in administrative rules. DBP Telekom's obligations and rights to render monopoly services have been described in the Telecommunications Regulation (*Telekommunikations-Verordnung*—TKV)[11] of June 1991 as amended in October 1992. On the basis of this legal framework DBP Telekom has published standard business conditions,[12] which are modified from time to time depending on the development of the various services rendered.

Agencies of Regulation, Competition Control

21-08   The regulatory agency is BMPT headed by the Federal Minister of Post and Telecommunication. All regulatory matters are dealt with by the ministry itself.

Competition control is implemented in two different ways. On the one hand the ministry reviews the matter under certain aspects regarding those services which are not monopoly services and assures that pricing and licensing conditions, as well as business conditions of DBP Telekom applicable in those areas, do not unfairly impair the competitiveness of other service providers.

Furthermore, the German law against Restraints of Competition (*Gesetz gegen Wettbewerbsbeschränkungen*—GWB)[13] has been modified as of January 1, 1990.[14] Since then the Federal Cartel Office (*Bundeskartellamt*) at Berlin has the possibility to review anti-trust practices in this area, which prior to 1989 because of the complete monopoly could not become subject to cartel law requiring the legal possibility of competition. However, as the German GWB does not provide like Article 86 of the EEC Treaty for a clause against the abuse of a dominant market position, these possibilities are much more limited than those of the European Commission's DG IV in this area.

---

[9] *ibid.* pp. 2373.
[10] July 21, 1989—BGBI I pp. 1456 *et seq.*
[11] Integrated text dated October 5, 1992—BGBI I pp. 1717 *et seq.*
[12] *cf. Amtliche Mitteilung der Deutschen Bundespost TELEKOM*, July 30, 1992, pp. 722 *et seq.*
[13] July 27, 1957—BGBI I pp. 1081 *et seq.*
[14] New publication of the full wording dated February 20, 1990—BGBI I pp. 235 *et seq.*

General Rules Governing Telecommunication Installations, Services and Apparatus

The law applicable to telecommunication installations is the Telecommuni- **21–09**
cation Installations Act (FAG) 1989 (amended in 1994) defining the monopoly rights of DBP Telekom in a more narrow way than before. Monopoly
rights are only transferred to the extent necessary to ensure the accomplish-
ment of DBP Telekom's legally defined tasks. The transfer has been made
by administrative rules of BMPT.

According to the FAG the network monopoly is the exclusive right of the
Federal Government to set up and operate transmission lines including
the network termination points related thereto (network monopoly) and
the right to set up and operate radio installations (radio monopoly).

Another important restriction is the telephone service monopoly which
is defined under the FAG as switching voices in real time for third parties.
All other telecommunication services are opened to competition and can
be provided by any third party via switched or fixed connections to be
obtained from DBP Telekom under its network monopoly. The liberalisation of telecommunication services is also defined in the FAG.

As to terminal equipment, such equipment has been liberalised com- **21–10**
pletely under the FAG subject to the proviso that neither the transmission
lines provided by DBP Telekom nor any terminal equipment or persons
may be harmed or endangered by the connection or operation of the equipment used. The equipment is subject to type approval by the Federal Office
for Approvals of Telecommunication (*Bundesamt für Zulassungen in der
Telekommunikation*—BZT). This name change which implies also a
change in substance indicating the more independent structure of this federal office became effective in March 1992. Prior to this the same office
was called *Zentralamt für Zulassungen im Fernmeldewesen* (ZZF) and even
earlier *Fernmeldetechnisches Zentralamt* (FTZ). Any type approval of
equipment is effected by a licence number and the abbreviation "BZT
approved", which is nowadays the only correct designation for type
approval.

Licensing Regime including Main Licence Categories

There are two types of licences under the amended FAG: **21–11**

   (i) individual licences (*Einzelgenehmigungen*)
   (ii) general licences (*Allgemeingenehmigungen*)

Several licences have been granted. They include licences for the operation of mobile radio communication systems, radio paging systems, satellite
networks and corporate networks. Each of those areas has different licensing conditions which will be described under sections 3 and 4 to the extent
necessary to understand the system applicable in Germany at present.

3. Regulation of Services

Analysis of Rights to Provide Telecommunications Services— Monopoly and Competitive Services

21–12   Under the Postal Constitution Act 1989 the following three categories of services have to be provided by DBP Telekom:

- monopoly services
- mandatory services
- optional services

(i) Monopoly Services

21–13   Monopoly services depend on the network monopoly, the voice monopoly and the radio monopoly. As radio communications will be dealt with separately, only the network and the voice monopoly will be analyzed in this section. Furthermore, as monopoly is the exception and competition is the rule there are only a few and very precisely defined monopoly services which the minister has transferred to DBP Telekom. The general definitions of the scope of the monopoly are part of the Telecommunications Installations Act 1989 and 1994 and have been detailed by the Federal Minister of Post and Telecommunications in administrative rules. DBP Telekom's monopoly services are subject to the Telecommunications Ordinance (*Telekommunikationsverordnung*—TKV) of June 1991 as amended in September 1992. It provides the legal basis for the general business conditions of DBP Telekom regarding monopoly services. Under the Ordinance DBP has the obligation to unbundle its monopoly services in accordance with market demand as far as reasonably possible. Furthermore, monopoly services have to be provided in accordance with Community law, in particular the ONP Directive. Another obligation is the repair of any telecommunication defect without delay, *i.e.* even at night and on sundays and holidays. Finally, it is stipulated that specific liability rules are applicable, if customers are suffering damages due to the use of a monopoly service.

(ii) Mandatory Services

21–14   Mandatory services (*Pflichtleistungen*) are defined to be services which DBP Telekom has to render in the public interest. According to the Postal Constitution Act 1989 such mandatory services can be imposed upon DBP Telekom by ordinance of the Federal Government.

(iii) Optional Services

21–15   Optional services (*Freie Leistungen*) imply that they can be provided by DBP Telekom and any competitor to the extent they wish to enter the market.

Fixed and Mobile Services

It is worth noting that cross-subsidies between different services of DBP   **21–16**
Telekom or between the three DBP enterprises by revenues generated for
monopoly services are legal under Article 28 of the Postal Constitution Act
1989 as well as section 7 of the new Law on the Regulation of Telecommunication and Postal Services. The criteria are the following:

> The network monopoly is defined as the Federal Government's exclusive
> right to set up and operate telecommunications transmission lines includ-
> ing the network termination points. The radio monopoly includes the
> right to set up radio and broadcasting installations (transmission of radio
> signals only).

> The scope of the monopoly is not defined in the FAG. As the definition
> is very important the BMPT has adopted administrative rules defining
> the network monopoly.[15]

> Under the network monopoly the holder has the right to set up and
> to operate telecommunications transmission lines including the network
> termination points related thereto as well as cable and radio communica-
> tion installations together with their transmission equipment.

> All terrestrial services—not only telephone but also cable television ser-
> vices—are subject to the network monopoly of DBP Telekom. The radio
> monopoly includes the right to set up radio and broadcasting installa-
> tions (transmission of radio signals only).

> The situation is different for mobile services. As of 1989 mobile com-
> munications have been opened to competition.

(i) Digital Cellular Radio

In 1989 a GSM licence was granted for the 900 MHz D 2 network operated   **21–17**
by Mannesmann Mobilfunk GmbH. This company competes with the D
1 mobile network of Deutsche Bundespost Telekom. Under the D 2 licence
it is necessary that all basic services are provided by the end of 1994 with
a coverage of 94 per cent of the population living in former West Germany
and 90 per cent in the New Federal States. The licence is valid until 2009.
    The D 1 network of DBP Telekom is subject to special administrative
rules which have been set by the Federal Minister of Post and Telecommunications with the aim of guaranteeing fair competition. One of the rules
is that internal prices paid by D 1 for monopoly services to Deutsche Bundespost Telekom have to be equal to those charged to D 2.
    Another licence has been issued in 1993 with respect to the new DCS
1800 standard to a consortium lead by Thyssen AG, the so-called E 1
licence.

---

[15] September 19, 1991, *cf.* Information Series on Regulation Issue no. 4 of BMPT, December 1991, Regulation of the Federal Government's Network Monopoly, pp. 163 *et seq.*

(ii) Trunked Mobile Radio

21–18   Trunked mobile radio is primarily destined for in-house communications of companies. Several types of licences are available.

Licence type A —for 14 city areas of Germany.

Licence type B —for all other areas in Germany not covered by licences of type A.

Licence type C —for large properties.

These services are operated within the 410 to 430 MHz frequency band. The licences are granted for 10 years. Whereas 28 type A licences have been awarded by tender in 1991 and 1992, type B and C licences are awarded by application. Any type B licence may cover up to 15,000 square kilometers (approx. 9,500 square miles). Type C licences have been awarded to airports, shopping-malls, etc.

DBP Telekom has itself 14 coverage areas competing with type A licence areas. There again, administrative rules have been issued by BMPT in order to guarantee fair competition between the trunked mobile radio operators regulating also the access of the private competitors to monopoly services.

(iii) Service Providers

21–19   The marketing of mobile communications has not been restricted to the network operators. In this part of the business they may cooperate or compete with service providers and special retailers. Service providers and retailers are neither subject to licensing requirements nor to section 2 of the FAG as they do neither set up nor operate their own telecommunications installations.

All type D and E licensees must give access to service providers whereas type A, B and C licensees do not have such an obligation. Service providers must not be network operators in a competing network, *e.g.* the D 2 operator cannot become a service provider for the E 1 network.

Satellite Services (Non Broadcasting)

(i) Satellite Licensing

21–20   Satellite communications also require a licence. Most of these licences cover several types of services, *e.g.* data, text, video and audio services. Most of the licences granted to date concern data transmission and business television, whereas others are in the field of satellite news gathering and broadcasting as well as land mobile services.

The commercial provision of interactive voice connections is not generally permitted. There is one exception for traffic within or to the New Federal States of Germany (former East Germany), where satellite licences for voice communications are granted, but only for a limited period due to the very special circumstances ending on December 31, 1997.

BMPT has published a standard satellite licence. France and Germany have agreed on the mutual recognition of their licences, there is also in a memorandum of understanding between Germany, France, the Netherlands and the U.K. aiming at the same goal.

(ii) Access to the Space Segment

Due to the competencies of the international satellite organisations 21–21 INTELSAT, EUTELSAT and INMARSAT no licence for the operation of satellite earth stations includes the right of direct access to the space segment of these organisations. Such satellite capacity must, at present, be obtained from one of the signatories to these conventions, *i.e.* national telecommunications operator.

As of May 1, 1992 competition among the signatories has been admitted in Germany, *i.e.* the option of the holders of a German satellite licence to obtain the necessary space segment from a foreign signatory. There is no requirement of participation or consent to be obtained from BMPT regarding the agreement between the user of the space segment and the foreign signatory.

Of course, the right to place satellites in orbit remains unaffected by any licence to operate an earth station or the agreement with the signatory. It depends on international space law which is also applicable in Germany. Most important are the radio regulations of the ITU and the statutes of INTELSAT, EUTELSAT and INMARSAT.

The German Government views the radio regulations of the ITU as nondiscriminatory, on the other hand it does not accept anymore the "economic harm rule" which is part of the INTELSAT, EUTELSAT and INMARSAT conventions. Such a rule does not seem to be any longer justified, the technical possibilities having fundamentally changed in the meantime.[16]

Radio Communications in General

Under the FAG the Federal Government has the exlusive right to set up 21–22 and operate radio installations. It is exercised by the Federal Minister of Post and Telecommunications. He may grant individual licences regarding radio installations. As part of the radio monopoly the Minister has extensive regulatory competencies including the right to assign frequencies.

4. REGULATION OF BROADCASTING

Basic Legislative Rules

The most important aspect of the regulation of broadcasting in Germany 21–23 is the difference in competencies between the Federal Government and the

---

[16] *cf.* Information Series on Regulation Issue no. 7 of BMPT, June 1992, "Licensing and Regulation in Mobile and Satellite Communications", p. 38.

Member States (*Laender*), the Laender being responsible for the programmes, *i.e.* the contents of the broadcasting, whereas the transmission
of the signals is part of the Federal Telecommunications Law. Under Article
73, no. 7 of the German Constitution the Federal Government has the
exclusive competence in matters of telecommunications. The distinction
between this competence and the competencies of the Member States
regarding the programmes was in dispute for many years. In the sixties the
Federal Governement tried to establish a national broadcasting station. In
a famous case before the Federal Constitutional Court[17] this law was
annulled and the Court defined the subject matter of the Federal Government's telecommunications competencies as follows:

> "Post and telecommunications comprise only the technical area of broad-
> casting including studio technique, but do not comprise broadcasting
> as such."[18]

21–24 The Federal Government is therefore only competent to regulate and
administrate the transmission technique. All other broadcasting matters are
protected under Article 5 of the German Constitution (right to freedom of
opinion) and are, according to the basic rule of Article 30, a subject matter
within the legal competencies of the Laender. The reason is that broadcasting is considered as a matter of cultural affairs.[19]

To describe this matter in brief it means that the Federal Government is
responsible for the transmission technique, and the Laender for the organisation and the contents of broadcasting. Since that decision this question
has no longer been in dispute. At the time the judgement was rendered
there existed only one German television programme (ARD) and several
radio stations. They owned the transponders for their broadcasting themselves. Under the judgement of the Federal Constitutional Court these
transponders would have had to be transferred to Deutsche Bundespost.
However, the Federal Constitutional Court ruled that existing transponders
would remain with the owners and as a consequence until today, the first
German TV programme ARD and the German radio stations do own their
transponders, whereas those stations broadcasting the second public programme (*Zweites Deutsches Fernsehen*—ZDF), as well as the so-called
third programmes (regional programmes) and the private tv stations do not
own any transponders, but have to use those of Deutsche Bundespost.

With regard to broadcasting telecommunications installations have only
a "service function"[20] which explains why a distinction could be made.

Terrestrial and other Delivery

21–25 Satellite broadcasting is regulated in the State Treaty for the Reorganisation
of Radio Matters (*Staatsvertrag zur Neuordnung des Rundfunkwesens*).

---

[17] 12 BVerfGE (Decisions of the Federal Constitutional Court), p. 205.
[18] *ibid.*, p. 25.
[19] 2 BVerfGE, pp. 25, 29; 12, pp. 2, 5.
[20] BVerfGE 12, p. 227.

According to another decision of the Federal Constitutional Court the use of direct broadcasting satellites belongs to the competencies of the Laender, but due to the specific nature of the task to all of them together as a joint competence.[21]

As far as broadcasting via cable or broadcasting programmes transmitted via telecommunication satellites are concerned they are also regarded as a subject matter under broadcasting law, *i.e.* a matter where the Laender are competent themselves.[22]

Regulatory Agencies

As far as telecommunications law is concerned the only regulatory agency is BMPT. As far as broadcasting is concerned the Laender have created *Landesmedienanstalten* (Media Agencies of the Laender) granting licences to private radio and tv stations and controlling their programmes. As to the ARD, ZDF and the so-called third programmes, which are also broadcast by ARD members, there are no regulatory agencies, but radio councils where all groups of society, in particular the political parties and the roman catholic as well as the protestant church, are represented and control to some degree the contents of the programmes.

21–26

5. REGULATION OF APPARATUS

Regulation of Apparatus, Supply and Installation

Under the present German telecommunications law everybody has the right to set up and operate type approved terminal equipment to the extent that telecommunications installations and transmissions are not disturbed (*cf.* section 2 of the FAG).

21–27

Type Approval and Testing Procedures

In the 1989 Telecommunications Installations Act 1989 it is stated that type approval will be granted, if it is assured that the transmission lines of DBP Telekom are neither harmed nor endangered by the connection or operation of the equipment to be approved. Furthermore, neither persons nor any other terminal equipment must be harmed or endangered by connecting or operating the equipment seeking type approval. There are different technical and operational standards which must be met. They are primarily destined to prevent any possible interference with radio installations. In order to ensure that service providers offering services to connect terminal equipment with the termination points of the telephone network are

21–28

---

[21] BVerfGE 73, pp. 118, 197.
[22] *cf.* Provisional Protocol of the Minister Presidents of the Laender of February 4, 1983.

reliable, such persons need a licence to render such services based on their technical and professional expertise.

The competent office for type approval is the Federal Office for Approvals in Telecommunications (*Bundesamt für Zulassungen in der Telekommunikation—BZT*). This office was set up in 1982 under the name of the Central Approvals Office for Telecommunications (*Zentralamt für Zulassungen im Fernmeldewesen—ZZF*). It was set up as an impartial approval authority with its headquarters at Saarbrücken. Prior to its creation there was a department for type approval within the Telecommunication Engineering Center (*Fernmeldetechnisches Zentralamt—FTZ*). Under the 1989 reform of telecommunications law the regulatory tasks were seperated from the operational ones. Whereas the ZZF was still a part of Deutsche Bundespost the BZT is nowadays an independent and impartial federal office under the direct supervision of the Federal Ministry of Post and Telecommunications.

21–29   Although the former abbreviations are still well known, any equipment marked "approved by FTZ or approved by ZZF" is no longer correct and marketable any more. Only a person having obtained a BZT type approval has the right to market such telecommunications equipment in Germany. With the mutual recognition of type approval procedures the situation will be eased when the approval granted by one of the other type approval offices within the E.U. will suffice for the free circulation of the equipment within the E.U. In Germany this rule has been incorporated with the amendment of the FAG as part of the Postreform II.

6. Technical Standards

21–30   Besides the conventional networks using euipment with analogue technique there is the new Euro ISDN network requiring special equipment for voice communications and data transmission, personal computer connections, etc. Under Euro ISDN the minimum offer is 64 Kbit/s as transmission rate without any restriction and a 3.1 kHz a/b transmission service.

In the future the Euro ISDN service will feature the transmission of the number of the caller to the user called transmission of the calling number.

Datex P is the service for data exchange offered by DBP Telekom, a digital network for professional use. The customer may choose transmission speeds between 300 bit/s and 64,000 bit/s, and in the future also 1.92 Mbit/s. The fault rate is very small with 10^{-9}.

There is also the DATEX L service which is a line switched data transmission service. This network is interesting particularly for trade banks, insurance companies and credit card organisations. There are four speeds available; 300 bit/s, 2,400 bit/s, 4,800 bit/s and 9,600 bit/s. Futhermore, there is a teledat service for data transmission. Such data can be transformed directly within computer systems. The transmission speeds varies between 300 and 23,000 bit/s.

There are also many other services not described here which are part of a handbook of DBP Telekom published annually.[23]

7. Outlook

Germany has accepted the 1998 deadline for liberalising voice communications. The privatisation of DBP Telekom is a preparatory step to make the German operator fit for competition. But telecommunications markets requiring very costly investments are by their nature changing from national monopolies to regional or even woldwide oligopolies. The real test may come when the network monopoly will be completely abolished, if the cost for providing the necessary infrastructure will have considerably decreased by then. At present nobody can say if and when this will be the case, although the German Minister of Post and Telecommunications has publicly announced that he wishes to abolish the network monopoly as well as the voice monopoly by 1998.[24]

21–31

On March 27, 1995 the Minister of BMPT published the key elements of a Future Regulatory Framework for the Telecommunications Sector.[25] The Minister intends to enact a new telecommunications law as of January 1, 1998. Subject to very intensive political discussions the new law shall open the German telecommunications market to competition by a liberal licensing policy. No limits in numbers are intended unless technically unavoidable (*e.g.* in the case of limited frequencies). Any telecommunications operator rendering commercial telecommunications services to the public needs a licence to the extent the actual voice and network monopolies are concerned by the intended service. The requirement of a licence for mobile and satellite communications will remain in force. As any licence can be limited in its scope (*e.g.* territory covered), licences will be accessible also to small and medium enterprises. Market dominating operators will be subject to a universal service provision.

In some areas smaller liberalisation measures will be enacted prior to 1998, *e.g.* for corporate networks and cable networks.

Whether this ambitious programme can become law depends to a large degree on the willingness of German politicians of all political parties. Quo vadis Postreform III?

---

[23] Das TELEKOM-Buch 1993/1994.
[24] *cf. Frankfurter Allgemeine Zeitung*, September 12, 1994, pp. 15, 17.
[25] BMPT, *Key Elements of a Future Regulatory Framework for the Telecommunications Sector*, March 27, 1995.

CHAPTER 22

Hong Kong

By Anne Hurley, Minter Ellison, Sydney

1. HISTORY OF DEVELOPMENT OF TELECOMMUNICATIONS IN HONG KONG

Industry Structure

Hong Kong is a major business and financial centre with a sophisticated   **22–01**
and well-developed telecommunications infrastructure. The Hong Kong
Government is committed to a policy which has, as one of its major objectives, the development of Hong Kong as the pre-eminent communications
hub for the Asian region both now and into the twenty-first century. Competition is seen to be the means to achieve the Government's policy objectives and the Government is thus pursuing the introduction of competition
in those industry sectors which are and have always been the preserve of
a monopolist provider and is seeking to introduce more competition in
those industry sectors which are already competitive.

In Hong Kong, the supply of domestic telephony services and the supply
of international services are separate. Until July 1995, local fixed public
voice services were provided under an exclusive franchise granted to the
Hong Kong Telephone Company Limited (HKTC) by the Telephone
Ordinance (Cap. 269). With the expiry of the franchise, four carriers
licensed under the Telecommunications Ordinance (Cap. 106)—HKTC,
New T&T Hong Kong Ltd, Hutchison Communications Ltd and New
World Telephone Ltd—will compete in the provision of fixed local services.
International services are provided under a licence granted to Hong Kong
Telecom International Limited (HKTI) which contains a schedule setting
out the services which HKTI has the exclusive right to provide. The exclusive licence will continue in force until its expiry in September 2006. In other
industry sectors—value added services (both domestic and international),
mobiles, paging, CT2, equipment supply—there is competition among
providers.

Both HKTC and HKTI are subsidiaries of a holding company, Hong-kong Telecommunications Ltd (HKTI). A unique feature of the monopoly provider in Hong Kong is that it is not a government-owned utility but a privately-owned and publicly-listed company. The shares in HKT are held by U.K.-based Cable & Wireless plc (approximately 57.5 per cent) and China International Trust & Investment Corporation Hong Kong (Holdings) Ltd (CITIC) (approximately 17.5 per cent) with the balance being held by public shareholders.

22–02   Domestic and international services have always been supplied by different entities in Hong Kong. Domestic services have been provided by HKTC since that company was formed in 1925, taking over the operations of the China and Japan Telephone and Electric Company. HKTC was initially granted an exclusive 50-year concession to supply and operate the domestic telephone service in Hong Kong, the concession being renewed for a further 20 years in 1975. Hong Kong's international services date from 1871 when a forerunner to Cable & Wireless plc laid the first submarine telegraph cable into Hong Kong. From 1937 to 1981 international services were provided by subsidiaries of Cable & Wireless plc, but in 1981 a separate Hong Kong subsidiary Cable & Wireless (Hong Kong) Ltd, in which the Hong Kong Government had a 20 per cent shareholding, was established to provide the international services (following the British Government's decision to privatise Cable and Wireless plc). In 1987 Cable & Wireless plc and HKTC merged; Cable & Wireless (Hong Kong) Ltd changed its name to HKTI, and the Hong Kong Government subsequently in 1990 sold its interest in HKT and Cable & Wireless plc sold 20 per cent to CITIC. Although the Hong Kong Government no longer is a shareholder, it continues to appoint two directors to the boards of HKTC and HKTI respectively.

The historic separation of domestic and international services produced an agreement, dating from 1949 with subsequent amendments, between HKTC and HKTI to share the international revenue. As it presently stands, HKTC receives 40 per cent of the net revenue for long haul traffic and 60 per cent for short haul.

22–03   Hong Kong has one of the highest telephone penetration rates in the world with a teledensity of approximately 50 per cent. The network developed by HKT has, since 1993, been fully digitalised and fibre optic cable is being incorporated in the network. Local calls in Hong Kong are free of charge and HKT claims to have some of the lowest international tariffs. HKT's operation in Hong Kong has been highly successful with a reported annual profit in March 1994 of more than HK$8 billion, which accounts for approximately 60 per cent of Cable & Wireless plc's worldwide profits.

22–04   There is a big demand in Hong Kong for mobile telecommunications services. Hong Kong has the highest usage of mobile phones in Asia, with more than four per cent of the population being mobile services customers. Cellular radio services were introduced into Hong Kong in 1984 following an arbitration decision that a public mobile radio telephone service did not come within the exclusive franchise granted to HKT under the Telephone Ordinance. There are currently four licensed operators of cellular networks in Hong Kong: Hongkong Telecom CSL Ltd (a subsidiary of HKT), Hutchison Telephone Company Ltd, Pacific Link Communications Ltd and SmarTone Mobile Communications Ltd. The Hong Kong Government has not put restrictions on the type of cellular technology used and a range of analogue and digital technologies are available, although the trend has been towards the adoption of digital technologies, particularly GSM. The prices for cellular handsets in Hong Kong are generally higher than prices elsewhere in the world, the reason for the higher prices being seen to be the huge and unsatisfied demand for cellular services. In November 1994 the Telecommunications Authority issued a statement outlining revisions to the licensing of mobile services. The statement invited applications for up to six licences for personal communications services (PCS) licences and up to four cordless access services (CAS) licences. PCS is defined as the category of services most closely related to the European personal communications networks or person communications services in the USA. CAS is defined as low mobility, cordless services which primarily provide public services for access to fixed telecommunications networks. The TA will not mandate which technology must be used in the provision of mobile services although analogue technology will not be permitted.

A huge success story in mobile telecommunications in Hong Kong is the second generation cordless telephone (CT2). This service provides less coverage and mobility than a cellular phone but is significantly cheaper. Originally, CT2 was only available for outgoing calls and was generally used in conjunction with a radio paging system, however the technology has been developed so that a limited two-way service will soon be available. Four CT2 operators have been licensed: Chevalier (Telepoint) Ltd, Hutchison Paging Ltd, Pacific Telelink Ltd and Hong Kong Callpoint Ltd (a subsidiary of HKT which, although holding a licence, is not yet in operation). CT2 has only been in operation in Hong Kong since 1992 but there is presently a total of more than 160,000 customers.

Radio paging in Hong Kong has always been competitive with more than **22–05** 30 licensed operators and in excess of one million customers (a population penetration rate of approximately 16 per cent). The average annual growth rate in the number of paging subscribers over the last 10 years was over 21 per cent.

There is one licensed cable subscription television provider in Hong Kong, Wharf Cable, which has been in operation since October 1993 with a three-year exclusivity period. In November 1993 HKT applied to the Hong Kong Government for a cable television licence to become effective when Wharf Cable's exclusivity expires in 1996.

One Hong Kong company, Asia Satellite Telecommunications Company Limited (AsiaSat) has so far been licensed in Hong Kong to launch and operate a satellite, the AsiaSat-1 satellite, with coverage over Asian countries. AsiaSat intends to launch AsiaSat-2 in early 1995 and another satellite operator, APT Satellite Company Ltd (APT), also has plans to launch two satellites.

22–06   One licence has been granted, to Hutchvision Hong Kong Ltd (Hutchvision) which is the operator of the STAR TV services, for the uplinking of programmes for satellite broadcasting. STAR TV's programmes are transmitted to Hong Kong and the Asian region via AsiaSatl. Originally restricted to free-to-air satellite television broadcasting, Hutchvision's licence was amended in November 1993 to provide for satellite subscription television programmes in Hong Kong and the region, with the restriction that throughout the three-year exclusivity period of Wharf Cable, its subscription satellite television programmes will have to be distributed via Wharf's network.

The supply of a Satellite Master Antenna Television (SMATV) System for the reception and distribution of satellite television broadcasting services is on a competitive basis. As at the end of February 1994, there were 72 SMATV operators licensed.

Pursuant to the exclusive rights conferred in respect of designated external services, only HKTI may establish earth stations for external telecommunications. The earth station established by HKTI on Hong Kong Island operates nine antenna dishes providing access to INTELSAT satellites over the Pacific Ocean and the Indian Ocean, Palapa B2-P and AsiaSat-1. An earth station planned for the future will allow access to INMARSAT satellites over the Pacific Ocean and the Indian Ocean. There are no regulatory restrictions over the satellites to be accessed by HKTI in the provision of its services. Government policy has dictated some exceptions to the exclusivity, which is further discussed below.

Institutional Structure

22–07   Telecommunications policy in Hong Kong is within the jurisdiction of the Economic Services Branch ("ESB") of the Government Secretariat. The Telecommunications Authority, a public officer appointed by the Governor pursuant to the Telecommunication Ordinance (Cap. 106), is responsible for the administration of telecommunications in Hong Kong as it is encapsulated in the Telecommunication Ordinance (Cap. 106), the Telephone Ordinance (Cap. 269), the Outer Space Act 1986 (U.K. legislation extended to Hong Kong covering the launch and operation of satellites as space objects) and related subsidiary legislation and instruments.

The Telecommunications Authority's responsibilities include the promotion of fair and effective competition in the telecommunications sector, the management and administration of the frequency spectrum, the development of technical standards and equipment testing and the protection of telecommunication consumer interests.

Since July 1, 1993 the Telecommunications Authority has been assisted by the Office of the Telecommunications Authority (OFTA). Prior to the establishment of OFTA, telecommunications administration was the responsibility of a branch of the Postmaster General's department. OFTA was set up in recognition of the fact that an independent regulatory body was necessary as a precondition to the achievement of the Government's policy of increased competition in telecommunications in Hong Kong.

Unlike most other countries which have de-regulated telecommunica-   22–08
tions, Hong Kong does not have an anti-trust or fair trading regime. Anti-competitive safeguards in licences are thus within the responsibility of the Telecommunications Authority to administer. Licences for the provision of public telecommunications services contain a condition prohibiting anti-competitive arrangements. The new FTNS licence has a more developed regime of anti-competitive safeguards, prohibiting horizontal and vertical anti-competitive conduct, abuse of a dominant position and price discrimination by a dominant carrier.

Regulation of broadcasting is separate from the regulation of telecommunications, although ESB and the Telecommunications Authority have input on the technical aspects of broadcasting and the Telecommunications Authority is a member of the Broadcasting Authority for this reason. Broadcasting policy in Hong Kong is formulated by the Recreation and Culture Branch (RCB) of the Government Secretariat. The Broadcasting Authority, a statutory body, assists RCB in the development of policy and the administration of the Television Ordinance and the Television and Entertainment Licensing Authority (TELA) monitors programme standards.

2. Existing Legislative and Regulatory Environment

The Telephone Ordinance (Cap. 269)

The Telephone Ordinance (Cap. 269) conferred on HKTC the exclusive   22–09
right to supply domestic telephone services until June 30, 1995 and imposed obligations in the provision of the services—the universal service obligation, tariff and interconnection obligations. Other provisions dealt with by the Ordinance included the control by the Telecommunications Authority of Hong Kong's numbering plan. Additional obligations as to charges and revision of charges were imposed on HKTC by the Telephone Regulation, which also gave legislative effect to an agreement reached between HKTC and the Hong Kong Government for a price cap mechanism until 1996 to limit price increases of a basket of HKTC services to CPI minus four.

With the introduction of competition as from July 1, 1995, HKTC's   22–10
provision of telecommunications services is regulated under the FTNS licence granted under the Telecommunications Ordinance. The universal service obligation is now imposed on HKTC by the licence. The Telephone Ordinance has been largely repealed, leaving only two operative sections. Section 3 confers on the Telecommunications Authority extensive powers

in respect of Hong Kong's numbering plan and section 4 provides for retention of the price control arrangements already applicable to HKTC.

The Fixed Telephone Network Services licence

22–11   The provision of local fixed services in Hong Kong is largely regulated by the conditions imposed on licensees under the FTNS licence. In addition to the provisions of the Telecommunications Ordinance, the general conditions of the FTNS licence impose on each licensee, amongst other things, obligations as to:

- control of interference and obstruction;
- provision of service;
- interconnection;
- compliance with Hong Kong's numbering plan;
- fair market conduct;
- accounting procedures; and
- tariffs.

The Telecommunication Ordinance (Cap. 106)

22–12   The Telecommunication Ordinance (Cap. 106), and its subsidiary legislation, is the principal legislative instrument for regulating the provision and operation of telecommunication systems, services and equipment in Hong Kong (other than public telephonic communication). The Ordinance was enacted in 1963 and although some legislative amendments were made in 1993 to enable domestic competition there has been no major change to the Ordinance since its enactment.

The Telecommunication Ordinance establishes the licensing regime for telecommunications in Hong Kong. Section 8 provides that:

"(1): Save under and in accordance with a licence granted by the Governor in Council or with the appropriate licence granted by the Authority, no person shall in Hong Kong or on board any ship or aircraft that is registered in Hong Kong—

(a) establish or maintain any means of telecommunication;
(b) possess or use any apparatus for radiocommunication or any apparatus of any kind that generates and emits radio waves notwithstanding that the apparatus is not intended for radiocommunication;
(c) deal in the course of trade or business in apparatus or material for radiocommunication or in any component part of any such apparatus or in apparatus of any kind that generates and emits radio waves whether or not the apparatus is intended, or capable or being used, for radiocommunication; or

(d) demonstrate, with a view to sale in the course of trade or business, any apparatus or material for radiocommunication."

The words "means of telecommunication" in section 8(1)(a) of the   **22–13**
Ordinance are taken to refer to telecommunication installations. The
Ordinance defines "telecommunication installation" as meaning any apparatus or equipment maintained for or in connection with a telecommunication service. A "telecommunication service" is defined to mean the provision of facilities for use by members of the public or by any person for the transmission or reception of messages or the provision on loan, lease or hire to members of the public or to any person of apparatus for telecommunication either within Hong Kong or with any place outside Hong Kong.

The Governor in Council is empowered to exempt any person or class of persons from the requirement to obtain a licence. The owner or user of apparatus connected to the public telephone system is exempted from the requirement to obtain a licence for the apparatus when it is not being used to provide a service to the public and similar exemptions apply in respect of apparatus used by the customers of other licensees under the Ordinance.

Licences under the Telecommunication Ordinance may be granted by   **22–14**
either the Governor-in-Council or by the Telecommunications Authority.
The Authority may grant any of the licences specified in Part 1 of Schedule 1 to the Telecommunication Regulations. Other licences may be granted by the Governor-in-Council. There are currently in force two licences granted by the Governor-in-Council conferring exclusive rights: the licence granted to HKTI in 1981 in respect of international services, which is further discussed below, and an exclusive licence granted to Chubb Electronics (Hong Kong) Limited (Chubb) to provide a fire alarm transmission system for relaying fire alarm signals from specified premises to the fire services department's control centre. In addition, in December 1990 Hutchvision Hong Kong Limited was granted a non-exclusive Satellite Television Uplink and Downlink licence by the Governor-in-Council as an exception to the exclusive rights with respect to external television and voice programme transmission services conferred on HKTI under its licence.

The licences specified in Part 1 of Schedule 1 to the Telecommunication Regulations for the provision of public telecommunications services are all non-exclusive licences. They are the Fixed Telecommunications Network Services (FTNS) licence, the Public Non-Exclusive Telecommunications Service (PNETS) licence, the Public Radiocommunication Service (PRS) licence and the Radio Paging System licence.

There are many other licences provided for in the Schedule, for example   **22–15**
the Closed Circuit Television licence, the Satellite Master Antenna Television licence, Radio Dealers licence, and various station licences (broadcast relay stations, ship stations, aircraft stations). All the licences are in a prescribed form containing a number of general conditions. The Telecommunications Authority may add further conditions to a licence. The fees for the licences which may be granted by the Authority are set out in the Telecom-

munications Regulations and the licence is valid for so long as it is specified in the Regulations.

A licence granted by either the Governor-in-Council or the Authority may be cancelled or withdrawn in the event of a contravention of the licence conditions or Ordinance. In addition, the Governor-in-Council may cancel or suspend a licence at any time if he considers that the public interest so requires.

A contravention of the requirement to obtain a licence renders a person liable to a fine of $5,000 and imprisonment for two years on summary conviction, and to a fine of $10,000 and imprisonment for five years on conviction on indictment.

22–16   Aside from establishing the licensing regime for telecommunications in Hong Kong, the Telecommunication Ordinance contains the following provisions for the control of telecommunications:

- the appointment of the Telecommunications Authority;
- control of the import and export of radiocommunication transmitting apparatus;
- controls on the use of radiocommunication apparatus on a vessel in Hong Kong waters or aircraft in Hong Kong;
- the power for the Government to take control of telecommunications stations in an emergency;
- the placement and installation of telecommunication lines and facilities;
- offences and penalties for contraventions of the provisions of the Ordinance, including conferring certain search and seizure provisions on the Telecommunications Authority in respect of offences under the Ordinance;
- provisions empowering the Broadcasting Authority to grant licences for sound broadcasting; and
- interconnection determinations and directions.

The amendments dealing with interconnection were made in 1993 to assist the introduction of competition in fixed services. The Telecommunications Authority is empowered to determine the terms and conditions of interconnection in respect of specified types of interconnection. Before making the determination the Authority must be satisfied that the parties to an interconnection arrangement have been afforded reasonable opportunity to make representations to him as to why a determination should not be made. The Authority is also empowered to issue directions in respect of interconnection, as well as compliance with licence conditions and Ordinance provisions, and to impose financial penalties for non-compliance with a direction.

The Licence for International Services

22–17   Pursuant to the powers conferred by the Telecommunication Ordinance for the granting of licences, the Governor-in-Council in 1981 issued a licence to

Cable & Wireless (Hong Kong) Ltd (now HKTI) to provide external services for 25 years. The licence expires on September 30, 2006.

The services which HKTI is licenced to provide are set out in two schedules to the licence: Schedule 1 confers the exclusive right to provide the circuits and carry on the services specified and Schedule 2 confers the non-exclusive right to carry on the services specified.

Schedule 1 confers exclusivity in the following: **22–18**

(a) Circuits by radio for the provision of external public telecommunications services.
(b) The operation of circuits by submarine cable for the provision of external public telecommunications services.
(c) External and internal Public Telegram Services.
(d) External and internal Public Telex Service.
(e) External public telephone services to subscribers to the Public Switched Telephone Network by radio, submarine cable and such overland cables as are authorised.
(f) External dedicated and leased telephone circuit services by radio, submarine cable and such overland cables as are authorised.
(g) External dedicated and leased circuits for—

    (i) telegraph
    (ii) data
    (iii) facsimile

(h) Hong Kong coast stations and coast earth stations of the Maritime Mobile Service and Maritime Mobile-Satellite Service.
(i) Hong Kong aeronautical stations of the Aeronautical Mobile Service and Aeronautical Mobile-Satellite Service for radiocommunications services between aircraft operating agencies and their aircraft in flight.
(j) International telecommunications services routed in transit via Hong Kong.
(k) Except to the extent that the Governor-in-Council may from time to time otherwise in writing direct, external television and voice programme transmission services to and from Hong Kong.
(l) Except to the extent that the Governor-in-Council may from time to time otherwise in writing direct, external private circuits between fixed points, to and from Hong Kong.
(m) External public data circuits for—

    (i) circuits-switched data
    (ii) facsimile and/or document transfer service
    (iii) packet-switched data.

Schedule 2 of Non-Exclusive Services includes all public telecommunica-   **22–19**
tions services not included in Schedule 1, or covered by any other exclusive licence or other right or franchise, as well as multidestination broadcast press radio reception and transmission services for press agencies.

The licence imposes on HKTI the obligation to provide services in a manner satisfactory to the Telecommunications Authority and to have the Authority's approval for its tariffs. In addition, HKTI has an agreement with the Government to reduce its IDD tariffs by a weighted average of 12 per cent (in nominal terms) over the period 1993–1996.

Other Regulatory Instruments

22–20   Other instruments which regulate the provision of telecommunications services in Hong Kong are:

- Guidelines on the Application and Interpretation of the Competition Provisions of the FTNS Licence;
- Guidelines on the Application and Interpretation of the Interconnection Provisions of the FTNS Licence;
- Statements by the Telecommunications Authority: as at the date of writing, eight statements have been issued between March and June 1995 dealing with interconnection and related competition issues.

Telecommunications Policy

22–21   It has already been mentioned that HKT's exclusive right to provide public telephonic communication will not be renewed when it expires on June 30, 1995. Non-exclusive fixed telecommunications network services licences will confer non-exclusive rights on HKT, and three new licensees, to provide voice and non-voice services over fixed networks.

The Hong Kong Government has announced its intention to honour the licence given to HKT1 in 1981 in respect of international services. However, it has also stated that it intends to pursue opportunities for further liberalisation in the provision of international telecommunications by, for example, widening resale opportunities.

Competition in the provision of mobile services is being increased. The Telecommunications Authority has stated that while technological developments are breaking down traditional distinctions between fixed and mobile services, the services will remain separate in Hong Kong for at least the next three years. One of the reasons for this approach is the recognition of the size of infrastructure investment required, particularly by the new FTNS licensees.

3. REGULATION OF SERVICES

Domestic Services

22–22   Public telecommunications services—fixed and mobile, voice and date, value-added—are open to competition. There is no legislative requirement that the applicant for a licence be a company registered in Hong Kong, although the Telecommunications Authority takes the view that it is desirable. There are no foreign ownership restrictions on licensees for public non-exclusive services.

For the provision of value-added services using licensed telecommunication service operators' networks, the value-added service provider currently needs a public non-exclusive telecommunications service (PNETS) licence. There is no limitation on the number of PNETS licences which may be issued, however, an applicant for such a licence is expected to provide certain information (such as financial capability and a detailed description of the service including technical details) and meet specified criteria before a licence will be granted. Charges for competitive services are not regulated.

For the provision of public non-exclusive services using radio, a public radiocommunications service licence (PRSL) is required, or, in the case of a public paging service, a radio paging system licence. For services whose operators must be limited because of spectrum constraints (*e.g.* cellular services), the Authority generally issues a *Gazette* notice to invite interested parties to submit applications by a set date. Guidelines for the submission of applications are generally issued, the applications are evaluated and the licences granted to the applicants considered most suitable by the Authority.

In respect of the provision of other types of public radiocommunication **22–23** services, the Authority has indicated that it is prepared to consider applications at any time for public radio paging services operating in the 279–281 Mhz band, community repeater (trunked radio) services operating in the VHF (Band III) and UHF (800 Mhz) bands, one-way data message services operating in the 922 MHz band and public mobile radio data services.

As with a PNETS licence, an applicant is required to provide information and meet criteria. There are more technical requirements for a PRSL licence relating to, for example, frequency band and channel width.

Where the value-added service provider or a PRSL operator interconnects with the public switched telephone network of HKTC, an interconnect charge as specified in the *Gazette* is payable to HKTC.

As mentioned earlier, HKTC is permitted to provide certain non-telephonic services on a competitive basis. Charges for those services are required to be specified by the Authority by order published in the *Gazette* and HKTC must not directly or indirectly impose a higher charge than that specified.

The supply of telecommunications equipment in Hong Kong is competitive with minimal licensing requirements but some technical requirements, as further discussed below.

Monopoly International Services

As already noted, HKTI has been granted exclusive rights to provide the **22–24** international services listed in Schedule 1 to its licence. With the development of new technologies and new services, questions of interpretation of HKTI's exclusivity are beginning to arise, that is, whether a new service falls within or outside the exclusivity conferred by Schedule 1. Examples of such services include call-back services, video-conferencing, video-telephony and mobile satellites. Consideration is currently being given to these issues.

The Governor-in-Council has made two exceptions to HKTI's exclusive rights in respect of satellite services:

- the Governor-in-Council has issued a licence to AsiaSat to operate an earth station for telemetry, tracking, control and monitoring of its satellites (TTC&M stations). Other satellite operators will be similarly licensed when their operations commence. Further, HKTI has agreed with the Hong Kong Government that earth stations for the provision of private circuits for intra-corporate or intra-organisational communications will be licensed under the "self-provision" licence currently being crafted;
- the Governor-in-Council has issued a licence to Hutchvision Hong Kong Ltd for the uplinking of STAR TV programmes. The Government's present policy is to consider the issue of further licences and to further Hong Kong as an uplinking centre for satellite television.

Competitive International Services

22–25   International services outside the exclusive rights granted to HKTI may be provided competitively. HKTI is also permitted to provide such competitive services subject to the terms of its licence, in particular the requirement to have the Authority's approval for charges for any services and not to charge more than the approved charge.

International value-added network services (IVANS) may be operated by any service provider licensed under a PNETS licence. The services may be operated over the public switched telecommunication networks of HKT (telephone, data, telex) or over international private leased circuits (IPLCs). In order to be operated over IPLCs between Hong Kong and another country, the other country must allow the services.

22–26   The scope of permitted IVANS excludes basic voice, telex and simple resale communications and includes:

(i) **Data communications:** code and/or format conversions, protocol conversion, store and forward, or store and retrieve. If value is added at one end, it need not be added at the other. A managed data network service (MDNS) comes within the scope of IVANS.

(ii) **Facsimile communications:** store and retrieve or store and forward. If value is added at one end, it need not be added at the other end.

(iii) **Voice communications:** value added voice service, which means any telecommunication service involving the conveyance of speech,

music or other sounds which does not amount to real-time telephonic conversation between humans.

Where access to IVANS by subscribers in Hong Kong is via the public
switched networks of HKT, a gazetted interconnect charge is payable by
the IVANS service provider. Where access is via dedicated circuits, the flat-
rate charges for the circuits are payable by the IVANS service provider and/
or its subscribers.

The provision of IPLCs is within the exclusive rights of HKTI, however,
legislation is currently being drafted to enable companies and organisations
to "self-provide" their own external circuits for intra-corporate traffic.

Satellite Services

In order to launch a satellite and operate it as a space object, a licence   22–27
under the Outer Space Act 1986 is required. A licencee is obliged to take
out adequate insurance to cover the liability of the governments of the U.K.
and Hong Kong arising from the outer space activities of the licensee. It is
intended to replace the U.K. legislation with Hong Kong legislation.

Any radio station on a space object registered in Hong Kong must be
licensed under the Telecommunications Ordinance.

As already noted, the provision of external telecommunications services
via satellite is within the exclusive right of HKTI, except to the extent that
AsiaSat has been licensed to operate an earth station in respect of its satellite AsiaSat1 and it is expected that other satellite operators will also be
so licensed. In addition, the self-provisioning legislation will cover satellite
communication.

The transmission of satellite television is also within the exclusive rights   22–28
of HKTI. The satellite television uplink and downlink licence, issued by
the Governor-General under the Telecommunications Ordinance, has been
introduced as an exception to that exclusivity. There is no restriction on
the number of uplinking licences, although currently only one has been
issued to Hutchvision Hong Kong Limited. A licensee under an uplinking
licence must not be more than 49 per cent foreign-owned. The uplinking
licence may impose some controls over the contents of the programmes
linked and may impose some language restrictions—for example, Hutchvision is presently not permitted to transmit Cantonese on two of its free-to-
air programmes. A programme supplier may enter into an arrangement
with HKTI to uplink its programmes, in which case the supplier will not
be required to obtain a licence as it will not be operating any telecommunication facility in Hong Kong.

To receive and distribute satellite television broadcasting services via a
SMATV systems requires the operator to take out a SMATV licence. There
are no foreign ownership restrictions and provision of SMATV systems is
competitive. There are financial and technical requirements imposed on a
licencee as well as restrictions on charges and the area that a SMATV
system may serve. In addition, the reception and distribution of satellite

broadcasting signals under a SMATV licence may not be used to distribute subscription-based satellite broadcasting signals, in order to protect the three-year exclusive franchise for cable subscription television broadcasting services granted to Wharf Cable from June 1, 1993.

Cable Television and Telecommunications

22–29 Cable television in Hong Kong was introduced in October 1993 when Wharf Cable began operation. There are no regulatory restrictions on cable companies carrying telephony and vice versa and current indications are that there will in future be direct competition on both fronts between at least Wharf Cable and HKT. Wharf Cable's licence confers a three-year exclusivity period. In November 1993 HKT applied to the Hong Kong Government for a cable television licence to become effective when Wharf Cable's exclusivity expires in 1996. In addition, HKT has announced its intention to trial video on demand over its network, a step which Wharf Cable sees as an infringement of its exclusivity for subscription television. Wharf Cable is originally delivering subscription television services through a multichannel multipoint distribution system (MMDS) while it rolls out its fibre-optic system, expected to be ready by 1995 to enable New T & T, a Wharf Cable sister company and new FTNS licensee, to use the network for competitive domestic telephony services.

4. REGULATION OF BROADCASTING

22–30 The regulation of broadcasting in Hong Kong is separate from the regulation of telecommunications, with broadcasting being the responsibility of the Recreation and Culture Branch (RCB) and the Broadcasting Authority and telecommunications being the responsibility of the Economic Services Branch (ESB) and the Telecommunications Authority.

The principal legislative instrument for the regulation of broadcasting is the Television Ordinance which provides for the licensing of commercial and subscription television and related matters such as programming, royalties, licensee restrictions and technical conditions. In addition, the regulation of sound broadcasting is rather anomalously found in Part IIIA of the Telecommunications Ordinance. The Television Ordinance was enacted in 1964 and has been amended as needed to cover new situations such as the introduction of subscription television.

There are two free-to-air terrestrial television broadcasters in Hong Kong—Asia Television Limited (ATV) and Television Broadcasts Limited (TVB) both of which provide a Cantonese channel and an English channel. There is one satellite subscription television broadcaster (STAR TV), which currently also provides free-to-air satellite broadcasting, and one cable subscription television broadcaster (Wharf). The terrestrial broadcasters and Wharf are licensed under the Television Ordinance. Hutchvision Hong

Kong Limited, which operates STAR TV, is licensed under the Telecommunications Ordinance to uplink STAR TV's programmes. Until Wharf's three-year exclusivity expires in 1996, STAR TV is required to provide its subscription satellite television programmes over Wharf's network.

Hutchvision's uplinking licence contains broadcasting controls over programme content and language restrictions because the uplinking of television programmes is classified for regulatory purposes as "broadcasting" services. For other telecommunications services which are "genuine" telecommunications services, broadcasting controls are not appropriate. The present policy is to draw a distinction, for regulatory purposes, between a satellite broadcasting service and a satellite telecommunications service on the following criteria: **22–31**

- if the uplinked programmes are unencrypted, and if the programme originator has no objection to the programmes being received by any member of the general public, the uplinking is a "broadcasting" service;
- where the uplinked programmes are encrypted, the uplinking is a "broadcasting service if decoders are made available to any member of the general public. If the decoders are made available to broadcasters or cable television headends only, the uplinking is a "telecommunications" service.

A licensee under the Television Ordinance is deemed to be a licensee under the Telecommunications Ordinance for the purposes of establishing and operating a television broadcasting service and associated telecommunication services and is authorised to possess or use radiocommunication equipment in connection with the operation of the broadcasting or associated service. The broadcaster is required to comply with the provisions of the Telecommunications Ordinance relating to the establishment and operation of the television broadcasting service and associated telecommunications services.

A licensee under the Television Ordinance must be a company formed   **22–32** and registered under Hong Kong's Companies Ordinance (Cap. 32) and must comply with conditions as to residence of its directors. The rules on cross-ownership vary as between the terrestrial broadcasters and the subscription broadcasters. Subject to the approval of the Broadcasting Authority, the terrestrial broadcasters are limited to broadcasting-relating investments and ownership in certain broadcasting-related companies is restricted to 15 per cent. There are no similar restrictions on the subscription broadcasters and no general limitations on cross-media ownership of newspapers or other media. The royalties payable to the Government also vary as between the terrestrial broadcasters and the subscription broadcasters, with the latter paying an additional subscription royalty.

The Television Ordinance makes provision for technical aspects of the transmission of broadcasting. The Telecommunications Authority is given the role of advising the Broadcasting Authority in respect of technical standards on the basis of which the Broadcasting Authority may issue Codes of Practice. In addition, the Telecommunications Authority is empowered to direct a licensee in respect of the provision of equipment and the sharing of facilities with other licensees.

22–33   The traditional separation between broadcasting and telecommunications based on the content/conveyance distinction is being challenged in Hong Kong by HKTC's proposal to offer a dial-up video-on-demand service over its telecommunications network. As the legislative framework currently stands, such a service is a "means of telecommunication" requiring licensing or exemption under the Telecommunications Ordinance. It does not come within the definitions of broadcasting in the Television Ordinance such as to require licensing under the Television Ordinance and bringing the content under regulatory control. The appropriate means of regulation of video-on-demand is currently an issue being debated in Hong Kong.

5. Regulation of Apparatus

22–34   The supply of telecommunications equipment in Hong Kong is competitive. There are no regulatory restrictions on network operators also supplying customer equipment, such as mobile handsets, except to the extent that anti-competitive arrangements are prohibited by a condition in the relevant licences conferring the right to offer public non-exclusive telecommunications services. HKTC's charges for apparatus and leased circuits are subject to the tariffing obligations under the Telephone Ordinance.

A licence is not required to import or sell terminal equipment such as PABX or facsimile equipment, but the equipment is required to meet technical specifications which aim to ensure that there are no electrical hazards to people and no technical harm to the networks to which they are connected.

The Telecommunications Ordinance requires a licence to deal in apparatus or material for radiocommunication, the relevant licence being the radio dealers licence. Radiocommunication equipment for use in Hong Kong is also required to be type-approved or type-accepted by the Telecommunications Authority.

6. Technical Standards

22–35   The Telecommunication Regulations makes provision for the Authority to make tests and measurements in respect of electrical and radiated interference and for the Authority at any time to make tests and measurements of any apparatus for telecommunication to determine whether it complies with any requirement applicable to it under the regulations or the conditions of any licence held. The Telecommunication (Control of Interference) Regulations also make provision in respect of the design, construction, and

use of designated apparatus for technical matters designed to control interference with a telecommunication system.

The Authority has issued a number of performance and safety specifications, including in relation to subscription television, subscriber equipment
connected to the PSTN and leased circuits, and radio equipment.

CHAPTER 23

Italy

By Livia Magrone Furlotti, Magrone e Ardito, Rome

---

1. HISTORY OF DEVELOPMENT OF TELECOMMUNICATIONS

Institutional Structure

At the beginning of this century, the liberal theories which, until then, had   **23–01**
governed the European states, gave way to state intervention and planning.
All public requirements, *i.e.* energy, transport (both by rail and road), post,
telecommunications, etc., were supposedly better served by the state than
by the private sector. Hence, the creation and organisation of monopolies
which will last until the end of this century, and which was sanctioned by
Article 43 of the Italian Constitution, enacted in 1948. This Article reads
as follows: "For reasons of public utility, the law can reserve from origin
or transfer, by means of expropriation and with due consideration, to the
state, to public agencies, to communities of workers or users, such enterprises or categories of enterprises that provide essential public services or
are connected with energy sources or monopolistic situations and that have
a character of pre-eminent general interest".

Under the pressure exerted by the E.U. Commission towards liberalisation of the market, we are now witnessing the counter-phenomenon of
the breaking down of all non-essential state monopolies.

Background to Legislation Presently in Force

In 1903,[1] the Italian Government was authorised to establish and operate   **23–02**
the telephone facilities for public service. The same year,[2] the first unified
text of telephone service regulations was enacted. It provided for the public

---

[1] Law no. 32 of February 15, 1903.
[2] Royal Decree of May 3, 1903.

telephone service to be dealt with on the basis of franchises ("concessions") granted by the State to applicants.

By means of a "concession", the State grants to a company, by governmental decree, the right to supply the public service on its behalf; the respective undertakings of the State and franchisee ("concessionaire") are governed by an agreement enclosed with the decree; the consideration due to the state consists of a percentage of the yearly turnover of the concessionaire company.

In 1907,[3] the Italian State revoked the existing concessions and assumed direct operation of the public service.

The first organised set of regulations concerning the entire telecommunications sector was contained in the Postal Code of 1936,[4] appealed in 1973 (see below, para. 23-05).

Article 2 of the Code provided for the exclusive right of the State to supply all telecommunications services to the public. The scope of the telecommunications services is defined in the article covering telegraph, telephone, broadcasting and optical telecomunications.

As regards the network and connected terminal equipment, the State monopoly is expressly provided for by Article 166 of the Postal Code, which submitted the establishment of telecommunications plants to concession by the State. Any disregard to the prohibition to establish telecommunications plants or equipment without either an authorisation for private connections, or a concession for services to be rendered to the general public is punishable with a criminal fine, according to Article 178 of the Code.

Industry Structure

23-03   Although having monopolised the service right from the beginning, the Italian State deals *directly* with only a part of these services (via the Ministry of Post and Telecommunications) and namely with a) telegraph; b) intercontinental and international telephone services; c) a minor part of the domestic telephone service. Domestic telephone and telegraph services effected by means of telephone transmission, have been provided to the community since 1925 by companies entrusted with a so-called "concession".[5]

The international and intercontinental connections by submarine cables are dealt with by Italcable, on the basis of a grant issued by Royal Legislative Decree and according to an agreement signed on August 6, 1935. The broadcasting is dealt with by *Italo Radio—Società Italiana per i Servizi Radioelettrici*—on a grant issued by Royal Legislative Decree, transformed

---

[3] Law no. 506 of July 15, 1907.
[4] Enacted by Royal Decree, February 27, 1936, no. 645.
[5] For this reason, and since 1925, domestic territory has been divided into five areas, each served by a company running the service on a special grant issued by Royal Decree, and according to an agreement drawn up by the Ministry of Post and Telecommunications. The companies are S.T.I.P.E.L. (Piemonte and Lombardia); T.E.L.V.E. (Tre Venezie); T.I.M.O. (Emilia, Marche, Umbria, Abruzzo & Molise); T.E.T.I. (Luguria, Toscana, Lazio & Sardegna) and S.E.T. (Southern Italy & Sicily).

by Parliament into Law no. 935 on April 4, 1935, and according to an agreement executed with the Ministry of Post and Telecommunications. Nautical telecommunications are granted to yet another company, namely SIRM.

The concession agreements set the standard of service and responsibilities imposed upon the company in view of expansion of service, including the building up of the local network. Royalties to the State are provided for in all agreements.

As for the State, a special agency belonging to the Ministry of Post, with separate assets and balance sheet, *Azienda di Stato per i Servizi Telefonici* ("National telephone service company") deals with the international and intercontinental telephone services, as well as with the telegram service not provided for by means of telephone transmission.

The telephone network—domestic, international and intercontinental— is built up and partly owned by the State and partly by "concessionaires".[6]

In 1964, the five concessions were discontinued and by Presidential Decree no. 1594, dated October 26, 1964, and a subsequent Decree dated March 6, 1968 regarding data service, a new concession was granted to SIP, on an exclusive basis for the whole of the Italian territory, with domestic service as an objective.[7]   23–04

On the same date and by the same Presidential Decree no. 497, dated March 6, 1968, Italcable (former concessionnaire for submarine cables— see above—belonging to the same group as SIP) was granted an exclusive right to provide part of the international telephone, telegraph, telex and data transmission services, as well as handle the entire intercontinental network.[8]

By Presidential Decree no. 1130 of February 12, 1965, the establishment and operation of satellite connections was granted to Telespazio, a company belonging to the same group as SIP.

In 1957,[9] the existing concession of telegraph services to the press with

---

[6] In 1957, the above concessions granted to five telephone companies, S.T.I.P.E.L., TEL.VE., T.I.M.O. S.E.T. and TE.TI. were renewed by Presidential Decrees dated respectively December 14, 1957 nos. 1405, 1406, 1407, 1409 for the first four and December 28, 1957 no. 1408 for the last one.

[7] Presidential Decree no. 374 of June 6, 1957, allows for the granting of a concession to companies owned through majority shareholding by the State, with no need for tendering. SIP is owned by the holding company of its group, STET, which is in turn owned by I.R.I.—*Istituto per la Ricostruzione Industriale*— at the time a State agency—now a company owned by the State managing one sector of the State shareholding (other sectors are owned by two different agencies, E.N.I.—for petroleum and chemical, and E.F.I.M. for manufacturing).

[8] The remaining part of international services,

    (a) Telephone in European countries and also in Algeria, Cyprus, Lybia, Morocco, Tunisia and Turkey are served by the *Azienda di Stato per i Servizi Telefonici*;

    (b) Data transmission in Algeria, Egypt, Lybia, Tunisia and Turkey are also served by the *Azienda di Stato per i Servizi Telefonici*;

    (c) Telegraph services in Albania, Algeria, Austria, Vatican City, Egypt, France, Greece, Yugoslavia, Lybia, Lichtenstein, Malta, Monaco, San Marino, Switzerland, Tunisia and Turkey are served by the Ministry of Post and Telecommunications;

    (d) Telex services in Algeria, Egypt, Lybia, Tunisia and Turkey are also served by the Ministry of Post and Telecommunications.

[9] Presidential Decree of May 4, 1957.

Radiostampa (belonging to the STET group) was renewed. Nautical telecommunications remain with SIRM and Telemar.

The most recent renewal of SIP, Italcable and Telespazio concessions was enacted in 1984[10] to expire in 2004. The concessions are now to be transferred to the new company, Telecom Italia SpA, resulting (see below, para. 23–08) from the merger of the above companies and SIRM and IRITEL (see below, paras. 23–06 and 23–08).

2. EXISTING LEGISLATION

Principal Legislation

23–05   Article 1 of Postal Code of 1973[11] provides for exclusive reserve to the state for all telecommunication services. Article 183 sets the same reserve over all telecommunication facilities and equipment, unless a specific concession is granted. According to Article 1, paragraph 2, private broadcasting equipment may be established on the basis of mere authorisation, as well as sound and image transmission by cable.

More recent legislation provides for complete restructuring of the TOs system on the one side, and for the progressive opening-up of the market in the equipment, and services' sectors on the other side, in due compliance with E.C. Directives.

The Restructuring of the TOs

23–06   In 1992[12] the *Azienda di Stato per i Servizi Telefonici* was dissolved. A new company was subsequently established, IRITEL, fully owned by IRI (*Istituto per la Ricostruzione Industriale*).[13]

IRITEL was granted the concession formerly held by the Azienda and also took over the services formerly dealt with by the Ministry itself, initially for a period of one year and then for a further period until the end of 1994.[14]

The Regulator, in this case the Ministry, and the telecom operator, in this case IRITEL, then became two separate entities (according to duties laid down upon Member States by E.C. Directive 388/90).

The manufacturer, ITALTEL, the network organiser, S.I.R.T.I., and various companies operating in other sectors also belonged to the IRI group through the holding company STET.

A new organisation of the telecommunication concessionnaires was subsequently planned[15] which established that the Ministry of Post and Tele-

---

[10] Presidential Decree of August 13, 1984, no. 523.
[11] Presidential Decree of March 29, 1973, no. 156.
[12] Law no. 58 of January 29, 1992, Art. 1, para. 3.
[13] A State agency until the Governmental Decree of July 11, 1992, confirmed by Law dated August 8, 1992, transforms it into a company, fully owned by the Treasury.
[14] Art. 1, para. 1 of the above Law no. 58/92.
[15] Art. 1, paras. 4 to 6 of Law no. 58 of 92.

communications and the Ministry of State shareholdings[16] must submit to C.I.P.E. (*Comitato Interministeriale per la Programmazione Economica, i.e.* Interministerial Committee for Economic Planning) a proposal for the complete restructuring of the telecommunications sector. After adoption of the final decision by CIPE, and approval of same by the special Parliamentary Commission, IRI would be in charge of putting it into operation. The proposal was aimed at "homogeneity of functions; efficient and economical management; transparency in the relationship of services provided under monopoly and those provided on the free market; compliance with E.C. regulations and a guarantee of coordination between services".

Article 2 of Law no. 58/92 also provided for new criteria for setting the   **23–07** tariffs, which had to be established on the basis of a proposal brought forward to the Interministerial Committee for Prices (CIP) by the Ministry of Post and Telecommunications subsequent to consultation with the Ministry of Treasury, the Ministry of Budget, Economic Planning and of State Shareholding (now re-named as Ministry for State Shareholding privatisation) within six months from the date when the law comes into force.

The draft proposal of CIPE, approved by a resolution dated December 30, 1992, based on the above provisions, sketched the following programme:

By December 31, 1994, a single operator of the telecommunications system would be established, by merging all existing operative companies of the STET group. The State shareholding in the company acting as a single telecom operator was to be progressively reduced to 50 per cent of the share capital.[17]

Activities not to do with the establishment and management of the network and provision of services, such as manufacturing of telecom products and plants, as well as management of the shareholdings owned, would be conferred upon holding companies which would not interfere with the activities of the single operator.

The new operator, by merging industrial and commercial activities, would maximise operative efficiency, and would have a positive effect on the level of tariffs, which would be set according to a multi-annual programme based on a "price cap" principle. This will also take into account indexes of efficiency, productivity, quality of service and reduction of indebtedness.

The merger has now taken place: The board of directors of SIP, Italcable,   **23–08** Telespazio, Iritel and SIRM, companies which operated the public services, met in March 1994, and set the exchange rate of the Telecom Italia, the new company's name-shares in relation to shares of the merging companies. Shareholder meetings followed to adopt the formal deed of merger. The

---

[16] (which has now been abolished and replaced by the Ministry for State shareholding privatisation).

[17] The responsibility of the single operator comprises: definition and timing of developments in system architecture; design and development of the equipment as a whole; procurement of financial resources; management of network and user relations; relation with the supervisory authority.

merger was executed in August 1994, to be effective retroactively as of January 1, 1994. Telecom Italia is at present the sole concessionnaire for the operation of the telecommunications system as a whole.

According to CIPE's resolution, the share capital of Telecom Italia, is planned to be largely sold to the public. IRI will maintain a significant shareholding but will cease to be the majority shareholder.

Also the companies carrying out production and installation activities in such a sector (also indirectly owned by the State through STET), will merge into a sole sub-holding of STET (see paragraph 23–06, above) and, be supposedly, privatised.

As a result of Governmental Decree no. 487 of December 1, 1993, the postal administration has been transformed into a State agency, denominated *Ente Poste Italiane*. The Decree provides for the State agency to be transformed into a company by December 31, 1996, and for the majority of shares to be sold to the public, according to a programme issued by CIPE on the same day as the Decree.

In line with Article 2 of the Decree, the newly established State agency acts instead of, and provides the services formerly provided by, the postal administration.

In compliance with Article 11 of the Decree, the Ministry of Post and Telecommunications remains the supervisor of all services in the postal sector, postal banking, telematic and telecommunications. The Ministry was conferred the functions of planning and regulation as well as the power to direct, co-ordinate, supervise and control all telecommunications and postal activities according to the provisions laid down by the law. The Ministry is also in charge of Italy's representation within the E.U. and in all international sessions; of analysis and research, both under the international and domestic scope of action as regards economic, technical and juridical evolution in the relevant sector. As we shall see further on, (see, paragraph 23–11) the Italian Government had alternatively proposed that the regulatory functions be entrusted to an authority independent of the Ministry.

The Deregulation of Equipment and Services

Services

23–09   In 1992,[18] the Parliament entrusted the Government with the task of implementing Directive 388/90, providing for deregulation of services.

As shall be described later, until April 1995 the relevant draft legislative decree has not yet been approved or enacted, due to the efforts of the State monopolist to retain privileges not in compliance with the E.C. Directive to be implemented, on the one side, and to the counter-effort of the Antitrust Authority to have the draft modified according to provisions of the Directive, on the other side (see paragraph 23–21). We have therefore been left for years with a legislative gap which prevented all intervention by the

---

[18] Law no. 142 of February 19, 1992.

Government in the field of deregulated services, notwithstanding the acknowledgement by the same that the deregulation itself had taken place, as enacted by the same Directive, due to force of self-execution (see the Antitrust Authority decision in the *3C v. SIP* case: paragraph 23–13, note 31).

Equipment

We refer to the following section 5, "Regulation of Apparatus", as there **23–10** is no general act governing deregulation of equipment, but the same has taken place by means of single provisions issued from time to time, each governing a different sector.

Agencies of Regulation and Competition Control

Agencies of Regulation

Under Italian telecommunication law, attention has traditionally been **23–11** focused on the regulation relating to production standards and homologations of terminal equipment. In fact, until recent times, only terminal equipment had been deregulated.

Such a task was, and is presently being, carried out by regulatory authorities immersed in the Ministry of Post and Telecommunications (the "P.T. Ministry"). However, the separation required by E.C. law between the regulatory authorities and the State agency directly providing telecommunication services has been achieved by the dissolution of the State agency, *Azienda di Stato per i Servizi Telefonici* (ASST), the establishment of the state owned company IRITEL (see paragraph 23–06 above) and execution of a Concession Agreement with the latter providing for the assignment of all services performed by ASST to IRITEL until December 31, 1995.

The following authorities are now entrusted with the task of regulating the telecommunications sector:

(a) *Consiglio Superiore Tecnico delle Poste, delle Telecomunicazioni e dell'Automazione* (Technical Council of Post, Telecoms and Automation)[19]: It has several consultative and proposition tasks: to prepare projects for the development of telecom services; to draft technical regulations and standards; to draft general and special technical contract specifications; to advise the Minister of Post on revision or amendment to the National Regulatory Plan of Telecommunication; to provide preliminary approval of any new telecommunications system.

(b) *Comitato per le Telecomunicazioni* (Telecommunications Committee)[20]: Among other commitments, it is in charge of preliminary assessment of projects relating to the installation of terrestrial fixed radio communication stations.

---

[19] Established by Law no. 433 of 1948, amended by Law no. 693 of 1975.
[20] Established by Law no. 281 of 1940.

(c) *Ispettorato Generale delle Telecomunicazioni* (**General Inspectorate of Telecoms, hereinafter "I.G.T."**)[21]: I.G.T. plays a pivotal role in regulating apparatus supply as well as in type-approval and testing procedures of telecom terminal equipment, since it is specifically charged with the provision of technical regulations, is responsible for accrediting test laboratories[22] and for issuing type approval certificates.[23] Finally, I.G.T. has been appointed as the Italian Authority to participate in to the ACTE Committee provided for by Directive 91/263.[24]

(d) A new Telecommunication Authority is going to be set up as a body independent from any political or administrative influence, in order to comply with E.C. requirements, and will be established according to the criteria laid down by C.I.P.E.'s decision of April 2, 1993. A bill of March 15, 1995,[25] is still pending before Parliament which will regulate the structure and the operation of such an Authority.

An outline of the tasks which are likely to be entrusted to the Authority includes: Supervision of general terms and conditions of contracts applied by the TO's; supervision of the regularity and efficiency of the services to be rendered by the TO's; assuring transparency of tariffs in order to avoid cross-subsidisation between the telecom services to be supplied; regulating tariffs structure, by applying the "price Cap" method, in compliance with the criteria adopted by the C.I.P. (Interministerial Committee for Prices) decision of December 30, 1992[26]; and, finally, it will act as arbitrator to settle any dispute between the TO's and their customers.

Competition Control

23–12   Competition control is performed by two authorities:

(a) The *Autorità Garante della Concorrenza e del Mercato* (Authority Guaranteeing Competition and Market, hereinafter the "Antitrust Authority"), and

(b) The *Garante per la Radiodiffusione e l'Editoria* (Authority for Broadcasting and Press Activities, hereinafter the "Broadcasting Authority").

The Antitrust Authority

23–13   The Antitrust Authority has been set up by Article 10 of Law no. 287, of 1990 (the "Antitrust Law"), and it has the general jurisdiction to apply Antitrust Law to antitrust issues such as agreements between competitors, abuses of a dominant position by one or more undertakings, and concentrations between undertakings.[27] The Antitrust Authority is also empowered

---

[21] Created by Law no. 432 of 1948, as amended.
[22] Art. 9, para. 1 of Law no. 220 of 1988 and Art. 6 of Legislative Decree no. 519 of 1992.
[23] Art. 8 of Law no. 220 of 1988 and Art. 2, para. 6 of Legislative Decree no. 519 of 1992.
[24] Art. 14 of Legislative Decree no. 519 of 1992.
[25] Bill of the Senate no. 2331 of 1995.
[26] Paras. d–f of the C.I.P.E. decision above.
[27] Arts. 2, 3 and 6 of the Antitrust Law.

to carry on general inquiries into economic sectors where trends of trade, price movements, or other circumstances suggest that competition is being distorted.[28] In addition, the Authority is entitled to bring to the attention of Parliament or the Government any distortion in competition deriving from legislative measures, providing an opinion upon actions to be taken in order to remove or prevent such distortions.[29] Finally, the Authority may render its opinions upon any bill of law and of regulations insofar as competition is concerned.[30]

The Antitrust Authority has played an important role in the liberalisation process by intervening on several occasions in the telecommunication sector, and finally opening some markets to competition.[31]

---

[28] Art. 12, para. 2, of the Antitrust Law.
[29] Art. 21, paras. 2–3, of the Antitrust Law.
[30] Art. 22, para. 1, of the Antitrust Law.
[31] Here follows an outline of the main cases:

  (i) Terminal Equipment (TE): in *Ducati/SIP* (6[1993] Boll.23) a distributor of T.E. claimed that the exclusive purchasing and distributing contracts for cellular telephones between SIP and its franchisees infringed the Antitrust Law. The Authority held that SIP had a dominant position on the relevant market and that it was abusing the same by imposing an exclusivity, since it prevented access to the market to its competitors, and ordered SIP to cease the infringement by renouncing the exclusivity clauses. See also *Assistal/SIP* (25–26 [1994] Boll. 5).

  (ii) Telecommunication Services: in *3C Communications/SIP* (5[1992] Boll. 6), SIP refused to supply the telephone lines requested by 3C for connection to equipment (duly homologated) that allowed it to provide the service of payment of telephone calls by credit cards. The Authority held that such service was not included within the services reserved to SIP since it had been liberalised by Directive 90/388 (which, it is worth mentioning, was held by the Authority to be directly applicable), since it was different from voice telephony or any other non deregulated services, as well as by Art. 3.3 of Ministerial Decree of April 6, 1990, as a value added service. The Authority held that SIP had a dominant position on the relevant market since it had an exclusive right to provide telecom services (*i.e.*, the supply of telephone lines) and that the unjustified refusal of SIP constituted an abuse of dominant position since it related to an "essential facility" necessary for 3C to provide its telecommunications services. See also *Sign/STET-SIP* (5 [1995] Boll. 17); *Telsystem/SIP* (25–26 [1994] Boll. 11).

  (iii) Cellular Mobile Telephony: In a communication to the Government, at the end of 1991, the Authority stressed for the first time the need for opening up to competition v cellular mobile telephony. Later, in June 1993, together with the publication of the results of a general inquiry on the telecom sector (Suppl. to 14 [1993] Boll. 1), the Authority held that the provision of GSM service had to be immediately opened to the competition of at least another provider. In fact: (i) GSM was not included in the 1984 Convention between the State and SIP (reference is made to para. 23–28, below); (ii) SIP already had the monopoly of the TACS mobile system and by providing GSM under monopoly it could discriminate between professional users (to which GSM would be reserved in the near future) and less sophisticated users (with which would saturate TACS); (iii) SIP had already invested a substantial amount of money and had already connected a large part of the Italian territory; (iv) In spite of the fact that it was only authorised to test the service, SIP was already marketing GSM and was selecting the best customers. Therefore, the Authority held that, without any step to open GSM to competition, SIP would irremediably increase its competitive advantage over possible competitors (Case no. 1265, of 23.6.1993, in 14 [1993] Boll. 55). The challenge to SIP exclusivity was successful since, a few months later, the Government decided to award by bid a second concession for GSM service (the service has been now awarded through tendering procedure, to the Omnitel Pronto Italia consortium, lead by De Benedetti).
  Finally, some months later, the Authority held that SIP was abusing its dominant position: (i) by extending its monopoly on fixed telephony and on TACS mobile telephony also to the GSM system (Reference was made to the ECJ cases *Télémarketing* and *RTT v. GB-INNO*), and (ii) by preventing competitors from entering into the market, since it already installed repeating stations for 95 per cent of the territory and was already marketing the service to the public, even if not yet authorised (Case no. 1532 of October 28, 1993, in 32 [1993] Boll. 5).

  (iv) Data Transmission: the Authority has recently decided to proceed to a general inquiry, pursuant to Art. 12.2 of the Antitrust Law, on the sector of data transmission. In fact, in view of the liberalisation following to the expected implementation of the Directive 90/388, the Authority held that "the development of competition in such sector may be impeded by the conditions of access to the network" (Case no. 1586 of November 10, 1993, in 34 [1993] Boll.29).

The Broadcasting Authority

23–14   The Broadcasting Authority has been established in order to apply the special provisions laid down by Law nos. 416 of 1981, 67 of 1987 and 223 of 1990, with the aim of preventing concentrations between subjects operating in the broadcasting or press field, which create dominant positions. A dominant position is deemed to exist where, as the result of the transfer of a newspaper title or television channel as well as of the shares or goodwill of any company, the undertaking concerned:

(a) controls more than 20 per cent of the national newspapers;

(b) controls more than 50 per cent of the regional newspapers;

(c) controls more than 50 per cent of the interregional newspapers;

(d) becomes connected with companies publishing more than 30 per cent of the national newspapers;

(e) owns one concession for nation-wide TV broadcasting, where it already controls more than 16 per cent of the national newspapers;

(f) owns more than one concession for nation-wide TV broadcasting, where it already controls more than eight per cent of the national newspapers; or

(g) owns more than two concessions for nation-wide TV broadcasting, where it already controls any newspapers.

Any such transaction bringing about a dominant position is null and void and the Broadcasting Authority may impose a time limit of between six and 12 months to eliminate such position.

The Broadcasting Authority has also exclusive jurisdiction to apply the Antitrust Law to those subjects carrying out broadcasting or publishing activity.[32]

Telecommunication Installations, Services and Apparatus

Telecommunication Installations.

23–15   As quoted under paragraph 23–05 above, Article 183 of the Post and Telecommunications Code (the "Postal Code")[33] states as a general rule that "no one can install or operate telecommunication equipment without the necessary concession".

Concession Agreements for both telecommunication services and installations were made with SIP, Italcable, Telespazio in 1984 and with Iritel in 1992 (extended in 1993).

SIP is granted an exclusive concession for the nation-wide installation and operation of telecommunication equipment.[34] Italcable was granted an exclusive concession for the installation and operation of telecommunication equipment necessary for connecting mainly extra-European countries

---

[32] Art. 21.1 of the Antitrust Law.
[33] Enacted by Decree of the President of the Republic no. 156 of 1973.
[34] Art. 1 of D.P.R. no. 523 of 1984.

(see below paragraph 23–16).[35] Telespazio is granted an exclusive concession, within the limits and conditions set in the annexed agreement, for the provision of space communications, as well as for the installation and the operation of the systems fit for realising telecommunication links through satellites.[36] Finally, Iritel was granted an exclusive concession for the installation and operation of the telecommunication equipment formerly granted to A.S.S.T. and, after the expiry of the concessions granted to SIRM and Telemar, also for the installation and operation of the telecommunication equipment of the latter two companies.[37]

As stated in para. 23–04 above, the four companies merged in June 1994 to become Telecom Italia, to which the four concessions are to be transferred. Therefore, installation of telecommunication equipment is in principle reserved to Telecom Italia SpA. However, the limits and extent to such a rule, as well as the steps towards deregulation of installation will be dealt with in paragraph 23–52 below.

Telecommunication Services

Article 1, paragraph 1 of the Postal Code (see above paragraph 23–05)   **23–16**
grants to the State the exclusive right to provide telecommunication services, with the exception of those referred to in paragraph 2 of the same Article, *i.e.* "a) private repeater installations for domestic and international radio and TV broadcasting; b) local installation for sound and images transmission by cable". The provision of the latter services is subject to authorisation.

However, according to Article 4 of the Postal Code, as we have seen above, the State may entrust other entities, through concession, with the actual provision of the telecommunication services reserved to the same.

At present, the services are provided under concessions as described below. Such concessions have been enacted by Presidential Decree no. 523 dated August 13, 1984 as regards SIP and Italcable, and by Ministerial Decree dated December 29, 1992, amended and prorogated by Ministerial Decree dated December 22, 1993, as regards IRITEL.

According to the above concessions:

Until December 31, 1994, IRITEL was to provide the following services:

(a) National service for European countries and North African countries belonging to the Mediterranean basin;
(b) national intercity telephone service together with SIP;
(c) telex service for European countries and North African countries belonging to the Mediterraneon basin (formerly dealt with by the PT Administration);
(d) nautical radio communications, including operation of coastal radio stations.

---

[35] Art. 2 of D.P.R. no. 523 of 1984.
[36] Art. 3 of D.P.R. no. 523 of 1984.
[37] Art. 1 of Ministerial Decree of December 29, 1992, as amended by Decree of December 22, 1993.

SIP (controlled by STET, the majority shareholder) continues to operate the following services[38] to the public[39]:

(a) Local telephone service;
(b) intercity telephone over short and medium wave/distance and national long distance telephone service together with IRITEL;
(c) mobile radio services;
(d) competitive auxiliary services.[40]

Italcable (controlled by the majority shareholder, STET) was to operate the following[41]:

(a) Telephone, telex and data transmission services with countries outside Europe, apart from the North African countries belonging to the Mediterranean basin;
(b) international telegraph service worldwide apart from Italian neighbouring countries;
(c) auxiliary services to the above mentioned services.

23–17   Telespazio (controlled equally by STET, Italcable and RAI—*Radio Audizioni Italiane*), apart from establishing and operating the satellite connections, including earth stations:

(a) was to operate competitive auxiliary services.

SIRM (36 per cent owned by STET):

(a) was to deal non-exclusively with nautical telecommunications, sharing the operation of this service with Telemar, a company owned by shipowners; and
(b) provides competitive auxiliary services and equipment on the free market.

Therefore, as a general rule, under Italian law, the provision of telecom services is reserved to the State concessionaire.

However, as we shall see in more detail at paragraph 23–21 below, the general rule referred to does not extend to all services.

Telecommunication Apparatus

23–18   The production of telecommunication terminal equipment has never been regulated, since the provisions of Article 183 of the Postal Code, which grants the state or its concessionnaires the exclusive right for apparatus installation and operation, but does not extend to the manufacture of apparatus which is not part of the network. However, as we shall see later, at paragraph 23–52, only apparatus type-approved may be put on the market and installed.

---

[38] With the exception of television broadcasting, maritime mobile radio communication, telegrams, telexes as well as the services reserved to any other concessionnaire.
[39] Art. 1, para. 1, of D.P.R. 523 of 1984.
[40] The institution and operation by SIP of the new telecom services described above shall be established and regulated by express provision of the Administration (Art. 8 of SIP Concession).
[41] Art. 2 of D.P.R. no. 523 of 1984.

On the contrary, as we shall see both installation and marketing of tele-com terminal equipment has been reserved for the State until recent times, when it was been liberalised (paragraphs 23–52 to 23–56, below).

As far as the network is concerned, according to the above mentioned Article 183, the installation of telecommunication plants has been tradi-tionally reserved to the State, and operated partly by the same, and partly by its concessionaires. As discussed below, paragraphs 23–49 to 23–50, Directive 90/388—recently implemented by a legislative decreee dated March 19, 1995, no. 103—has deregulated the use of the public network for providing all telecom services different from voice telephony, and has allowed leasing of switched private networks from the TO's by private operators.

Licensing Regime, including Main Licence Categories

We shall first outline the general rules governing licensing in the Italian system and then provide detailed information about requirements set by the law in order to perform the activities relating to telecommunication services and equipment as well as to broadcasting.   23–19

The State may entrust individuals, by a so-called "concession", with the right to perform activities reserved to itself. Concessions are therefore defined as administrative acts granting rights to the applicants.

For other activities, which are not reserved to the State, the law can, for reasons of public interest, provide that they be carried out only by indi-viduals complying with pre-determined requirements or having pre-determined characteristics. Such compliance, or the existence of such char-acteristics, will be ascertained by the State, which will grant an "authorisation" to the applicant. The right to carry out the activity which has been authorised is, in this case, pre-existent to the granting, but its use is conditioned by the same.

The licensing regime for carrying out activities in the telecommunication sector is laid down in the Postal Code now in force (see above under para-graph 23–05). The Code provides for the procedures to be followed for the granting of concessions aimed either to perform a public service[42] or to perform private telecommunications connections for one user, only between different premises of the same.[43] The former is granted by a Decree of the President of the Republic, upon proposal of the Minister for Post and Telecommunications,[44] after a public bid unless the concessionnaire is a company controlled by the State.[45] The latter is granted by Decree of the Minister of Post and Telecommunications. The regime is connected with the exclusive reserve to the State of all telecommunications services, now deregulated, apart from voice telephony, telex, paging and satellite services.

[42] Arts. 196–211.
[43] Arts. 213–218. The Postal Code provides also, as we shall see, for the various authorisations to be requested on a case by case basis.
[44] Art. 196.
[45] Art. 198.

23–20   Also the frequencies for providing radiocommunication services are reserved to the State, which may grant them to private operators by concession, as we shall see in more detail under paragraph 23–30, below. As a consequence, also broadcasting is subject to prior concession by the State.

Installation and maintenance of telecom terminal equipment is now subject to the granting of an authorisation to the interested party. Further details on the applications and the condition for the granting of such authorisations are given in paragraphs 23–48 *et seq*.

Finally, as far as the provision of liberalised services is concerned, no authorisation is required to carry out any deregulated services on the public network, apart from circuit and packet-switched data transmission (see below paragraph 23–21); the latter, as well as all deregulated services performed by means of leased lines, have to be authorised by the Ministry of Post and Telecommunications. The department of the Ministry in charge of granting authorisation is the Inspectorate General for Telecommunications.

3. REGULATION OF SERVICES

Detailed Analysis of Rights to Provide Telecommunications Services: Those Subject to Monopoly and those Open to Competition

23–21   In 1990, a new National Regulatory Plan of Telecommunication (the "Telecom Plan" or "Plan")[46] was enacted to consolidate and supersede any previous provision. As far as telecommunication services are concerned, the Telecom Plan defines the technical conditions for the utilisation of the main services of the public network by the providers of applicative or value-added services."[47]

Article 2 of the Telecom Plan defines telecommunication as "any sign, writing, image or sounds of any nature transmitted via cable, radio electric, optical or any other electromagnetic system" and then defines and classifies the different telecommunication services into four groups, as follows:

(a) Main Services (M.S.). Such are those services offered by a telecommunication network for transferring information between termination points of the network (annex I to the Plan, gives the following examples of M.S.: analogue circuit switched service for telephony; analogue mobile circuit switched service for telephony; digital direct circuit point-to-point service at 9600 bit/s; digital direct circuit "no restriction" service at 64 Kbit/s, etc.);

(b) Teleservices (T.S.). These are services which allow communication between termination points according to internationally normalised procedures, relating to communication standard and information

---

[46] Enacted by Ministerial Decree of April 6, 1990.
[47] Art. 1, para. 2, of the Plan.

treatment (annex I to the Plan, gives the following examples of T.S.: telephone service and telex);

(c) Supplementary Services (S.S.) to the M.S. and T.S. Such are those services which may not be performed independently for M.S. or T.S. and which can be distinguished from the latter since they constitute a further service (annex I to the Plan, gives the following examples of S.S. to M.S.: identification of calling number; closed user group, hot-line, etc., and of S.S. to T.S.: conference calls, etc.);

(d) Applicative and/or value-added services (V.A.S.). Those are services which are not included within the services mentioned under points (a)–(c) above, either normalised or not, which provide a further and distinct service from those provided by M.S. (annex I to the Plan, gives the following examples of V.A.S.: electronic mail; conversion of protocol; fax, telex and video text; optional services such as alarm-clock; video phone, etc.).

Article 3, paragraph 1, of the Telecom Plan confirms that the services **23–22** "described under points (a)–(c) above are under the monopoly of the State and are provided directly or through a public concessionaire". However, article 3, paragraph 3, of the Plan expressly states that the "Plan does not regulate the provision of the services described under point (d), which are offered under competitive regime, save that any V.A.S. shall use the M.S. described under point (a) above, and which are supplied under a monopoly regime, by using the interfaces regulated by the Minister of Posts".

Therefore, only value-added services were liberalised telecom services according to Italian law until Legislative Decree no. 103 of 1995 was enacted. Other services, such as data transmission and, last but not least, the mere resale of capacity, were liberalised first by the self-executing Directive 90/388 and subsequently by the above mentioned Decree.

It is interesting to see guidelines leading to the Government regulation:

All services, apart from telex, voice telephony (including mobile), satellite services and paging would have been deregulated, by the Ministry of Post and Telecommunications' Decree of September 23, 1992 and also by the draft legislative Decree filed with Parliament for approval at the beginning of 1993, had they been enacted.

In order to supply the public with deregulated telecommunication services—apart from circuit-switched and packet-switched data transmission, which undergo a special discipline—the draft of the Ministerial Decree provided clearance on the use of switched links in the public network, under the sole condition that notice of start-up of the service, together with a description of the service to be supplied and the means to be used to that end, was transmitted to the Ministry (General Inspectorate for Post and Telecommunications) by the new operator, 60 days beforehand; when the service is to be provided through direct links of the same, only the private operator, not the public operator, although the competitor, had to apply for a licence. As regards data transmission, the new operator was obliged to apply for a licence from the Ministry (General Inspectorate of Telecommunications), both for transmittal through switched and direct connections. Resale of more capacity was prohibited.

23–23   Access to the network, and the granting of licences, were subject to the same requirement provided for in Article 3 of Directive 388/90.

By a later draft legislative Decree, filed on September 13, 1993, which took into account the remarks of the Antitrust Authority, any difference between the public and the private operator for licensing in the sector of leased links usage was abolished.

Moreover, complying even further with the Antitrust Authority remarks, the mere resale of capacity was allowed for data transmission and the prohibition was maintained only for resale meant for voice telephony, mobile telephony, telex and satellite communications.

A lump sum would have to be paid as upon the granting and renewal of the licence, set by the Ministry of Post and Telecommunications Ministerial Decree issued in agreement with the Ministry of Treasury (the amount used to be 6 million lire) and a fee would have to be paid each year according to specific instruction to be issued by the same Ministry (it used to be 50 million lire).

23–24   By Law no. 146 dated February 22, 1994, Parliament again entrusted the Government with the authority to implement by means of legislative decrees E.C. Directive 90/388 on deregulation of services, and E.C. Directive 92/44 on Open Network Provision.

On March 17, 1995 the Government finally enacted the above mentioned Legislative Decree, no. 103. The Decree deregulates all services provided by means of switched or direct links of the public network, apart from voice telephony as defined in Article 3.1 of the same Decree.

The scope of the Decree does not cover telex, mobile telephony, paging and satellite communications.

In compliance with the Antitrust Authority's remarks relating to the previous draft bill, the Decreee submits to authorise only packet and circuit-switched data transmission and data transmission on leased lines. The text of the Decree governing the granting of authorisation has been criticised by one of our most prominent authors because it leaves with the Ministry of Post and Telecommunication a full discretionary power by providing that authorisations shall be granted on the basis of "objective and non discriminatory criteria" without defining the same.

Regulation of each single telecommunication service, as well as an outline of the draft legislation for implementation of Directive 90/388, will be discussed in more detail at paragraphs 23–25 et seq.

Fixed (including Cable TV where Relevant to Telecommunications) and Mobile Services

Fixed Services

23–25   As far as services are concerned, we have already drawn a distinction between the services subjected to State monopoly and those which are liberalised (at paragraph 23–21 et seq., above). However, a more detailed examination of the extension and the limits of reserved services now appears necessary.

Data transmission, resale of mere capacity and voice telephony for closed users group are now liberalised in Italy within the limits and under the conditions sets by Decree no. 103, 1995.

As far as voice telephony services rendered to the public are concerned, without doubt they are still reserved by the State.

Mobile Services

Paging

Paging services have been organised in 1985[51] and are reserved to SIP (now   **23–26** Telecom) by the concession granted for supply of public domestic telephone service.[52] The regulations presently in force[53] state that terminal equipment for a paging service may be freely purchased on the market, while connection to network and maintenance are reserved by SIP (Telecom).

In May 1995, the General Meeting of Shareholders of Telecom Italia SpA resolved to establish a new company, Telecom Italia Mobile, to which assets and infrastructures related to mobile car telephony and mobile personal telephony would be conferred, together with the relevant concession (see below).

Mobile car telephony

Organised in 1985 as a public service,[54] this service is reserved to SIP, in   **23–27** the same way as is paging, by the concession agreement for supply of public domestic telephone service. The service is limited to terrestrial means of transport. Terminal equipment may be freely purchased and maintained by the user; connection to the public network is reserved by SIP.

Mobile personal telephony

This was organised as a public service in 1990.[55] SIP has maintained from   **23–28** start-up of the service that, being concessionnaire of the domestic telephone service, it has the exclusive right also to supply mobile service under the provisions of the concession agreement enclosed with the 1984 Decree granting the concession.

The Italian Antitrust Authority (see above n. 31) has opposed SIP's alleged right since November, 1991, when its first advice was sent to Parliament, suggesting the Italian Government should open up the market in the field of mobile telephony, at least to a second operator. After other

[48-51] Ministerial Decree dated August 3, 1985.
[52] Approved by Decree of the President of the Republic dated August 13, 1984 no. 523. The Agreement enclosed with the Decree provides for the sole mobile maritime service to be excepted.
[53] Ministerial Decree no. 43 January 8, 1992.
[54] Ministerial Decree dated August 3, 1985.
[55] Ministerial Decree no. 33 of February 13, 1990, subsequently modified by Ministerial Decree no. 512 of November 8, 1993; tariffs are set by Ministerial Decree dated February 14, 1990.

interventions of the Authority in the ensuing years, the above advice was finally complied with by the Italian Government, and a bid for a second operator to provide mobile telephone service, making use of new the GSM system, was won in March 1994 by a joint venture between Olivetti (51.9 per cent), Bell Atlantic (16.6 per cent), Cellular Communication (14.7 per cent), Telia Int. (9.7 per cent) Lehman Brothers (8.5 per cent), which represent altogether 70 per cent of the consortium, and Pactel (34 per cent), Mannesmann (15 per cent), Banca di Roma (15 per cent) and others (36 per cent), representing the remaining 30 per cent whose name is OMNI-TEL—Pronto Italia. Criticism about the bidding procedure is still going on in Italy.

According to the splitting mentioned above, Telecom Italia Mobile SpA has become the exclusive concessionaire of analogue mobile telephony (TACS) and one of the non exclusive concessionaire of the digital mobile telephony (GSM). The latter concession had been granted to Telecom Italia SpA by a Decree of the President of the Republic dated December 22, 1994.

Satellite Services

23–29   These services are still reserved to the State, as Directive 90/388 does not apply to it, and our Postal Code, providing for the exclusive right of the state to supply telecommunications services, has not yet been amended, apart from value-added services, which have been deregulated by the Telephone and Telegraph Regulatory Plan of 1990 (see above).

The TO which has been granted the exclusive right to establish and maintain satellite connections is Telespazio. Terminal equipment is deregulated only for VSAT receivers.

Deregulation is to take place in the near future by means of an E.C. Directive which has up until now only been a draft.

Radio Communication Generally

23–30   We make reference to paragraphs 23–01 to 23–04 above regarding historical information. Means of broadcasting are and have always been reserved to the State. The public service has been dealt with by concessionnaires. The most recent concession was granted to RAI in March 1994.[56]

The exclusive right reserved to the State for Broadcasting services was limited by the Italian Constitutional Court in 1976 to national broadcasting,[57] liberalising local (regional and city broadcasting). After a long dormant period, during which no regulation governing local broadcasting was enacted, which consequently gave way on a large scale to abusive use

---

[56] The relevant agreement bears the date of 15 March 1994.
[57] By its decision no. 202 of July 28, 1976 the Court stated that the reserve to the state of radio and TV broadcasting within "local" areas—to be determined on the basis of "reasonable" criteria having regard to geographical, urban and socio-economical conditions—did not comply with Art. 3 and 21 of our Constitutional Chart; the said Art. provide for right to equal treatment of the citizens and freedom of speech and thought.

of the available frequencies, the matter was finally fully regulated in 1990.[58]
The 1990 law provides for the public service to be dealt with by the State
through a concession and for concessions to be granted to applicants for
private service both on a national and on local scale.

Reference is made to the following paragraphs 23–47 *et seq.* for detailed
information on the present regulation.

4. REGULATION OF BROADCASTING

Basic Legislative Rules

The two main features of the radio and television system actually in force   **23–31**
in Italy are (i) the State exclusive control of use of wavelengths and frequency bands and (ii) the running of the public information service via
different sorts of concessions, both private and public. Broadcasting is governed by Law no. 223, of August 6, 1990, ("Regulation of public and
private broadcasting system"). This system aims at achieving pluralism,
and impartiality of information (Article 1).

The concessions are granted by the Government (Minister of Posts and
Telecommunications) to companies entrusted with public (RAI) and private
services.[59]

Recently, the concession to concessionaires of private service, which represented the main innovation of the above Law, has been specifically regulated by Law no. 422 of October 27, 1993. Previously, the private service
operated on the basis of temporary authorisations.[60] Besides these two main
texts, there are several other laws governing partial aspects of the matter.[61]
The multiplicity of the rules concerning radio and TV broadcasting also
reflects the lively political discussion about the issuing of these provisions.

The broadcasting system provided by Law no. 223, 1990 has not yet
been totally accomplished. Because of this state of the regulation, the latest
laws refer to a new reform act which should reorganise the whole matter.[62]

The concessions are granted according to a national plan of distribution
of frequency bands, drawn up by the Ministry.[63] This plan divides the
national territory into many areas in accordance with the density of population, the geological structure of the territory involved, town planning and
the prevailing social conditions of the site. An adequate plurality of networks must be granted in all areas. According to this plan, which is to be
revised every five years, the Minister of Posts and Telecommunications
assigns the frequency bands to the concessionaires on the basis of criteria
set down by law. In all cases, the frequencies necessary for carrying out

---

[58] Law no. 223 of August 6, 1990.
[59] *ibid.* Art. 16.
[60] *ibid.* Art. 32.
[61] The most relevant laws are: Law no. 206 of 1993; Law no. 515 of 1993; Law no. 483 of 1992; Law
no. 482 of 1992; Decree 73/1991.
[62] According to Art. 2 of Law no. 206 of 1993 the new broadcasting system regulation should be carried
out within 1995.
[63] Art. 3 of Law no. 223 of 1990.

the objectives of the public telecommunications network must be made available to the public concessionnaire.

Sources of Financing

23–32   Sources of financing for the transmission of public telecommunications are subscription charges,[64] paid by users of the service, and the proceeds of advertising. Among the private transmitters, there are networks transmitting at no charge, which finance themselves by means of advertising, and Pay TVs as well, whose consumers are supplied with a decoder for messages transmitted.

The Public Concessionnaire

23–33   The concessionnaire of public service is a state-owned corporation (RAI—*Radiotelevisione Italiana SpA*), under Parliament's control; such control is exercised for the public interest of the broadcasting activity carried out. The law provides for a series of particular rules concerning the operation and the duties of the Board of Directors, and the appointment of its members, by the Presidents of the House of Deputies and the Senate. On March 15, 1994, an agreement was executed between the Ministry of Post and Telecommunications and the RAI, for the exclusive concession throughout national territory of public radio and television broadcasting. The concessions concern both installation of systems and transmission of programmes and have a duration of 20 years.

A specific discipline has also been imposed upon private concessionaires.[65] There can be two different types of concessions: a) commercial, and b) community radio broadcasting. The latter is characterised by absence of aims of profitmaking and for cultural, ethnic, political and religious aims. The concessionnaires should therefore be non-profit organisations.

The concession for national commercial broadcasting can be granted exclusively to limited companies or cooperatives whose statements of objects comprise the carrying out of telecommunications and editorial activities or other activities which are however related to information and entertainment. The majority shareholding of these companies cannot belong to a foreign physical person, or to a foreign company, or to a trustee company.[66] Any transfer of the concessionaire companies' shares to such persons or companies is void. On a regional level the concessions can also be granted to Italian or E.C. physical persons, organisations and companies, provided a guarantee of 300 million lire is paid.[67]

Market Concentrations

23–34   Law no. 223 of 1990 contains important provisions for the control of market concentration. This Law guarantees the maximum transparency of

---

[64] Art. 15 of Law no. 103 of 1975.
[65] Art. 16.1 of Law no. 223 of 1990 and Law no. 422 of 1993.
[66] Art. 17 of Law no. 223 of 1990. For other requisites of concessionaire companies see also Arts. 4 and 5 of Law no. 422 of 1993.
[67] Art. 16.8 of Law no. 223 of 1990.

the structure and management of radio and television broadcasting companies, in order to ensure the necessary conditions for the realisation of a pluralistic system of information.

For this purpose, the regulatory law has instituted a national register of radio and television broadcasting companies[68] entrusted to the *Garante per l'editoria e la radiodiffusione* ("Publishing and broadcasting authority", see paragraph 23–39) and the concessionnaires, both the public concessionnaire and the private ones, as well as the companies carrying out production and distribution of the programmes and the advertising concessionaries, all have to be registered. All information regarding the shareholders must also be registered, including ownership of shares, and controlling companies.

Besides the register of the radio and television companies, the radio and television companies are obliged to notify the *Garante*, for antitrust purposes, of the following occurrences: (i) transfer of shares belonging to the concessionnaire companies; (ii) if the companies are linked with or controlled by other entities holding at least 10 per cent of their capital; (iii) execution of shareholders' agreement.[69]

Also the accounts[70] of the registered companies must be submitted every year to the "Garante" and should contain information on transmitted programs.

Article 15 of the regulatory law, which is the main provision on control of market concentration, gives the notion of a dominant positions in the mass media industry. Pursuant to a Constitutional Court decision of December 7, 1994, no. 420, Article 15 of Law no. 223 of 1990 must be modified by the Parliament to reflect that no operator can own more than 25 per cent of national networks.

Terrestrial and other Delivery

The regulatory law of broadcasting has, principally, as its aim the transmission of radio and television programmes via electromagnetic frequencies. **23–35**

As far as other means of transmission are concerned, some specific provisions for a) broadcasting via cable and b) on codified transmissions are in force.[71]

The installation of the broadcasting cable system, mono and multi-channel, is carried out by the State. Such activity is performed either directly by the State or with the help of telecommunications network and service concessionnaires.

The distribution of television programmes and sound via cable is however subject to authorisation by the Ministry of Post and Telecommunications. This authorisation is different from the concessions described above.

Distribution authorisation for programmes via cable is also subject to Antitrust rules.

[68] *ibid.* Art. 12.
[69] *ibid.* Art. 13.
[70] *ibid.* Art. 14.
[71] Art. 29 of Law no. 223 of 1990; Decree no. 73 of 1991; Art. 11 of Law no. 422 of 1993.

For codified transmissions mentioned above, a gradual move towards a system of reception exclusively via cable or satellite is foreseen. For satellite transmissions, there is no specific norm. The exclusive concession to RAI is however regulated by a convention with the State.[72]

Regulatory Agencies

23–36   *Ministero delle Poste e Telecommunicazioni* (**"Ministry of Post and Telecommunications"**)[73]: The Ministry is the authority supervising the entire radio and television activity. Besides issuing concessions and authorisations necessary for the radio and television companies, the Ministry controls planning of the frequency network, installation of systems and has also, together with the *Garante*, punitive powers in case of infringements of the rules concerning the contents of programmes.

23–37   *Commissione Parlamentare* (**"Parliament Committee"**)[74]: At the top of the public apparatus of control is the Parliament committee for general guidelines and the supervision of radio and television services. It consists of 40 Members of Parliament and is set up with the intent of entrusting to Parliament the management of television in order to guarantee fundamental values of impartiality and pluralism.

This committee has the task of providing guidelines, surveillance, and controlling the entire matter of radio and television broadcasting, including aspects of content of programmes.

23–38   *Garante per la Radiodiffusione e l'Editoria* (**"Publishing and Broadcasting Authority"**)[75]: The Law has instituted an authority for broadcasting and publishing. This figure assumes, within the provisions of law, a central and specific position. It is a monocratic administrative authority, independent of government, with antitrust powers (see also paragraph 23–12) in relation to the press and at the same time controls the activities of the concessionnaires and of the broadcasting system in general.

Amongst the powers of the authority, as far as radio and television are concerned, besides those above mentioned, the authority is in charge of keeping the register of radio and television companies, it also reviews the accounts of these companies and supervises the indexes of viewers/listeners and the report with the regional committee for radio and television services.[76]

The *Garante* is also in charge of (i) carrying out investigations and (ii) applying fines in the event the parties do not comply with their obligations in connection with advertising and programming, as well as (iii) preventing the formation of dominant positions in the mass communication sector.[77]

[72] Art. 1 of D.P.R. no. 367 of 1988.
[73] Arts. 1, 3, 18, 31 of Law no. 223 of 1990; Art. 1 of Law no. 422 of 1993.
[74] Arts. 1–4 of Law no. 103 of 1975.
[75] Arts. 6, 31, and 34.7 of Law no. 223 of 1990.
[76] *ibid*. Arts. 8, 9, 20, 21, and 26.
[77] *ibid*. Art. 15.

Every year the *Garante* submits a report to Parliament on activities carried out and on the state of application of law.

Amongst the functions of the authority, the power to rectify is extremely important.[78] Whoever feels to have suffered material or immaterial damage, from transmissions contrary to the truth, has the right to ask the concessionnaires to broadcast a special amendment. If the amendment is not broadcasted within 48 hours, upon request of the interested party, the Broadcasting Authority will provide thereto.

Consiglio Consultivo degli Utenti ("Consumers' Consultation Counsel")[79]:   23–39
Within the office of the *Garante*, a counsel has been established with the institutional aim of liaising *Garante* and community. Such a body, formed by representatives of both consumers and associations of experts, has been appointed with the task to ensure protection of consumer interests. The regulations of this body are established by the Broadcasting Authority.

Comitato per i Servizi Radiotelevisivi ("Regional Committee for Radio   23–40
and Television Services")[80]: At each regional council, a committee for radio and television services is elected, composed of experts in radio and television communications. This body is in charge of providing consultancy services to the region on such matters and of making proposals to the public concessionnaire regarding regional programming. The functioning of this body is regulated by the regions.

Regulation of Transmission v. Content

Besides control on the radio and television companies, the regulation of   23–41
radio and television broadcasting has the aim of regulating broadcasted programmes. Here follows a brief overview of the principal aspects of regulation of the content of programme.

Advertising[81]

The law foresees first a maximum hourly and daily and weekly percentage   23–42
for advertising (for the public concessionnaire: 12 per cent per hour and four per cent per week; for private national broadcasting stations: 18 per cent per hour and 15 per cent of daily; for private stations: 20 per cent per hour and 15 per cent daily).

Concerning content, it is forbidden to evoke in whatever manner any form of discrimination, to offend personal dignity, religious beliefs, induce harmful behaviour in respect of health and environment, harm minors, or to introduce advertising during a cartoon programme. It is also forbidden to advertise drugs or medical treatment.

[78] *ibid.* Art. 7.
[79] *ibid.* Art. 28.
[80] *ibid.* Art. 7.
[81] Art. 8 of Law no. 223 of 1990; Decree of the Ministry of Post and Telecommunications dated December 9, 1993.

The law also foresees a limit to the duration of interruptions of cinemato-graphic works, by advertising spots. In addition, an absolute prohibition may be imposed by the *Garante* where works of high artistic nature or of religious and educational value are broadcasted.

23–43   Sponsorship is also regulated in detail. Sponsorship is only allowed when it may be recognised and distinguished. Any kind of subliminal sponsorship is absolutely forbidden. Sponsorship is defined as a contribution from a public or private company, not involved in television or radio activities, to the financing programmes, with the aim of promoting its own name, trademark, image, activity and product. It may consist exclusively in the previews of the programme, as well as on billboards, in promos and in spot-jingles. Sponsorship is forbidden to manufacturers or sellers of tobacco, superalcoholic drinks, drugs or medical treatment available by doctor's order. The news and political and economic information cannot be sponsored either.

Electorial Propaganda[82]

23–44   The recent general regulation of radio and television programmes also extends to electorial propaganda. The private concessionnaires who intend transmitting electorial programmes near to election time, should first pre-pare a self-conduct code which fixes conditions of access to electorial pro-paganda spots. In any case, the electorial programming should allow access to all interested parties and guarantee impartiality and pluralism of political information.

The self-conduct code should be presented to the *Garante* 35 days before polling day. Also, electorial programming should be announced, during main viewing time, at least 30 days before the elections and should be carried out according to the code of self-conduct of the broadcasting com-panies and according to the regulations set down by the *Garante* (n. 24). This rule also distinguishes between electorial publicity, which is prohib-ited, and propaganda, which is allowed. Both transmissions and electorial spots portraying politics in a "show-biz" light, without any presentation of candidates or political programmes are not allowed.

Obligations Concerning Programming[83]

23–45   The concessionnaires must broadcast the same programme throughout their territory. It is forbidden to broadcast codified or subliminal messages, violent or immoral programmes, which could damage the growth of minors, or films prohibited to under 18's. Furthermore, in order to protect the cinema, cinematographic works cannot be broadcast before a period of two years has elapsed from the date of their first showing, unless there is a different agreement with the copyright holders.

Upon private concessionnaires a minimum daily and weekly programme

[82] Law no. 515 of 1993; Law no. 81 of 1993; Regulation of *Garante* January 26, 1994.
[83] Arts. 10, 15, 20 of Law no. 223 of 1990.

threshold is also imposed, as well as the obligation to broadcast the daily news.

As far as control on programme content is concerned, the obligation imposed upon the concessionnaires to rectify untrue news broadcasts has to be stressed. Such rectification should be made within 48 hours of the request, and must be broadcast in the same hourly band and with the same importance as in the transmission which damaged the interested parties.

The concessionaries are obliged to broadcast, free of charge and immediately, communications from public bodies, should the need arise.

Moreover, in accordance with European regulations, the law provides[84] that European cinematographic works shall be not less than 40 per cent, for the first three years from the issuing of the concession, and 51 per cent, for successive years, of all cinematographic works broadcasted.

Monopoly and Competitive Services

In spite of the competitive aims of the new regulations, the new public and private free enterprises have yet to be realised in the broadcasting sector.   23–46

As already seen above, the legal discipline grants exclusive rights to the State on the use of wavelengths and provides for pluralistic supply of information. However, a double market has been created. On the national level, there is a duopoly between RAI and the Fininvest Group, while on the regional level we assist competition between many different networks.

5. REGULATION OF APPARATUS

We have already outlined above at paragraph 23–18, in very broad terms, the regulations relating to telecommunication apparatus. However, we   23–47 shall now give more details on their supply, the right to market them, their installation, and maintenance.

Regulation of Apparatus Installation and Supply

Installation and Supply of the Telecommunication Network

Article 183 of the Postal Code reserves the installation and the operation   23–48 of the telecommunication network to the State, which entrusts third parties by concession without carrying out this activity itself. We shall deal with the ground network first and then with the satellite network.

The ground network

The domestic network was owned and operated partly by SIP and partly   23–49 by Iritel (see above), while the international network was owned and oper-

---

[84] ibid. Art. 26.

ated partly by Italcable and partly by Iritel, according to the countries to be connected (see paragraphs 23–21 *et seq.* above). Apart from mobile, the overall network is now operated by Telecom Italia SpA. Therefore, as far as the installation of the ground telecom network is concerned, there is no doubt that according to Article 183 of the Postal Code, it is expressly reserved by the State.

In any case, the provisions of Directive 90/388 firstly and subsequently those contained in Legislative Decree no. 103 of 1995 have substantially amended the above mentioned provisions of the Postal Code in so far as the supply of the network is concerned. As a matter of fact private operators may lease lines from the TO entrusted with the network and are now entitled to simply resell the capacity of such leased lines for the provision of data services provided an authorisation licence to that end is granted by the Ministry of Post and Telecommunications.

However, Italy has not yet taken the measures necessary for the licensing procedures provided for by the above mentioned Decree. In the wake of such regulations, we shall give an outline of the present regulation relating to the leased lines.

23–50   Article 295 of the same Code expressly provides, at paragraph 1, that "anyone may apply to be granted the use of a direct point-to-point telephone link to make telephone conversations between two different premises located within the same city network or within different city networks" and, at paragraph 2, that "direct point-to-point telephone links may also be used for connecting premises belonging to different persons".

Moreover, according to Article 296, paragraph 1, of the Postal Code "The provider [TO] may allow the connection between several point-to-point links which are being used by the same user; in this case the operator shall provide all the necessary for the organization and the functioning of the link" but, according to paragraph 2 of the mentioned provision "the link may not be made available to third parties and shall be exclusively used for calls relating to the user".

Therefore, anyone may apply to the TO actually entrusted with the network to be granted point-to-point or multi-point dedicated connections, and may create its own circuit, also having more than one termination point and reaching different persons. Only the following private networks have been realised through switched links from the public network: Alitalia, FFSS, Reuter, SWIFT.[85]

However, under these regulations it is not possible to connect several leased circuits and realise a private network, for what is called a "Closed Users Group". In fact, it is expressly provided that any connection between the different multipoint direct circuits shall be performed without any switch.[86]

---

[85] The concession of the private networks to Reuter's and SWIFT have been granted by Ministerial Decree dated December 22, 1989, and renewed by Decree of December 31, 1990.
[86] Ministerial Decree dated August 4, 1982.

Satellite network

Finally, as far as the satellite network is concerned, a distinction should be   23–51
drawn between the satellite segments and the transmit-receive antennas. In
fact, the former are granted to Telespazio on an exclusive basis[87] while the
latter, (with the exception of the stations of Fucino, Lario and Scansano,
which are granted to Telespazio) will be granted by decision of the Minister
of Posts and Telecommunication.[88]

It is worth mentioning that, according to Article 9 of the Telespazio
Convention Agreement of 1984, "consistently with the needs of the service,
links may be assigned by Telespazio, with the prior authorisation of the
Administration, to other State agencies as well as to other persons which
have been duly granted a concession or, in urgent cases and for limited
period of time, a special authorization by the same Administration". Therefore, under special circumstances, satellite links may also be assigned to
private persons, which may thus build up a satellite telecom network to
provide services.

Installation of Telecommunication Terminal Equipment

The very first step towards liberalisation of the installation of telecom   23–52
equipment was taken by Articles 284 and 285 of the Postal Code, according
to which, users may request the concessionnaire to connect their terminal
apparatus to each other and may also entrust authorised companies with
the task of performing such connection.

The trend towards the liberalisation of installation has been accelerated
during the last years. In fact, "the installation, testing, connection to public
networks and maintenance of terminal equipment is carried out by the
public service operator or by enterprises which have been granted a special
authorisation for this purpose,[89] on behalf of the users". However, users
may directly install simple equipment.[90] Any infringement of such provision
will carry a fine of up to 10 million lire (Article 2.1).

Finally, Article 1, paragraph 5 of Law no. 109 of 1991 empowers the
Minister of Posts to enact by decree the requirements for being granted the
authorisation to instal, as well as the technical requirements and procedures
for connections to the public network. These provisions have been enacted
by Decree no. 314 of May 23, 1992.

Therefore, since 1991, installation of telecom equipment has been liberalised and is no longer reserved to the national telecommunication oper-

---

[87] In fact, pursuant to Art. 1, para. 2 (b), of the Agreement between the Minister of Post and Telecommunication and Telespazio, enacted by D.P.R. no. 523 of 1984 Telespazio is granted an exclusive concession
"for the installation and the operation: . . . b) of the space segments systems fit for realising telecom
links through artificial satellites".

[88] According to Art. 1, para. 2 (a) of the above mentioned Concession Agreement with Telespazio, "the
installation and the other ground stations [different from Fucino, Lario and Scanzano] must be authorized by the Administration".

[89] Art. 1, para. 3, of law no. 109 of 1991.

[90] ibid. para. 5.d.

ator. However, only type-approved terminal equipment may be installed.[91] Installation of terminal equipment which has not been submitted to type approval is prohibited,[92] and the equipment may be seized.[93]

Supply and Right to Market Telecommunication Terminal Equipment

23–53   No special provisions have ever regulated the manufacture of telecommunication apparatus, therefore, as a general rule, the manufacture of such apparatus has always been free. The different types of equipment are produced by various manufacturing companies and their installation is only subject to the previous approval of the Minister of Post and Telecommunication.

On the other side, the right to market telecom terminal equipment has been characterised by a progressive liberalisation and opening to competition. As a matter of fact, the supply of terminal equipment has been implicitly liberalised since August 1984.[94]

This is the state of liberalisation with respect to each kind of terminal apparatus:

23–54   **Telephones and telecopiers:** The sale of telephones and telecopiers has been expressly liberalised by Article 1, paragraph 1, of Law no. 109 of 1991. This states that "users may supply themselves, directly or through the public TO, with terminal equipment allowed to be connected to the public network". Even if the mentioned law is misleadingly headed "New Provisions on the Connection and Testing of Internal Telephone Equipment", its provisions should apply to any terminal equipment.

23–55   **Modems, PABX and V.A.S. terminal equipment:** The liberalisation of the supply of modem and other data trasmission equipment has been provided by Article 3 of the Concession Agreeement between the State and SIP of 1982 to regulate the activity of data and signal transmission services.[95]

Finally, the full liberalisation of any other value-added services terminal equipment (*e.g.* equipment to allow use of credit cards to pay phone calls) has been definitely achieved as from January 1, 1989.[96]

23–56   **Satellite Antennas:** The provisions applicable to satellite antennas differ according to whether (a) VSAT receive-transmit or (b) VSAT receive-only are concerned.

(a) Installation and operation of VSAT receive-transmit has exclusively

---

[91] Arts. 1.2 and 2 of the Postal Code and Art. 20 of the National Telecom Plan.
[92] Art. 1.2 of Law no. 109 of 1991.
[93] *ibid.* Art. 2.2.
[94] Art. 12 of the Concession Agreement of August 1, 1984 between the Minister of Post and Telecommunication and SIP enacted by D.P.R. no. 523 of 1984.
[95] Concession Agreement of July 1, 1982, enacted by Ministerial Decree of August 4, 1982.
[96] By Ministerial Decree of December 30, 1988.

been granted to Telespazio, which provides such terminal equipment to the operators.

(b) As for VSAT receive-only, according to Article 3 of the E.C. Directive 88/301 on competition in the markets of telecommunication terminal equipment, economic operators have the right to "import, market, connect, bring into service and maintain terminal equipment". Annex I to such Directive evidences that the definition of terminal equipment is also extended to "receive-only satellite stations not reconnected to the public network of a Member State".[97]

Moreover, the same result is achieved if it is assumed that the provisions of Law no. 109 of 1991 apply to any telecom terminal equipment.

Maintenance of Telecommunication Terminal Equipment

Maintenance has followed the same steps towards liberalisation adopted with regard to the installation activity referred to at paragraph 23–52 above. Therefore, at present, maintenance of telecommunication terminal equipment is performed for the user by SIP or by enterprises which have been granted a special authorisation.[98]   **23–57**

Type Approval and Testing Procedures

Electromagnetic Compatibility

Legislative Decree no. 476 of 1992 implemented Directive 89/336, on harmonisation of Member States legislation relating to electromagnetic compatibility.[99]   **23–58**

Only electric or electronic equipment complying with the harmonised technical rules (N.E.T.) or, failing such rules, with the rules of the Member States (Article 6) may be put on the market (Article 11).[99a]

The conformity of equipment shall be attested by the E.C. declaration of conformity and the related E.C. mark released by the manufacturer or the importer within the E.C. to be kept at disposal of the Minister of Post and Telecoms (Article 7).

As far as radio-transmitting equipment is concerned, the conformity shall be attested by declaration of conformity of the manufacturer on the basis of the E.C. type-examination certificate issued by a notified body of a Member State (Article 8).

The Inspectorate General of Telecommunication is the notified body to

---

[97] The mentioned provision by no doubt has direct effect in Member States, since the deadline for implementing Directive 88/301 has expired, and it provides for sufficiently clear and specified provisions which do not impose further obligations on the Member States.

[98] Art. 1, para. 3 of Law no. 109 of 1991.

[99] Even if the Decree refers to Directive 92/31, the latter has not yet been implemented in Italy.

[99a] Ministerial Decree of September 15, 1994 laid down a list of the Italian rules (enacted by *Comitato Elettronico Italiano*, or CEI rules) implementing the European technical harmonised rules on electromagnetic compatibility. Moreover, FM and HM radio broadcasting equipment shall satisfy the minimum requirements laid down by Ministerial Decree no. 311 of 1994.

issue the E.C. type-examination certificates in Italy. The latter shall be issued within 90 days from the communication of the conformity tests made by the appointed laboratories (Article 9).

Conformity of Terminal Equipment to Essential Requirements (Homologation of T.E.)

23–59   The procedure relating to homologation[1] of terminal equipment[2] is now regulated by Annex 11 to Ministerial Decree no. 314 of 1992.

The General Inspectorate of Telecommunication (I.G.T.) is the authority entrusted with the granting of the homologations.[3] The requests for homologation shall be filed before I.G.T. on paper bearing stamp duties and providing certain information[4] together with the documentation provided for by Article 5 of Annex 11.[5]

I.G.T. will assess the documentation and submit the equipment to technical examination[6] in order to assess whether the equipment is compatible with the network and complies with the E.T.S. Finally, I.G.T. will also assess whether the equipment complies with safety regulations (Article 6).

Within 180 days of the submission of the application and the filing of the requested documentation I.G.T. shall grant the certificate of homologation or shall communicate the reasons for rejecting the application. Against the refusal an appeal may be filed before the Administrative Courts of Rome.

Recognition of Conformity of Equipment Manufactured within other Member States

23–60   Legislative Decree no. 519 of 1992 implemented Directive 91/263, on recognition of conformity of other Member States telecom terminal equipment.

Terminal equipment which (i) has been authorised (granted the conformity assessment) for connection to the public network by a notified body of another E.C. Member State and (ii) has been marked with the E.C. mark, may be freely marketed in Italy and used if duly connected to the network and maintained (Article 10).

[1] *i.e.* a declaration that equipment may be connected to the public telecommunication network.
[2] *i.e.* according to Art. 2 of Annex 11 to Ministerial Decree no. 314 of 1992, any equipment to be connected to the public network, excluded those for radio and tv broadcasting as well as those relating to mobile maritime and air service. The procedure for the homologation of equipment to be part of the network substantially differs from the one referred to and has been recently amended by D.P.R. no. 395 of 1994.
[3] Art. 3 of Annex 11 to Ministerial Decree no. 314 of 1992.
[4] Such as names and place of residence of the interested party, type, trademark and model of the terminal equipment, intented use of the equipment, information of any homologation or certificate granted by any other E.C. authority.
[5] Such as documents describing the characteristics of the equipment as well as its interface and its circuits and any possible additional equipment; any certification evidencing the electromagnetic compatibility as well as the conformity to common specifications of the equipment, etc.
[6] Either through the appointed laboratories or through the Istituto Superiore delle Poste e Telecomunicazioni, a body with technical purposes created within the Ministry of Posts and Telecommunication (Art. 6 of Annex 11).

The Inspectorate General of Telecommunications is the notified body to issue the E.C. type-examination certificates and to control terminal equipment in Italy (Article 2.6).

The administrative approval may also be requested to the I.G.T., which shall grant it on the basis of either:

(a) an E.C. type-examination certificate[7], together with the manufacturer's declaration of conformity to the type[8], or
(b) a certificate relevant to the full quality assurance system[9] (Article 7).

Finally, I.G.T. shall supervise and control both the marketing and the use of telecom terminal equipment, by random checks (Article 11), and may adopt interim measures, such as seizure, if the equipment is not properly used or disturbs the public network (Article 12).

6. TECHNICAL STANDARDS

Apparatus

Directive 86/361 has been complied with in 1988.[10] Directive 91/263[11] has been complied with in 1992[12] (see above paragraph 23–58).   **23–61**

Network and Interfaces

According to Directive 90/388 (Article 4), conditions governing access to the public network have to be made objective, non-discriminatory and public.   **23–62**

According to Directive 90/387, Member States are obliged to establish and supply an open network (ONP).

In 1993,[13] the matter began to be regulated within the Italian system of law, by setting the procedures needed in order to comply with the above E.C. Directives. It was provided that:

(a) Harmonised conditions for access and use of the public network will be set by decree of the Ministry of Post and Telecommunications (the decree has still to be published in the Official Journal of the

---

[7] To be granted by the I.G.T. itself: a) upon a special application, b) provided that no other application has been filed before another is notified, and, c) previous testing of the conformity of the equipment to the essential requirements by one of the laboratories accredited in accordance with Annex 5 to Legislative Decree no. 519 has taken place.

[8] To be rendered in accordance with Annex 2 to Legislative Decree no. 519.

[9] *ibid.*

[10] Ministerial Decree no. 220 of May 28, 1988.

[11] Please note that pursuant to Art. 2 of Legislative Decree no. 519 of 1992 (implementing the mentioned Directive), "Technical Basic for Regulation" nos. TBR5 and TBR9 have been adopted as national technical standards relating to the requirements for connection as far as the approval of terminal equipment of radio mobile stations for the GSM System is concerned.

[12] Legislative Decree no. 519 of December 19, 1992.

[13] Legislative Decree no. 55 of February 9, 1993.

Italian Republic and, to our knowledge it has not yet been drafted).
(b) The above conditions: 1) must guarantee objectivity, transparency and equal access to everybody; 2) will be phased according to progressive normalisation and harmonisation of the technical standards, with regard to the different sectors of use, defined as follows:

- leased lines;
- packet and circuit switched data transmission;
- digital network integrated with services;
- voice telephony service;
- telex service;
- mobile services;
- new type of access to the network as well as access to new intelligent functions of the network;
- access to large band network.

23–63   While operators wait for the decree of the Ministry of Post and Telecommunications, which will actually make public the harmonised conditions of access to the Italian network, Legislative Decree no. 55 dated February 9, 1993, whose title is "Implementation of Directive 90/387/CEE on institution of an internal market for telecommunications services by means of provision of an open network of telecommunications" states (Article 3) that: "Access and use of the public network of telecommunications and, when necessary, to the public service of telecommunications, is allowed to all, in order to supply the above services to the public, under the conditions, in the ways and with the limits provided for by the law, in compliance with Directive 90/388/CEE, regarding competition in the telecommunications markets. Observance of domestic technical standards reproducing the corresponding harmonised European standards concerning the technical interfaces, as published in the Official Journal of the European Communities, leads to the presumption of compliance with basic requirements consisting of: a) safety of network operation; b) safety of network integrity; c) for reasons of general public interest, interoperability of telecommunications services and data protection." Directive 92/44 relating to the application of the ONP to leased lines has been implemented by Legislative Decree no. 289 of May 2, 1994. However, surprisingly (and in spite of "whereas" no. 14 of the preamble of Directive 92/44 which conferred to each national Regulatory Authority—the referee—the responsibility of the implementation of the Directive), Article 3 of the Decree empowers the telecommunication Operator, *i.e.* the market player, to determine the condition of authorisation for leasing lines.[13a] However, "any user may have access to the leased lines for its own use or for providing services to third parties . . . within the limits set out by the [not yet enacted] Decreee implementing Dir. 90/388" (Article 5.1) and no restriction to their use is allowed (Article 5.3). Fees for leased lines in the Union, the one-stop procedure for a single

---

[13a] However although the conditions of access to the Italian PTN are not clear enough, in the event of either Telecom Italia's refusal to supply leased lines or of excessive charging, any private operator may start an antitrust procedure against Telecom Italia for abuse of its dominant position on the lines, an essential facility for the oeprator.

billing, will be set out by a further Decree of the Minister of Post and Telecommunications (Article 10).

Finally, in 1994,[14] Directive 91/287 was implemented, dealing with frequency bands to be attributed to co-ordinated introduction into the Community of a Digital European Cordless Telecommunications (DECT). The system is defined in the Decree as a technical system complying with European standards of telecommunications ETSI ETS 300 175. The frequency band reserved for that purpose is 1880 to 1900 MHz.

[14] Ministerial Decree dated March 18, 1994.

CHAPTER 24

Japan

By Greg L. Pictrell, Coudert Brothers, Tokyo and
Masayuki Okamoto, Tanaka & Takahashi, Tokyo

1. DEVELOPMENT OF THE TELECOMMUNICATIONS INDUSTRY

Overview

NTT/KDD Former Monopoly

Modern telecommunications in Japan was established during the period of   **24–01**
the 1870's following the Meiji reformation which brought western influ-
ences to Japan. The communications infrastructure remained in the sole
control of the Japanese government. Following World War II, until 1985,
the telecommunications business in Japan was monopolised by Nippon
Telegraph and Telephone Public Corporation (NTT Public Corp.). NTT
Public Corp. was established in 1952 to assume control from the Government to centralise the reconstruction of Japan's telecommunications after
the war. Its initial goals of rebuilding the infrastructure, meeting the
increased demand for telephones (resulting from Japan's rapid economic
growth following the war) and, finally, providing nationwide automatic
direct dial network service were substantially accomplished during the late
1970s. The international telecommunications market had been monopolised by Kokusai Denshin Denwa Co., Ltd. (KDD), created by the Government in 1953 also as part of post-war reconstruction reforms. Subsequent developments, including an increase in demand for sophisticated
and diversified services, technological progress in network development and
interconnection and the growing economic and technological capability of
private industry in telecommunications, led to a recognition by the Japanese
Government that centralised telecommunications operations were no
longer required. As a result, in April 1985 the Government reformed
Japan's telecommunications industry, accomplishing a transition from
regulation to liberalisation and from monopoly to competition.

1985 Reforms

24–02   The major reformation of Japan's regulatory framework in 1985 was achieved through the enactment of two laws—the Nippon Telegraph and Telephone Corporation Law (NTT Law) and the Telecommunications Business Law (TBL). The first law resulted in the privatisation of NTT Public Corp. as NTT Corp. and provides the basis for certain restrictions on NTT. The second law provides the foundation for the reforms and the currently existing framework for the telecommunications business. Telecommunications business is now categorised as either Type I (providing telecommunications services via one's own telecommunications circuit facilities) or Type II (providing services through circuits leased from entities operating Type I businesses). For the purposes of this Chapter, the entities operating such telecommunications businesses are referred to as Type I or Type II carriers.

Japan's market is now characterised by active competition in most service areas in the Type I businesses, the emergence of a large number of Type II businesses offering a diverse range of services, more flexibility in rate setting and the investment of foreign capital in Japan's telecommunications sector. Liberalisations permitting ownership by users of telephone sets (and more recently, in 1994, cellular phones), rather than lease from NTT (or the cellular telephone service providers) and gradual reforms of equipment procurement policies have also increased competition by encouraging consumer electronics manufacturers to enter this manufacturing sector and by diversifying supplier participation.

Further NTT Reforms

24–03   The Japanese Government has continued to explore and to implement methods for fair and effective competition between NTT and the new common carriers (NCCs) that emerged as a result of the reforms. These efforts are intended to improve market conditions for the NCCs, which must still access NTT's monopolised local networks. In response to a report by the Telecommunications Council of the Ministry of Posts and Telecommunications (MPT) on a proposed restructuring of NTT, the Government promulgated the Measures to be Taken in Accordance with Article 2 of the Supplementary Provisions of the Nippon Telegraph and Telephone Corporation Law (1990 Governmental Measures). As a result, in April 1992, NTT was separated into a Long-Distance Communications Division and a Regional Communications Division. In July 1992, NTT's mobile communications business, which included paging and cellular telephone service, was established as a separate subsidiary (NTT DoCoMo), and subsequently, in July 1993, the service area for such separate entity was divided into nine regional areas covering Japan. Interconnection (as further described in paragraphs 24–50 to 24–51), and open network issues were also addressed. The Government's continuing attempts to provide competition with NTT in new markets were recently evidenced by the requirement by the MPT that NTT establish a separate corporation to apply for licenses to provide personal handy phone service. However, pressure appears to continue for a formal separation of NTT into separate regional companies, similar to the creation of separate regional telephone companies in the U.S. following the breakup of AT&T. In 1995, pursuant to the 1990 Governmental Measures, the Government is required to review the status of NTT and to make recommendations, taking into account the results of such measures.

Industry Structure

Overview

Japan's telecommunications industry can be divided into two principal cat-   **24–04**
egories: Telecommunications Services (Domestic Type I, International Type I, Special Type II and General Type II) and Broadcasting Services (public and private radio and television terrestrial broadcasting, satellite broadcasting, cable television and cable sound broadcasting). Additionally, a special wire broadcast telephone service exists, primarily in rural areas, where cable sound broadcasting facilities are used for both community sound broadcasts and telephone service, which is governed under a separate law. This subsection briefly describes the current environment for primary categories of the telecommunications service business and the broadcast industry. Statistical and factual information, as well as the state of the law, is provided as of March 1995, unless another date is referenced.

Telecommunications Industry

Type I Telecommunication Businesses

There are 109 NCCs, in addition to NTT and KDD (which were the only   **24–05**
common carriers prior to April 1985).

Long Distance Domestic Carriers: These carriers provide long distance tele-   **24–06**
phone and data services using their own transit trunks, but use NTT local circuits at a point of interface for access to the caller and the receiver. The three carriers in this group—Japan Telecom Co., Ltd (Japan Telecom) DDI Corp. (DDI), and Teleway Japan Corp. (Teleway Japan)—started leased circuit services in 1986 and started telephone services in 1987. NTT also provides telegraph, telegram and ISDN services. Japan Telecom, whose initial primary investors were three of the Japan Railway companies, sold shares to the public in 1994. In 1993, DDI sold shares to the public. The largest investors in DDI and Teleway Japan (which is still private) are Kyocera and Toyota Motor Corp., respectively.

Regional Domestic Carriers: These carriers provide leased circuit services   **24–07**
via their own communications networks in limited regions. Eleven companies provide, or are scheduled to provide, services in separate regions, with TT Net, the initial entrant in this group, also providing telephone services.

24-08   **Satellite Carriers:** These carriers provide leased circuit satellite communications services via their own communications satellites. In addition to NTT, there are two NCCs: Space Communications Corp. (SCC), in which Mitsubishi Corp. and Mitsubishi Electric Corp. are the largest investors, and Japan Satellite Systems, Inc. (JSAT), which is jointly owned by four of Japan's largest trading companies and resulted from a merger of Japan Communications Satellite Co., Inc. and Satellite Japan Corp. in August 1993. SCC owns the Superbird-a and Superbird-b satellites (both launched in 1992), having transponders in both the Ku band and Ka band. JSAT has JCSAT-1 and JCSAT-2 (launched in 1989 and 1990, respectively), having transponders in the Ku band. The National Space Development Agency, NTT and certain other companies own CS-3a and CS-3b, which include both Ka band and C band transponders. JSAT is planning to launch JCSAT-3, which will include Ku band and C band transponders, and NTT is planning to launch two additional satellites (N-STARa and N-STARb), which will include Ka band, Ku band, S band and C band transponders. In conjunction with June 1994 amendments to the TBL and Radio Law permitting international satellite owners to provide international telecommunications service into Japan, policy changes were adopted to allow the satellite NCCs, previously limited to domestic services, to provide services outside of Japan. With the significant increase in the number of other Asian satellites, both telecommunications and broadcasting services are expected to be increasingly provided through these other satellites.

24-09   **International Carriers:** These carriers provide international telecommunications services, such as leased circuit and telephone services, via their own submarine cables and by linking with overseas carriers. In addition to KDD, there are two NCCs: International Telecom Japan Inc. (ITJ), known by the 0041 access code, and International Digital Communications Inc. (IDC), known by the 0061 access code, both of which commenced leased circuit and telephone services in 1989. ITJ, whose primary investors are three of the major Japanese trading companies, has foreign investment from France Telecom and British Telecommunications plc. IDC, whose major investors are Itochu and Toyota Motor Corp., includes Cable & Wireless plc as a major foreign investor.

24-10   **Cellular Phone Carriers:** These carriers provide cellular phone services by establishing their own base stations in service areas. There are 26 carriers (including the nine NTT DoCoMo regional subsidiaries and companies mostly affiliated either with DDI or Japan Telecom) who provide, or are scheduled to provide, regional services, with generally two carriers licensed per area. Although analogue systems were previously used in all areas, certain carriers have recently commenced digital services using either the 800 Mhz or 1.5 Ghz band. Nippon IDOU Tsushin Corp. (IDO), which has investment from Teleway Japan and Toyota Motor Corp., is the largest of these NCCs and serves the Kanto area (which includes Tokyo), and Chubu area. They have introduced separate analogue systems using both the NTT system and an incompatible Motorola system.

Person Handy Phone: Personal handy phone service was allowed by a 1993   **24–10a**
amendment to the MPT regulations. The MPT granted licenses to the separate NTT subsidiary, a subsidiary of DDI and a consortium which
includes Japan Telecom. Service is expected to commence in July
1995.

Radio Paging Carriers: These carriers provide radio paging services by   **24–11**
establishing their own base stations in service areas. There are 40 carriers
(including the nine NTT DoCoMo regional subsidiaries) providing regional
services, with two carriers licensed per area.

Type I Telecommunications Business Market Changes

The entry of the NCCs into the international and domestic telecommunica-   **24–12**
tions business has resulted in both a decrease in market share by the established carriers in each area and a reduction in rates. According to 1993
fiscal year information, the market shares of the NCCs were 23.0 per cent
and 29.5 per cent in the international leased circuit and telephone areas
respectively, in competition with KDD. Competition with NTT (and its
affiliated companies) within the domestic market has shown a similar trend,
with the scope of the effect varying by service type. The NCCs' market
share ranged from 10.1 per cent for telephone services to 36.3 per cent for
cellular service.
  Liberalisation of the rate-setting process and the introduction of competition by the NCCs has resulted in rate reductions of as much as two-thirds,
depending on the service sector, compared with rates immediately prior to
the reforms. The NCCs entered the markets with rates below those of the
established carriers. Subsequent progressive reductions by both the established carriers and the NCCs resulted in still lower rates, but the difference
between the competitors' rates has greatly decreased due in part to the
MPT's limits on approval for further reductions in rates and fees. However,
since rates still remain high compared with other countries (particularly the
U.S.), somewhat controversial "call-back services" in the telephone service
sector are proliferating to allow users to take advantage of charges at significantly lower rates in other countries.

Special Type II Telecommunications Business

There are 44 Special Type II carriers, which lease circuits from Type I   **24–13**
carriers but provide large scale (to many and unspecified users) and/or
international systems. The wide-ranging scope of services provided by these
carriers includes (i) communications-related services, which can be divided
into basic communication services (including virtual private networks,
resale of leased circuits, packet-switching, circuit-switching) and value-
added communication services (including E-mail, electronic data exchange,
file transferral, voice mail) and (ii) info-communications services (including
online information processing and online databases). Of these carriers, 31
provide international VAN services, in conjunction with bilateral agreements between Japan and each of the currently 23 countries.

General Type II Telecommunications Business

24–14 The General Type II carriers, which number 2,063 and whose numbers are growing rapidly, provide similar services in the domestic sector and to a limited user group, mostly in conjunction with, or in support of, other types of business activities.

Broadcast Industry

Terrestrial Broadcasters

24–15 **NHK:** The Japan Broadcasting Corporation (NHK) is a public-interest corporation established in 1950 under the Broadcast Law to provide AM or FM sound broadcasting and television broadcasting reception throughout Japan.

NHK provides two terrestrial television broadcasting channels (one general and one educational), two BS television broadcasting channels, two AM sound broadcasting channels and one FM sound broadcasting channel. NHK also provides international short-wave broadcasting services from Japan (as Radio Japan).

NHK's revenues are derived mostly from annual subscription fees which are required to be paid under the Broadcast Law by any person who has equipment capable of receiving NHK television broadcasting. NHK is prohibited from broadcasting advertisements for other businesses and, therefore, receives no advertising revenues.

The Broadcast Law was amended in 1994 to allow television broadcasting to commence from Japan to other countries. NHK will now be able to broadcast television programmes overseas through satellite carriers and will be able to enter into partnerships to offer NHK programs to foreign cable television broadcasters.

24–16 **Commercial Terrestrial Broadcasters:** Commercial terrestrial broadcasters, which are private corporations established under the Commercial Law, there are 203 companies operating 7,736 television stations, 240 AM stations, 239 FM stations, three FM mutiplex stations, two shortwave stations, 7,194 television sound multiplex stations and 5,570 television teletext multiplex stations. Revenues of commercial terrestrial broadcasters are derived mostly from advertising.

While NHK has authority to provide nationwide ground-based networks, commercial terrestrial television broadcasters are generally licensed by region. However, commercial terrestrial television broadcasters are loosely organised into networks centered on five flagship broadcasters in Tokyo for procuring a stable supply of television programmes. The five flagship television broadcasters of the networks have very close ties to the major Japanese newspaper publishers. Similarly, the local television broadcasters have close ties to the local newspaper publishers.

As a result of the previously referenced 1994 amendment to the Broadcast Law, commercial broadcasters will be able to entrust their television programs to "facility supplying broadcasters" for broadcasting intended to be received within Japan and overseas.

Since 1966, the MPT had conditioned the grant of a provisional licence for a broadcasting radio station on the applicant's Articles of Incorporation providing that a transfer of its shares was subject to the approval of its Board of the Directors. This condition, which resulted in difficulties for the applicant to list its shares on a Japanese stock exchange, was terminated by the MPT in February 1994.

The University of the Air: The University of the Air is a fully accredited   **24–17**
university, maintained by the University of the Air Foundation, a public-interest corporation established in 1981 under the University of the Air Foundation Law (law no. 80 of 1981) to utilise broadcasting to promote a new university education system where all courses are broadcast.

Satellite Broadcasters

BS Broadcasters: Broadcast Satellite (BS) broadcasters both establish and   **24–18**
operate satellites and supply programmes. The BS services are offered via the BS-3a satellite (launched in 1990) and the BS-3b satellite (launched in 1991). The BS services offered are television broadcasting services provided by NHK and Japan Satellite Broadcasting Inc. (JSB) and television sound multiplex broadcasting services provided by Satellite Digital Audio Broadcasting Co. Ltd. (SDAB). Since November 1991, the BS-3b satellite also has been used for high-definition TV (HDTV) test broadcasting. As of December 1994, NHK and JSB had a total of 8 million subscribers for their BS satellite broadcasting services, while SDAB had only 81,000 subscribers.

CS Broadcasters: Communication Satellite (CS) broadcasting services are   **24–19**
provided by communication satellites originally intended for use for telecommunications which are now also being used for broadcasting pursuant to a 1989 amendment of the Broadcast Law. Before this amendment, to qualify as a broadcaster an entity had to both (i) establish and operate a broadcasting station and (ii) supply and have responsibility for the compilation of broadcasting programs. The amended law now allows the entity establishing and operating a CS satellite to be different from the entity supplying and compiling the programmes. The former is defined as a "facility supplying broadcaster" and the latter as a "program supplying broadcaster".

CS broadcasting services are provided via the JCSAT-2 and the Superbird-b satellites owned by JSAT and SCC, respectively. CS services include 14 pulse code modulation (PCM) sound broadcasting channels and 11 television broadcasting channels which are provided, or are scheduled to be provided, by programmes supplying broadcasters. In 1993, two PCM sound broadcasters ceased operations, but four new television broadcasters were approved.

As of December 1994, there were about 22,000 subscribers to PCM sound broadcasting services and about 239,000 subscribers to CS television broadcasting services. Programmes supplying broadcasters, whose revenues are derived mostly from subscription fees, provide revenues to the facility supplying broadcasters by payment of charges for use of the broadcasting facilities.

Cable Broadcasters

24–20 Cable Television Broadcasters: Cable television broadcasting in Japan was originally intended to serve areas with poor terrestrial broadcasting reception. The MPT limited the grant of licences to owners in the operational area, resulting in a proliferation of very small facilities, in contrast to the U.S. As of March 1994, there were over 55,000 cable television facilities serving around 9 million subscribers, over 90 per cent of which were facilities with less than 500 drop terminals and about 70 per cent of which conducted only retransmitting business. "Urban cable television facilities", originally established in 1987, are now becoming wide-spread. An urban cable television facility is a cable television facility that has (i) 10,000 or more drop terminals, (ii) five or more channels for broadcasting its original programmes, and (iii) interactive capability. As of March 1994, there were about 1.5 million subscribers to urban cable televisions and 158 urban cable television facilities. The MPT's policy that a cable television broadcaster must have its operational basis (such as its shareholders and officers) in a region where its facility is established was abolished in December 1993. This policy change is expected to result in a consolidation in the cable television market, leading to more efficient service and a decrease in the currently high charges.

The development of urban cable television has been facilitated by space cable network systems in which a provider of cable television programmes (regulated under the TBL as a Type II carrier) supplies a cable television broadcaster with programmes via a CS satellite, rather than by the previous means of delivery of video tapes, making a wider selection of programmes possible.

The Cable Television Broadcast Law distinguishes between two types of cable television broadcasters—the type which establishes and operates a cable television broadcasting facility, and the other which leases a cable television broadcasting facility established by others.

24–21 Cable Sound Broadcasters: As of March 1994, there were about 12,000 cable sound broadcasting facilities, over half of which were used for broadcasting community information in rural areas. Of these facilities, 462 facilities were used also for telephone services.

Continuing Technology Development

24–22 Japan's industry and Government continue to investigate and implement new developments in telecommunications technology. By the end of fiscal year 1995, NTT is expected to complete an ISDN infrastructure covering the whole of Japan with either INS Net 64 or INS Net 1500 (the two ISDN standards in Japan). As previously mentioned, personal handy phone service will commence in July 1995. In the broadcast area, Japan's HDTV broadcasting, which had used an analogue system called MUSE, is expected to change to a digital system. Increasing attention is being focused on multimedia and the integration of various digital information and telecommunications services, partly in response to globalisation in this area.

2. CURRENT REGULATORY AND LEGISLATIVE ENVIRONMENT

Institutional Structure

Ministry of Posts and Telecommunications

Since its establishment in 1949 as one of the primary ministries of the 24–23 Japanese Government and successor to the Ministry of Communications, the MPT has been responsible for all aspects of telecommunications, as well as the postal system. Its minister is appointed by the Prime Minister and is a member of his cabinet. Although the minister changes with the political party in power, the bureaucracy controlling policies and their administration remains stable, like most other ministries in Japan.

The MPT is divided into a number of bureaux responsible for various aspects of postal services and telecommunications. The telecommunications bureaux include the Communications Policy Bureau, the Broadcasting Bureau and the Telecommunications Bureau (the latter further being divided into the Telecommunications Business Department and Radio Department, each responsible for the administration of the laws governing these separate areas). In addition to a number of institutions established under the MPT, including the Institute for Posts and Telecommunications Policy and the Communications Research Laboratory, the MPT has a number of councils which have important policy responsibility.

Telecommunications Regulatory Organisations

The Telecommunications Technology Council of the MPT has overall 24–24 responsibility for coordinating with the ITU to set and conform to international standards and to coordinate with outside administrative bodies and various manufacturers, users, institutes and other organisations in establishing standards for Japan. Detailed procedures have been developed for the standardisation process.

Responsibility for wired communication systems rests with the Telecommunications Technology Committee (TTC) and for radio communications with the Research and Development Center for Radio Systems (RCR), which were both established as public interest corporations for setting domestic standards.

The Radio Regulatory Council (RRC), which is composed of five members appointed by the MPT (with consent of the Diet), is established under the Radio Law. The RRC is provided with authority to investigate and make recommendations to the MPT on a broad range of matters relating to the radio waves and broadcasting and to examine and adopt resolutions on complaints filed against administrative dispositions by the MPT. The RRC must give interested parties an opportunity to express their opinions in its review of the MPT's establishment, modification or abolishment of MPT ordinances relating to specified provisions of the Radio Law.

Authority for approving compliance with the technical standards has been given by the MPT to the Japan Approvals Institute for Telecommunications Equipment (JATE) for terminal equipment and to the Radio Equipment Inspection and Certification Institute (*Musen-setsubi Kensa-kentei Kyokai*, or MKK) for radio equipment (as discussed in paragraphs 24–67 *et seq.*).

Principal Legislative Instruments

24–25   Described below are the principal laws which govern telecommunications in Japan, as grouped by the function of the law.

General Laws

24–26   These laws provide the administrative framework for the principal regulatory body or frameworks applicable to all types of business.

- Ministry of Posts and Telecommunications Establishment Law (law no. 244 of 1948). This law provides for the establishment, duties, jurisdiction and organisation of the MPT.
- Law concerning Anti-trust and Assurance of Fair Trading (law no. 54 of 1947). This law provides the general framework of antitrust and fair trade regulation applicable to all types of businesses and transactions.
- Administrative Procedure Law (law no. 88 of 1993). Although not directly oriented toward telecommunications, this law is intended to provide a uniform process and framework for dealing with governmental administrative agencies. These procedures could assist in minimising the frustrations present in the licensing process. Pursuant to the Administrative Procedure Law, the MPT issued examination standards and standard review periods for administrative acts relating to telecommunications (including licence, permission and approval). The examination standards and the standard review periods are available at the MPT and the regional bureaux of telecommunications.

Facilities Laws

24–27   These laws provide the framework for establishment and operation of telecommunications facilities applicable to all types of business (including the broadcasting business).

- Cable Telecommunications Law (law no. 96 of 1953). The basic law in the field of cable telecommunications, this law regulates mainly the establishment and operation of facilities (exchanges, equipment, lines and other electrical facilities) used for cable telecommunications and broadcasting.

- Radio Law (law no. 131 of 1950). The basic law in the field of radio communications, this law regulates mainly the establishment and operation of facilities used for radio telecommunications and broadcasting.

Business Operations Law

Because Japanese law distinguishes between the concepts of "broadcasting" and "telecommunications", businesses in these two areas are governed by different laws. **24–28**

- Telecommunications Business Law (law no. 86 of 1984) (TBL). This law, which replaced the Public Telecommunications Law (law no. 97 of 1953), regulates the conduct of most types of telecommunications business in Japan. The TBL sets forth the basic rules to be applied to telecommunications carriers.
- Broadcast Law (law no. 132 of 1950). This law regulates the operation of the broadcasting business (excluding the cable television broadcasting business and the cable sound broadcasting business), and the establishment and scope of NHK.

Further, businesses governed by the following group of laws are excluded from the "telecommunications business" governed by the TBL and relate to the type of services provided:

- Cable Television Broadcast Law (law no. 114 of 1972). This law regulates the operation of the cable television broadcasting business, and the establishment and operation of certain cable television broadcasting facilities.
- Law to Regulate the Operation of the Cable Sound Broadcasting Services (law no. 135 of 1951). This law regulates the operation of the cable sound broadcasting business.
- Law Regarding Wire Broadcast Telephones (law no. 152 of 1957). This law regulates the operation of the cable broadcast telephone business, which utilises cables in primarily rural areas for both community cable sound broadcast and for telephone service.

Specific Telecommunications Operations Organisation Laws

These laws were enacted to regulate specific telecommunications carriers. **24–29**

- Nippon Telegraph and Telephone Corporation Law (law no. 85 of 1984). The NTT Law was enacted to place certain obligations on NTT considering the public nature of its business (although the general commercial law and corporation law apply to its operations as a non-governmental company).
- KDD (Kokusai Denshin Denwa Co., Ltd.) Law (law no. 301 of 1952).

The KDD Law was enacted to privatise the international sector of the telecommunications business.

Regulations; Guidance

24–30   The basic laws provide a very general framework in each applicable area and are implemented by the promulgation of Cabinet Orders and MPT Ministerial Ordinances, and the issuance of MPT Ministerial Notices which together constitute the body of detailed regulations that interpret the laws. Also relevant are ministerial guidelines, which are generally not issued in writing but are important to provide direction on the approach which could be expected to be applied by the MPT in certain situations. Ministerial Guidance is issued in writing to specific parties to provide practical direction on specific issues. Additionally, information will be communicated during the informal discussions with MPT officials that are usually part of the licensing process.

Foreign Ownership Restrictions

24–31   The restrictions imposed on foreign ownership and participation in tele-communications-related companies vary by the type of facility and service governed under the different laws. With the implementation of the 1985 reforms, foreign investment in the telecommunications business in Japan became possible.

TBL Restrictions

24–32   Under the TBL, there are foreign ownership restrictions on Type I carriers, but not on either of the Type II carriers. The following foreign ownership or management restrictions are applicable to Type I carriers. A telecommunications business carrier cannot be granted a licence by the MPT to carry on a Type I Telecommunications Business if such carrier is:

   (i)   an individual of non-Japanese nationality;
   (ii) a foreign government or its representative;
   (iii) a foreign entity or organisation; or
   (iv) any entity or organisation which is represented by any person or body set forth in the preceding three items, or one-third or more of whose officers are such persons, or one-third of whose total voting rights is exceeded by the aggregate of voting rights directly held by such persons or bodies and voting rights defined as indirectly controlled by such persons or bodies through any other entity or organisation by the Ordinance for Executing the TBL. (With regard to the application of indirectly controlled voting rights for this item, the Ordinance sets forth an attribution approach that uses a multiplication method if a person or body coming within items (i) to (iii) above holds at least 10 per cent of an intermediate entity or organisation which holds at least 10 per cent of the voting rights of such carrier.)

Pursuant to the TBL, a Type I carrier, whose shares are listed on the Japanese stock exchanges or registered by the Japan Securities Dealers Association, may refuse to register the name and address regarding shares acquired by any person or body mentioned in items (i) to (iii) above or any entity or organisation at least 10 per cent of whose voting rights are held by such person or body, if compliance with the request for registration would result in the Type I carrier falling within the provisions of item (iv) above. The restrictions on the registration of shareholders of NTT and KDD provided in the NTT Law and the KDD Law are more stringent, prohibiting shareholder registration which would result in NTT or KDD falling within the provisions of item (iv) and limiting ownership of voting rights to one-fifth of the total voting rights.

However, pursuant to amendments to the TBL which became effective in June 1994, international satellite carriers are now permitted to apply for licences as Type I carriers to supply international satellite services, with no restrictions on foreign ownership and the registration of shareholders.

Radio Law Restrictions

With certain limited exceptions, under the Radio Law foreign ownership   **24–33** restrictions are placed on the holders of licences granted by the MPT for establishing radio stations not used for broadcasting (including facility supplying broadcasters). Although the restrictions under the Radio Law are the same as those under the TBL with respect to items (i), (ii) and (iii), paragraph 24–32, along with the management limitations in item (iv), there is an important difference with regard to voting rights. Indirectly controlled voting rights are not counted and therefore no attribution approach is used with regard to ownership of voting rights, resulting in different calculations for investments by Japanese corporations that have foreign ownership or which are intermediate entities.

Foreign nationals can now acquire multichannel access radio and amateur radio licenses (pursuant to June 1993 amendments) and can hold licences for certain other limited applications. Pursuant to amendments to the Radio Law effective June 1994, concurrently with the amendments to the TBL referred to above, international satellite carriers are now permitted to apply for licences for establishing satellite stations, with no restrictions on foreign ownership.

The restrictions for licences for establishing radio stations used for broadcasting (other than for relay broadcasting and for facility supplying broadcasting) are more stringent. Any foreign person in the position of an officer is precluded from executing the licensee's business (rather than the less than one-third of the officers under the TBL) and ownership of voting rights is limited to one-fifth of the total voting rights, but without including indirectly controlled voting rights, which would therefore use no attribution approach.

Broadcast Law Restrictions

Under the Broadcast Law, the same foreign ownership restrictions applic-   **24–34** able to licences for establishing radio stations used for broadcasting apply to the approval by the MPT for conducting a programme supplying broadcasting business. Furthermore, similar restrictions on the registration of shareholders to those applicable to Type I carriers apply to broadcasters (other than NHK and The University of the Air).

Cable Television Broadcast Law Restrictions

24-35   Under the Cable Television Broadcast Law, foreign ownership restrictions apply to the grant by the MPT of licences for establishing cable television broadcasting facilities. Although the restrictions are similar to those applicable to licences for establishing radio stations used for broadcasting, there is an important difference with regard to the MPT's discretion. The MPT is granted discretion in, and is not prohibited from, issuing a licence if a company falls within the listed conditions. Using its discretion, the MPT is now allowing direct foreign ownership of less than one-third of the voting rights of a cable television carrier, rather than the one-fifth stated in the Cable Television Broadcast Law. However, a television cable carrier that only requires a notification under the Cable Telecommunications Law, as described in paragraph 24-61 *et seq.*, is not subject to any foreign ownership restrictions.

Cable Sound Broadcast Law Restrictions

24-36   Under the Law to Regulate the Operation of the Cable Sound Broadcasting Service, there are no foreign ownership restrictions on a cable sound broadcaster.

Wire Broadcast Telephones Law Restrictions

24-37   Under the Law Regarding Wire Broadcast Telephones, there are no foreign ownership restrictions on a wire broadcast telephone business carrier.

Penal Provisions

24-38   Each of the various laws includes penalty sections specifying maximum prison terms or fines for violation of the applicable provisions.

3. TELECOMMUNICATIONS SERVICES REGULATION

Licensing Regime

Overview

24-39   The telecommunications business in Japan is regulated by the TBL. A "telecommunications business" provides telecommunications services, defined as intermediating communications of others through the use of telecommunications facilities, or any other provision of telecommunications facilities for the use of communications of others. Specifically excluded are the following businesses: (i) facility supplying broadcasting business (governed by the Broadcast Law), (ii) cable sound broadcasting business (governed by the Law to Regulate the Operation of the Cable Sound Broadcasting Services), (iii) wire broadcast telephone business (governed by the Law Regarding Wire Broadcast Telephones), and (iv) cable television broadcasting business and the business of leasing cable television broadcasting facilities (governed by the Cable Television Broadcast Law). As stated in paragraphs 24–12 to 24–14, telecommunications business is divided into Type I and Type II, according to whether the carrier establishes or only uses the "telecommunications circuit facilities". Further, the provisions of the TBL (except for general provisions relating to censorship and secrecy, discussed below) do not apply to the following telecommunications businesses: (i) any telecommunications business which exclusively provides telecommunications services to a single person which is not a telecommunications carrier; (ii) any telecommunications business which provides telecommunications services through telecommunications facilities established in the same premises, or through telecommunications facilities which do not satisfy in scale the standards set forth in the Ordinance for Executing the TBL (five kilometer lines in total); and (iii) any Type II telecommunications business which provides telecommunications services other than intermediating communications of others through the use of telecommunications facilities.

Type I Telecommunications Business

A Type I telecommunications business provides telecommunications services by establishing telecommunications circuit facilities. The term "telecommunications circuit facilities" is defined to mean transmission line facilities connecting transmitting points with receiving points, switching facilities installed as inseparable units, and other accessory facilities. **24–40**

Type II Telecommunications Business

A Type II Telecommunications Business, further subdivided into special and general, is a telecommunications business other than a Type I Telecommunications Business. **24–41**

(i) **Special Type II Telecommunications Business**: There are two kinds of Special Type II telecommunications business: (a) a telecommunications business that provides a number of unspecified people with the use of telecommunications facilities that exceed in scale the capacity standards (500 circuits for 1200 b/s conversion) for accommodating telecommunications circuits set forth in the Order for Executing the TBL (domestic special Type II telecommunications business); and (b) a telecommunications business that provides telecommunications facilities designed for communications between Japan and foreign points for the use of communications of others (international special Type II telecommunications business). A carrier who carries on the latter business actually provides its services by transmitting data from an access point in Japan to a node via a private circuit leased from the Type I carrier who establishes such circuit.

(ii) **General Type II Telecommunications Business**: A general Type II telecommunications business is a Type II telecommunications business which is not a special Type I telecommunications business.

General Provisions

24–42   Several general principles apply to all carriers. Communications handled by a telecommunications carrier are not to be censored and their secrecy is to be protected. Carriers cannot discriminate unfairly in providing telecommunications services. A carrier is required to give priority to necessary emergency communications in the event of a disaster or similar emergency. Furthermore, an international carrier must comply with international treaties and agreements relating to the international telecommunications business.

Licensing Procedures

Type I Carrier Permission

24–43   A person intending to operate as a Type I carrier must file an application to obtain permission from the MPT. The application shall include (i) the name and address of the applicant, (ii) the category of telecommunications service in accordance with classifications listed in the Ordinance for Executing the TBL and a description of the service, (iii) the service area, and (iv) an outline of telecommunications facilities, in addition to a business plan and other specified documents. An applicant will not qualify if (i) the person or entity (or any of its officers) was fined or penalised or its license was revoked within the prior two years under specified telecommunications laws or (ii) the person does not satisfy the foreign ownership restrictions.

In addition to determinations that the applicant has an adequate financial basis and technical capability to perform properly the requested telecommunications business and that the plan for such business is feasible, the MPT will grant permission with reference to the following additional criteria: (i) the telecommunications service to be provided by the carrier must be appropriate in light of the demand in the service area, (ii) the introduction of such telecommunications business should not result in a significant excess of telecommunications circuit facilities to be used for such business in all or part of the area or route to be covered, and (iii) the introduction of such business should contribute to the sound development of telecommunications. Based on the foregoing, the MPT can exercise significant discretion in granting permission to Type I carriers. The standard review periods issued by the MPT following the implementation of the Administrative Procedure Law specify periods of one to two months as a general guideline for the review period after filing the formal application. However, the informal discussion prior to the filing generally requires a long period.

Upon obtaining permission, a Type I carrier is required to commence service within a prescribed period and, prior to commencement, obtain MPT confirmation that the facilities conform to technical standards. Any material changes in the matters described in an application require prior MPT permission. The permission granted is for an indefinite term, unless the MPT attaches conditions to the permission or unless revoked by the MPT upon certain conditions. Prior approval is also required before the carrier entrusts any part of its facilities to another person.

A merger or amalgamation of a Type I carrier (unless such carrier is the survivor in a merger with a person who is not a Type I carrier) or a transfer, inheritance, suspension or discontinuance of its business is not allowed without MPT approval.

Special Type II Carrier Registration

A person intending to operate as a Special Type II carrier must register by **24–44** filing an application with the MPT. The application must include information similar to that required for a Type I carrier, except that the service area need not be included and the number of required documents for attachment (which similarly includes a business plan) is fewer.

The MPT shall register the carrier on the Special Type II telecommunications carrier register unless (i) the person or entity (or any of its officers) was fined or penalised or its registration was revoked within the prior two years under specific telecommunications laws, or (ii) the person does not have an adequate financial basis and technical capability to perform properly the requested telecommunications business.

Although the term "registration" connotes a simple process, before filing an application, an applicant would in fact be required to visit and explain the proposed services to the relevant officer of the applicable division of the MPT, and would be subject to informal advice from such officer (*i.e.* the guidance referred to in paragraph 24–30). In many cases, the completion of the registration may require a number of visits to the relevant officer.

Any material changes in the registered matters require the amendment of the registration by application. Similar to the Type I carrier, the registration is effective indefinitely, unless the MPT revokes the registration for specified reasons. A merger or amalgamation of such carrier or transfer, inheritance, suspension or discontinuance of its business is treated in the same way as that of a General Type II carrier, with only a notification required.

General Type II Notification

A General Type II carrier is not required to be registered and is required **24–45** only to submit a notification to the MPT using a designated form. Any merger or amalgamation of such carrier or transfer, inheritance, suspension or discontinuance of its business also requires notification only.

Tariff and Pricing Policies

Type I Tariff Approval

24–46   A Type I carrier is required to establish a tariff which sets forth charges and the other terms relating to its services and obtain approval of the tariff from the MPT. A tariff approved by the MPT must meet the following standards: (i) charges must be fair and reasonable, (ii) calculating methods of charges must be stipulated, (iii) discussion of the respective responsibilities of the carrier and user and cost allocations related to installation of facilities and other works must be clear, (iv) the tariff cannot unreasonably restrict utilisation of the facilities, (v) no discriminatory provision can be included and (vi) emergency communication must be addressed. Any change in the items set forth in the tariff requires prior approval by the MPT.

Neither a Type I carrier nor a Special Type II carrier can provide telecommunications services except pursuant to the applicable approved or notified tariff; except that a Type I carrier can provide Type II carriers with approved non-tariff services, as discussed below. However, a Type I carrier or a Special Type II carrier can reduce or exempt charges to the extent permitted in the applicable ordinance. Pursuant to 1995 amendments to the TBL, certain charges, to be specified by MPT ministerial oridinance, will only need to be notified, rather than be approved. Furthermore, use by a Type I carrier of a "standard tariff" will also only require notification.

Special Type II Tariff Notification

24–47   A Special Type II carrier must also submit to the MPT a notification of the tariff setting forth the following items: (i) a description and contents of the proposed services; (ii) the area in which the proposed services are to be provided; (iii) a price schedule; (iv) provisions related to liabilities; (v) the method of important communications in an emergency; and (vi) certain other designated matters.

The tariff may be made effective from the date of acceptance. As with registration procedures, the normal practice is to submit a draft tariff to the appropriate division of the MPT for its review and, after pre-clearance on the draft is obtained, a formal notification will then be submitted. Any change in the items set forth in the tariff must be notified to the MPT. The procedural steps described above will apply to this notification procedure for changes.

General Type II Services

24–48   A General Type II carrier is not restricted in the pricing policies for its services, nor is it required to file any notification.

Non-Tariff Based Services

24–49   A Type I carrier may enter into a contract to provide services to a Type II carrier on different conditions from those stated in its tariff, but it must first obtain the approval of the MPT for the non-tariff based services. The main criterion for such approval is whether the non-tariff based contract will benefit the public by providing services with added value (*e.g.* electronic mail and EDI services). Applications normally have been approved if the prescribed criteria are met.

Interconnection of Telecommunications Facilities

Legal Framework

A Type I or a Special Type II carrier must obtain approval from the MPT   **24-50** to enter into or amend an agreement with another Type I or a Special Type II carrier for interconnection or sharing of telecommunications facilities. However, only notification to the MPT is required for a Type I or Special Type II carrier to enter into or amend an agreement with a General Type II carrier, or for two domestic Special Type II carriers to enter into or amend an agreement relating to facility interconnection or sharing. A Type I or Special Type II carrier must obtain approval from the MPT to enter into, amend or terminate an agreement with a foreign government or a foreign person or entity with regard to the material matters specified in the Ordinance for Executing the TBL.

Where negotiations between carriers for the interconnection or sharing of telecommunications facilities (excluding situations where only MPT notification is required), or for provision of non-tariff based service by a Type I carrier to a Special Type II carrier, fail to be conducted or to lead to an agreement, a party may request the MPT to order execution of an agreement to promote the public interest. Disputes with regard to the terms and conditions of the agreement arising after issuance of such order can be resolved through an arbitration conducted by the MPT upon a party's request, with ultimate recourse to court allowed if a party is dissatisfied with the financial terms awarded.

NTT Network Access

After the 1985 reforms, the emergent NCCs needed to use the functions   **24-51** and information of NTT's networks. The 1990 Governmental Measures were to implement a decision to clarify the specific access conditions related to the mode of connection and technical requirements and to allow other carriers to utilise NTT's networks under conditions equivalent to those of NTT. The MPT has had "Open Network Policy Study Meetings" which proposed ongoing dialogue between NTT and Type II carriers for opening networks and clarification by NTT of its plans in this area. In March 1992, NTT presented its plan to the MPT for implementation of open networks, which was revised in March 1993 and March 1994.

In October 1994, in response to MPT pressure, NTT agreed to allow Japan Telecom to use NTT local networks for frame relay systems, a service NTT had previously begun to provide. Also in late 1994, pursuant to arbitration proceedings with the MPT by DDI, Teleway Japan and Japan Telecom, MPT ordered NTT to open its local networks to allow them to provide virtual private network service (a system of codes linking offices globally), a service NTT had begun to operate in February 1994. In response to continuing controversy relating to access to NTT's local communication network, the MPT in February 1995 issued administrative guidance to NTT on clarification of procedures and related matters concerning negotiations on interconnection by other carriers to NTT's local network. The NTT responded to such guidance with the promulgation of a plan containing standards that NTT would apply, as to which the MPT solicited public opinions from interested parties to fully evaluate the implementation.

Connection of Terminal Equipment to Circuits

24–52   General principles regarding connection of a user's terminal equipment to a Type I carrier's circuit facilities are set forth in the TBL and the specific regulations promulgated by the MPT. A Type I carrier may not refuse a request from a user to connect the user's terminal facilities with the carrier's telecommunications circuit facilities except as specified by ordinance or where such connection does not conform to technical conditions established by the MPT and technical requirements established by the carrier with approval of the MPT.

Before a user may use terminal equipment connected to the Type I carrier's circuit facilities, the Type I carrier must inspect the user's facilities to ensure that the technical conditions and requirements are met. However, if the user's equipment has already received technical standards compliance approval from the MPT's designated agency, JATE, or if otherwise specified by MPT, then the Type I carrier will not inspect the connection (with limited exceptions if there is evidence of interference or malfunction).

In addition, a Type I carrier shall not refuse a request to connect its circuits to customer-provided telecommunications facilities (*i.e.* a user's facilities other than terminal facilities) unless such connections will not conform to technical conditions and requirements or the carrier receives certification from MPT that such connection would make it financially difficult for the carrier to maintain telecommunications circuit facilities.

A licensed installation technician must execute or supervise the connection of the user's facilities with the Type I carrier's circuit facilities, unless the terminal equipment had received technical standards compliance approval and is connected in the manner specified by the applicable MPT notice. The TBL provides the framework for licensing requirements and allows MPT to designate an examination agency to conduct qualification examinations for applicants for licences for installation technicians and chief telecommunications engineers (to supervise the installation, maintenance and operation of the carriers' own facilities).

Telecommunications Facilities Maintenance

Methods of services and maintenance and control of telecommunications 24–53
facilities are under the supervision of the MPT. Type I and Special Type
II carriers are required to establish administrative rules for their facilities,
submitted in advance to the MPT, to insure reliable and stable services.
Such carriers are also required to appoint properly licensed chief telecommunications engineers.

Use of Land and Water

A Type I carrier may request the owner of land or buildings to negotiate 24–54
a use right for the carrier's wires, cables, antennas and other facilities if
such use is reasonable and necessary, subject to the approval of the prefectural governor of the area. Any use may not seriously interfere with the
ordinary use of the land and buildings and the use of a building is limited
to the support of lines. The term of the right is 15 years (or 50 years for
the establishment of underground cables or structures or ground structures
made of steel or concrete). Notification is submitted to the prefectural governor upon agreement.

If negotiations do not result in agreement, the Type I carrier may apply
to the prefectural governor for arbitration, which would involve public
notice and the presentation of opinions from interested parties and would
result in a decision regarding the details of the use right, including compensation (the standard amounts for compensation for typical uses and
lands are set forth in the Ordinance for Executing the TBL).

Provisions are also included for temporary use of land during the process
of establishing facilities or lines, entry into or passage through land, cutting
trees and relocation of lines. After termination of use by a Type I carrier,
the land is required to be restored to its original state or appropriate compensation provided.

Procedures are specified for obtaining use of public waters for laying an
underground cable (with participation in the process by the Ministry of
Agriculture, Forestry and Fishing if fishing rights are involved and such
ministry has authority), and protection of such cables.

4. RADIO COMMUNICATIONS REGULATION

Overview

Even if a telecommunications or broadcasting business is governed by the 24–55
TBL or the Broadcast Law, respectively, if the business uses radio waves
in its operation, separate licensing is required under the Radio Law. The
Radio Law and the various implementing regulations thereunder provide
the framework and detailed provisions for the licensing of radio stations

and operators, approval and certification of equipment, and requirements for station operation and inspection. Except for very limited cases, equipment used only for reception is not subject to MPT regulation.

Licensing Requirements

24–56 Except for certain low power stations, a radio station cannot be established without obtaining a licence in advance from the MPT.

A licence is obtained by submission to the MPT of an application and documents containing information specified in the Radio Law (such as purpose, reasons, counterparty of communications, kind of communications, equipment location, emission type, frequency and power, operational focus, construction details and scheduled date of commencement of operations), with additional information required for radio station licences for broadcasting, ships, aircraft and satellites. Upon submission, if the application complies with (i) technical standards (ii) the feasibility of frequencies, and (iii) the Essential Standards for the Establishment of Radio Stations (other than Broadcasting Stations), the MPT will issue a provisional licence for the station. The MPT may request a personal appearance and submission of additional information as part of the process. The application period will vary depending on the type of facilities to be licensed. After installation is completed, the MPT will inspect the radio facilities to check for conformity with the provisional licence and the qualifications of the station operators and will grant a full licence upon passing inspection. For equipment which obtained technical standards compliance certification, radio station licences can be granted using simplified procedures (as further described in paragraph 24–69). The licence will be valid for the term fixed in the Ordinance for Executing the TBL, with a maximum limit of five years for general purposes (with certain exceptions) and will specify in detail the limitations on operations. However, the initial term may be shorter than the prescribed term because the MPT specifies the same expiration date of licences for certain types of radio stations. After expiration of the licence term, an application to the MPT for regranting a licence is necessary. The simplified licensing procedures can be used for the regrant of the licence. Any changes relating to the items in the licence will require prior MPT permission. Licences may be inherited by a person or assigned to a successor of a company formed by amalgamation or into which such company is merged with permission by the MPT.

The radio licensing process for broadcasting stations is generally similar to the standard process. The documents required to be attached to the application additionally include a business plan, types of broadcasting and broadcast area. Further, the MPT will initially review whether the application complies with (i) technical standards, (ii) the feasibility of the frequencies requested in accordance with the Plan for the Available Frequencies Allocated for Broadcasting, (iii) adequate financial basis requirements and (iv) the Essential Standards for the Establishment of Broadcasting Stations. Since radio broadcasting station equipment cannot receive technical standards compliance certification, a simplified licence application cannot be used for the initial application.

An ordinance under the Radio Law implements a fee schedule for activities governed by the Radio Law (such as station and operator licensing, technical standards compliance and inspection). The Radio Law also provides for the annual payment of specific "Spectrum User Fees" to fund certain administrative expenses.

Radio Operators

With certain exceptions (which include use of low power stations which   24–57
obtained technical standards compliance certification), a radio station can only be operated under the supervision of an appropriately licensed radio operator. The Radio Law establishes a classification system for radio operators and provides for designation of a training agency and application and examination procedures for operator licensing.

Operational Issues

Operational requirements for various types of radio stations are specified   24–58
in the Radio Law, along with general prohibitions on interfering with other stations and intercepting or divulging communications directed to a particular person.

The Radio Law also provides authority for the MPT to issue call signs for operation, and to change radio equipment frequencies, location or power, with appropriate compensation for losses. The MPT must inspect certain stations at regular intervals for compliance and has designated an inspection agency for this purpose, and may revoke station and operator licences if violations occur. The MPT has authority to limit the construction of structures within "radio propagation blockage prevention areas" for routes of important radio communications.

Administrative Decision Disputes

The Radio Law provides procedures for handling protests against adminis-   24–59
trative dispositions made by the MPT. Objections submitted to the MPT are initially referred to the RRC, which is required to hold a hearing. An action may be brought in a court with appropriate jurisdiction to reverse the final MPT decision.

5. BROADCASTING REGULATION

Basic Legislative Rules

Overview

Under Japanese law, the concept of "broadcasting" is distinguished from   24–60
that of "telecommunications" by the intention for the communications to

be directly received by the general public (unspecified persons), rather than specified persons. In determining the class of persons intended for receipt, the following factors are considered: the relationship between sender and receiver; whether contents of communications are based on the nature of such relationship; whether the communications are intended to be secret; whether the receiving equipment is controlled by a sender; and whether advertisements are placed in communications.

Service and Regulatory Distinctions

24–61 Broadcast communications are separately regulated based upon whether the medium of communication is radio or cable. Cable broadcasting is further separated for regulatory purposes into television and sound broadcasts. For each of these three categories, the operation of broadcasting business, including the compilation of programmes (the software side), is separately regulated under separate laws from those governing the establishment and operation of the broadcasting facilities (the technical and hardware side). The laws applicable to each category are listed in the table below.

If a cable broadcaster intends to conduct a telecommunications business (such as providing telephone services) by using its cable facility, in addition to a broadcast service business, such broadcaster will be subject to the Cable Telecommunications Law with respect to the establishment and operation of the facilities and the TBL with respect to the operation of telecommunications business. However, if a Type I carrier establishes cable facilities, the notification requirement under the Cable Telecommunications Law is exempt. In November 1994 the MPT issued guidelines on the commercialisation of cable telephony.

Laws Regulating Broadcast Business and Facilities

Type of Broadcasting	Broadcast Business	Facilities
Radio Wave (excluding programme supplying broadcasters)	Broadcast Law	Radio Law
Radio Wave (Programme supplying broadcasters)	Broadcast Law	Not applicable
Cable Television (More than 500 drop terminals)	Cable Television Broadcast Law	Cable Television Broadcast Law; Cable Telecommunications Law
Cable Television (500 or fewer drop terminals)	Cable Television Broadcast Law[1]	Cable Telecommunications Law
Cable Sound	Law to Regulate the Operation of the Cable Sound Broadcasting Services	Cable Telecommunications Law

[1] Cable Television Broadcast Law does not apply to rebroadcast through a facility that has 50 or less terminals.

Licence Procedures

Radio Wave Broadcasters

A broadcaster who uses radio waves must obtain a licence from the MPT   **24–62**
for establishing a broadcasting station in accordance with the Radio Law.
A separate licence under the Broadcast Law is not required. The licensing
procedures for establishing a broadcasting station are similar to those for
establishing other radio stations and are described in paragraph 24–56.

In addition to the licence requirements under the Radio Law, a facility
supplying broadcaster must notify the MPT of the charge for broadcasting
services for programme supplied by the programme supplying broadcasters
and other conditions for providing the services in accordance with the
Broadcast Law, and cannot vary those conditions unless such broadcaster
again notifies the MPT.

Programme Supplying Broadcasters   **24–63**

A programme supplying broadcaster must obtain approval from the MPT
for conducting programme supplying broadcasting business in accordance
with the Broadcast Law. The MPT will issue a certificate of approval which
will include the type of broadcast, broadcasting frequency and kind of
broadcast. If the broadcaster intends to change any of the kinds of broadcasting listed in the certificate, it must obtain permission from the MPT.
The MPT approval is effective for a term of five years and may be renewed
upon application to the MPT.

Cable Television Broadcasters

A cable television broadcaster who intends to establish a cable television   **24–64**
broadcast facility that has over 500 drop terminals must obtain a licence
from the MPT for establishing the facility in accordance with the Cable
Television Broadcast Law. Material changes to the facilities plan (including
the service area and location of facility) or the frequency to be used or
facility require prior permission from the MPT and other non-material
changes require notification. The standard review periods issued by the
MPT following the implementation of the Administrative Procedure Law
specify periods of two months as a general guideline for the review period
after formal approval. However, the discussion prior to the filing may take
an additional period.

If a cable television facility will have 500 or less drop terminals, a cable
television broadcaster must only submit a notification to the MPT in
accordance with the Cable Telecommunications Law to establish the facility and to make any changes in the notified matters.

In addition to the licence requirement under the Cable Television Broadcast Law or notification requirement under the Cable Telecommunications
Law, a cable television broadcaster must submit a notification to the MPT in
accordance with the Cable Television Broadcast Law to conduct cable

television broadcasting business and to make any changes in the notified matters.

If a cable television broadcaster intends to broadcast its original programmes and charge subscription fees, the broadcaster must prepare a tariff setting forth subscription fees and submit a notification of the tariff to the MPT in accordance with the Cable Television Broadcast Law and any changes in the tariff.

Cable Sound Broadcasters

24-65   A cable sound broadcaster must submit separate notifications to the MPT for establishing a cable sound broadcasting facility in accordance with the Cable Telecommunications Law, and for conducting cable sound broadcasting business in accordance with the Law to Regulate the Operation of the Cable Sound Broadcasting Service, along with changes in the matters in either notice.

Programme Regulation

24-66   The Broadcast Law provides for general minimum standards for programming, requires broadcasters to establish standards for the compilation of broadcast programmes and requires the establishment of a Consultative Organisation on Broadcast Programmes to maintain the appropriateness of broadcasting programmes. Similar provisions also apply to cable television broadcasting and cable sound broadcasting. However, if a cable television broadcaster receives television programmes from a broadcaster, and retransmits such programmes without any change at the same time they are received, rather than broadcasting its own original programmes, the provisions with respect to programming in the Broadcast Law will not apply. Further, such provisions also will not apply to facility supplying broadcasters because they do not supply or have responsibilities for the compilation of programmes.

6. Equipment Certification

Introduction

24-67   Certification for radio and terminal equipment used in Japan is issued by the MPT or, in the majority of cases, by two certification agencies designated by MPT. There are restrictions on the use of certain types of non-certified equipment; in most cases, using certified equipment will save cost and steps in the licensing or installation process for users. Therefore, manufacturers and importers benefit from seeking compliance certification from the MPT or its designated agencies. The two certification agencies publish English language guides to assist in the certification process.

Terminal Equipment

The general requirements for certification of terminal equipment are set out   **24–68**
in the TBL and in applicable MPT regulations.

Technical compliance certification is granted on a "type approval" basis for terminal equipment. JATE has been designated by the MPT as the compliance approval agency for terminal equipment. If compliance approval has been granted for the type of terminal equipment to be used, no inspection by a Type I carrier is necessary for the connection of the equipment to its network unless there is evidence of malfunction. Further, the Type I carrier cannot refuse to have equipment of the type approved connected to its network.

Specific compliance standards are set forth in technical conditions and technical requirements. Technical conditions are set forth in the applicable MPT ordinance. There are several categories of technical conditions: general technical conditions, conditions specific to analogue telephone terminals and conditions specific to mobile telephone terminals.

Technical requirements are the electric and network control requirements that must be met by terminal equipment to be connected to other networks. They are issued by the Type I carriers and approved by the MPT. Separate requirements are issued by each carrier for connecting different types of terminals, such as telex, circuit switching, facsimile and leased circuit terminals. There are also separate requirements for domestic and international terminals. Technical conditions and technical requirements, according to principles stated in the TBL, must preserve telecommunications circuits and functions thereof, prevent interference to other users, and clearly divide responsibility between telecommunications circuit facilities established by a Type I carrier and terminal facilities connected to them by a user. Depending on the type, equipment may have to comply with multiple sets of conditions and requirements.

Applications for compliance approval are first made by submitting documents to JATE. Foreign companies may submit approval requests directly or by proxy with a power of attorney, but all documents must be submitted in Japanese. JATE may require the submission of additional documents before issuing approval, or, if approval cannot be granted on the basis of the documents, the applicant may request examination of the equipment. If JATE grants approval, an approval label must be affixed to the equipment, which can be prepared either by the applicant according to JATE's specifications or by JATE.

If the equipment is modified, another compliance approval application must be submitted but the application fees will be reduced in certain cases.

Equipment using "low power" electric waves must be approved both by JATE and MKK, the radio equipment certification agency discussed below; in such a case, if documents are not submitted to both agencies, the relevant parts of documents will be forwarded by one agency to the other. Equipment using "weak" electric waves, however, need be approved only by JATE.

Radio Equipment

24–69   Technical standards for radio equipment and the equipment certification regime are set out in the Radio Law and in specific MPT regulations.

Certain radio apparatus cannot be installed unless it meets the "type approval" test of the MPT (such as radar and other equipment to be used on ships and aircraft). Within this category, however, specified types of apparatus need only pass a certification test for that type equivalent to the one conducted by the MPT.

Specified radio equipment to be used for small-scale radio stations may receive "technical standards compliance certification" from the MPT or its designated agency, MKK. Radio equipment certified by MKK includes that used by portable mobile stations aboard aircraft, by land mobile radio stations for cellular radio telephone communications and MCA mobile radio communications, by radio stations using specified frequencies of radio waves, citizen radio and cordless telephone, and by VSAT stations for satellite communications. Different regional MKK offices are designated to handle each equipment category. Technical compliance certification may be sought by "written application" (submitting test data and documents only) or "test application" (submitting the equipment itself as well as certain documents). Manufacturers or importers can apply to MKK to receive technical standards compliance certification for units of their equipment. Certification is less expensive for additional units of a type of equipment that has already passed the "type approval" test.

Radio stations that use only certified radio equipment may eliminate both the provisional licensing stage and inspection after construction. Further, certain equipment, such as citizen radio and cordless telephones, may only be used if certified, in which case no licence is necessary for persons using such certified equipment.

Biographical Information

24–70   Greg L. Pickrell is the managing partner of the Tokyo office of the international law firm of Coudert Brothers. Mr Pickrell received a B.S. degree in Physics from Iowa State University and a J.D. degree from The University of Michigan Law School. Prior to heading the firm's Tokyo office, Mr Pickrell was a partner in the firm's Silicon Valley office in California, focusing on the representation of high technology companies, including telecommunications companies.

Masayuki Okamoto is an attorney with the Japanese law firm of Tanaka & Takahashi. Mr Okamoto received his L.L.B from Tokyo University.

Julie N. Mack, an associate in the Tokyo office of Coudert Brothers, contributed to this chapter. Ms Mack received a B.A. and J.D. degree from Harvard University.

CHAPTER 25

The Netherlands

By Marjolein Geus, Buruma Maris, The Hague

1. HISTORY OF DEVELOPMENT OF TELECOMMUNICATIONS IN THE NETHERLANDS.

Industry Structure

Telecommunications in the Netherlands started with the telegraph service 25-01
of the Dutch Railway Company[1] in 1845. Some years thereafter, when the
telegraph service had gained in importance, the Government decided that
telegraphy should become a public service. Subsequently voice telephony
was being developed; initially this development was left to private enterprises. Nevertheless, as voice telephony grew more important, private telephone networks were taken over by local authorities and subsequently by
the Dutch State. In 1913 the State Enterprise PTT was established to provide the public postal, telegraph and telephone services and to control and
operate the corresponding telecommunication network.

Apart from the telecommunication network of the State Enterprise PTT
other networks were developed for different purposes such as the cable
broadcasting networks which were constructed by the municipalities.[2] The
construction and operation of transmitters for the national broadcasts were
entrusted to the Dutch company NoZeMa. Finally other large private telecommunications networks were constructed for internal use such as the
networks of the Dutch railway company, the electricity companies and several large private companies. Although all of these networks have been
constructed for a specific purpose other than the provision of telecom services they have become increasingly important in view of the discussions
about the liberalisation of telecommunications infrastructure and services.

---

[1] *De Hollandse IJzeren Spoorwegmaatschappij* (H.IJ.S.M.).
[2] It should be noted that Royal PTT Netherlands NV has taken over several cable broadcasting networks.
Presently the PTT owns the largest cable broadcasting network operator in the Netherlands (CASEMA).

25–02   In 1989 the Dutch PTT was transformed from a State enterprise into a private company, in order to create a more flexible national telecom operator to cope with the expansive development of the telecommunication market. In June 1994 the shares in the Dutch PTT were listed on the Amsterdam Stock Exchange and the Dutch Government sold a first tranche of shares (approximately 30 per cent). It is likely that the Dutch Government will lose its controlling interest in Royal PTT Netherlands NV in the course of 1995 when it is expected that the next tranche of shares will be sold.

The participation of private companies on the Dutch telecom market has increased since 1989. Licences have been granted to private companies to operate various networks such as Mobitex and Euromessage. The participation of the private sector will increase in the near future. As of September 1994, competition has been introduced in the field of public mobile communications.

Finally it should be noted that it is the intention of the Government to introduce competition with respect to the fixed telecommunications infrastructure by the end of 1995. On May 22, 1995, two Bill's were presented to the Second Chamber of the Dutch Parliament allowing for a second national fixed network operator. Such operators should preferably be formed out of the present owners of large private networks such as the Dutch railway company, the electricity companies and cable tv operators.

Institutional Structure

25–03   The fact that in the Netherlands telecommunications were considered a public task which was therefore almost exclusively entrusted to the State enterprise PTT had consequences for the institutional structure as well. Until the privatisation of the PTT in 1989 almost all tasks such as regulation, supervision and operation were centralised with the State enterprise PTT acting on behalf of the Minister of Transport.[3]

Background to Legislation

25–04   The first Act which dealt with the various aspects of telecommunications was the Telegraph and Telephone Act 1904.[4] According to this Act other parties than the Dutch State (*i.e.* the State Enterprise PTT) were not allowed to construct, maintain or operate public or private telecom networks or to provide telecom services without a concession or a licence of the Minister of Transport. Furthermore the market for terminal equipment was the exclusive domain of the PTT.

On the occasion of the privatisation of the PTT in 1989 the Telegraph

---

[3] The development of the broadcasting sector took place along different lines in which the PTT was not involved. Therefore the institutional structure in this area has always been very different.
[4] *Telegraaf en Telefoonwet* 1904.

and Telephone Act 1904 was replaced by the Dutch Telecom Act 1988[5] and several other Acts.[6] New regulations were needed to provide for a suitable company structure, the performance of public services, the transfer of personnel and the transfer of assets and liabilities. In view of the developments on a European level the opportunity has been taken to liberalise part of the telecommunications market and to separate the regulatory and the operational functions.

Traditionally the broadcasting area is not entrusted to the Minister of Transport but to the Minister of Culture. Broadcasting has partly been regulated in the Media Act 1987[7] and partly in the Broadcasting Transmitters Act 1935.[8]

2. Existing Legislative and Regulatory Environment

Principal Legislative Instruments

The field of telecommunications has mainly been regulated in the Dutch Telecom Act 1988. A major amendment to this Act has entered into force on September 1, 1994.[9] On the basis of the 1988 Act Governmental Decrees and Ministerial Regulations have been issued such as: 25–05

- the Governmental Decree on Mandatory Services[10];
- the Governmental Decree on Radio Transmitters[11];
- the Governmental Decree on Cable Broadcasting; Networks and Cables/Cableworks[12];
- the Governmental Decree on Terminal Equipment[13]; and
- the General Ministerial Directives.[14]

It should be noted that E.U. regulations and other E.U. developments have rather frequently been implemented by means of so-called "Notifications of the Minister of Transport". In such notifications it is announced that a certain article in the Dutch legislation will no longer be applied or will be interpreted in another way.

The Minister of Transport is authorised to take the various decisions that have to be taken on the basis of the 1988 Act, such as with respect to licence applications for telecommunications installations, applications for type approval, applications for exemptions, frequency allocations, bind-   25–06

[5] Wet op de Telecommunicatievoorzieningen 1988 [1988] O.J. L520.
[6] The Machtigingswet PTT Nederland NV Act 1988 [1988] O.J. L521; The Personeelswet PTT Nederland N.V. Act 1988 [1988] O.J. L519; The Postwet, Act 1988 [1988] O.J. L522.
[7] [1987] O.J. L249.
[8] Radio Omroep Zenderwet Act 1935; [1935] O.J. L403.
[9] Wet Mobiele Communicatie 1994; [1994] O.J. L628.
[10] Besluit Opgedragen Telecommunicatiediensten.
[11] Besluit Radio Electrische Inrichtingen.
[12] Besluit Draadomroepinrichtingen en kabelinrichtingen.
[13] Besluit Randapparatuur.
[14] Besluit Algemene Richtlijnen Telecommunicaties.

ing directives addressed to the concessionaire (Royal PTT Netherlands NV) and administrative sanctions. Such ministerial decisions are subject to appeal at the Administrative Court.

Apart from the Telecom Act 1988, the Media Act 1987 and the Act on Broadcasting Transmitters 1935 should be mentioned. Certain telecommunications services such as teletext and datacasting have partly been regulated in the Media Act and partly in the Broadcasting Transmitters Act.

Finally the Dutch Competition Act is applicable with respect to the telecommunications sector. On the basis of this Act the Minister of Economic Affairs may act against deeds of unfair competition.

Agencies of Regulation

25–07   Since the privatisation of the Dutch PTT almost all regulatory and supervisory functions are centralised with the Ministry of Transport (the Department on Telecommunications and Post). The Ministry of Transport is entrusted with the preparation of Bills and Governmental Decrees as well as the issuance of Ministerial Decrees. Furthermore the Minister of Transport has to decide on applications for licences and exemptions. The Minister of Transport is also entrusted with monitoring whether the Royal PTT Netherlands NV and other companies and organisations comply with their obligations under the Telecom Act 1988 and with their licence conditions. Finally the Minister of Transport may address binding directives to PTT Netherlands NV and may impose administrative sanctions in case of violation of the provisions of the Telecom Act.[15]

With respect to type approval procedures, notified bodies and testing laboratories have been appointed (see below paragraphs 25–73 to 25–77.

25–08   Apart from the regulatory and supervisory institutions several advisory and consulting bodies have been established:

- The Advisory Board for Post and Telecommunication.[16] This organisation is entrusted with an advisory task either at the request of the Minister of Transport or at its own initiative.
- The Consulting Committee for Post and Telecommunications.[17] This Committee has been established in order to allow the Minister to consult with the representative organisations of the various interested parties.

[15] It has been decided during a hearing in the Second Chamber of Parliament on March 21, 1994 that the supervisory tasks had to be transferred from the Department on Telecommunications and Post to another department of the Ministry of Transport. As of April 1, 1995 this decision was effected by the constitution of a separate supervisory department for telecommunications and post, the Directie Toezicht Netwerken Diensten (TND). The decision whether the supervision should altogether be separated from the Ministry of Transport will probably take several years.   It should be noted that apart from the forementioned administrative tasks the Minister of Transport (acting on behalf of the Dutch State) is also the majority share owner in Royal PTT Netherlands NV, (see below). It has been argued that the Ministry of Transport should transfer the management with respect to the shares to the Ministry of Finance to avoid conflicting interests.

[16] *Commissie van Advies voor Post en Telecommunicatie van de Adviesraad voor Verkeer en Waterstaat.*

[17] *Overlegorgaan Post en Telecommunicatie.*

- The Consulting Committee for PTT.[18] This Committee has been established for the consultation by the concessionaire (Royal PTT Netherlands NV) of representative organisations of the various interested parties.

Furthermore the Dutch Government (the Minister of Transport and the Government) has been granted specific rights in the Articles of Association of Royal PTT Netherlands NV. For instance various protective measures have been introduced in the Articles of Association, including a so called "Golden Share" and two types of preference shares. The Golden Share empowers the Minister of Transport to veto a range of important decisions that might affect the future of the concession owner and the supply of concessionary tasks.

Competition Control

Competition control is partly entrusted to the Minister of Transport and partly to the Minister of Economic Affairs. The Minister of Transport has to supervise whether Royal PTT Netherlands NV complies with the prohibitions with regard to cross subsidies which have been laid down in the General Ministerial Directives. The Minister of Economic Affairs is entrusted with the application of the Dutch Competition Act,[19] which also applies to the telecommunications sector.   25–09

General Rules Governing Telecommunications Installations, Services and Apparatus

The structure of the Telecom Act 1988, will be completely reviewed. However, it will probably take several years before the new Telecom Act will enter into force. Therefore the Netherlands will have to cope with their present Act for quite some time. However, it should be noted that on May 22, 1995 two Bills have been presented to Parliament in which major "interim-Amendments" to the Telecom Act 1988 are proposed in view of a further liberalisation of the fixed telecommunications infrastructure. These Bills, which are seriously criticised at present, will be discussed briefly in paragraph 25–21. According to the Government the Bills should have been passed by the end of 1995. For a good understanding of the regulation of installations, services and apparatus, it is important to give a general outline of the structure of the Telecom Act 1988:   25–10

Chapter I: Definitions.

Chapter II: Regulations with respect to the telecommunications infrastruc-   25–11

---

[18] *Overlegorgaan PTT.*
[19] *Wet Economische Mededinging* (WEM) 1958; [1958] O.J. L491. The latest Amendement to the WEM was published in [1995] O.J. L287 and will enter into force on September 1, 1995.

ture of the concessionaire. The Telecom Act provides for an exclusive concession. The concessionaire has the following obligations:

- The provision of leased lines throughout the country against uniform tarrifts and conditions.
- The provision of the so-called mandatory telecommunications services throughout the country against uniform tariffs and conditions. These mandatory services are fixed voice telephony, telex, telegraphy and mobile maritime telephone services.
- The provision of certain interconnection facilities.

On the other hand the concessionaire has the exclusive right with regard to the construction and the maintenance of the telecommunications infrastructure pertaining to the mandatory services, leased lines and interconnection facilities. Futhermore the concessionaire has a priviliged position with regard to the supply of mandatory services. Royal PTT Netherlands NV has been appointed exclusive concession owner under the Telecom Act 1988. Royal PTT Netherlands NV has entrusted the performance of the concessionary tasks to her subsidiary PTT Telecom BV.

25–12   Chapter II of the Telecom Act 1988 refers to General Directives issued by the Minister of Transport with respect to the concessionnaire. In these General Directives rules have been laid down with respect to for instance:

- The quality of the universal service, the quality of the network and the quality of leased lines.
- Change of the tariffs; tariffs of leased lines.
- The obligation to provide information to the Minister of Transport.
- The general terms and conditions applied by Royal PTT Netherlands NV
- A prohibition of cross subsidies.
- An obligation to discuss important concession matters with interested parties in the Consulting Committee for PTT.
- A conciliation procedure open to users of the public services.

25–13   **Chapter IIA:** Regulations with respect to designated technical systems for public mobile telecommunications. This Chapter came into force on September 1, 1994.[20] Chapter IIA provides for the issuance of a specific kind of licences for certain innovative technical systems for public mobile telecommunications. A technical system will be appointed on the basis of Chapter IIA if the development of such a system is considered in the general economic interest. Accordingly Chapter IIA contains provisions in order to guarantee the construction of the designated networks and the supply of the corresponding services.

25–14   **Chapter III:** Private installations. According to Chapter III licences may be granted for the installation, the use and the operation of telecommunica-

[20] Mobile Communications Act 1994; [1994] O.J. L628.

tions installations and networks other than the telecommunications infrastructures referred to in Chapter II and IIA.

Chapter IV: Terminal Equipment. The terminal equipment market has been   **25–15**
liberalised in the Netherlands in accordance with the relevant European directives. It should be noted that the provisions of Chapter IV of the Telecom Act 1989 with respect to type approval and testing procedures have been set aside by a Notification of the Minister of Transport in order to comply with the Telecommunications Terminal Equipment Directive.

Chapter V: Suppression of certain electric and electronical installations and apparatus.

Chapter VI: Regulation of the obligation to allow the construction, the maintenance and the removal of cables and cableworks by the concessionaire.

Chapter VIA (not yet in force): Conciliation by the Minister of Transport. Interested parties may apply for a judgement by the Minister of Transport with respect to the measures of the concessionaire within the ONP Directives.[20a]

Chapter VII: Fees charged by the Minister of Transport with respect to for instance the grant of licences and exemptions, type approval, and supervision.

Chapter VIII: Appeal of Ministerial decisions.

Chapter IX: Supervision and administrative sanctions.

Chapter X: Criminal provisions

Chapter XI, XII, XIII: Various provisions with respect to for instance the   **25–16**
supply of telecommunications services in exceptional circumstances, legal tapping and transitional provisions.

   It should be noted that telecommunications services such as teletext and cable tv information services have been regulated in the Media Act 1987. Furthermore the use of the remaining capacity of the broadcasting transmitters of NoZeMa[21] for telecommunications services such as datacasting has been regulated in the Media Act 1987 and the Broadcasting Transmitters Act 1935. These Acts will be discussed in paras. 25–55 *et seq.*

[20a] This new Chapter VIA to the Telecom Act was adopted by the First Chamber of Parliament on November 1, 1994 (Bill 23632) and will probably enter into force by the end of 1995.
[21] See below para. 25–55.

Licensing Regime, including Main Licence Categories

25–17   As indicated above Royal PTT Netherlands NV has been granted the exclusive concession with respect to the construction, the maintenance and the exploitation of the telecommunications infrastructure pertaining to the mandatory services, leased lines and certain interconnection facilities. Other parties, including Royal PTT Netherlands NV if acting outside the concession, need a licence of the Minister of Transport to install, maintain, use or exploit telecommunications installations, networks and apparatus. An exception to this rule are cables and cableworks which are exclusively constructed on private properties. Furthermore a specific regime applies to terminal equipment and certain types of radio receivers which only need to be approved as to type.

The following main licence categories can be distinguished:

25–18   **Category 1:** Licensed infrastructures on the basis of Chapter IIA. This category contains the licences for the construction, maintenance and operation of certain designated innovative infrastructures for mobile telecommunications. At this moment GSM, ERMES and DCS 1800 have been designated.

25–19   **Category 2:** Private installations in the sense of Chapter III. This category comprises the licences for private installations in accordance with Chapter III of the Telecom Act. The following categories of licences for private installations can be distinguished:

(i) Radio communications installations:

— Radio communications generally:

- radio transmitters not intended for telecommunications between fixed points (Article 17 of the Telecom Act 1988);
- Radio transmitters intended for telecommunications between fixed points (Article 18 of the 1988 Act);
- Radio receivers (Article 19 of the 1988 Act).

— mobile public networks other than the technical systems which have been designated on the basis of Chapter IIA (Article 17 of the 1988 Act),

— mobile communications within closed user groups and trunking networks (Articles 17 and 14, paragraph 6, of the 1988 Act),

— satellite ground stations (Articles 17 and 18 of the 1988 Act).

(ii) Cable broadcasting networks using cable systems and/or radio communications intended for the distribution of programmes including:

— a basis licence for the distribution of programmes in the sense of the Media Act (Article 21 of the Telecom Act);

— an additional licence for the supply of telecommunications services (Article 22 of the Telecom Act);
— an exemption for the provision of fixed connections on behalf of Chapter IIA networks (Article 22, para. 2, of the Telecom Act).

(iii) Telecommunications installations using cables or cableworks, such as Local Area Networks and Wide Area Networks (Article 23 of the Telecom Act), including:

— a licence for the construction, maintainance and use;
— an exemption for the provision of fixed connections on behalf of Chapter IIA networks.

Category 3: transitional licences. A transitional licence on the basis of Chapter III of the Telecom Act 1988 has been granted to Royal PTT Netherlands NV with respect to the existing analogue mobile systems (ATF, ATF2, ATF3 and SMF3). However, as a deviation from Chapter III specific transitional regulations apply to these systems.   25–20

Future Regulation

The present regulation in the field of telecommunications will be amended in the near future. A total revision of the Telecom Act is expected before 1998. Apart from this revision it is expected that by the end of 1995 two interim Acts will come into force, mainly in order to introduce competition in the fields of fixed telecommunications infrastructures and telecommunications services (except for public fixed voice telephony and the telex service which will be liberalised in 1998 at the latest). On May 22, 1995 the relevant Bills were sent to the Second Chamber of Parliament.[22] It should be noted that the Bills are still under discussion and that the final Acts may include changes to various (major) aspects.   25–21

According to these Bills the following "interim Amendments" will be made with respect to the existing regulation.

(a) Competition will be introduced in the field of fixed telecommunications infrastructure. Apart from the telecommunications infrastructure of the concessionaire, a licence will be granted to a second operator of national fixed telecommunications infrastructure. The second national infrastructural licence will only be granted if the applicant operates licensed cable broadcasting networks and/or licensed cables/cableworks that cover a substantial part of the Netherlands to be

---

[22] Bill with respect to the amendment of the Telecom Act 1988, the Media Act 1987, the Broadcasting Transmitters Act 1935 and the Code of Criminal Procedure relative to the liberalisation of fixed telecommunications infrastructure (Fixed Telecommunications), TK24163, 1994/1995, nos. 1, 2; Bill with respect to the regulation of licences for the construction, maintenance and exploitation of fixed telecommunications infrastructure (Licences for Fixed Telecommunications Infrastructure), TK24164, 1994/1995, nos. 1, 2.

determined by Ministerial Decree. Furthermore, additional regional infrastructural licences may be granted, preferably to existing operators of licensed cable broadcasting networks and/or licensed cables/cableworks, such as the Dutch railway company and energy utilities, of a certain minimal size to be determined by Ministerial Decree. The Bills provide for a set of rights and obligations with respect to the owners of infrastructural licences. Main elements are that the owner of an infrastructual licence has the right to expand his fixed network within his territory and that he has the obligation to provide leased lines to anyone upon request against a compensation within this territory.

(b) In addition cable tv operators will have the choice to opt for a less regulated liberalisation by means of a simple registration in accordance with the draft Directive of the European Commission of December 21, 1994 ([1995] O.J. C76/8).

(c) A licensing/registration regime will be adopted with respect to the provision of telecommunications services. At present telecommunications services are mainly regulated as a part of the regulation of the infrastructure (see below).

(d) The implementation of the Services Directive of the European Commission will be completed and the additional E.C. Directive with respect to satellite communications[23] will be implemented.

3. REGULATION OF SERVICES: DETAILED ANALYSIS OF RIGHTS TO PROVIDE TELECOM SERVICES; THOSE SUBJECT TO MONOPOLY AND THOSE OPEN TO COMPETITION

General Remarks

25–22   As indicated above the Telecom Act 1989 does not encompass a specific licensing or registration regime for telecommunications services. In the present Telecom Act telecommunications services have mainly been regulated as a part of the regulation with respect to telecommunications installations. As a consequence the monopolistic nature of certain services follows mainly from restrictions to provide such services via certain infrastructures.

A legal system which provides for licensing or registration of services is envisaged in the imminent revision of the Telecom Act.

Services Subject to the Monopoly of the Concessionaire

25–23   Article 4, paragraph 1, of the Telecom Act 1988 includes the obligation of Royal PTT Netherlands N.V. to provide certain mandatory services to be

---

[23] E.C. Directive 94/46 (1994) O.J. L268.

defined by Governmental Decree. As of September 1, 1994, the mandatory services encompass the following:

- fixed public voice telephony,
- mobile maritime services,
- telex, and
- telegraphy.

Initially the mandatory services encompassed fixed data-transportation, paging services and public mobile telephony. However, in view of the Services Directive of the European Commission[24] fixed data transportation has been deleted from the Decree on Mandatory Services.[25] As of September 1, 1994, the public mobile services (except for mobile maritime services) have been deleted from the mandatory services.[26]

The mandatory services are subject to a monopoly of the concessionaire, because the offering of such services by means of other infrastructures such as leased lines and networks and installations in the sense of Chapters IIA and III is prohibited.

Services Offered by Means of Leased Lines

The use of leased lines is regulated in Article 5 of the Telecom Act 1988.   25–24
According to this provision the following regime is applicable:

- leased lines may not be used for the provision of public voice telephony services to third parties.[27]
- an exception has been made for public mobile telephony. Leased lines may be used by or on behalf of operators of public mobile telephony and their service providers. The exception applies only to the mobile service for which the licence has been granted.
- Furthermore leased lines may be used for the provision of telephony services if such services are part of a value added service.

Former restrictions with respect to the offering of fixed and mobile data-transportation services by means of leased lines no longer apply.

With respect to prohibition to provide public voice telephony services to third parties the E.C. Services Directive must be taken into account. According to the Commission the Member States may not prohibit the provision of the following voice telephony services by means of leased lines:

- voice telephony services on a non-commercial basis;
- voice telephony services that are offered to closed user groups;
- voice telephony services which do not connect two network termina-

---

[24] Commission Directive 90/388 on competition in the markets for telecommunications services.
[25] Decree of December 20, 1993; [1994] O.J. L21.
[26] Decree of December 15, 1994; [1995] O.J. L94.
[27] Leased lines may not be used either for the provision of the other mandatory services (telex, telegraphy and maritime mobile telephony).

tion points of the switched network at the same time (only dial-in or dial-out services);

- voice telephony services which do not constitute direct transport or switching of speech in real time (for instance store and forward and voice mail applications and least cost routing); and
- services meeting demand which is not satisfied by the current telephone service or which does not consist of a significant part of the transport of speech such as facsimile, video conferencing and video phone.

According to the Bills which have been discussed in paragraph 25–21 above the Minister of Transport has now accepted this interpretation.

25–25   Leased lines may be used to offer voice telephony to closed user groups. The Minister of Transport has issued a notification with respect to closed user groups in order to comply with the Service Directive.[28] The notification gives the following definition of a closed user group: companies which belong to the same economic entity and groups whose members have a continuous economic or professional relationship among themselves from which the communications requirements derive. The following examples are given:

(a) a single organisation encompassing distinct legal entities, such as a company and its subsidiaries or its branches in other Member States incorporated under the relevant domestic company law (often referred to as "corporate networks");

(b) different institutions or services of international and intergovernmental organisations;

(c) a common activity network in which the link between the members of the group is a common business activity. In this case the communications requirements are part of the services concerned such as fund transfers for the banking industry, reservation systems for airlines, information transfers between universities involved in a common research project, re-insurance for the insurance industry and interlibrary activities;

(d) an "integrated business community" or "business web" encompassing a corporation and its major business partners, suppliers, dealers and in some cases major customers such as shipping companies and their insurance companies and other carriers or airlines and tour operators.

25–26   An additional notification has been published with respect to dial-in and dial-out services.[28a]

The implementation of the Services Directive will be complete with the introduction of the "interim Amendment" to the telecom regulation referred to in paragraph 25–21 above.

---

[28] Notification of May 30, 1994; *Gazette*, June 3, 1994.
[28a] Notification of the Minister of Transport of May 1, 1995, *Gazette* no. 84.

Services Offered by Means of Licensed Networks in the Sense of Chapter IIA

Licensed installations on the basis of Chapter IIA of the Telecom Act 1988   25–27
may only be used for the public mobile services pertaining to the technical
system for which the licence has been granted. Therefore it is not allowed
to provide any other telecommunications services by means of a licensed
network. Licence criteria and licence conditions are discussed below.

Services Offered by Means of Private Installations

The offering of services by means of a private installation is restricted by: 25–28

- the licence criteria;
- the licence conditions;
- general prohibitions in the Telecom Act 1988.

These will be discussed in more detail below.
Nevertheless some general rules are applicable with respect to the
offering of services by means of private installations[29]:

- Private installations may not be used to provide voice telephony services.
- Private installations may not be used to provide fixed public data
  transportation services.
- Private installations may be used for value-added services if the tele-
  communications infrastructure of the concessionaire is used as well.
  With respect to cable broadcasting networks a different regime is
  applicable (see below).
- A specific regime is applicable with respect to mobile services (see below).
- It is not allowed to connect a private installation to any other private
  installation unless by means of the telecommunications infrastructure
  of the concessionaire (for example using leased lines of Royal PTT
  Netherlands NV). With respect to radio transmitters and cable/
  cableworks the Minister of Transport may grant an exemption from
  this prohibition if the concessionaire is not able or not prepared to
  provide a similar connection within a reasonable term and against
  reasonable conditions.

Fixed Services (Fixed Voice Telephony, Fixed Data Transportation, Cable Broadcasting Networks and Cables/Cableworks)

Fixed voice telephony

As indicated above fixed voice telephony services are part of the mandatory   25–29
services which have to be provided by the concessionaire. Fixed voice tele-

---

[29] Art. 14, 22 and 25 of the Telecom Act 1988.

phony may not be supplied by other parties with the exception of fixed voice telephony within closed user groups which may be supplied by means of leased lines.

Fixed data transportation

25–30 Fixed data transportation services have been excepted from the Decree on Mandatory services. As indicated above fixed data transportation services may be offered by means of leased lines but not by means of licensed infrastructures in the sense of Chapter IIA or private installations in the sense of Chapter III. With respect to the fixed data transportation service of the PTT an agreement has been reached between the Minister of Transport and Royal PTT Netherlands NV according to which PTT will continue to offer its fixed data transportation service in accordance with the Council Recommendation 92/382.[30] The packet-switched data network of the PTT no longer belongs to the concessionary's infrastructure. The packet-switched data network has to be considered as a set of leased lines which are leased by the commercial department of PTT from the concessionary department of PTT.

Services by means of cable broadcasting networks:

25–31 The Telecom Act 1988 only deals with the infrastructural aspects of cable broadcasting networks and with telecom services which are offered by means of cable broadcasting networks. The question whether a broadcasting program (including teletext and datacasting services) may be distributed by means of a cable network is governed by the Media Act 1987 (see below paragraph 25–55).

According to Article 21, paragraph 3, of the Telecom Act the basic licence for the construction and the operation of a cable broadcasting network may be granted if the size of the cable network is limited to the territory of a municipality.[31] Apart from procedural objections a basic licence may only be refused:

(a) if the applicant does not have sufficient technical and financial means to safeguard the future operation of the cable network;[31a]
(b) if someone else already obtained a basic cable network licence for the same area.

---

[30] Council Reccomendation of June 5, 1992 on the harmonised provision of a minimum set of packet-switched data services (PSDS) in accordance with open network provision (ONP) principles.

[31] Very small cable networks designated by the Minister may be exempted from the licence obligations (*e.g.* cable networks connecting a few buildings or apartments).

[31a] Discussion is taking place concerning these criteria with respect to the so-called "white spots", *i.e.* industrial and rural areas in municipalities. Licences to construct and operate a cable tv network in these areas used to be refused by the Minister of Transport on the basis of these criteria. However, in view of the proposed liberalisation of existing fixed telecommunications infrastructure a licence to construct and operate a cable tv network in these areas became increasingly important. As a result various legal administrative proceedings are pending in which the refusals to construct and operate a cable tv network in industrial areas are being disputed. At the same time the Minister of Transport issued a Notification stating that existing cable tv operators can apply for a licence to exapnd their networks into industrial areas within the municipalities in which their networks are located on the basis of an Article 23 licence (Notification of May 1, 1995, *Gazette*, No. 84). This Notification is also being disputed.

The owner of a basic licence for a cable network may not use the network to provide other services than the distribution of programmes in the sense of the Media Act 1987. For telecom services an additional licence or an exemption on the basis of Article 22 of the Telecom Act is required.

An additional licence may be granted on the basis of Article 22, para-   **25–32** graph 3, with respect to the provision of other services than the distribution of programmes in the sense of the Media Act 1987. However, no additional licence can be granted with respect to point to point services such as voice telephony and data transportation services between subscribers connected to the cable network. In addition to the refusal grounds for basic licences, an additional licence will be refused if the granting is contrary to an efficient provision of telecommunications as determined by the general social and economic interest.

An exemption from the prohibition to provide point-to-point services may be granted on the basis of Article 22, paragraph 2, with respect to the provision of services on behalf of the owner of a Chapter IIA licence. This exemption entered into force on September 1, 1994, and is considered as an anticipatory provision with respect to the liberalisation of fixed telecommunications infrastructure.

Article 25, paragraph 1, of the Telecom Act 1988 prohibits connection of the cable broadcasting network to another cable broadcasting network (or any other private installation) except by means of the concessionaire's telecommunications infrastructure.

Finally it should be noted that in anticipation of the liberalisation of fixed infrastructures and telecommunications services, operators of cable broadcasting networks are enabled to benefit from the following exceptions which have been published in the *Gazette*:[31b]

- technical adaptations of their network in preparation of the liberalisation;
- technical experiments to provide telecommunications services between subscribers.

The adaptations and experiments are only allowed if:

- they do not relate to the provision of the public voice telephony which is still reserved on behalf of the concessionaire;
- approved terminal equipment is being used;
- the primary task of the cable operators, to distribute broadcasting programmes, is not disturbed;
- services are offered on a non-commercial basis;
- services are limited to the area for which the licence has been granted; the number of subscribers involved follows from the technique which is being used; and
- the cable tv operators have registered the experiment with the Ministry of Transport.

[31b] Notification of the Minister of Transport, November 30, 1994, *Gazette*, No. 233.

Services Using Cables and Cableworks

25–33   Cables and cableworks may not be used to provide mandatory services and datatransportation services to third parties (Article 14, paragraph 2). Cables and cableworks may be used for the provision of value added services if the concessionare's infrastructure (for instance leased lines) is used as well. Furthermore it is not allowed to connect cables and cableworks to other private installations unless by means of the concessionare's infrastructure (Article 25, paragraph 1). The Minister of Transport may grant an exemption from this prohibition if the concessionaire is not able or not prepared to provide such a connection within a reasonable term and against reasonable conditions.

An exemption from the prohibition of Article 14, paragraph 2, may be granted for the provision of services on behalf of the owner of a Chapter IIA licence (Article 14, paragraph 5, of the Telecom Act 1988).

Apart from the general restrictions referred to above and apart from procedural objections, a licence for a cable or a cable work shall be refused on the basis of Article 23, paragraph 2 of the Telecom Act 1988:

(a) if Royal PTT Netherlands NV is able and prepared to provide a similar connection—as applied for—within a reasonable term and against reasonable conditions;
(b) if otherwise the general, social and economic interest relating to efficient telecommunications services is opposed to the granting of a licence.

The technical aspects of the cable work and the intended use thereof will be prescribed in the licence conditions.

It should be noted that according to notification of the Minister of Transport of April 29, 1994[32] the concessionaire shall not invoke its right of first refusal with respect to cables not exceeding 500 meters.

Radio Communications Generally

25–34   Radio Communications installations are regulated by the Articles 17–20 of the Telecom Act.[33] The following categories can be distinguished:

- Transmitters not intended for telecommunications between fixed points (Article 17, Telecom Act).
- Transmitters intended for telecommunications between fixed points (Article 18, Telecom Act).
- Radio receivers (Article 19, Telecom Act).

With respect to satellite services and mobile services specific rules are applicable. These will be discussed separately.

According to Article 14, paragraph 1, radio communications installa-

---

[32] *Gazette*, May 20, 1994.
[33] The NoZeMa radio transmitters for broadcasting purposes follow a different legal regime which will be discussed in section 4.

tions may not be used to provide mandatory services nor any other direct transport of data. Value-added services are allowed on the condition that the concessionare's telecommunications infrastructure is used as well.

According to Article 25 private radio communications installations may   **25–35**
not be connected to other private radio communications installations nor any other private installations unless by means of the concessionare's infrastructure. The Minister of Transport may grant an exemption if the concessionaire is not able or not prepared to provide such a connection within a reasonable term and against reasonable conditions.

Categories of transmitters designated by the Minister of Transport may be exempt from the licence requirement. An exemption has been issued for terminal equipment such as mobile telephones and cordless telephones and for trunking apparatus.

Apart from various procedural objections, a licence for a radio-   **25–36**
transmitter not intended for telecommunications between fixed points (Article 17 of the Telecom Act 1988) will be refused:

(a) If an efficient use of the terrestrial frequencies requires so.
(b) If it precludes an efficient telecommunications service as determined by the general social and economic interest.
(c) If a licence would be contrary to the Media Act 1987 or the Act on Broadcasting Transmitters 1935.

In past decisions the Minister of Transport has, *inter alia*, refused a licence on the grounds as indicated in (a) and (b) if the applicant intended to provide a service that would be contrary to the general prohibition of Article 14, paragraph 1.

The nature of the radio transmitter and the intended purpose use thereof will be prescribed in the licence conditions.

Article 18 formulates in addition to Article 17 slightly different grounds   **25–37**
for the refusal of a licence. Instead of the grounds referred to under (a) and (b) above, a licence may be refused:

(a) if Royal PTT Netherlands NV is able and prepared to provide a similar connection—as applied for—within a reasonable term and against reasonable conditions;
(b) if otherwise the general, social and economic interest relating to efficient telecommunications services is opposed to the granting of a licence.

Parliamentary debate on Article 18 of the WTV clarified that the terms "requires", used in Article 17, paragraph 7, subsection c, and "opposed", used in Article 18, subsection b, of the WTV, indicate that the first has a more restrictive meaning than the latter. This leaves more discretionary power under Article 18 to the Minister of Transport in comparison to Article 17.

In May, 1994 the Minister of Transport has issued a notification[34] in which it is stated that the concessionaire will not invoke the right of first refusal referred to under (a) with respect to certain beam transmitters for a distance up to 500 meters and for certain transmitters for mobile and temporary applications.

25–38   With respect to radio-receivers the principle of free flow of information applies to receivers intended for the receipt of broadcasting programmes (programmes intended for the general public). As a consequence of the ITU obligation to safeguard the secrecy of messages, the receipt of messages not intended for the general public and the corresponding receivers may be regulated for this purpose. The Governmental Decree on Transmitters submits certain radio receivers to a licence duty, such as satellite data receivers and receivers using a decoder. Apart from procedural objections, according to Article 19, paragraph 4, of the Telecom Act 1988 a licence may be refused:

(a) if the protection of the rights of third parties in radio communications requires so;
(b) if the fulfilment of the State's obligations under international treaties requires so.

Satellite Services

25–39   Private satellite earth stations are regulated by the Articles 17–19 of the Dutch Telecom Act 1988, which have been discussed previously. The following categories of earth stations should be distinguished:

• mobile earth stations (Article 17 of the Dutch Telecom Act);
• fixed earth stations (Article 18 of the Dutch Telecom Act);
• Receive Only Stations (ROES) (Article 19 of the Dutch Telecom Act).

Apart from the general rules referred to above the Minister of Transport has issued a notification in which it has been announced that the concessionaire will not invoke its right of first refusal with respect to fixed VSAT's, VSAT-hub stations and satellite news gathering earth stations of a rather small size which have been specified in the annex to the notification.[35]

It should be noted that no legal monopoly on behalf of the concessionaire exists with respect to the space segment.

As of May 1, 1993 a one stop shopping procedure has been introduced with respect to the applications for licences for a satellite network in the Netherlands, the U.K., France and Germany.[36] This procedure applies to VSAT's and earth stations for "satellite news gathering". The Head Depart-

[34] *Gazette*, May 20, 1994.
[35] *Gazette*, May 20, 1994.
[36] Protocol of March 26, 1993.

ment for Telecommunications and Post of the Ministry of Transport has been appointed as the national coordinator for the Netherlands.

Mobile Services

With respect to mobile services a specific regime is applicable. The follow-   25–40
ing categories of mobile services should be distinguished:

- The designated innovative public mobile services in the sense of Chapter IIA (GSM, ERMES and DCS 188);
- Other public mobile telecommunications services;
- Mobile telecommunications services for closed user groups;
- The existing analogue public mobile services of Royal PTT Netherlands N V (ATF 1, ATF 2, ATF 3, SMF).

Innovative Public Mobile Telecommunications in the Sense of Chapter IIA of the Telecom Act 1988

General Remarks

As of September 1, 1994, the Act on Mobile Communications entered into   25–41
force. According to this Act a new Chapter IIA has been added to the Telecom Act 1988 which provides for the issuance of specific licences for certain innovative types of public mobile telecommunications. The innovative mobile telecommunication systems which are governed by Chapter IIA are designated by Governmental Decree. A technical system will be designated on the basis of Chapter IIA if the system is considered important in the general economic interest. Accordingly Chapter IIA contains certain provisions in order to guarantee the construction of the designated telecommunications systems and the provision of the corresponding services.

Licences on the basis of Chapter IIA will be granted for a fixed period to be established by Governmental Decree. The Minister of Transport will decide not later than two years before the expiration whether a licence can be renewed and under which conditions (Article 13k of the Telecom Act 1988).

Already the systems GSM, ERMES and DCS 1800 have been designated   25–42
as technical systems in the sense of Chapter IIA of the Telecom Act.[37] Two licences will be issued for the GSM-system.[38] The first GSM licence has been issued to Royal PTT Netherlands N V. With respect to the second GSM licence a tender has been conducted. Pursuant to this tender the second GSM licence has been granted to a consortium, "MT2", consisting of IN6, Vodaphone, Vendex International, Internatio Müller, LIOF-Limburg and Telecombinalie. With respect to ERMES three licences will

---

[37] Governmental Decrees of August 4, 1994 on GSM [1994] O.J. No. 629 (GSM) and [1994] O.J. No. 630 (ERMES and DCS 1800).
[38] The Government may decide to grant further GSM licences in the future.

be issued. The first ERMES licence will be issued to Royal PTT Netherlands. With respect to the other ERMES licences the tender procedure applies. The tender procedure for ERMES will probably be started by the end of 1995. With respect to DCS 1800 it has not yet been decided how many licences will be issued and whether such licences will be nationwide, regional or local. Initially it was intended to start the tender procedure by the end of 1996. However, the Second Chamber of Parliament urged the Minister of Transport to start the tender for DCS 1800 earlier. With respect to DCS 1800 (and any other Chapter IIA systems which may be appointed in the future) Royal PTT Netherlands N V is not ensured of a licence and will have to tender like the other applicants.

The GSM licences will be granted for an initial period of 15 years. The ERMES licences will be granted for an initial period of 10 years.

The Selection Criteria

25–43   According to Article 13h of the Telecom Act 1988 the following selection criteria are applicable with respect to tender procedures on the basis of Chapter IIA of the Act.

- The application should fulfil the minimum legal requirements concerning legal status and ability to comply with the licence regulations.[39] Other requirements concern liquidity, solvency, technical means, know how and experience.
- Furthermore the applications which are in accordance with the minimum requirements will be subject to comparison on several aspects including:

   — the degree in which the application exceeds the minimum requirements with respect to liquidity, solvency, technical means, know-how and experience,
   — the quality of the telecommunications infrastructure,
   — the use of frequencies,
   — the services that will be provided and the proposed tariffs.

Licence Obligations

25–44   Because the development of the technical systems and the corresponding services which are designated under Chapter IIA are considered a matter of public interest, the Telecom Act 1988 and the licence conditions provide for several obligations similar to the obligations of the concessionaire such as:[40]

- A universal service obligation with respect to the services to be described in the licence (Article 13c, paragraph 2, of the Telecom Act 1988);
- Obligations with respect to the capacity, the quality and the further characteristics, including the possibility of legal tapping, of the licensed infrastructure;

---

[39] Such as the provision of GSM services and facilities which are harmonised within the GSM Memorandum of Understanding.
[40] Arts. 13g and 13l of the Telecom Act 1988.

- the use of frequencies;
- an innovative service level of high quality, corresponding to the "state of the art"; tariffs should be published; furthermore obligations are imposed with respect to the security of the network and the services and privacy protection;
- A conciliation procedure open to specific categories of customers (consumers);
- The obligation to comply with international agreements with respect to the technical systems and the corresponding services;
- An obligation to provide information to the Minister of Transport required for the purpose of a proper supervision;
- The obligation to grant access to service providers on objective and non discriminatory conditions with respect to tariffs and technical specifications (Article 13s).
- According to Article 13l, paragraph 2e, of the Telecom Act maximum tariffs will be prescribed in the licence conditions for a period of six months from the date of the service provision.

According to Article 13d the network of the licensee may not be used otherwise than to provide the services which can be offered by means of the technical system for which the licence has been granted.

The Network

On the basis of a Chapter IIA licence the licensee is entitled to construct and   **25–45**
maintain his own telecommunications infrastructure with the exception of fixed connections, other than beam transmitters. According to Article 13c Chapter IIA licensees may use shared infrastructure for the offering of their services. However a minimum of self-installed infra-structure is required (Article 13l of the Telecom Act 1988). The licensee is not entitled to install his own cables, cableworks and fixed satellite-connections. With respect to such fixed connections the following options are available;

(a) Leased lines of the concessionaire (Royal PTT Netherlands NV) either supplied directly by the concessionaire or subleased from a third party.
(b) Private cable broadcasting networks or private cable-works if an exemption has been obtained by the operator of such networks from the Minister of Transport. An exemption will be granted if the cable broadcasting network or the cable work has a certain minimal size (at least 500 subscribers for cable broadcasting networks and a provincial or nationwide coverage for cableworks).
(c) If the concessionaire is not able or not prepared to supply the cable, cablework or satellite connection within a reasonable term and against reasonable conditions, the Minister of Transport may consent to the installation of such a connection by the licensee.

The licensee is entitled to interconnect his infrastructure with: **25–46**

- the telecommunications infrastructure of the concessionaire;

- the telecommunications infrastructure of another Chapter IIA licensee; and
- foreign installations and infrastructures.

The interconnection with private installations is prohibited unless an exemption has been granted by the Minister of Transport.

Interconnection

25–47   The concessionaire is obliged on the basis of Article 4, paragraphs 4 and 5, to supply facilities for the interconnection of the network of the licensee to the public switched telephone network. If possible, the concessionaire is obliged to supply the interconnection facilities which are requested by the licensee. The tariffs for interconnection should be cost based. If no agreement can be reached between the licensee and the concessionaire, the Telecom Act 1989 provides for an arbitration procedure before the Minister of Transport. The Minister will examine whether basic principles with respect to non discrimination and cost based tariffs are being violated by the concessionaire (Article 13q of the 1989 Act).

Licensed infrastructures should be made suitable for legal tapping by the licensee. Investment costs have to be borne by the licensee. With respect to GSM the network must be suitable for legal tapping before January 1, 1996.

Asymmetrical Regulations

25–48   In view of the advantages of Royal PTT Netherlands NV resulting from the unequal start of the exploitation of GSM and ERMES the second (or third) operator may apply for the following options (Article 13x of the 1989 Act):

(a) the right to offer services as a service provider on the network of Royal PTT Netherlands NV with the use of an own number block;

(b) the right to use the network of Royal PTT Netherlands NV using his own number block as soon as his own network is partly operational (roaming);

(c) number portability, allowing users to retain their original subscription number when transferring from Royal PTT Netherlands NV to the other operator.

According to the Minister the costs of these provisions should be borne by the applicant.

Entrance Fee

25–49   Extensive discussions took place with respect to the question whether Chapter lla licensees should pay an entrance fee or a percentage of their profits. As a result of these discussions this issue will be discussed by the Parliament within the framework of the future frequency policy. On February 17, 1995 the Minister of Transport issued a White Paper containing proposals for further discussion.[40a] The implementation of the new policy, which still has to be determined, will have to be incorporated in the revision of the Telecom Act 1988, which will probably take quite some time.

Other public mobile telecommunications services

The Mobile Communications Act 1994 which entered into force on September 1, 1994 has liberalised other public mobile telecommunications services as well. Articles 14 and 17 of the 1989 Act have been amended in order to allow licences for radiotransmitters on the basis of Chapter III for public mobile telecommunications services.

25–50

With respect to licences for public mobile telecommunications services the licensing regime for transmitters on the basis of Article 17 of the 1988 Act is applicable (see above). However, the following specific rules apply:

- The prohibition of article 14 of the Telecom Act with respect to the supply of mandatory services and direct transport of data is not applicable with respect to the public mobile service for which the licence has been granted.
- Apart from the procedural grounds, the grounds for refusal of the licence have been limited to the following instances:

  — If an efficient use of the terrestrial frequencies requires the refusal;
  — If a licence would be contrary to the Media Act 1987 or the Broadcasting Transmitters Act 1935.

- Leased lines of the concessionaire may be used by or on behalf of the licensee or his service providers with respect to the supply of the public mobile communications services for which the licence has been granted.

Fixed connections which are needed by the licence owner for the construction of his public mobile telecommunications infrastructure must be leased lines of the concessionaire. Other private installations may not be used for this purpose.

25–51

The operation of a public mobile service on the basis of Chapter III is regulated to a lesser extent than the operation of such a service on the basis of Chapter IIA. However, the licence conditions may contain certain obligations in order to ensure that the network is is constructed within a certain period and that the services will in fact be provided.

Mobile Telecommunications Services for Closed User Groups

With respect to mobile telecommunications services on behalf of closed user groups by means of radio transmitters the general licence criteria of Article 17 are applicable as well. However, according to Article 14, paragraph 6, the

25–52

[40a] TK 1994/1995, 24095, No. 2.

Minister of Transport may grant an exemption from the prohibition of Article 14, paragraph 1, with respect to mobile telecommunications services within closed user groups if the concessionaire is not able or not prepared to provide a similar provision within a reasonable term and against reasonable conditions.

The exemption on the basis of Article 14, paragraph 6, is used to enable for instance radiotelephone installations and trunking networks.[41]

Analogue Mobile Systems ATF1, ATF2, ATF3 and SMF3

25–53   Because the public mobile services have been deleted from the mandatory services, the Mobile Communications Act 1994 had to reclassify the existing systems for public analogue mobile telecommunications networks of Royal PTT Netherlands NV (ATF1, ATF2, ATF3 with respect to mobile telephony and SMF3 with respect to paging). The transitional regulations of the 1994 Act provide for a licence on the basis of Article 17 of the Telecom Act 1988 with respect to these systems. However specific conditions are applicable with respect to these systems.

According to the transitional provisions, the universal service obligation with respect to ATF3 is maintained until one of the GSM operators is able to offer a similar service. After this period Royal PTT Netherlands NV will be entitled to continue the ATF3 service although the Minister of Transport may terminate the ATF3 licence upon a five years notice if the development of the GSM services is affected.

The universal service obligation with respect to ATF1, ATF2 and SMF3 is not maintained. However a licence to operate these systems is granted for a period of five years. After this term the licence will be renewed, unless the Minister of Transport is of the opinion that an efficient and appropriate use of frequencies requires otherwise.

25–54   As long as Royal PTT Netherlands NV is obliged to offer ATF3 services, most of the concession obligations continue to apply. An exception is made for the regulations in the General Directives which control the change of tariffs. Instead Royal PTT Netherlands NV is obliged to charge a uniform tariff during this period. As soon as Royal PTT Netherlands NV is no longer obliged to offer ATF3, the offering of these services will become less regulated. For instance general directives with respect to the quality of the services are no longer applicable. However, other obligations such as the obligation to offer uniform tariffs and conditions and the prohibition of cross subsidies are maintained during this period. The same applies for the offering of SMF3 services. With respect to the offering of ATF1 and ATF2 very little has been regulated. The general directives no longer apply to the offering of these systems.

---

[41] *Gazette*, March 21, 1991, No. 57.

4. REGULATION OF BROADCASTING

Basic Statutory Rules

Broadcasting has been regulated in the following Acts: 25–55

- **The Media Act 1987**[42]: The Media Act regulates the various aspects of the Dutch broadcasting system such as public and commercial broadcasting organisations, the contents of broadcasting programmes, which programmes may be distributed by means of transmitters and cable broadcasting networks, technical facilities, the financing of the Dutch broadcasting system and the relationship between broadcasting organisations and the press.
- **The Broadcasting Transmitters Act 1935**[43]: This Act regulates the construction and the exploitation of broadcasting transmitters which are operated by the Netherlands broadcast transmitter company, NoZeMa, on behalf of the public broadcasting organisations.
- **The Telecom Act 1988**[44]: This Act regulates the construction and the operation of cable broadcasting networks.

Traditionally the Dutch public broadcasting system is based on the provision of programmes on a non-commercial basis by associations which represent various social, cultural and ideological groups in the Dutch society. Nevertheless, in view of international regulations commercial broadcasting has been introduced in the Netherlands in 1992. The amendment to the Media Act 1987 which entered into force on January 1, 1995 has the aim to strengthen the position of the public broadcasting organisations in relation to the commercial broadcasting organisations.

According to the Media Act 1994 the following categories of programmes should be distinguished:

Public Broadcasting Programmes

(a) The national broadcasting programmes of the public broadcasting   25–56
    associations.
        Public broadcasting associations must represent various social, cultural or ideological groups in the Dutch society. Furthermore a certain minimum number of members is required. As of September 1, 1995 public broadcasting associations need a concession granted by the Minister of Culture.[45]
(b) The national broadcasting programmes of the National Broadcasting

---

[42] The Media Act 1987 has been amended by the Act of April 28th 1994, [1994] O.J. L385. The amended Act has entered into force on January 1, 1995. Additional amendments were made by the Act of December 21, 1994, O.J. 1945, in force as of January 1, 1995, and by the Act of December 23, 1994, O.J. 1946, in force as of December 30, 1994. Hereafter the text of the amended Act will be discussed.
[43] *Radio Omroep Zender Wet 1935* [1935] O.J. L403.
[44] [1988] O.J. L520.
[45] Articles 31–38 of the Media Act 1994 [1994] O.J. No. L386.

Organisation.[46] As of January 1, 1995 the National Broadcasting Organisation has been divided in two seperate legal entities: the National Broadcasting Organisation[47] entrusted with the supply of broadcasting programmes which should be jointly provided[48] and the National Broadcasting Programmes Organisation[49] entrusted with the provision of broadcasting programmes which are not provided by the public broadcasting associates.[50]

The National Broadcasting Organisation is also entrusted with the provision of the national teletext programme.

(c) The national broadcasting programmes of educational broadcasting organisations, religious and ideological denominations and political parties. The Authority for the Media[51] is entrusted with the allocation of broadcasting time to such organisations.[52]

(d) The national broadcasting programmes for public information.[53]

(e) The national broadcasting programme of the Organisation for the Broadcasting of Advertisements.[54]

(f) The broadcasting programmes of the regional and local broadcasting organisations. The Authority for the Media is entrusted with the allocation of broadcasting time to regional and local broadcasting organisations.

The regional and local broadcasting organisations are also entitled to supply a teletext programme.

Commercial Broadcasting Programmes

25–57   (g) National commercial broadcasting programmes which are being distributed by means of cable broadcasting networks. With respect to these programmes a licence should be obtained from the Authority for the Media.[55] Such a licence will be granted if certain criteria are met, for instance transmission agreements must be concluded with cable operators up to at least a certain minimum coverage.[56]

The commercial broadcasting organisation is also entitled to transmit a teletext programme.

(h) National commercial broadcasting programmes which are distributed by means of cable broadcasting networks and by means of transmitters (terrestrial or satellite).

A licence from the Minister of Culture is required with respect to the distribution of a commercial broadcasting programme by means of transmitters (terrestrial or satellite).

---

[46] *De Nederlandse Omroepstichting* (NOS).
[47] *Nederlandse Omroepstichting* (NOS).
[48] Art. 51d of the Media Act 1994.
[49] *Nederlandse Omroepprogramma Stichting* (NOS).
[50] Art. 51b Media Act 1994.
[51] Arts. 39e–39g of the Media Act 1994.
[52] *Commissariaat voor de Media.*
[53] Art. 39h of the Media Act 1994.
[54] *Stichting Etherreclame* (STER); Art. 39b of the Media Act 1994.
[55] During the first three years the licence will be granted by the Minister of Culture.
[56] Art. 71 c and 71 h of the Media Act 1994.

Pay Television Programmes[57]

> (i) Pay television programmes which are distributed by means of cable 25–58
> broadcasting networks. With respect to these programmes a licence
> from the Authority for the Media is required.[58]
>
> (j) Pay television programmes which are distributed by means of trans-
> mitters (terrestrial or satellite). With respect to these programmes a
> licence is required from the Minister of Culture.[59]

Other Programmes

> (k) Datacasting programmes (programmes consisting of pictures, of 25–59
> alphanumerical data and other stationary pictures such as cable tv
> information services and teletext programmes which are distributed
> on a separate channel).
>
> For the distribution of these programmes by means of a cable
> broadcasting network no licence is required. If distributed by means
> of transmitters a licence from the Minister of Culture on the basis
> of Article 167 of the Media Act is required.
>
> (l) The broadcasting programme of Netherlands Radio Worldservice.
>
> (m) Foreign broadcasting programmes which are distributed in accord-
> ance with the applicable foreign broadcasting regulation.
>
> (n) Broadcasting programmes consisting of the full-length sound repro-
> duction of the parliamentary debates and Council meetings.

The Media Act 1994 encompasses certain of obligations with respect to 25–60
public broadcasting organisations and public broadcasting programmes in
order to ensure a multiform supply of broadcasting programmes and to
strengthen the position of the public broadcasting organisations. Regulations are applicable with respect to the allocation of broadcasting time,[60]
restrictions with respect to advertisements, the coordination of programmes
on a certain channel, the content of the programmes including minimum
amounts of cultural, instructive and educational programmes, the financing
of the public broadcasting system, etc. With respect to commercial broadcasting the regulations focus on the protection of the public broadcasting
system and the relation to the newspaper market.

Terrestrial and other Delivery (*e.g.* Satellite, Cable)

Terrestrial and Satellite

According to Article 167a of the Media Act 1994 the following broad- 25–61
casting programmes may be distributed by means of transmitters (terrestrial
and/or satellite) in the Netherlands:

---

[57] Including pay radio programmes.
[58] Art. 73 of the Media Act 1994.
[59] In the Media Act 1994 an exception has been made with respect to teletext pay television programmes
which are being distributed via a seperate channel.
[60] With respect to public broadcasting three television channels and five radio channels are available.

- The public broadcasting programmes;
- foreign broadcasting programmes which are distributed in accordance with applicable broadcasting regulations in the country of origin;
- broadcasting programmes consisting of the full-length sound reproduction of the parliamentary debates and Council meetings;
- commercial broadcasting programmes if the Minister of Culture has consented to the distribution of such programmes by means of transmitters (terrestrial and/or satellite);
- datacasting programmes; and
- pay television programmes if the Minister of Culture has consented to the distribution of such programmes by means of transmitters (terrestrial and/or satellite).

It should be noted that in the event of satellite transmission, the uplink is not considered as the distribution of a programme in the sense of Article 167a of the Media Act 1994.

With respect to the construction and operation of the transmitters the provisions of the Broadcasting Transmitters Act 1935, and the Telecom Act 1988 apply (see below):

Cable Broadcasting Networks

25–62   With respect to the distribution of programmes by means of cable broadcasting networks the Articles 65 and 66 of the Media Act 1994 are applicable.

According to Article 65 the cable operator is obliged to distribute certain broadcasting programmes. The main programmes for which this obligation applies are:

- National public broadcasting programmes;
- local and regional broadcasting programmes which are intended for the municipality or the region where the cable broadcasting network is located;
- television programmes of the Belgium national broadcast organisation, if such programmes can reasonably be received with a standard antenna by the cable tv operator;

25–63   Furthermore Article 66 determines which other broadcasting programmes may be distributed by the cable operator:

- Local and regional public broadcasting programmes other than the programmes mentioned in Article 65.
- Foreign broadcasting programmes which are distributed in accordance with applicable broadcasting regulations in the country of origin.
- Broadcasting programmes consisting of the full-length sound reproduction of the parliamentary debates and Council meetings.
- Datacasting programmes.
- Pay television programmes for which a licence has been granted by the Authority for the Media.

- Commercial broadcasting programmes for which a licence has been granted.

The construction and the operation of cable broadcasting networks has been regulated in the Telecom Act 1988 and has been discussed in sections 2 and 3.

Regulatory Agencies for Broadcasting

As of January, 1 1995 regulatory tasks in the field of broadcasting have been   **25–64** shared out between the Minister of Culture, the Minister of Transport, the Authority for the Media and the National Broadcasting Organisation.

The Minister of Culture is entrusted with those tasks which are considered essential with respect to the governmental media policy such as:

- the granting of concessions to the national public broadcasting associations and the allocation of broadcasting channels;
- the determination of the total amount of broadcasting time;
- the determination of the broadcasting budget.

The Minister of Transport is entrusted with the granting of licences for cable broadcasting networks and transmitters for broadcasting purposes on the basis of the Telecom Act 1988. Especially with respect to the allocation of frequencies to commercial broadcasting organisations, the Minister of Transport plays an important (though disputable) role.[61]

The Authority for the Media is entrusted with: **25–65**

- the supervision with respect to the compliance with and the implementation of the Media Act. Furthermore the Authority may impose administrative penalties.
- The granting of licences and permits on the basis of the 1994 Act.

The National Broadcasting Organisation is the organisation for cooperation and coordination of the public broadcasting organisations. The tasks of the National Broadcasting Organisation have become more important on the basis of the amendment which entered into force on January, 1 1995. The National Broadcasting Organisation is entrusted with:

- the distribution of broadcasting time among the national public broadcasting organisations;
- the coordination of the programmes of the public broadcasting organisations.

Regulation of Transmission

Transmitters (Terrestrial and Satellite)

NoZeMa is the Netherlands broadcast transmitter company through which   **25–66** the Dutch State and the national public broadcasting organisations are

---

[61] The role of the Minister of Culture with respect to the allocation of frequencies to commercial broadcasting organisations is disputed as well.

working together with respect to the transmission of the Dutch public broadcasting programmes. According to the Broadcasting Transmitters Act 1935 NoZeMa has the exclusive right to constuct and exploit transmitters intended for the distribution of the Dutch public broadcasting programmes. As long as NoZeMa fulfills this task, no licence can be granted to any other party on the basis of article 17/18 of the Telecom Act 1988 (see above sections 2 and 3).[62]

With respect to the distribution of "other broadcasting programmes", NoZeMa has been granted a right of first refusal if such distribution can be considered as an "object of cooperation" between the public broadcasting organisations. If NoZeMa is able and willing to construct and operate such broadcasting transmitters no licence will be granted to any other party on the basis of Articles 17 and 18 of the Telecom Act 1988. A decision of the Administrative Court of January 27 1993[63] made it clear that the right of first refusal does not apply to foreign programmes because the distribution of such programmes can not be considered as an "object of cooperation". The same applies with respect to commercial programmes. The situation with respect to pay television is not clear.

25–67   Transmitters intended for the distribution of programmes which are not covered by the Act on Broadcasting Transmitters such as foreign programmes and commercial programmes have to comply with the licence requirements of Articles 17 and 18 of the Telecom Act 1988 (see above sections 2 and 3). NoZeMa has no preferred status in this area.

Some questions have been raised with respect to the use of the remaining capacity of the NoZeMa transmitters for the distribution of public broadcasting programmes. NoZeMa is using the remaining capacity for datacasting services.[64] At present it is assumed that such use of the remaining capacity is covered by the Broadcasting Transmitters Act 1935 and no licence in the sense of Article 17 is required.

The exclusive position and the right of first refusal of NoZeMa on the basis of the Broadcasting Transmitters Act 1935 also covers the distribution by satellite. However, until now the Dutch public broadcasting programmes are not transmitted by means of satellites. Furthermore it should be noted that the earth station and the uplink do not fall within the scope of NoZeMa's exclusive status.

Cable Broadcasting Network

25–68   With respect to the construction, maintenance and operation of cable broadcasting networks a licence on the basis of Article 21 of the Telecom Act 1988 is required. See above paras. 25–05 *et seq.* and 25–29 *et seq.*

---

[62] With respect to the programmes of local broadcasting organisations NoZeMa has given up its monopoly.
[63] *College van Beroep voor het Bedrijfsleven*, January 27, 1993, AB 1994, 31.
[64] With a licence of the Minister of Culture on the basis of Article 167 of the Media Act 1994.

Monopoly and Competitive Service

As indicated above NoZeMa has an exclusive position with respect to the   **25–69**
construction, the maintenance and the operation of transmitters (terrestrial
and satellite) intended for the distribution of the Dutch public broadcasting
programmes.

With respect to cable broadcasting networks the Telecom Act 1988
determines that for each area only one operator will be licenced. Licences
in this area are granted on the basis of the first come first serve principle.

With respect to the provision of broadcasting programmes the Organisation for the Broadcasting of Advertisements (STER) has an exclusive position among the national public broadcasting organisations. Apart from this
legal monopoly no exclusive rights have been granted with respect to the
provision of programmes, although competition in the broadcasting area
is regulated to a large extent (see above).

5. REGULATION OF APPARATUS

Regulation of Apparatus Supply and Installation

Terminal Equipment

The definition of terminal equipment in the Telecom Act 1988 is not very   **25–70**
clear as a result of the Mobile Communications Act 1994 in which a new
category of licensed telecommunications infrastructure has been introduced. However in practice the following categories of terminal equipment
are distinguished by the Ministry of Transport:

- apparatus intended to be connected to the telecommunications infrastructure (including leased lines) of the concessionaire in the sense of
  Chapter II of the 1988 Act;
- apparatus intended to be connected to licensed telecommunications
  infrastructures in the sense of Chapter IIA of the 1988 Act;
- apparatus intended to be connected to the existing telecommunications infrastructures for analogue mobile telephony and
  paging (ATF1, ATF2, ATF3 and SMF3) of Royal PTT Netherlands
  NV for which systems a transitional licence has been given on the
  basis of Chapter III.

Apparatus intended to be connected to public mobile networks in the
sense of Chapter III (such as the Euromessage System) are not considered
to be terminal equipment.

The Telecom Act 1988 regulates the connection to the public tele-   **25–71**
communications network as well as the marketing, the installation and the
maintenance of terminal equipment. According to Article 29, paragraph 3,
of the Telecom Act 1988, it is prohibited to connect terminal equipment

to the public telecommunications network unless such equipment has been properly type approved. According to Article 29, paragraph 8, it is not allowed to market (sell, lease, offer, etc.), or to store terminal equipment which has not been properly marked as being type approved. The approval and testing procedures will be discussed below. With respect to the installation and maintenance of terminal equipment on a professional basis, Article 29, paragraph 9 of the 1988 Act, and Article 10 of the Decree on Terminal Equipment require a professional education up to a certain level in order to warrant sufficient professional skills. Furthermore a special licence by the Minister of transport is required with respect to the production, marketing, export, installation and maintenance of radio transmitters.

Other Apparatus

25–72   Other apparatus must fulfill the general requirements with respect to for instance the EMC Directive[65] and the Low Voltage Directive.[66] Moreover a separate type approval procedure applies with respect to transmitters and radio receivers on the basis of Article 17 of the 1988 Act (see below). Terminal equipment has been exempted from this separate type approval requirement. However, with respect to apparatus (solely) intended to be connected to mobile networks on the basis of Chapter III, such as Euromessage, this type approval procedure is applicable.

Type Approval and Testing Procedures

Terminal Equipment

25–73   In view of the Telecommunications Terminal Equipment Directive[67] the regulation in the 1988 Act with respect to the approval of terminal equipment has to be reviewed. The Act with respect to the revised regulation of terminal equipment will probably be presented to the Parliament in the course of 1995. In the meantime the Directive of April 29, 1991, is implemented by means of a Notification of the Minister of Transport.[68] According to the Notification the following procedure for type approval is applicable.

25–74   With respect to terminal equipment for which a Common Technical Regulation (CTR) has been established, the procedures for type approval referred to in the Directive apply. However, at this moment only a few CTR's are available. For those types of terminal equipment for which no CTR has been established, the national technical requirements apply and a separate national type approval for the Netherlands is required. However, according

---

[65] Directive 89/336.
[66] Directive 73/23.
[67] Directive 91/263 [1991] O.J. No. L128 and the amendment to this Directive of July 22, 1993 [1993] O.J. No. L220.
[68] Notification of April 1, 1994, published in the *Gazette*, April 14, 1994, No. 72.

to the Notification of April 14, 1994, the procedures for type approval given by the Directive are also applicable with respect to the national type approval procedure. Consequently the applicant may choose one of the following procedures regardless whether CTR's have been established:

(i) An (E.C.) type-examination followed by an (E.C.) declaration of conformity to type; the product checks are carried out by the HDTP[69] or a designated test laboratory.
(ii) An (E.C.) type-examination followed by a production quality assurance;
(iii) An (E.C.) full quality assurance.

With respect to the type-examination and the declaration of conformity   25–75
to type the HDTP[70] has been designated as notified body in the sense of the Telecommunications Terminal Equipment Directive. With respect to the production quality assurance and the full quality assurance the N.V. KEMA has been designated as notified body. These notified bodies are also entrusted with the national type approval for those area's where no CTR's have been established.

Furthermore the following test laboratories have been designated in the sense of the Telecommunications Terminal Equipment Directive to carry out tests pertaining to the procedures referred to above:

• NMi Certin BV with respect to terminal equipment, radio transmitters and EMC;
• Telefication BV with respect to terminal equipment, radio transmitters and EMC;
• N.V. KEMA with respect to EMC.

Other Apparatus: Radio Transmitters and Radio Receivers

With respect to transmitters and radio receivers a seperate approval proced-   25–76
ure applies on the basis of Articles 17 and 19 of the Telecom Act 1988. According to this procedure a declaration of conformity of one of the forementioned test laboratories should be filed with the Minister of Transport (the HDTP) together with the application for type approval. Transmitters and radio receivers which qualify as terminal equipment have been exempted from this separate type approval procedure.

Marking

The marking of terminal equipment for which CTR's have been established   25–77
should be in accordance with the Telecommunications Terminal Equipment Directive 91/263. With respect to terminal equipment for which no CTR's have been established the national marking requirements apply. The same applies with respect to transmitters and radio receivers which have to be type approved on the basis of Article 17 and 19 of the Telecom Act 1988.

---

[69] The Department on Telecommunications and Post of the Ministry of Transport.
[70] ibid.

Rights to Market

25–78   As indicated above, it is not allowed to market (offer, sell, lease, or provide otherwise) or to store terminal equipment which has not been properly marked as being type approved. Furthermore a licence is required with respect to the production, marketing, export, maintenance and installation of radio transmitters. Finally it is not allowed to market radio receivers which do not comply with the technical requirements or which are not properly marked. Finally with respect to the marketing of transmitters and radio receivers certain conditions with respect to the registration and storage are applicable.

6. TECHNICAL STANDARDS

Apparatus

25–79   As indicated above apparatus have to comply with technical requirements. With respect to terminal equipment CTR's or national technical requirements are applicable. With respect to other apparatus (transmitters and radio receivers) technical requirements have been determined by the Minister of Transport as well. However, apart from these technical requirements which relate to the "essential requirements" that apply to terminal equipment and similar requirements for other apparatus there is no limitation with respect to specific technical standards.

Networks, Interfaces

The Telecommunications Infrastructure of the Concessionaire

25–80   According to the General Directives, the interfaces of the network of the concessionaire should comply with European and international recommendations, directives, standards and specifications as far as possible.[71] In addition leased lines should be in accordance with "international recommendations".[72] Furthermore the telecommunications infrastructure of the concessionaire should comply with "modern standards".[73] Apart from these general obligations, no concrete technical standards are prescribed by the Minister of Transport.

Licensed Telecommunications Infrastructures in the Sense of Chapter IIA

25–81   Licensed telecommunications infrastructures (inclusive of the interfaces) should comply with the technical standards of the technical system for which the licence has been granted.

---

[71] Art. 2.6 of the General Directives.
[72] Art. 2.4 of the General Directives.
[73] ibid., Art. 2.1.

Private Networks in the Sense of Chapter III

Private radio communications installations should comply with the tech-   25–82
nical requirements which have been determined by the Minister of Transport in order to safeguard that they fulfil certain essential requirements
(see above Section 5). However, no specific technical standards have been
prescribed.

With respect to cables and cableworks no specific technical requirements
have been determined by the Minister of Transport.

With respect to cable broadcasting networks a set of technical requirements has been determined by the Minister of Transport, which are in
accordance with the former PTT standard. However, as these technical
requirements are considered more or less outdated, a licence can be given
if certain essential technical requirements are met.

The Packet Switched Data Service

At the occasion of the removal of the packet switched data service from   25–83
the mandatory services, an agreement was concluded between the Minister
of Transport and Royal PTT Netherlands NV with respect to the continuation of the provision of a packet switched data service which is in accordance with the technical standard as indicated in the Council Recommendation of June 5, 1992.[74] According to the agreement Royal PTT Netherlands
will continue this service until sufficient competing services are available.[75]

ONA, ONP and Similar Open Architecture/Access Principles

The Concessionare's Telecommunications Infrastructure and Mandatory Services

With respect to the mandatory services and the supply of leased lines ONA   25–84
and ONP principles apply such as:

- a universal service obligation;
- tariffs and conditions should be non-discriminatory and based on
  objective criteria;
- tariffs and conditions should be made public.

With respect to the tariffs for the mandatory services the General Directives encompass a system to control the changes of tariffs with the intention
to prevent excessive increases for the "average" user.[76] With respect to
leased lines the ONP Directive 92/44[77] has been implemented in the General
Directives. Provisions with respect to cost oriented tariffs have been
added.[78]

---

[74] Council Recommendation 92/382 on the harmonised provision of a minimum set of packet-switched
data services (PSDS) in accordance with open network provision (ONP) principles.
[75] Governmental Decree [1993] O.J. No. 21.
[76] Art. 5 of the General Directives.
[77] Council Directive 92/44 on the application of open network provision to leased lines.
[78] Amendment of January 27, 1993, *Gazette*, February 9, 1993.

Furthermore in view of the introduction of Chapter IIA with respect to licensed infrastructures a regulation with respect to interconnection has been added in Article 4, paragraphs 4 to 6, and Article 13q of the Telecom Act 1988. According to these provisions the concessionaire is obliged to provide interconnection facilities. If no agreement is reached between the parties, the Minister of Transport will examine whether the tariffs and conditions which are offered by the concessionaire are non-discriminatory and objective and whether the offered tariffs are costbased.

Licensed Telecommunications Infrastructures in the Sense of Chapter IIA

25–85   With respect to licensed telecomunications infrastructures certain ONP and ONA principles apply as well, such as:

- an obligation to provide a nationwide service upon uniform conditions;
- an obligation to publish tariffs.

However, the licensee is not obliged to supply his services to end users against uniform tariffs. Furthermore the licensee owner should give access to his network to service providers upon objective and non-discriminatory conditions relating to tariffs and technical specifications. These conditions should be made public.

Private Networks in the Sense of Chapter III

25–86   With respect to private networks the following ONP and ONA related principles apply:

- The standard licence conditions for cable broadcasting networks encompass a universal service obligation on behalf of households for the cable broadcasting operator within his territory. Furthermore conditions and tariffs should be made public.
- The exemption by the Minister of Transport on behalf of the operators of cable broadcasting networks and cableworks on the basis of Article 14/22 of the Telecom Act 1988 with respect to the supply of fixed connections to Chapter IIA licensees encompasses ONP/ONA provisions. If such operators apply for an exemption, they are obliged to offer similar connections against objective and non-discriminatory tariffs and conditions to all licensees. Furthermore the private operators are obliged to offer their fixed connections to all applicants of a Chapter IIA licence against non-discriminatory conditions and tariffs.
- Furthermore ONP/ONA principles apply with respect to the mobile analogue systems ATF3 and SMF3 of Royal PTT Netherlands NV As long as these services are offered, they must be offered to anyone against uniform tariffs and conditions which should be made public.
- With respect to public mobile networks for which a licence has been granted on the basis of Chapter III no ONP/ONA principles are applicable. However the licence conditions may encompass certain

ONP/ONA related provisions such as the obligation to construct and operate a nationwide network.

The Packet-switched Data Service

According to the agreement between the Minister of Transport and Royal   **25–87**
PTT Netherlands NV referred to above, the PTT will continue to supply the packet-switched data service in accordance with the Council Recommendation on the harmonised provision of PSDS in accordance with ONP principles until a working competition has been reached.

CHAPTER 26

New Zealand

By Jim Stevenson, Buddle Findlay, Wellington

1. HISTORY OF DEVELOPMENT OF TELECOMMUNICATIONS IN NEW ZEALAND

Former Industry and Institutional Structure

Up until major legislative reforms in the period 1986–1989 the communica-   26–01
tions services sector in New Zealand was characterised by a very high
degree of government intervention through both economic regulation and
ownership. In part, this reflected the necessary involvement of successive
governments, since European settlement in the nineteenth century, in the
funding and development of an infrastructure of essential services.

Heavy government involvement in telecommunications and broadcasting
was also driven by social and political factors. While New Zealand's small
population (now some 3.5 million) has tended to concentrate in a small
number of major urban centres, a significant proportion live in rural areas
and related servicing towns. The existence of hilly and mountainous terrain
in many regions compounds the difficulties of establishing and maintaining
telecommunications and broadcasting services in these areas and throughout the country (which is about the size of the U.K.). Yet the rural sector
is the source of the great majority of New Zealand exports namely agricul-
tural and horticultural products. This factor helped lead to the evolution
of government policies to provide both universal telephone service and
access to basic sound radio and television broadcasting services.

New Zealand developed a monopoly PTT, latterly known as the New
Zealand Post Office (NZPO), responsible to a government Minister. The
NZPO combined its original postal services functions with responsibilities
for telegraphy and telephony including control of the "wireless" use of the
radio spectrum.

Other government owned entities responsible for railways, electricity transmission and broadcasting transmission as well as a few government agencies also developed their own private telecommunication networks. However, resale of their telecommunication capacity was effectively prohibited as was resale of NZPO licensed radio services.

Broadcasting services were also essentially developed by a succession of state-owned broadcasting entities of various descriptions. Market entry for private broadcasting firms, while eventually opened up, was restricted by a broadcasting warrant system.

Background to Legislation

26–02   The current regulatory environment for telecommunication and broadcasting markets in New Zealand largely derives from a series of government policy decisions between 1985 and 1990. Underpinning all these decisions was the then Government's goal of making the New Zealand economy more internationally competitive. A competitive telecommunications infrastructure was an important requirement for achieving that goal.

The initial impetus for reform was the Government's desire to improve the performance of the public sector (including government owned telecommunication and broadcasting facilities). To this end certain policy principles were implemented by the Government *viz*:

- Commercial functions were separated from non-commercial functions such as regulatory functions and policy advice to the Government.
- Commercial functions were re-organised as businesses in the form of companies under national companies law.
- State-owned commercial activities were exposed to competition in their relevant product markets unless there were overriding social or practical reasons (*i.e.* the principle of competitive neutrality).
- Government social objectives for public enterprises were to be distinguished from, and implemented in a manner which did not conflict with, commercial objectives.

Of all the Government agencies operating in the mid 1980s, the NZPO provided the most comprehensive example of the mix of commercial and non-commercial functions hitherto undertaken within a single Government department. Policy advisory functions of the NZPO ranged across national and international telecommunications services, postal services and banking services. The NZPO also operated commercially in those sectors. Moreover the NZPO administered the all-pervasive regulation of the telecommunications and postal services sector. The NZPO also regulated the very important resource of the radio spectrum which provided the basis for reinforcement of market entry restrictions in telecommunication markets.

Legislative and Administrative Reform 1987–1990

26–03   In part, implementation of the above principles, the telecommunication business activities of the NZPO were corporatised, as at April 1, 1987, into a state-owned enterprise called the Telecom Corporation of New Zealand Ltd.

Other corporations were formed for postal services and banking activities. The Telecommunications Act 1987 (see para. 26–19 *et seq.*), was enacted as part of legislation implementing wide ranging liberalisation of markets in which state-owned enterprises operated. This provided for the progressive freeing-up, between 1987 and 1989, of customer premises equipment and related services markets and for some degree of liberalisation of value added services markets. Policy and regulatory functions were transferred to a separate government department (now the Ministry of Commerce).

The social and regulatory framework in which the former NZPO operated was so highly entrenched that it was not readily susceptible to immediate liberalisation of network services. The Government was cautious, too, about further liberalisation especially since there was no real precedent, internationally, for complete deregulation of network services. It therefore commissioned further studies into the economic and social effects of further liberalisation and options for reform.

The further official studies, including advice from the independent consultants Touche Ross commissioned by Government in 1987, suggested that significant economic benefits would result from network service deregulation. As a consequence of these studies the Government introduced the Telecommunications Amendment Act 1988 which removed the residual statutory restrictions on the supply of telecommunication services from April 1, 1989.

The Government had decided that it would establish neither a special   **26–04** industry regulatory regime nor any special ad hoc authority (like an OFTEL) to promote the entry of new competitors. It chose instead to rely on the then recently strengthened anti-trust statute, the Commerce Act 1986 (see paras. 26–14 to 26–18 and 26–40 to 26–45 below), which includes provision for enforcement of prohibitions of anti-competitive behaviour either by an agency called the Commerce Commission or by private legal action. Telecom NZ also gave certain undertakings to the Government in June 1988 that charges for interconnection would be based on costs and that dealings as between Telecom subsidiaries and competitors would be non-discriminatory and on an arms length basis. A similar undertaking was given in July 1989. However the Government also warned that greater intervention might occur, if warranted, following experience of the effectiveness of the regime.

The establishment of fully competitive entry in telecommunication markets was accompanied by similar moves in 1988/89 in the related broadcasting sector. State-owned commercial activities in television and radio were corporatised and entry restrictions into broadcasting removed along with the abolition of the former Broadcasting Tribunal (see paragraph 26–49). In accordance with government policy favouring a company structure for state broadcasting assets, the Broadcasting Corporation of New Zealand (BCNZ) was abolished and replaced by two state-owned enterprises. Specific statutory privileges enjoyed by the BCNZ were also removed. Policy and regulatory functions were separated.

Restrictions on the provision of telecommunications links for broadcasting were removed. Conversely, deregulation also presented an opportunity to utilise broadcasting transmission assets and services for other tele-communication services.

Access to radio frequencies was liberalised subsequently under a new radio spectrum management regime enacted under the Radiocommunications Act 1989 (see below). This paved the way for the tendering of telecommunication and broadcasting frequencies subject to constraints under the Commerce Act 1986 on anti-competitive acquisitions of spectrum property rights. Social policies to ensure adequate behavioural standards in broadcasting as well as to promote New Zealand programming were implemented through a new Broadcasting Act 1989. (See paras. 26–49 and 26–50).

Subsequent Developments

26–05   The period leading up to and after network service deregulation witnessed expressions of interest by national and international parties in establishing telecommunication businesses in New Zealand. In particular, two consortiums were formed linking foreign telecommunication interests to state railways and broadcasting organisations which had established telecommunication infrastructure. This led in turn to the commencement of interconnection negotiations between those consortiums and Telecom in 1989/90 and the subsequent merger in 1990 of those consortiums into a major fixed services network competitor now known as CLEAR Communications Limited.

In line with the then government policy of selling state-owned assets to reduce the level of public debt, Telecom Corporation of New Zealand Limited was sold in 1990. Following a review of the regulatory environment the Government decided that there should be no substantive changes in the environment except for the imposition of new information disclosure requirements on Telecom and regulation making powers with respect to international services (see paras. 26–29 to 26–31).

Kiwi Share Principles

26–06   The Government decided to secure Telecom's early commitments, when a state-owned enterprise, to certain measures affecting residential service. These were enshrined in the form of pledges to be accepted by the new owners of Telecom. These were subsequently given contractual effect by their inclusion in Telecom's Articles of Association and made enforceable by way of a government owned preference share (the Kiwi Share). The pledges may be summarised as follows:

    (a) The local "free" calling options are to be maintained for all residential customers although Telecom is given the right to develop optional tariff packages which entail local charges for those subscribers who elect to take them as an alternative;

    (b) Telecom is permitted to charge no more than the standard residential rental for ordinary residential telephone services subject to increases no greater than the New Zealand consumer price index unless the overall profitability of the Telecom regional operating companies[1] is unreasonably impaired;

(c) the line rental for residential users in rural areas is to be no higher than the standard residential rental;

(d) Telecom is obliged to make ordinary residential telephone service as widely available as it was at the date of the adoption of the pledges.

Provision is made within the Articles for amendments to the principles and modifications with the agreement of the Kiwi shareholder (*i.e.* the Minister of Finance).

Limitations on foreign shareholding of Telecom NZ were also introduced 26–07 through modifications to its Articles of Association at the time of privatisation. No person may have a "relevant interest" in 10 per cent or more of the total voting shares without the approval of the Government and the Board. No person, who is not a New Zealand national, may have a relevant interest in more than 49.9 per cent of the total voting shares for the time being without the Government's written approval. There are also powers for the sale of shares acquired without the requisite approvals.

Under an agreement with Government, the American purchasers of Telecom were obliged to reduce their aggregate ownership in the company to not more than 49.9 per cent of the company's then outstanding ordinary shares by September 12, 1993 (or under certain circumstances with consent by the New Zealand Government by September 12, 1994). Reduction in share ownership had to include offers to the public and institutions in New Zealand of shares having an aggregate offer price of at least NZ$500 million and could be further accomplished by public sales or private placements of shares on terms and conditions, in New Zealand or other jurisdictions as determined by the new owners. In accordance with these obligations, various offerings have been made with a view to reducing the shareholding to the prescribed limits.

The Articles state that their provisions were not intended to confer any benefit on and are not enforceable by any person other than the Government. The Kiwi share may be converted to an ordinary share at any time by the holder at which time all rights and powers attaching to the Kiwi share will cease.

2. Existing Legislative and Regulatory Environment for Telecommunications

Principal Legislative Instruments

The principal legislative instruments comprising the regulatory environ- 26–08 ment are:

---

[1] In 1993 these ROCs were combined into a single company Telecom New Zealand Limited. For convenience the Telecom Group of Companies is referred to in this chapter as Telecom NZ.

(i) Competition Legislation

26–09   • The Commerce Act 1986.

(ii) Telecommunications Specific Legislation

26–10   • The Telecommunications Act 1987 incorporating Amendment Acts
        • The Telecommunications (International Services) Regulations 1994.
        • Telecommunications (Disclosure) Regulations 1990.

(iii) Radiocommunications Legislation

26–11   • The Radiocommunications Act 1989.
        • The Radiocommunications (Radio) Regulations 1993.

(iv) Fair Trading and Consumer Legislation

26–12   • The Fair Trading Act 1987.
        • Consumer Guarantees Act 1993

(v) Privacy and Law Relating to Misuse of the Telephone

26–13   • The Privacy Act 1993.
        • The Telecommunications Act 1987.
        • The Radiocommunications Act 1989.

The above legislation and the Broadcasting Act 1989 (see paras. 26–49 to 26–50) are administered in the New Zealand Ministry of Commerce, Wellington. The administration of the specific communications statutes (ii) and (iii) above) and related policy advice are the responsibility of the Communications Division of the Ministry. The Division reports, through the Secretary of Commerce, to the Minister holding the portfolios of Communications and Broadcasting.

The regulatory environment in New Zealand is also affected by obligations arising under, or from, international agreements to which New Zealand is a party notably ITU Conventions, INTELSAT and INMARSAT and, more recently, the General Agreement on Trade in Services negotiated in the Uruguay Round. The Communications Division also has administrative responsibility for these agreements.

Competition Legislation

26–14   New Zealand has no industry specific regulation of entry into markets for the supply of telecommunication goods or services. As noted above, restrictions were abolished progressively from 1987 to 1990. The original Telecommunications Act 1987, *inter alia*, provided for the phased relaxation of restrictions on customer premises equipment and related services. The Telecommunications Amendment Act 1988 removed the remaining restrictions on the supply of telecommunication services of all kinds on April 1, 1989. The remaining provisions of that Act, as further amended by the Telecommunications Act 1990, mainly comprise measures to facilitate competition. This legislation and regulations thereunder complement the Commerce Act 1986.

The Commerce Act 1986, which has the object of promoting competition in New Zealand markets, is the general law of competition (anti-trust) in New Zealand. The Commerce Act 1986 interfaces directly with the Telecommunications Act 1987 by virtue of section 6 of that latter Act which confirms the primacy of the Commerce Act 1986.

The major parts of the Commerce Act 1986 of relevance to the telecommunications sector are Part II—Restrictive Trade Practices; Part III—Business Acquisitions; Part IV—Control of Prices.

(i) Commerce Act 1986, Part II—Restrictive Trade Practices

This part prohibits a range of restrictive trade practices notably: **26–15**

- Arrangements between competitors that substantially lessen competition in a market (section 27).
- Arrangements between competitors that exclude supply to or purchases from rivals (section 29).
- Arrangements that lead to prices being fixed amongst competitors (section 30).
- A dominant firm in a market using its position purposefully to restrict competitive conduct, etc. (section 36). This section is particularly relevant to interconnection with Telecom New Zealand and its application is discussed in more detail in paragraphs 26–40 to 26–45.
- Suppliers fixing the price at which goods may be sold by other traders (resale price maintenance) (sections 37 and 38).

The Commerce Act 1986 establishes a Commerce Commission which is **26–16** responsible, *inter alia*, for enforcement of prohibitions against restrictive trade practices. It may seek injunctions in the New Zealand High Court against such practices or seek pecuniary penalties for contraventions. Part VI provides for pecuniary penalties of up to NZ$5 million for companies, and up to $500,000 for individuals, to be imposed. Individuals or companies affected by a restrictive trade practice may also seek injunctions and take actions for damages through the High Court. Private parties may not seek pecuniary penalties.

Authorisation of prohibited practices, except the misuse of a dominant position (Section 36), may be sought from the Commerce Commission in accordance with Part V of the Commerce Act 1986 on the grounds that the practice has public benefits overriding relevant anti-competitive detriments.

The Commerce Commission does not have a general right of inquiry or report except in accordance with its express statutory functions.[2]

---

[2] *Telecom Corporation of NZ Ltd v. Commerce Commission* (1994) 5 NZBLC 293, H.C.; (1994) 5 NZBLC 482, C.A. Between November 1991 and June 1992, the Commission undertook an inquiry at its own initiative into the development of competition in the telecommunications industry in New Zealand and the efficacy of the regulatory framework. It issued a report in June 1992 critical of both the Telecommunications (Disclosure) Regulations 1990 (See paras. 26–23—26–28) and Telecom NZ. The New Zealand Court of Appeal found, confirming the finding of the High Court, that the Commission acted *ultra vires* its powers under the Commerce Act 1986 and that it was improbable that Parliament meant to give by implication, authority to make and publish unappealable findings.

Section 26 of the 1986 Act requires the Commission in the exercise of its powers to have regard to the economic policies of the Government as transmitted to the Commission by the Minister of Commerce. These are also formally Gazetted. No such formal statement has been transmitted with respect to telecommunications.[3]

(ii) Commerce Act 1986, Part III—Business Acquisitions

26–17   Section 47 of the 1986 Act, as well as section 27, (above) applies to business acquisitions.

Section 47 provides that no person can acquire the assets or shares of a business if that acquisition will result in:

* The acquisition of a dominant position in a market;
* the strengthening of a dominant position in a market.

Section 47 does not apply to a business acquisition where there is a bare transfer of market dominance, *i.e.* where dominance is transferred from one party to another without being strengthened in the process. Neither section 47 nor section 27 applies to business acquisitions which have been

---

[3] However the following statement of existing policy was published by the Minister of Communications on December 9 1991:

"The Government sees competition as the best regulator of telecommunications markets. Accordingly, there will continue to be no statutory or regulatory barriers to competitive entry into telecommunications markets in New Zealand.
To maintain conditions of effective competition, the Government places primary reliance on the operations of the Commerce Act 1986. In particular, it relies on the enforcement of the statutory prohibitions against anti-competitive practices, including misuse by any person of a dominant position in a market and the prohibition against business acquisitions which create or strengthen dominance.
The following supplementary measures will continue to apply:

(a) Telecommunications (Disclosure) Regulations 1990; and
(b) Telecommunications (International Services) Regulations 1989.

If it proves to be necessary, the Government will consider the introduction of other statutory measures or regulation. It will take particular care to ensure that it is not seen to be acting merely to enhance the commercial position of one firm or group within society at the expense of another.

Interconnection

The interconnection of competing networks with the Telecom network is the critical competition issue. This was recognised from the outset when the decision to deregulate was first announced.
On 6 July 1989 the then Chairman of Telecom, Sir Ronald Trotter gave the following undertaking, that it was "Telecom's policy to ensure that interconnection will be provided to competitors on a fair and reasonable basis, and that the relationships between Telecom companies will not unfairly disadvantage competitors."
The Government expects this undertaking will continue to be honoured.
   It is essential that the interconnection is achieved between Telecom and other competitors in telecommunications markets, including international, domestic long distance, domestic local, and cellular services.
   To this end, the Government will expect all parties:

to act in good faith;
to expedite negotiations, and any court actions;
to recognise the unique regulatory features of New Zealand's telecommunications market.

Despite the progress which has been made, further progress is necessary in the introduction of new, competitive services. The Government's overriding concern is to ensure that the regulatory environment within which negotiations are occurring contributes to its growth and employment strategy".

cleared or authorised by the Commission under Part 5 of the 1986 Act. As with restrictive trade practices, the Commission may seek penalties in respect of acquisitions contravening the 1986 Act. It may also seek an order for the divestiture of specified assets or shares in respect of offending transactions. Injunctive proceedings are also available both to the Commerce Commission and to private parties. Individuals or companies may seek damages if they have suffered loss as a result of the acquisition.

The Commerce Commission has indicated in a statement of practice that telecommunications and television broadcasting are sensitive "industries" in which business acquisitions are more likely to raise competition concerns and that it expects those involved to seek clearance or an authorisation. Section 47 is also cross-linked to acquisitions of radio spectrum property rights by the Radiocommunications Act 1989.[4]

(iii) Commerce Act 1986, Part IV—Control of Prices

Part IV makes provision for the imposition of price control, by Order in   **26–18** Council generally, or in respect of particular businesses or specific products and services in circumstances where the Minister of Commerce[5] is satisfied that conditions of effective competition do not exist and that control is necessary to protect users, or consumers or, as the case may be suppliers.

Presently there are no price controlled items under the Commerce Act 1986. Nor is there any Government policy statement on the circumstances in which control would be considered in relation to telecommunication goods or services. The Commerce Commission may initiate the reports to the Minister on whether or not price control should be imposed. The Minister of Commerce may also require the Commission to report.

The Telecommunications Act 1987

General

The Telecommunication Act 1987 does not provide a framework for regu-   **26–19** lation of the telecommunication sector or interconnection between carriers. Rather it provides for:

(a) statutory protection of the right of any network operator to prevent interference with its network;

(b) measures to facilitate the establishment and maintenance of telecommunication networks (including the transmission of broadcasts);

(c) regulations to be made under the Act facilitating competition but

---

[4] Radiocommunications Act 1989, s. 138 (as amended by s. 8(2) of the Radiocommunications Amendment Act 1990). For an application of this section see, *e.g.* judicial review of Commerce Commission decision on Telecom NZ acquisition of TACS-B management rights for cellular telephone frequencies, *Broadcast Communications Ltd v. Commerce Commission* (1991) 4 TCLR 537; Appeal by Telecom NZ against Commission decision on acquisition of AMPs A cellular telephone frequencies *Telecom Corp of NZ Ltd v. CC Judgment of High Court* (1991) 4 TCLR 473, C.A. judgments (1992) 3 N.Z.L.R. 429.

[5] The Minister of Commerce is advised by the Business Policy Division of the Ministry of Commerce. It can be assumed that the Minister would also consult closely with the Minister of Communications over any proposal for the imposition of control with respect to telecommunications.

only for the imposition of information disclosure requirements on
Telecom and in respect of international services;

(d) residual provisions brought forward (with modification) from previous Post Office legislation. These are:

   (i) the offences relating to offensive and indecent use of a telephone
       (sections 8 and 8A);

   (ii) the admissibility in proceedings of "telegrams" and computer
        records of telex and toll calls (section 9).

   (iii) authorisation of the interception of telecommunications by
         employees of network operators for the purposes of main-
         taining telecommunication services.

Network Operator Status

26–20   A network operator is defined as either Telecom NZ or a person declared
as such by Order in Council under section 2A of the 1987 Act. While
there is no statutory requirement to secure a declaration before providing
telecommunication services in New Zealand, a network operator may take
advantage of certain statutory rights or protections notably with respect to
special rights of access to land and the road reserve to lay or construct
lines (see below).

In line with the stated purpose of the section "to facilitate entry into and
competition in telecommunication markets" the requirements to secure a
declaration are relatively straightforward. In order to qualify for the status
the Minister of Communications (who has the power to recommend that
declarations be made) must be satisfied that a declaration is "necessary"[6]
for the applicant to carry on a business providing—

   (i) Facilities for telecommunication between 10 or more other persons,
       being facilities enabling at least 10 of those persons to communicate
       with each other; or

   (ii) Facilities for broadcasting to 500 or more other parties, being facilit-
        ies that enable programmes to be transmitted along a line or lines
        to each of those persons.[7]

---

[6] The Ministry currently requires (Telecommunications Information Leaflet No. 2) that applications
include:

* details of the actual business to be provided, including anticipated customer levels;
* a copy or details of the company's business plan, showing how the business development is to proceed;
* details of costings and financial planning regarding the development and implementation of the proposal; and
* details of geographic areas in which cable laying will be necessary and the number of customers that
  will be reached by that cable.

In addition, in order to obtain assurance that an application for network operator is substantial the
Ministry seeks:

* a copy of the company's certificate of incorporation (this is essential to establish that the company is
  a legal entity);
* a list of the company's managers, directors and major shareholders; and
* a brief company history.

[7] The extension of coverage to organisations providing broadcasting services by lines was enacted by s.
86(2) of the Broadcasting Act 1989.

Declared network operators therefore include both telecommunications providers and broadcast or interests.[8]

Under Telecom NZ policy, such status is also a pre-requisite to Telecom NZ entering into negotiations with carriers on trunkside interconnection with Telecom's network (although Telecom NZ will also negotiate with registered operators under the Telecommunications (International Services) Regulations 1994.

Prevention of Interference

Section 6 of the 1987 Act requires the agreement of a network operator **26–21** for connection of any additional line apparatus or equipment to:

(a) any part of the operator's network; or
(b) to any line, apparatus or equipment connected to the network.

Injunctive proceedings and actions in damages are available to the network operator (sections 20C and 20D).

The section 6 provides authority for the imposition by Telecom NZ and, ostensibly, other network operators, of standards for network interconnection.

Section 6(2) of the 1987 Act makes plain that whatever conduct is taken under or purporting to be under, the authority of the section such as a refusal to agree to connection of line apparatus and equipment, is subject to the Commerce Act 1986 (which includes the prohibitions against restrictive trade practices).

Special Rights Relating to Land and Road Reserve

The Amendment Act 1989 re-defined rights of access to land for network **26–22** operators to ensure all operators were placed on an equal footing to the maximum extent possible. Under previous law Telecom NZ enjoyed special rights of access to land, including rights, to construct lines on private land.

The Amendment Act 1989 minimises special access rights for Telecom and other operators although, for practical reasons, it allows Telecom NZ to retain rights of access to existing works and lines for maintenance and repairs (section 12). While all network operators have to seek the agreement of land owners to construct, erect or lay new lines (and works) section 11 empowers a District Court, on application by an operator, to make orders for entry and related purposes. The court must be satisfied that the operator has taken all reasonable steps to negotiate an entry agreement and that no practical alternative route exists.

All network operators have rights to construct lines and minor fixtures

---

[8] The following organisations have been declared network operators (current Orders in Council only)— Ashburton CA. TV Ltd; S.R. 1992 No. 15; BellSouth New Zealand Ltd; S.R. 1991 No. 69; Broadcast Communications Ltd; S.R. 1990 No. 285; Civic Enterprises Ltd; S.R. 1989 No. 299; Clear Communications Ltd; (formerly called the Alternate Telecommunications Company Ltd, S.R. 1990 No. 285; Comsys Ltd; S.R. 1992 No. 15; Kiwi Cable Company Ltd; S.R. 1990 No. 92; Multi Band Television Ltd; S.R. 1990 No. 114; New Zealand Post Ltd; S.R. 1993 No. 94; New Zealand Rail Ltd; S.R. 1993 No. 16; Pacificom Networks Ltd; S.R. 1992 No. 15; Sky Network Television Ltd; S.R. 1990 No. 76; Telecable Holdings Ltd; S.R. 1989 No. 228; TransPower New Zealand Ltd; S.R. 1992 No. 205.

on, along, across, over or under any road provided they meet reasonable conditions of the relevant authority (sections 15 and 15A).[9] An appeal lies to the District Court if the local authority imposes unreasonable conditions.

A network operator may construct and maintain telephone cabinets and such like (section 18).

Section 20 provides protection of title to line and works. Section 19 provides for compensation for injurious affection in respect of damage arising from the exercise of statutory powers under the 1987 Act. Provision is also made for the removal and trimming of trees interfering with lines (sections 13 and 14).

Resource Management

26–23   Further, albeit limited provision is made in New Zealand planning law for the facilitation of essential services such as telecommunication and radiocommunication networks.[10] A network utility operator, as defined in the Resource Management Act 1991, may apply for ministerial approval of the operator as a "requiring authority" for a particular project or work or network utility operation. The Minister for the Environment must be satisfied that approval of this special status is appropriate for the operation concerned and that the operator will satisfactorily carry out all the responsibilities of an authority as well as give proper regard to interests of those affected and the environment.

Telecommunications (Disclosure) Regulations 1990

26–24   Sections 5C, 5D and 5E of the Telecommunications Act 1987 enable regulations to be made to promote disclosure of certain financial accounting and pricing information of Telecom NZ. Section 5C enables the imposition of information disclosure requirements on Telecom. Section 5D requires Telecom to provide the Ministry of Commerce with information including full details of contracts on request to ensure compliance with the information disclosure requirements. Section 5E establishes offence provisions and penalties for non-compliance with the Regulations.

The following information is required to be disclosed under the current Regulations:[11]

(a) financial statements; and
(b) the prices, terms and the conditions of certain prescribed goods and services.

---

[9] For a discussion on the meaning of terms, notably the term "road", in the context of ss. 2, 11, 15 and 15A see *Clear Communications Ltd v. Morrison NZ* (1993) unreported, November 15, 1993, C.A.
[10] Resource Management Act 1991, Pt 8. Network operators which have received project approval include Broadcast Communications Limited; Telecom NZ and its subsidiaries Telecom NZ International Ltd and Telecom Mobile Communications Ltd.
[11] S.R. 1990 No. 120; Amendment No. 1 S.R. 1993 No. 380.

Financial Statements

The Regulations require Telecom's network operating company, Telecom   26–25
NZ Ltd, to publish half-yearly financial statements (profit and loss
accounts, balance sheet and a statement of accounting principles).

Telecom is required to have statements available on request at Telecom's
three principal metropolitan offices within three months of the end of both
the first half and the full financial year.

Pricing Information

Telecom is required to publish, quarterly, the prices, terms and conditions   26–26
for the supply of certain prescribed telecommunication goods and services.
The current services have been prescribed—

- access to the public switched network operated by Telecom;
- interconnection to a network owned and operated by Telecom for the
  purpose of operating any other public switched network, whether or
  not owned or operated by Telecom;
- leased circuits; and
- telecommunication links that enable the making of—

  — telephone calls within Telecom's free calling area; and
  — telephone calls to places outside Telecom's free calling area including national and international toll calls.

Telecom must disclose the *standard* contract prices, terms and conditions   26–27
of the five prescribed service categories: monthly rentals charges, local call
charges, domestic, long distance and international (toll) call charges, leased
circuits, and interconnection charges. Telecom must also disclose the principles or guidelines it applies in determining discounts (including maximum
discounts). Where a discount of 10 per cent or more is allowed it must disclose the size of such discount and the principles used in its determination.

Where Telecom supplies a basket of services which includes a prescribed
service and a discount of 10 per cent or more is given across the board to
all the services, the Corporation must disclose the contents of the basket
and the discount allowed. It should be noted that this disclosure requirement not only applies to transactions between Telecom and its customers
but also applies to transactions between Telecom and its subsidiaries.

Within 30 days at the end of each calendar quarter Telecom is obliged
to publish an update of any changes which have been made to prices, terms
and conditions during a quarter. Every two years Telecom is required to
publish a complete summary of the prices, terms and conditions of the
prescribed services. Publication is effected by making the information available at Telecom's principal offices.

Additional Information Requirements of Government

The Ministry of Commerce may request any statements, reports, agree-   26–28
ments, particulars and other information for the purposes of monitoring

compliance with the Regulations (section 5D). The Ministry may also request information to ascertain the scope of information required to be made available to the public to facilitate competition in the supply of tele-communications goods and services. The Ministry also considers, incident-ally, that these financial statements are relevant to the Government in its monitoring of Telecom's Kiwi share principles contained in Telecom's Art-icles of Association.

Publication of Interconnection Agreements

26–29   Since December 1993, following a review of the Regulations, Telecom NZ is now required to publish the full text of its actual interconnection agree-ments except any details relating to the location of physical links for service delivery between networks.

International Services

Telecommunications (International Services) Regulations 1994

26–30   Section 5 of the Telecommunications Act 1987 provides for a degree of regulation of international telecommunications but for the purpose of pro-moting a competitive market in international telecommunication services in New Zealand. Provision is made for registration of persons establishing, operating or maintaining facilities in New Zealand for the purpose of pro-viding services in New Zealand to or from territories outside New Zealand. There are fines for non-registration and for failure to comply with any term or condition of registration.

Under Regulations[12] coming into force on March 1, 1995, registration by the Secretary of Commerce is required in respect of operators providing the following services:

(a) Leased circuits that are connected both with public networks in New Zealand and with public networks in the territory of an overseas operator;

(b) public switched telecommunications services to or from territories outside New Zealand.

The Regulations are not intended generally to apply to "callback" services.

Registration for the latter services simply requires conformity with information and application fee requirements (regulation 4(2)). Applica-tions in respect of leased circuits will be reviewed by the Ministry against additional considerations relating to the conditions of resale in the overseas territory and the risk of and level of harm to New Zealand users form one way accounting rate by-pass (regulation 5(1)).

---

[12] S.R. 1994 No. 280. Registration under the former 1989 Regulations provides transitional registration under the new regulations but only up to April 30, 1995. Application and annual registration fees are payable (NZ$1,000 and $10,000). Six weeks processing time respectively is the Ministry's target.

The Ministry of Commerce, in assessing these criteria, is expected to be **26–31** concerned to ensure that relevant regulatory conditions in the overseas country are broadly equivalent with those of New Zealand, *e.g.* resale to third parties is permitted; interconnection to the PSTN is possible, etc. Applications relating to services between New Zealand and Australia, Sweden, U.K. and the USA respectively will be considered favourably.[13]

Under newly liberalised provisions, registered operators[14] will be able to negotiate terms with overseas operators without compulsorily being subject to requirements relating to parallel accounting and proportionate return of traffic. However the Ministry of Commerce is given discretionary power to impose such conditions having regard to the desirability of promoting a competitive international services market in New Zealand and the interests of New Zealand based users (regulation 7(2)).

The imposition of conditions will normally be considered after complaint. The Ministry is expected to take account of a range of situations involving abuse of market power. "Similar proportion" indicators for intervention will be taken to mean a proportion that is—

- within five per cent of expected traffic share within any six-month period within the first two years of service;
- within two per cent in any six-month period after two years service commencement.

International Agreements

New Zealand is a member of the International Telecommunication Union **26–32** (ITU). New Zealand is also a party to INTELSAT and INMARSAT satellite systems. The New Zealand Government is a party to the agreements, and as noted (above), the Ministry of Commerce is the responsible Government department. The Government has designated Telecom New Zealand as the New Zealand signatory to the operating agreements.

Telecom is represented in respect of INTELSAT by its wholly-owned subsidiary Telecom New Zealand International Limited (TNZI) and has an ownership share in INTELSAT along with other signatories. To meet competition objectives, the Government has arranged with TNZI that bona fide potential operators for INTELSAT capacity, including entities registered under the Telecommunications (International Services) Regulations will be permitted by TNZI to seek a direct contractual relationship with INTELSAT for the provision of services if this is so requested. The arrangement is known as direct access.

Entities qualifying for direct access to INTELSAT but who choose not to exercise this option or only part of this option are entitled to have their

---

[13] Ministry of Commerce Draft Compliance Statement (December 1994); Telecommunications (International Services) Regulations 1994.
[14] The following organisations have been registered as at March 1994—Telecom New Zealand International Ltd (effective August 1989); Optus Pty Ltd (effective October 1990); Clear Communications Ltd (effective October 1990); Television New Zealand Ltd (effective April 1991); Global Telecom Systems Ltd (effective September 1992).

business dealings conducted in such a manner that ensures that TNZI does not derive any commercial advantage from the knowledge of these dealings. This arrangement is known as equal access.

In order to facilitate both these arrangements an office exists within TNZI, but independent of TNZI's commercial operations, with the function of managing all INTELSAT matters whether they be for Telecom or for competitors. This is known as the New Zealand Office of Signatory Affairs (NZOSA).[15]

Radiocommunications[16]

26-33   As indicated above, New Zealand's traditional administrative system for the regulation of radio spectrum was inadequate to cope with market led economic policies in the downstream markets of telecommunications and broadcasting. Following a detailed review of existing legislation and practice a new scheme was introduced under the Radiocommunications Act 1989 as further amended.[17] This comprises the following features:

- Rights to spectrum use within defined radio engineering parameters may be created in the form of management rights with a duration of up to 20 years.
- These rights have, by statute, certain attributes of property including the right of transfer, aggregation or division and uses of financial security.
- Original management rights are established by and registered in the name of the Crown of New Zealand acting through the Secretary of the Ministry of Commerce. Provision was made for protection of incumbency interests.
- The Crown as holder of the original rights, or other persons to whom management rights have been transferred, are called managers.
- Managers may confer various forms of licences on other persons (right holders) for use of the management rights within prescribed parameters of those rights. Licences may be transferred or mortgaged.
- The exercise of both management and licence rights is constrained by statutory obligations and minimum conditions primarily designed to avoid radio interference.
- In addition to enforcement by the Secretary of Commerce against contraventions of the 1989 Act, provision is made for private legal remedies for protection of rights, including new actions in tort.
- A formal scheme of registration has been established for the recording of details of management rights and licences as well as other transactions such as transfers, aggregation, mortgages and caveats.
- The former radio apparatus licensing scheme was continued and

---

[15] The detailed procedures for arranging access are set out in the NZ Ministry of Commerce document INTELSAT Satellite System: Procedures for Access: September 24, 1993.

[16] For more detailed review of New Zealand experience under its Radiocommunications law see the author's commentary: Proceedings CIRCIT Conference on Radio Spectrum Management November 1993. (Editor Chee-Wah, 1994).

[17] In May 1994 the Ministry of Commerce released a discussion document on the Act inviting comment on various issues which may lead to amendments to some provisions of the Act.

complements the property rights scheme. Licensing[18] is subject to regulatory criteria including Government policy statements.[19] Currently these facilitate competition in telecommunications. Although the legislation does not say so, the radio apparatus licensing scheme remains, in part, as a transitional measure, for certain areas of spectrum intended to be converted into property rights and sold by tender. It was envisaged, nevertheless that for some time to come, the creation of new management rights would be limited to the frequency range of 44–1400MHz because of uncertainty about future international usage for frequencies outside that range.

Interference management in relation to licences is addressed by prevention, prohibition and resolution mechanisms. Prevention includes minimum requirements as to licences such as out-of-band emission limits on transmissions and compliance with interference minimisation measures. There are express prohibitions against endangering the functioning of radio navigation services along the lines of the international radio regulations and unlawful transmission. Receivers can take actions in tort with respect to interference by licence right holders and there is a "first-in-time" mechanism based on maximum permitted interfering signals. **26–34**

For radio apparatus licensing, the old style co-ordination and supervision as well as prosecution of offences is the responsibility of the Crown through the Ministry of Commerce.

The manner in which the Crown transfers its original management rights for licences created under the Act is neither prescribed in statute nor in regulations. However, public administrative law principles apply to the Crown's conduct in the exercise of Crown powers to create and transfer management rights.[20]

Following Government approved allocation and transfer policies and administrative processes, formal procedures have been devised and refined for the transfer of spectrum. These include the initiation of a public consultation process, the establishment of any new management rights including expressions of interest and an engineering plan for the spectrum usually involving external consultations, the announcement of policy decisions on those plans and planned allocations (including social uses) and disposition by tender. **26–35**

Since late 1989–90 tradeable property rights (*i.e.* management rights or licence rights) have been created and disposed of by tender under the following bands:

(a) 806–890MHz—management rights for cellular telephone known as AMPs-A (held by Telecom NZ), AMPs-B (held by Telecom NZ under an incumbency), TACS-A (held by BellSouth) and TACS-B **26–36**

---

[18] Radiocommunications (Radio) Regulations 1993; S.R. 1993 No. 340.
[19] Such statements are subject to review in accordance with public administrative law principles; *Professional Promotions & Services Ltd v. Attorney General & Anor* (1990) 1 N.Z.L.R. 501.
[20] *Mirelle Pty Limited v. Attorney General* (1992) unreported, November 27, 1992, H.C. The particular order arising under the judgment was overturned after an uncontested appeal by the Crown.

(held by Telstra following a resale of a right originally held by Tele-com NZ but denied on competition grounds under the Commerce Act). In early 1994 the Ministry of Commerce announced a proposal for a modification of rights as between these parties.

(b) 525–1625kHz—a Crown management right under which licences have been issued for MF-AM sound broadcasting.

(c) 988.8–100.2MHz—Crown management right with licences for VHF-FM sound broadcasting.

(d) 518–582MHz and 646–806MHz—Crown management rights and licences suitable for UHF television broadcasting.

(e) 2.3–2.396 GHz.

As at May 1994, the Ministry had under review planning and establish-ment of management rights and licences in land mobile frequencies and in the 1.8–2.1 GHz band for local cellular services.

Fair Trading Act 1987 and Consumer Law

26–37   The Fair Trading Act 1987 is the counterpart of the Commerce Act 1986, designed to enable the consumers to take advantage of increased competi-tion. It comprises a comprehensive set of measures covering misleading or deceptive conduct in trade giving rise to civil remedies,[21] a law against false and misleading representations, and prohibitions against unconscionable selling practices. Although, the 1987 Act also provides for the promulga-tion of consumer information requirements for product and services and product safety standards, these are unlikely to be applied to telecommuni-cations equipment and services given the special standards setting process in existence (see paras. 26–56 et seq.).

Intending network operators should also consider the provisions of the recently enacted Consumer Guarantee Act 1993. In relation to services, this legislation provides statutory guarantees as to reasonable care and skill, fitness for a particular purpose, time of completion (if no time is specified) and, reasonableness of price (if no price is specified). The 1993 Act only applies to the supply of services to consumers, for personal, domestic or household use. Suppliers of services are not able to contract out of the statutory guarantees except in very limited circumstances. Suppliers can contract out where the consumer is a business.

Law relating to privacy and misuse of the telephone

Privacy of Telecommunications

26–38   The Radiocommunications (Radio) Regulations 1993[22] prohibit any receiver of a radiocommunication, not intended for that person, from

---

[21] For an example of proceedings involving telecommunications advertising see *Telecom Corporation of NZ Ltd v. Clear Communications Ltd* (1992) 4 NZBLC 102, 839.

[22] Reg. 28. Authorised under s. 134(d) of the Radiocommunications Act 1989.

making use of the radio communication or information derived, to reproduce it, or to disclose its existence.

Section 6 prohibits interference with a network (discussed above). Section 216B of the Crimes Act 1961 prohibits unlawful interception of private communications. Interception of telecommunications by a network
operator employee (within his authority) is expressly authorised for the
purpose of maintaining telecommunication services[23] as is interception lawfully carried out under criminal or security statutes.[24]

The recently enacted Privacy Act 1993 has application to providers of
telecommunications services. In particular this Act establishes principles for
the collection, use and disclosure of and access to personal information
held by any "agency" such as a services provider. A "privacy officer" must
be appointed to encourage compliance with the principles which are
enforced by way of complaint to the Privacy Commissioner. The Privacy
Act 1993 has raised issues surrounding the use of information and information services vital for effective competition in network services[25] which are
under review by the Ministry of Commerce. Meanwhile the major network
operators have been preparing a draft privacy code on telecommunications
services. This is expected to be published for comment in mid 1995.

Misuse of Telephone

Section 8 creates offences relating to the use of offensive language and other   **26–39**
disturbing uses of the telephone.[26] Section 8A creates an offence for indecent or obscene telephone calls for pecuniary gain.

3. REGULATION OF SERVICES

Analysis of Rights to Provide Telecommunications Services

There are no statutory entry restrictions on the provision of any telecom-   **26–40**
munication services.

Telecommunication Interconnection Policies

However the ability of network operators to provide competitive network   **26–41**
services is affected by the interconnection policies and practices of Telecom
NZ which owns and operates the PSTN. The "regulatory" framework
under which Telecom develops its policies and practices and conducts interconnection arrangements principally comprises the anti-competitive proscriptions of section 36 of the Commerce Act 1986.

---

[23] s. 10, Telecommunications Act 1987,
[24] NZ Security Intelligence Services Act 1969, ss. 4A–4B. The Misuse of Drugs Amendment Act 1978, ss. 14–29.
[25] See paper commissioned by Ministry of Commerce (Longworth Associates 1992).
[26] For interpretation of s. 8 see *Whittaker v. Police*, unreported, December 8, 1993, Tipping J.; *Spooner v. Police*, unreported, June 29, 1992, Fisher J.

For section 36 to have application, it must be established that Telecom:

(a) is in a dominant position in a market; and
(b) has used its dominant position for a "substantial" purpose which falls into one of the defined proscribed purposes.

26–42   The defined proscribed purposes (361D (a), (b) and (c)) are:

(a) restricting the entry of any person into a market in which Telecom is dominant; or
(b) preventing or deterring any person from engaging in competitive conduct in that or in any other market; or
(c) eliminating any person from that or any other market.

26–43   Network operators seeking to establish Telecom interconnection policies will require:

(a) Telecom's current Interconnection Guidebook;
(b) relevant permit to connect specifications (PTCs) and Telecom network access specifications TNA (see below);
(c) Interconnection Agreements with existing network operators which are publicly available from Telecom under the Disclosure Regulations; and
(d) information on Telecom's prices and terms generally. See information provided by Telecom under the Disclosure Regulations and Telecom's current list of charges for its services (TLOC).

The responsibility for providing information on Telecom policies and negotiating interconnection agreements with Telecom is confined to the Corporate Policy Group of Telecom New Zealand.

26–44   Telecom's interconnection policies and conduct in respect of interconnection negotiations have been challenged under the Commerce Act 1986 in the following litigation:

(a) Re terms for interconnection for competing local network services.[27]
(b) Re Telecom's conduct over a requirement for the customers of competitors' toll services to dial an access code.[28]
(c) Re Telecom's terms for the provision of additional Points of Interconnection.[29]
(d) Re Telecom's terms for interconnection with respect to 0800 services to be provided by CLEAR to its customers.[30]
(e) Telecom's conduct with respect to modification of Telecom managed

---

[27] *CLEAR Communications Ltd v. Telecom Corporation of NZ Ltd* (1992) 5 TCLR 166, H.C.; 413, C.A.; (1994) 6 TCLR 138, P.C.
[28] *CLEAR Communications Ltd v. Telecom Corp of NZ Ltd and others.* Interim award of Arbitrator May 26, 1994.
[29] *CLEAR Communications Ltd v. Telecom Corporation of NZ Ltd.* CP 25/94, H.C., Wellington Registry (proceedings suspended by agreement).
[30] *CLEAR Communications Ltd v. Telecom Corporation of NZ Ltd.* (1992), CP 373/92, H.C., Wellington Registry (proceedings suspended by agreement).

or leased PABX equipment from customers for CLEAR's toll services.[31]

Of the above proceedings, only the local service case and the non-code access cases proceeded to full hearing. Both resulted in findings under section 36 against Telecom and orders for enquiries into damages.[32]

In the local service case the High Court held, in December 1992, that **26–45** Telecom New Zealand had breached the 1986 Act by asking CLEAR too much for interconnection up to its stance at trial and for not providing Direct Dial In (DDI) service but found that Telecom was entitled to adopt, under provisions set by the court, a pricing thesis known as the Baumol/ Willig rule to recover forgone revenue. An enquiry for damages was ordered in respect of the DDI breach.

The Court of Appeal, in December 1993, upheld CLEAR's appeal that the adoption of the Baumol/Willig rule was also a breach of the Commerce Act 1986 and it also enunciated certain principles for interconnection pricing by the parties. The Court of Appeal also awarded CLEAR damages (for the DDI breach as also ordered by the High Court) and costs.

The Judicial Committee of the Privy Council (currently New Zealand's highest court and based in London) allowed Telecom's appeal in part, finding that the final position adopted by Telecom at trial based on the Baumol/ Willig rule did not breach the 1986 Act since it did not involve the "use" by Telecom of its dominant position although it found anti-competitive purpose established on the part of Telecom in terms of section 36 of the 1986 Act. Neither party was awarded costs. However, the Privy Council, responding to concerns about the risk of monopoly rents in Telecom's interconnection prices, left open the prospect of price control of Telecom's interconnection price by the Government under the Commerce Act 1986.

The Government announced on November 9 1994 that it had instructed **26–46** officials to examine the public policy implications of the Privy Council judgment, including consideration of the Baumol/Willig rule for interconnection in telecommunications and the wider implications of the Privy Council's judgment on section 36 of the 1986 Act. Government Ministers stated that the Government was keen to ensure that the dynamic benefits of competition continue to develop to the advantage of users. At the same time Ministers emphasised their expectation that the parties should resume negotiation.

The outcome of the officials' examination is expected in 1995.

The Privy Council interpretation of the Commerce Act 1986 with respect to the Baumol/Willig rule and the related Kahn competitive parity rule has

---

[31] Proceedings withdrawn by agreement.

[32] The local service case required the courts to access a package of terms including the requirement that Telecom customers dial an access code to reach CLEAR's local network and pay a different price than for calls within the Telecom network. The case also concerned Telecom's demand for an "assess levy" as part of local service interconnection pricing which included as one justification the use of economic principles propounded by American Economists Professor William Baumol, Dr Robert Willig and Dr Alfred Kahn.

been criticised by CLEAR, *inter alia*, as incompatible with U.S. anti trust and regulatory decisions, and the views of its authors. The London judgment is also at odds with the subsequent OFTEL Consultative Document "A Framework for Effective Competition" of December 1994 in which the "efficient component pricing rule" (ECPR) is found to be unattractive on competition and other grounds.

The judgment has also prompted further calls by the major network operators for further Government intervention incorporating a compulsory framework for dispute resolution (including the formal adoption of unbundled and incremental cost based interconnection principles).

In the non-code access case, the Arbitrator found Telecom had breached section 36 in respect of its conduct surrounding the implementation of obligations to provide a non-code access feature to CLEAR under its Interconnection Agreement. He ordered an enquiry into damages which is expected to be concluded in 1995.

Fixed and Mobile Services

26–47   There are no special regulations depending on the class of service.

Satellite Services (Non-Broadcasting)

26–48   There are no restrictions on the provision of services by satellite other than the requirement to obtain radio apparatus licences under the Radio Regulations 1987. The issuing of licences is subject to the criteria under the Radio Regulations 1993, international agreements under the auspices of the ITU and, in respect of broadcasting, bilateral Ministerial arrangement with Australia.

Radiocommunications

26–49   See paragraphs 26–08 to 26–39.

4. REGULATION OF BROADCASTING

Basic Legislative Rules

26–50   There are no special restrictions on entry into broadcasting markets. Subject to any applicable general law, any person may broadcast subject to any relevant requirements of the Radiocommunications Act 1987 and the Broadcasting Act 1989. Apart from disestablishing the former Broadcasting Tribunal and the system of warrants, the Act:

(a) establishes the Broadcasting Commission (now known as New Zealand on Air) to collect the public broadcasting fee prescribed under the Broadcasting (Public Broadcasting Fees) Regulations 1989[33] and disburse the proceeds in grants aimed at promoting defined social objectives in broadcasting *viz*;

  (i) to reflect and develop New Zealand identity and culture by promoting programmes about New Zealand and New Zealand interests and promoting Maori language and culture.
 (ii) to maintain and extend where the Commission considers appropriate, coverage of television and sound radio broadcasting not otherwise commercially viable;
(iii) to ensure a range of broadcasts are available to provide for the interests of women, children, persons with disabilities and minorities.
 (iv) to encourage the establishment and operation of archives of programmes which are likely to be of historical interest.

(b) Establishes the Broadcasting Standard Authority and requires broadcasters to adhere to certain behavioural standards in their programming and develop codes of practice in relation to programming;
(c) establishes arrangements for election broadcasting administered by the Authority;
(d) Imposes minimal restrictions on advertising hours. All broadcasters are restricted from advertising on certain days of religious observance or commemoration.

On July 1, 1993 an additional agency was established called Te Reo **26–51** Whakapuaki Irirangi to promote Maori language and Maori culture by making grants available for broadcasting and programmes drawn both from the Broadcasting Commission and the Government.[34]

The Radiocommunications Act 1989

A substantial number of tradeable property rights in AM and FM sound **26–52** radio and UHF TV broadcasting have already been sold by the Crown to the private sector.

The 1989 Act also makes express provision for reservation of certain radio frequencies for non-commercial use or for specified social purposes. The 7th schedule of the 1989 Act specifically protects the former warrant rights of some 30 non-commercial broadcasting organisations which had obtained their warrants to broadcast prior to July 1989.

Incumbent broadcasters at the commencement of the 1989 Act were also issued with property rights representing their warrant frequencies, subject to the payment of an annual or lump sum levy relating to either the resale value or the market value of their previous licences.

---

[33] S.R. 1989 No. 123; Amendment No. 1 S.R. 1990 No. 303
[34] Broadcasting Amendment Act 1993.

Overseas Ownership Controls

26–53 There are no statutory controls on cross media ownership other than the general controls over anti-competitive business acquisitions under Part 3 of the Commerce Act 1986.

Although the original Broadcasting Act 1989 contained restrictions on foreign ownership specific to broadcasting media, these controls were removed in 1991 by an amendment to the Act. Acquisitions involving overseas ownership are subject, like all other industries to the Overseas Investment Act 1973, and related regulations. As at May 1995, government policies are very liberal.[35]

It should be noted however in this context that the major free to air television service provider, Television New Zealand Limited, is a wholly owned state owned enterprise. TVNZ operates two nationwide VHF channels. A third nationwide VHF channel is operated by TV3. Regional, music and sport channels are emerging. Five pay to view UHF channels have also been established with coverage in many centres along with a regional cable TV and free to air UHF channels enterprise.

Terrestrial and other delivery mechanisms

26–54 There are no restrictions on the transmission technologies employed in transmission services. The Broadcasting Act 1989 and the Radiocommunications Act 1989 permit the introduction of new technologies or services without regulatory restrictions specific to the technologies of services concerned.

Regulation of Content

26–55 There are no controls such as quotas on the amount of New Zealand content. New Zealand content and other defined social objectives is promoted through the transparent funding scheme administered by the Broadcasting Commission (NZ on Air).

5. REGULATION OF APPARATUS AND STANDARDS

Regulation of Telecommunication Equipment Standards

26–56 Provision is made for the setting of standards under the Telecommunications Act 1987 and the Radiocommunications Act 1989.[36]

---

[35] There are no special requirements with respect to telecommunications except for Telecom Corporation of New Zealand Ltd (*supra*, paras. 26–03 *et seq.*).
[36] For a detailed review of the standards setting process in New Zealand see "Telecommunications, Standards Setting Process in New Zealand" (August 1993) Information Paper Ministry of Commerce.

Telecommunication Act 1987

Section 6 of the Telecommunications Act 1987 provides that the agreement   **26–57**
of the network operator is required before the connection of any apparatus
line or equipment to a network. The operation of this provision is subject
to the Commerce Act 1986. As the operator of the PSTN, Telecom New
Zealand, upon the commencement of the phased deregulation of Customer
Premises Equipment (CPE) markets, established a "Permit to Connect"
(PTC) certification system including a consultation process.
    Current PTC and TNA specifications fall under the following three series:

- 100 series on general specification and codes of practice covering such
  matters as the general conditions of PTC, electrical safety requirements of PTC and Telecom's network interface characteristics.
- 200 series on product certification covering such matters as technical
  requirements for all types of customer CPE including telephone instruments customer premise wiring and ISD and terminal equipment.
- 300 series on network interconnection specifications covering such
  matters as general requirements for network interconnection, telephone network interconnection with MFC signalling, and telephone
  network interconnection using CCITT number seven signalling.

TNA specifications deal with the characteristics of Telecom's network   **26–58**
and are not subject to prior consultation. Testing of telecommunications
equipment is undertaken by Telecom accredited laboratories both in New
Zealand and overseas or by Telecom's own laboratories where the laboratory facilities are not available. Alternatively complex equipment such as
PABX systems are subject to limited permit trials based on compliance
standards by the suppliers. If equipment is found to be in compliance with
the specifications, it is permitted to carry the "telepermit" label.
    While other network operators have the ability to establish standards
for equipment interfacing their network, there are no published standards
as yet.

Radiocommunications Act 1989

Sections 116, 133 and 134 of the Radiocommunications Act 1989 make   **26–59**
various provision for regulation of radio apparatus.
    Section 133 of the 1989 Act enables the Ministry of Commerce to set
standards or specifications relating to the performance of radio apparatus
or any system for effecting of radiocommunication. Relevant reference
standards must be complied with as part of the safety certification required
for registration of licences.
    Section 134(i) enables regulations to be made prescribing requirements
for the standardisation of:

(i) technical systems for radiocommunications;
(ii) technical formats for radiocommunications.

This may be regarded as a residual power only to be exercised when any necessary standards action is not achieved by voluntary means nationally or by international agreement.

Compliance for the Standards for Interfering Equipment (section 134(g))

26-60   Part V of the Radiocommunications (Radio) Regulations 1993 states that persons are prohibited from installing, using, manufacturing or selling interfering equipment in breach of a standard notified by the Ministry except for the purpose of testing for compliance.

Intentional emission specifications for radio transmitting are based on international standards with ETSI Standards. The unintentional emissions and specifications which cover the emission of radio interference by electrical equipment such as hairdryers and computer terminals are based on IEC Standards. Regulations in 1993 have introduced a system based on a declaration of conformity with labelling requirements similar to that adopted by the IEC. Exemptions from licensing have been issued for restricted radiation devices (including cordless telephone) under the Radio Regulations 1993.

Electrical Safety

26-61   Safety legislation is administered by the Energy and Resource Division of the Ministry of Commerce. The Electricity Act 1992 established a new electrical safety regime in New Zealand from April 1, 1993. The Electricity Regulations 1993 require that all fittings, electrical appliances, electrical installations, works and associated equipment shall be designed, constructed, maintained, installed and used so as to be electrically safe. The Regulations specify offences relating to non-electrically safe equipment or works. The Regulations are supplemented by 19 electrical Codes of Practice.

Voluntary Standards

26-62   Voluntary standard setting is available through the Standards New Zealand (SNZ) which operates in accordance with the Standards Act 1988. SNZ has established a telecommunication subcommittee with the objective of setting national standards for telecommunications. With the emergence of more public and private networks using different technologies Telecom no longer controls the end to end telecommunications network so the SNZ subcommittee will contribute to the establishment of those areas in which Telecom does not have an interest.

Numbering

26-63   There are no direct statutory controls over numbering. However, following a review by the Ministry of Commerce of Telecom's management of the national numbering system, the Minister of Communications decided in 1993 to establish an advisory committee. This committee called the Telecommunications Numbering Advisory Group (TNAG) comprises all carrier groups together with user and consumer organisations and is chaired by the Ministry of Commerce. The group has persuasive rather than any mandatory powers.

CHAPTER 27

Russia

1. DEVELOPMENT OF THE TELECOMMUNICATIONS INDUSTRY IN RUSSIA

The Russian economy is currently at the final stage of transition from a centrally-planned economic system with virtually no room for private entrepreneurship to a market economy with freedom of enterprise and emphasis on private initiative. The new Constitution of the Russian Federation enacted December 12, 1983 provides in Article 34 that each individual "has the right freely to use his or her abilities and property for entrepreneurial and other economic activity not prohibited by law". Articles 8 and 35 of the Constitution for the first time in the constitutional history of Russia proclaim that private property is protected by law. To accomplish the objective of transforming a command economy into a market model, the Russian Government has been pursuing, for more than four years, the policy of privatisation of the public sector of the national economy. According to official statistics, more than 70 per cent of the economy has already been transferred to private ownership.

27-01

The communications industry also has undergone significant changes. Although privatisation of this sector of the economy is subject to certain restrictions, private initiative has also found its way into the provision of communication services. Under Russian law, "telecommunications enterprises" (*i.e.* State owned) were subject to privatisation but only pursuant to a decision of the Government of the Russian Federation in each individual case and with the State retaining a certain percentage of the stock of the privatised entity for three years. As of now, virtually all of the former State telecommunications enterprises have been corporatised and, subject to the above restrictions, shares of the newly formed joint stock companies have been sold to the public. Also, there have been created a significant number of private operators which, pursuant to licences of the Russian Ministry of Communications, provide a wide variety of telecommunications services to a growing number of clients throughout the Russian Federation. According to the Ministry, more than 600 licences have been

27-02

granted to private telecommunications operators in Russia which represents an approximate annual sales volume of U.S. $100–200 million.

In general licences have been granted for specific services in specific territories for periods of between three and ten years depending upon the scope of the commitments undertaken by the licensee. The content of licences has been relatively brief in comparison with other European models, typically including obligations with respect to the provision of services, network penetration and roll-out as well as quality of service commitments. Tariffs have not been regulated by the Ministry although a few socially sensitive tariffs are directly regulated by the State; licences have normally simply included restrictions on price discrimination as between customers.

2. EXISTING LEGISLATIVE and REGULATORY ENVIRONMENT

Telecommunications Policy

27–03   Until recently, telecommunications regulatory policy had developed somewhat piecemeal. Nevertheless the Ministry of Communications has been working hard to create a coherent body of principles and to formulate these principles into what has become a considerable body of draft legislation. Equally, in terms of network infrastructure development, detailed plans have been formulated for expanding and upgrading this infrastructure and for interconnection of the multiplicity of networks in Russia. One, a "General Plan of Development of Telecommunications", has been adopted by the Ministry and is available to the telecommunication companies. The other is the "Main Provisions of the Third Stage of Development of an Interconnected System of Telecommunications in Russia"; this document contains the plan up to the year 2000 in detail and up to 2005 in general.

The Government's strategy is to encourage investment through a variety of fund-raising mechanisms, the emphasis now being on seeking non-state financing guaranteed by the Ministry and private enterprise. Clearly the Government is hoping to attract investment in the stocks of the newly privatised public operators. Rostelecom, the Russian national long-distance operator, recently offered 513,000 shares, almost a quarter of its stock, at a voucher auction.

27–04   In addition to privatisation, a major aspect of the Government's reform policy is encouraging fair competition in most industries of the national economy. One of the strategic objectives of the Ministry of Communications and the Government in the communications sector is the demonopolisation of the sector as set forth in the "Programme of Measures to demonopolise the Communications Industry for 1993 and Subsequent Years" approved by the President of the Government of the Russian Federation on December 9, 1993 (the "Programme"). The Programme acknowledges that the current state of the communications market reflects a shortage of most types of services (demand by far exceeds supply) and the dominant position of the enterprises currently operating within it. Reforming the sector will include supporting the existing (*ie.* former State) enterprises as well as gradually building up privately-funded alternative infrastructure. The objectives of the Programme include the following:

- restriction of monopoly activities, creation of a control mechanism in the sector, as well as regulation of the operations of enterprises having a dominant position in the communications services markets;
- promotion of competition in the industry and ensuring sufficient supply of communications services on the market, improving the investment climate, conversion of defence-oriented enterprises, use of departmental networks for these purposes and adequate licensing policy;
- protection of consumers' rights, certification of equipment and standardisation of communications services, creation of a control and liability mechanism for poor quality services, and adequate tariff policy; and
- privatisation of telecommunications enterprises, diminishing direct State regulation of such enterprises' operations, restructuring the sector.

The Programme sets broad standards for determining the monopoly activities of enterprises which have a dominant position in their respective markets; the criteria for specifically determining dominance and monopolistic activities are to be defined by a joint commission of the State Committee on Anti-Monopoly Policy and the Ministry of Communications. The Programme goes on to describe mechanisms for restructuring the industry, supporting "new structures" in the sector and regulating activities including: 27–05

- principles for full, partial or no privatisation of communications enterprises;
- a licensing system to protect consumers' rights and ensure that all business entities, irrespective of their form of ownership, should have an opportunity to receive licences;
- licence fees;
- certification of equipment intended for use in the Integrated Communications Network;
- certification of public communication services;
- regulation of tariffs, including a maximum rate for "socially significant" communication services;
- grants and subsidies out of the republican budget to support federal, regional and local development programmes;
- providing tax preferences and exemptions to communication enterprises; and
- regulating profitability of monopolistic enterprises.

The State Committee on Anti-Monopoly Policy and the Ministry of

Communications are to analyse the implementation of the Programme and make any necessary adjustments.

Telecommunications Legislation

(i) Federal Law on Communications 1995

27–06   On February 23, 1995 the new Federal Law on Communications ("the Law") was published and thus at long last set out a comprehensive legal and regulatory framework for the sector. The Law lays down general principles for the right to carry on communications activities, describes Government involvement in communications regulation and operation, establishes the institutional agencies involved in regulation and administration of communications and deals with the more operational aspects such as ownership of networks, protection of fair competition, interconnection, privacy and liability. These principles include:

- Communications enterprises shall be established and operate in an integrated economic environment, in diverse forms of ownership and in a competitive environment.
- Communications networks and facilities may be federally owned, the property of the Russian Federation's constituent entities, municipally owned, or the property of natural persons or legal entities, including non-Russian organisations or nationals.
- There will be freedom of conveyance of messages via telecommunications networks and facilities throughout the Russian Federation.
- Communications enterprises are to be supervised by and accountable to the Federal Executive authorities.
- Tariffs for communications services are to be set on a contractual basis (*i.e.* by the operators and services providers themselves) except in the case of certain tariffs (essentially those which are "socially significant") which may be regulated by the State.
- Network interconnection payments are to be negotiated and agreed by the communications enterprises concerned.

27–07   At an operational level, one of the hallmarks of Russian telecommunications in the past has been the proliferation of what the Law describes as "departmental networks", the networks run by various federal ministries. These are expressly covered by the Law and are given the right to provide communication services to the public or other communications users provided the operators have duly obtained a licence in the same way as any other operator/service provider whether public or private. Procedures for interconnecting such departmental networks are to be "defined" by the Russian Government.

Communications activities are to be managed through an integrated system of "federal communications authorities" (discussed under "Regulatory Institutions" below) but the Law itself is vague as to who and what these authorities actually are. The Ministry is expressly given responsibility for the issue of telecommunications, television and radio broadcasting licences but no mention is made in the Law of the State Radio Frequencies Commission, the State Telecommunications Communication or the Service of the State Supervision over Communications.

(ii) Statute on Licensing of Communications Activities 1994

The Statute on Licensing of Communications Activities was adopted by 27–08 the Russian Government by Decree 642 of June 5, 1994. This statute, which deals mainly with the procedural aspects of licence applications and grants, is discussed in detail in paragraph 27–21 below. Although enacted before the Law, it is still in effect and has not been repealed or amended.

Regulatory Institutions

The Ministry of Communications is the central body of federal authority 27–09 in the Russian Federation, having responsibility for State management of the communications industry and supervising responsibility for the condition and development of all types of communications.

Other executive authorities that exist to assist the Ministry in its regulatory responsibilities are discussed in more detail in paragraphs 27–15 to 27–21 below, but currently comprise a State Telecommunications Commission, State Radio Frequencies Commission and a new supervisory body called (in translation at least) the "Service for State Supervision of Communications". Each of the Ministry, the State Radio Frequencies Commission and the State Telecommunications Commission operate pursuant to a special statute enacted either by the President or the Government of the Russian Federation.[1]

The structure of the federal communications authorities referred to in the Law has not yet been finalised. The Government has wished to preserve the flexibility to adjust such structure as the reform of the communications sector is implemented. Accordingly the only regulatory authority mentioned in the Law is the Ministry itself, which for the time being will be at the core of the regulatory system.

Telecommunications Networks

Chapter 2 of the Law confirms the existing mixture of networks of which 27–10 the Russian communication system is composed. At the heart of the Law is the concept of an Interlinked Communications Network[2] for the whole of the Russian Federation. This will be made up of the state owned and

[1] Ministry: Governmental Decree no. 1022, December 25, 1992; amended by Decree no. 541, May 25, 1994.
[2] Defined in Art. 2 as "the totality of technologically integrated and centrally controlled telecommunication networks in the Russian Federation".

newly privatised public networks and Government departmental networks. The Law states that the development and improvement of the Interlinked Communications Network will be based on the principle of technological integrity of all networks and communications equipment. Furthermore, the Government will support measures for the improvement of the Interlinked Communications Network from the federal budget.

The existing networks of the Russian telecommunications system are as follows:

(i) The Public Communications Network

27–11   The public communications network is an integral part of the Interlinked Communications Network and includes trunk, regional and local telecommunications lines which are under the jurisdiction of the Russian Federation irrespective of their owner and form of ownership. This begs the question somewhat as to the extent of the jurisdiction of the Russian Federation, a constitutional matter. The Law does not make it clear whether there is any room left for the constituent entities of the Russian Federation in regulating local portions of the communications sector. Paragraph 2 of Article 4 of the Law provides that certain issues outside the scope of the Law may be regulated by the legislation of the constituent entities of the Russian Federation "to the extent of their regulatory jurisdiction". The scope of such local powers is not defined in the Law or, at present, anywhere else. Responsibility for the functioning and development of the public network lies with the federal communications authorities.

(ii) Departmental Communications Networks

27–12   The "departmental" communications networks serve the special needs of the federal ministries and departments. These networks can also be (and are) used for offering communications services to the public. The departmental networks are to be linked with the public network on a contractual basis once the technical equipment of the departmental networks conforms to the rules and technical norms established for the public network, and (where communications activities are to be pursued) after a licence has been obtained in the proper way.

(iii) Dedicated Communications Networks

27–12a   Dedicated communications networks are defined in the Law as networks operated by actual persons or legal entities, which have no access to the public communications network. However, if connected to the public communications network, dedicated communications networks are to change in their status and become components of the public communications network, with all that that entails.

In practice dedicated networks should include what are generally known as "bypass" or "overlay" networks, except where these are interconnected with the public network.

Commercial communications networks can be set up by any legal or physical person, including foreign investors who have recognised legal status. Upon connection with the public network, owners of commercial communications networks must satisfy the conditions of connection, and equally the demands of ensuring the compatibility of the connected networks with established rules and technical norms.

(iv) Communications Networks for Management, Defence, Security and the Protection of Law and Order of the Russian Federation

Article 9 of the Law states that a part of the public communications network should be set aside to secure communication needs for the purposes of government, defence, security and the protection of law and order in the Russian Federation. The Law states that such networks should be provided under leasing arrangements as provided by Russian Law.   27–13

Specially authorised agencies will be empowered by the President of the Russian Federation to secure government communications. These agencies will ensure the confidentiality of the information conveyed on the networks under their responsibility. The rights and obligations of these agencies will be defined by Russian law.

The federal communications authorities, the communications agencies of other federal executive authorities, and communications operators, irrespective of their form of ownership, are obliged to secure the immediate availability of communications channels for the purposes of government, defence, security and the protection of law and order in the State and to take immediate measures to replace and repair damaged links.

Status of Public Operators; Network Ownership

The general principles governing communications activities, as set out in Article 5 of the Law—for example, the equality of all communications operators, the limitation of monopolistic activity, the freedom to convey communications, would all seem to indicate that anyone can set up as an operator providing public telecommunication services in the Russian Federation. This must, however, be in accordance with the licensing provisions, set out in Article 15 of the Law and also in the Statute on Licensing of Communications Activities. It is also confirmed in Article 17 which states that communication networks may be federally owned, owned by constituent entities of the Russian Federation, by local municipal authorities or otherwise be the property of natural persons or legal entities, including non-Russians.   27–14

Future Draft Legislation

I understand that there are two draft laws which are pending: 27–15

    (i) a law on the radio frequency spectrum; and

(ii) a law on television and radio broadcasting.

The Ministry has also to consider important regulations on interconnection, prepared by the State Telecommunications Commission for the Ministry's approval (see below, "Network Interconnection").

3. REGULATION OF SERVICES and EQUIPMENT SUPPLY

Regulatory Authorities

27–16   Chapter 3 of the Law stipulates (Article 11) that communications shall be "managed" (*i.e.* in the regulatory sense) through "an integrated system of federal communications authorities". The organisation and structure of these authorities is to be defined by separate statutes approved by the Government. The federal authorities are responsible for matters such as:

- development of State policies on communications;
- "intersectoral" co-ordination of telecommunications operations;
- "arrangements and support" for the supervision of communications activities by the State; and
- development of State policies on radio frequency allocation and use.

Turning to each of the current authorities:

(i) The Ministry of Communications of the Russian Federation

27–17   The Ministry of Communications is the central organ of the federal executive authorities in the field of communications. Its role is to secure the fulfilment of Government policy and to co-ordinate the activities of communications enterprises under the jurisdiction of the Russian Federation. The structure and organisation of the activities of this Ministry are defined in a specific Decree, confirmed by the Government.

27–18   The basic functions of the Ministry of Communications include:

- the regulation of the provision of communications services, with the principal aims of satisfying the needs of: the population; State authorities; defence, security and law-enforcement services; business organisations and other legal bodies;
- drafting Federal laws;
- preparation of State programmes for the development of communications infrastructure;
- communications licensing;
- organisation of the procedure for certification of communications equipment and services;
- organisation of State supervision of communications equipment and networks;
- allocation of radio frequencies and orbital positions for civilian communications satellites;

- the resolution of disputes on technical questions;
- in circumstances envisaged by Russian Law, carrying out the function of central management of all communications networks on the territory of the country;
- the management of State academic institutions concerned with the training of communications specialists;
- implementing policy directed towards the restriction of monopolistic activity and the support and encouragement of genuine and effective competition between communications enterprises;
- representing Russia in international organisations and securing the protection in international law of Russian interests in the field of communications and the fulfilment of obligations, as accepted by statute, of the Convention, Acts and Regulations of the International Telecommunication Union and of the Universal Postal Union;
- the fulfilment of other functions, defined by the Government of the Russian Federation.

(ii) State Telecommunications Commission; Service for State Supervision of Communication

For the present specific authorities formed by and under the Ministry of Communications to date with respect to telecommunications include the following: **27–19**

- the State Telecommunications Commission. Its structure, activities and composition are established by specific Regulation, confirmed by the Russian Government. The Commission supervises the main functions of the Ministry as a whole and additionally has responsibility for agreeing to certain technical tasks and for setting technical conditions in the construction, reconstruction and technical adaptation of most communications networks (essentially all networks except those of certain Ministries and Federal Agencies);
- the Service for State Supervision of Communications. This Service supervises standards, conditions of certification and the state of telecommunications and postal communications networks and equipment (with the exception of the certain Ministries and Government Agencies referred to above). It also supervises compliance with licensing conditions, including control of the use of the radio frequency spectrum and of the orbital positions of communications satellites.

(iii) The State Radio Frequencies Commission

The State Radio Frequencies Commission is a managerial body, of which the structure, activities and composition are again defined by a specific Regulation. Its main tasks are as follows: **27–20**

- working out a long-term policy for the allocation and usage of the radio frequency spectrum by various types of radio equipment and

working out tables for the allocation of the radio frequency bands between the radio services of the Russian Federation;
- working out a technical policy to ensure compatibility of radio equipment and, for this aim, promoting scientific investigation.

Licensing

27–21   In June 1994 the Russian Government approved a new Statute on Licensing of Communications Activities.[3] This instrument is designed to deal with the more procedural aspects of telecommunication services licensing. The Statute—

- sets out a structure for considering licence applications;
- details the requirements to be followed by applicants and the information they should give;
- provides for how and when licences are to be accepted or refused;
- specifies maximum terms for the duration of licences;
- establishes the right to charge licence fees (though without detailing a fee scheme); and
- stipulates the grounds on which licences can be modified, suspended or revoked.

Telecommunication licences are expressed to be issued for a term of three to ten years, in line with current practice. It can be anticipated that many operators and investors will be seeking longer licence periods, to give them security for pay-back of their investment.

27–22   Article 15 of the Law deals with the general legal framework for licensing of communications activity. The issuing, variation, extension, or termination of licences to provide telecommunications services are the responsibility of the Ministry of Communications. However, licences will not be required in certain cases. Exempt activities comprise networks operated exclusively for the purposes of government, defence, security and the protection of law and order as well as for "industrial process support"; and also networks confined to one group of premises or several adjacent groups of premises or installed in vehicles, ships, planes or some other mode of transport.

The types and duration of licences, the conditions attached to them, suspensions and terminations of licences, and equally any other question of licensing will be regulated by the laws of the Russian Federation. A licence or any right attached to it can only be transferred, either in whole or in part, from one legal person to another, after those legal persons have received new licences. Anyone providing communications services without a licence or infringing the rules governing its use, will be liable under Russian law, unless such activity is exempt. There is a right of appeal to the

[3] Resolution no. 642, June 5, 1994.

courts or to the arbitration court on the licensing decisions made by the Federal Communications Authority.

On May 12, 1995 the Ministry of Communications adopted a "Regulation on the procedure for payment and the amount of the fee for the issue of licences in the field of telecommunications in the Russian Federation". This establishes a scale of fees for various telecommunications activities based on multiples of the minimum wage.

Network Interconnection

According to Article 8 of the Law, procedures for interconnecting a depart-   27–23
mental or dedicated communications network and the public network shall be defined by the Government of the Russian Federation. To this end, by Resolution No. 107 of January 25, 1995, the State Telecommunications Commission approved regulations "for the connection of telecommunications networks to public telecommunications networks, and management of telephone traffic via public telecommunications networks in the Russian Federation". These potentially very important regulations, which are in draft form, are, at the time of writing, with the Ministry for its approval. If so approved, they will:

- lay down detailed arrangements for levels and points of connection, to be further defined in the special provisions of the licence issued to the operator of the network to be connected (which appears to mean the network seeking connection with the public network): the regulations envisage connection at both the local and trunk (junction) level;
- provide that technical details for the interconnection should be defined in specifications to be provided by the public communications network operator: guiding principles are given in the regulations;
- require authorisation for operation of the interconnection by an agency of the Service for State Supervision of Communications;
- with respect to international telephone traffic, provide that such traffic to, from and via the public network must be carried only through the switching facilities of the gateway exchanges of Russian public telephone network operators (*e.g.* Rostelecom). This appears to be a reiteration of the "rule" practised by the Ministry which violated the Russian law on "Competition and Anti-Monopoly Activities in Commodities Markets 1991"; this was the subject of a proceeding before the State Anti-Monopoly Committee which in June 1994 issued a stop order requesting the Ministry to rescind the rule and release operators (*e.g.* overlay operators having interconnection with the public network) from the requirement. The Ministry has not, however, as yet rescinded the rule.

According to the proposed regulations, authorisations for connections are   27–24
not to be withheld for any reasons other than those defined in the regulations and such action by a public communications network operator would be treated as a breach of the terms of its licence. In addition, the terms

and conditions of interconnection should provide for the parties' mutual responsibility for the quality of communication services and their liability to users.

According to Article 21 of the Law network interconnection fees are to be set on the basis of agreements negotiated by the connecting operators. Any disputes over such matters are to be resolved by a court of law or arbitration court; it is unlikely such bodies would have the technical, financial or other relevant competence to deal properly with such disputes.

Fair Competition

27–25   The Government's programme for anti-monopoly policy is discussed above (see para. 27–03). Article 20 of the Law provides that communications operators shall be liable under Russian law "for any abuse of their position or for any other action which may prevent or restrict competition".

Tariffs

27–26   Article 21 of the Law sets out the situation as regards the regulation of tariffs for communications services. Tariffs are in general to be unregulated, with operators fixing their prices unilaterally or on a contractual basis. However, the Government reserves the right where necessary to control prices for specific communications services, namely the "socially significant" services referred to in the Programme.

Emergency calls for fire, police, ambulance and other "emergency services" are to be provided free of charge.

Privatisation

27–27   If there is to be a change in the form of ownership of communications networks and equipment under Federal State ownership, this must be carried out according to the procedure envisaged by Russian legislation. The procedure is the same in the case of a change of ownership of networks and equipment remaining in the ownership of State "constituent entities", except that notification must be given to the Ministry of Communications. Privatisation of State enterprises must conform to the Law "on the Privatisation of State and Municipal Undertakings". If a State-controlled communications enterprise is transformed into a joint-stock company, the State shall be the majority shareholder in such company for a period to be determined by the Government of the Russian Federation. Foreign investors can take part in the privatisation of State and municipal communications enterprises on conditions defined by the legislation of the Russian Federation.

In general, a change in the form of ownership of communications networks and equipment is permitted if it does not infringe any technological rules for the functioning of communications networks.

Investment Financing

Article 19 of the Law provides that in general terms investments in the   27–28
communications industry are to be made pursuant to the laws of the Russian Federation on investment programmes. Such programmes, to be drawn
up by the federal communications authorities, should accommodate suggestions of the Russian Federation's constituent entities and are to be subject
to approval by the Government of the Russian Federation.

Communication activities supported by the federal budget are to be financed as envisaged by the federal state communications development programmes and to be specifically included in the federal budget.

Natural persons or legal entities involved in projects to develop and sup-   27–29
port communications may, according to Article 19, be eligible for guarantees, loans on favourable terms, tax and other privileges.

Relations of Communications Enterprises with Regulatory
Authorities and with other Enterprises and Organisations

Chapter 5 of the Law is devoted to the relations of communications enter-   27–30
prises with regulatory authorities and with other organisations. It sets out
guidelines as to the rights and obligations of communications enterprises
in various areas of their activity, particularly the construction of buildings
and equipment, the use of land, and special rights in the event of accidents.

The Law states that in the planning and development of towns and residential areas, regard must be had to the buildings and equipment necessary
for communications enterprises. Town authorities are under an obligation
to give communications enterprises a right to lease separate accommoda-
tion which is suitable for their technical needs.

Communications undertakings have a right to construct buildings and
equipment on land in accordance with the law, and also on roofs, bridges,
railways, etc., with the consent of the owner or tenant of this land. They
also have a right to set up and to service communications apparatus on
such land. Owners, users and tenants have the right to refuse only for
reasons "defined by the laws and other legal acts passed in the Russian
Federation". In the event of dispute the matter may be submitted to a court
of law or arbitration court. Communications undertakings can enforce this
right against owners and tenants only after they have given written notice
of their intentions. If the owner or tenant objects, the undertaking can
take the question to the competent executive authority or to the arbitration
court.

The Law also provides for compensation to be paid by a communications
undertaking to the owner or tenant of land in the event that the value of
that land decreases as a result of construction works undertaken by the
communications enterprise. The amount is to be set according to the
(unspecified) laws of the Russian Federation. Communications undertak-

ings are under a general obligation to leave land in its original condition when their use of it terminates.

27–31   The Government is referred to as being responsible for the protection of communications equipment and facilities, the radio frequency spectrum and the positions of communications satellites in orbit. This could include restrictions on the production and import into the Russian Federation of radio equipment which creates irregular interference in the functioning of electro-magnetic systems. Any persons who permit harm to any communications equipment or who infringe any rules relating to the acquisition, import, use and registration of radio equipment, commit an offence under Russian law and are liable to compensate communications enterprises who have suffered loss of profit as a consequence.

One Article of the Law is devoted to the elimination of the consequences of accidents on communications networks. All executive authorities, both federal and local, are required to give immediate help to communications undertakings in the event of some accident to the communications lines. Such help is to be paid for, in the first instance, by the appropriate communications enterprise; however, executive authorities would appear to have a discretion to pay for it out of State resources. Such works can be carried out on any lands without requiring the consent of the landowner. Compensation for the damage caused to the communications operator, or to other legal and physical persons, resulting from the accident, will be paid for out of insurance funds, or at the expense of the guilty party. Again the State authorities, both federal and local, have a discretion to pay this compensation out of relevant budgets.

Other rights are afforded to operators in the event of such a breakdown, for example the priority use of local transport facilities.

Rights of Communications Users

27–32   Chapter 6 of the Law establishes a framework of rights to which communications users are to be entitled. All consumers are to have the right to convey communications over the telecommunications networks. Nobody can be refused access to the public communications network. Further, both consumers and operators have the right to demand connection with the public communications network of their own networks or terminal points provided they fulfil the conditions of interconnection, as defined for various operators or as agreed in licences. Only equipment certified in the standard way can be connected to the public networks.

Operators are obliged to supply users with communications services that meet with "applicable" quality and technical standards. It is unclear, however, exactly what these standards are.

Owners of networks and equipment are under an obligation to provide priority services for emergency communications.

Certain categories of officials will be entitled to privileged and advantageous use of communications services. Such categories include Russian State officials, diplomatic and consular representatives of foreign states, representatives of foreign organisations, and also separate groups of citizens. Privileges conferred can include priority usage as well as preferential rates.

The categories of such privileged persons are defined by Russian law   27–33
and the regulations of "constituent entities", as well as by international agreements. A procedure for compensating those State-owned communications enterprises which suffer as a result of these privileges will be defined by the Government of the Russian Federation and by the executive authorities of the "constituent entities".

The rights of consumers generally in relation to the supply of communications services of a reliable quality are to be protected by the laws of the Russian Federation, *e.g.* the Law "on the Protection of the Rights of Consumers". The rights and guarantees conferred in this law cannot be restricted in any way by regulations passed by the "constituent entities".

The secrecy of communications is a principle enshrined in the Russian Constitution. The Law also imposes upon communications operators the obligation to secure the secrecy of all communications. Derogations from this principle can only be permitted on the grounds and in the manner established under Russian law. Phone tapping is illegal unless authorised by a court of law.

Dispute Resolution

In the case of default by a communications operator, the "user" (customer)   27–34
may claim compensation and Article 38 of the Law sets out a detailed procedure for the filing of such claims. Disputes relating to communication activities which arise between communications enterprises or between communications users and enterprises are to be resolved according to the laws of the Russian Federation.

Disputes which arise between communications undertakings based in Russia and the undertakings of other states will be decided by reference to the relevant international conventions and agreements.

4. RADIOCOMMUNICATIONS

The Ministry of Communications has as one of its functions the allocation   27–35
of radio frequencies and orbital positions for civilian communications satellites. Equally, the State Radio Frequencies Commission has, at present, responsibility for forming policy in the area of the allocation and usage of the radio frequency spectrum, as well as for working out a basis for the allocation of the radio frequency bands between the radio services of the Russian Federation. The Commission also develops technical policy to ensure the compatibility of radio equipment. Finally, the Service for State Supervision of Communications supervises control of the use of the radio

frequency spectrum and of the orbital positions of communications satellites.

Article 12 of the Law is entirely devoted to the regulation of the usage of the radio frequency spectrum and of the positions of communications satellites. It is stated that the development and implementation of policies and procedures securing the allocation of the radio frequency spectrum and the effective use of radio frequencies and of the positions of communication satellites in accordance with the interests of the Russian Federation and with regard to international treaties and conventions, will be organised by the Government of the Russian Federation.

In order to ensure the electro-magnetic compatibility of radio equipment, the Government of the Russian Federation will establish a procedure for the allocation of radio frequencies, the special conditions of planning, construction, acquisition, exploitation and import from abroad of radio equipment and high-frequency apparatus. The Government will also define the measures to be taken to protect radio reception from industrial sources of interference.

Frequencies, once allocated to particular enterprises, can be changed in the interests of the protection of State activity, defence, security and the protection of law and order. The Law provides, in the event of such a transfer to another frequency, for compensation for any loss caused.

5. Technical Standards: Certification

27–36 All communications equipment and facilities used on the Interlinked Communications Network of the Russian Federation must be subjected to a compulsory process of testing and "certification" for compliance with the "established standards and other norms and technical requirements". Communications services which are supplied on the public communications network "may" also be subject to such "certification".

This process is to be carried out by the Federal Communications Authority with the help of test centres empowered to do this and accredited within the established framework of Russian organs for standardisation, metrology and certification. On completion of the process of certification for each model of communications apparatus, the Federal Communications Authority will issue a certificate of the established standard. Any party guilty of infringing the rules for the certification of telecommunications equipment will incur the penalties defined in Russian Law.

In Article 28 of the Law communications users are given the right to expect certain standards of service. Operators are obliged to supply to consumers communications services which meet with the requisite standards of quality, technical norms, certification, and to supply conditions of contract for the supply of such services.

6 BROADCASTING

Television and radio broadcasting activities are authorised by the Ministry   27–37
of Communications of the Russian Federation in co-ordination with the
Ministry of Press and Information and the All-Russian State Television and
Radio Broadcasting Company. Pursuant to Presidential Decree No. 2255
dated December 12, 1993 the Ministry of Press and Information was dissolved and its functions with respect to television and radio broadcasting
licensing were assigned to the Federal Commission on Television and Radio
Broadcasting ("the Federal Commission").

The licence issued by the Ministry of Communications gives to a licensee
the right to use technical means for broadcasting, while a second type of
licence, issued by the Federal Commission and described below, permits
the holder to broadcast on a certain TV channel. The new Licensing
Statute, Decree No. 642, applies to licences issued by the Ministry with
respect to broadcasting systems. The Ministry of Communications is also
responsible for issuing certificates for communications equipment through
specially created "enterprises" accredited with Goskomstandart.

The Law of the Russian Federation on Means of Mass Information ("the
Broadcasting Law") No. 2124–1 dated December 27, 1991 (sections 30–
34) provides that the State policy in the area of broadcasting shall be determined by the Federal Broadcasting Commission and local broadcasting
commissions.

Pursuant to Article 31 of the Broadcasting Law 1991, "broadcast
licences" are issued by the Federal Commission. Article 31 provides that
the holder of such a licence has the right to broadcast within the limits of
the licence. Nowhere in the Broadcasting Law 1991 is it stated explicitly
that it is forbidden to broadcast without a licence, although the requirement of a licence is clearly implied.

Article 8 of the Broadcasting Law requires the registration of "mass
media" with the Ministry of Press and Information.

Section 3 of the Statute "On the procedure for the manufacture, importation into the Russian Federation and use on the territory of the Russian
Federation of High Frequency Radio-electronic Equipment" (Decree 643)
provides that the assignment of broadcast frequencies is carried out by the
State Committee on Radio Frequencies under the Ministry of Communications. Section 3 also provides that the production, importation or use on
the territory of the Russian Federation of radioelectronic equipment is subject to licensing by the main department of the Service of State Supervision
of Communications of the Russian Federation.

CHAPTER 28

Spain

By Almudena Arpon de Mendivil y de Aldama,
Gomez-Acebo & Pombo, Madrid

1. HISTORY OF THE DEVELOPMENT OF TELECOMMUNICATIONS IN SPAIN

Industry Structure

In Spain, telecommunication is regarded as an important sector, given its   28–01
contribution to the country's generation of wealth in terms of employment,
technological innovations, capital investment, tax contributions and
improved management of the country's resources. On the other hand, the
country depends heavily on the service due to a combination of the lack
of an existing transport infrastructure, its vast and complex territory and
its dispersed population. It may therefore be assumed that it is a basic
instrument for social cohesion and territorial stability in the country.

Today in Spain telecommunication is regarded as the driving force of
the next cycle of economic development. A sound indicator of its potential
for growth is the interest shown by foreign companies in the national
market, as much as with the trade in equipment as with services. It is a
sector which, in spite of not being large, is important; its contribution in
1992 of 2.4 per cent to the GNP is far from insignificant.

In order to briefly describe the structure of the industry in Spain a distinction should be made between the sectors of the supply of services and that
of the manufacture of equipment. The former is distinctly led by "TELE-
FONICA DE ESPAÑA, SA" (henceforth TELEFONICA), as a result of the
monopolistic situation which shall be discussed further on; TELEFONICA
occupies fifth place amongst operators at a European level and eleventh in
the world, and is undergoing a significant process of internationalisation. In
addition to TELEFONICA, there exist another four state-owned operators
(RETEVISION, the Autonomous Postal and Telegraph Organ (OACT),
HISPASAT and RTVE), and new private entities which have been awarded

the concessions recently granted by the Ministry of Public Works, Tourism and Environment (MOPTMA): there are three licensees for paging within a national scope and 49 within a local scope; 16 licensees for regional trunking; eight for data transmission; one for automatic mobile telephony; and lastly, three licensees each for both private and satellite TV respectively.

28–02   With regard to the industrial subsector of equipment, its volume of production from 1990 to 1992 was sixth amongst the developed countries and fourth in the world. On this last point, in terms of contribution to the GNP, Spain was ahead of countries like the USA, Japan, Germany and France. For Spain, this subsector should therefore be very important as it is the only sector of information technology which holds a relatively important position.

The most significant factor which has set the industrial framework of the sector in Spain has been, without a doubt, the monopolistic situation granted by law to TELEFONICA as highlighted by the Defence of Competition Tribunal in its 1993 Report which points out the different problems faced by the industry derived from such a monopoly, namely the obligation to buy liberalised services from TELEFONICA in order to receive others that are monopolised, the slow supply of services, etc. The situation is such that the Tribunal considers that in Spain, in this industry, prohibition is the norm; they mention up to 18 prohibited activities.[1]

Institutional Structure

28–03   The institutions which are involved in telecommunications in Spain are:

- The Parliament, insofar as it promulgates the rules within a legal hierarchy of law that regulate the sector;
- The Government, which has the legislative initiative to draft laws and to proceed with the regulatory development of the same while exercising its legislative powers and all its executive competences over what is foreseen in such rules;
- MOPTMA, charged with the duty of fulfilling all the functions assigned to it by the rules on telecommunication; amongst other duties, it must ensure the cooperation and interconnection of telecommunication networks and infrastructures in order to endeavour to attain the optimum output from the services and rationalise investment, draft and supervise the implementation of the National Plan for Telecommunications, propose and develop technical laws relating to the provision of services as well as those concerning the granting of administrative concessions, authorisations and licences.

[1] Report on "Political Remedies which can favour freedom of competition in the Services Sectors and Halt Damage Caused by Monopolies" (1993), Spanish Defence of Competition Tribunal, in compliance with functions commended by the Law on Defence of Competition, to submit to the Government proposals for legislative changes in order to increase competition together with a concrete request to the Government to this respect.

- The General Secretariat of Communications, which is part of   **28-04**
MOPTMA, has the duty of proposing legislation for communications
and TV and of managing the relations between the State Administration
and TELEFONICA. The General-Directorate of Telecommunications,
the OACT and also the Delegate of the Government in TELEFONICA
all depend on the General Secretariat. According to an announcement
by MOPTMA it is foreseen that in the future this department will be
split into two: the General-Directorate for the Management of this
sector entrusted with the preparation and review of the National Telecommunications Plan, development policies, investments, establishment of the tariffs framework, approval and homologation of terminals,
and the General-Directorate for the Regulation entrusted with the
supervision of contracts, prices and offer of services under monopoly,
implementation of European rules, sanctions, etc.[2]
- The Telecommunication Consultation Committee, assigned to the
General-Directorate of Telecommunications, is the Government's
highest consultative organ on telecommunication matters. It is chaired
by the Minister of Public Works and Transport and is composed of
administrative representatives, users, suppliers of services, equipment
producers and of the most representative trade unions in the sector.
- The Defence of Competition Tribunal and the General-Directorate for   **28-05**
the Defence of Competition, both assigned to the Treasury Ministry, are
in charge of watching over the application of competition laws in Spain.
- The public service operators, TELEFONICA, RETEVISION, OACT,
HISPASAT or RTVE. At this point it is sufficient to note that TELE-
FONICA, created in 1924, is a private company that exploits the telecommunication services under rules of concession, under strong public
control; the State owns approximately 30 per cent of the capital, can
appoint up to a maximum of five members to the Board of Directors,
in addition to those which the General Meeting of Shareholders may
name, and appoint a delegate of the Government who has, amongst
other faculties, the power of veto over any agreement reached by
TELEFONICA on any matter relating to public interests. The delegate, who depends on the General Secretariat of Telecommunications,
is actually the Director General of Telecommunications so there is
some coincidence between the regulator of services and the operator
of the same.

Background to Legislation

Until 1987 telecommunication activities in Spain were controlled by out-   **28-06**
dated legislation: Acts dating back to April 22, 1885, October 26, 1907

---

[2] MOPTMA has recently adopted an "Agreement on the Telecommunications Policy for the period 1994
to 1998" which amongst other issues, foresees an arbitration entity assigned to MOPTMA of which the
functions are: to give a report on the proposals of tariffs on telecommunication services offered under
an exclusive régime or subject to obligations of public service, watch over the compliance of prerequisites
of competition, arbitrate and resolve conflicts between operators of networks and services and assess
MOPTMA over the development of regulations which may affect markets subject to competition.

and June 26, 1934. It became evident that this legislation was not able to cope with either the technological advances in telecommunications or the necessary changes in Spanish legislation arising after its accession to the European Communities in 1986, consequently forcing Spanish authorities to create up to date legislation.

At present, the basic legislative instrument is the Spanish Telecommunication Law (LOT) which came into force on December 19, 1987, subsequently modified in 1992.[3] However, the laws which regulate the concession made to TELEFONICA for the supply of telecommunication services are also fundamental. These provisions were originally enacted in 1924 and are currently governed by the contract approved in January 1992, between the State and TELEFONICA.[4]

2. Existing Legislative and Regulatory Environment

Principal Legislative Instruments

28-07   A more fundamental legal instrument having an impact on telecommunication is the Spanish Constitution of 1978[5] which establishes the following principles:

- Freedom of expression and of communicating or receiving true information by any means are protected, within the bounds of the corresponding rights to privacy, integrity and youth and child protection. (Article 20).
- Public initiative in economic activity is acknowledged by provisions which reserve to the public sector essential resources or services (Article 128.2).
- The responsibility for the legislation and general regulation of communications, post office and telecommunications is exclusively reserved to the State (Article 149.1.21).[6]

In order to implement these constitutional principles as well as the E.U. directives on telecommunications, Spain enacted the main legislative instrument that rules telecommunications in Spain, which is LOT.[7] This law defines the different kinds of telecommunication services and the legal régime for the provision of each either in competition or as a monopoly; it determines the intervention of the administration for the regulation of

---

[3] Law 31 of December 18, 1987 on the Regulation of Telecommunications modified by Law 32 of December 3, 1992 for the modification of LOT.
[4] Contract of December 26, 1991 approved by resolution of January 14, 1992, preceding the one dated October 31, 1946.
[5] The Constitution was approved by referendum of December 6, 1978; it entered into force on December 29, 1978, date of its publication in the Official Gazette.
[6] Please note that Spain recognises local governments who have the authority to enact their own legislation on certain matters.
[7] See n. 3.

the sector; and defines the breaches of the telecommunication rules and their corresponding sanctions. Conceived in such broad terms the provisions are brought into effect by a series of implementing regulations, the power to make such regulations having been entrusted to the Government by Royal Decree. Therefore, the possible provision of services under competition becomes delayed and there is legal uncertainty: until the regulation is enacted the corresponding services cannot be provided and the rules are unknown.

The LOT has been developed by several regulations, some of them of a general nature such as the Regulation for the development of the LOT in relation to the use of radioelectrical public domain and the value-added services which use that domain[8] and other specific regulations for particular services, amongst which the last one has been the Technical Regulation for the Provision of the Value-Added Telecommunication Service of Automatic Mobile Telephony of 1994.[9]   **28–08**

The rules on television and radio broadcasting can also be considered as basic legislative instruments. Of particular relevance is the Law on the Statute of Radio and Television and its technical regulation.[10] There are also other rules within the hierarchy of law approved or in the process of being passed by the Parliament which regulate and liberalise certain services referred to further on.

Though not a legislative instrument, due to its importance for the sector, it is pertinent to mention the contract between the Government and TELEFONICA of 1992. The term of this contract is 30 years and it appoints TELEFONICA as an operator under monopoly for basic and carrier services.[11] One must also take into consideration the Law for the Defence of Competition[12] which even though not specifically aimed at the telecommunications sector, is particularly relevant as through its application the monopoly of TELEFONICA is surveyed.

Agencies of Regulation, Competition Control

(i) Agencies of regulation

MOPTMA is the main agency with responsibility for the regulation of telecommunications. Amongst the different bodies which form part of MOPTMA, the following are particularly relevant: the General Secretariat of Communications and the Directorate-General of Telecommunications. As explained above, the General Secretariat appoints the delegate of the Government in TELEFONICA. The principal functions of the regulator are   **28–09**

---

[8] Royal Decree 844 of July 7, 1989.
[9] The GSM Regulation was approved through Royal Decree 1486 of July 1, 1994.
[10] Law 4 of January 10, 1980 on the Statute of Radio and Television and Royal Decree 1160 of September 22, 1989 approving its Regulation.
[11] See n. 4.
[12] Law 16 of July 17, 1989.

those of determination of the nature of the services that companies wish to provide, the establishment of the conditions for such provision and the rules setting out technical specifications. Finally, MOPTMA has been assigned the tasks of law enforcement and inspection of services, equipment, and of civil telecommunication systems and plants. An interesting point to note is that the officers who perform inspections may request police protection when conducting their task.

(ii) Competition control

28–10   Regarding competition control, the authorities in charge of this area are the Competition Defence Tribunal and the Competition Defence Service. They are administrative entities organically attached to the Treasury Ministry and are entrusted with the application of the competition rules contained in the Competition Defence Law in addition to those relating to Unfair Competition[13] (provided that the unfair competition practices affect public interest) and also of the competition rules of the EU.

The primary function of the Tribunal is to take any action relating to competition law issues. However it also provides advisory opinions and reports at the request of Government officials such as the one quoted.[14] The Competition Defence Service is in charge of the preliminary procedures that will ultimately lead to Tribunal proceedings.

The Competition Authorities are currently playing an interesting role in the liberalisation of the sector as reflected in the above report as well as in the judgments pronounced following the claims submitted against TELEFONICA for the abuse of its dominant position.

28–11   In those cases, the Tribunal has declared:

- That TELEFONICA was carrying out abusive practices consisting in the irregular modification of the tariffs of its services between certain urban areas. Such practices moving a harmful and unjustified impact on the national economy and on the users of the said public service.[15]
- That TELEFONICA was abusing its dominant position in the telephony equipment and services market by tying to the lease contracts an option to purchase certain equipment. The Tribunal condemned TELEFONICA imposing a fine of 45,000,000 pts.[16]
- Lastly, the Tribunal declared that TELEFONICA abused its dominant position by unjustifiably refusing to supply and by delaying the supply of telephone lines to 3C Communications de España, S.A.. Such action was held to infringe both the Spanish and European Union anti-trust rules, and resulted in the imposition of a penalty amounting to 124,000,000. pts.[17]

[13] Law 3 of January 10, 1991 on Unfair Competition.
[14] See n. 1.
[15] Case 167/80 of March 26, 1981.
[16] Case 328/93 of October 1, 1993.
[17] Case 350/94 of February 1, 1995.

General Rules Governing Telecommunication Installations Services and Apparatus

The general rules governing telecommunications installations, services and apparatus are the following: 28–12

- Telecommunication installations are to be provided by the operators of basic and carrier services. Any other telecommunication installation requires the granting of a licence by MOPTMA which will be denied if there are basic or carrier services which provide an effective substitute for the special telecommunication installation foreseen by the applicant.
- Regarding services, the basic rules establish four types: basic services to be provided under monopoly by TELEFONICA; carrier services also provided under monopoly by TELEFONICA and other service providers, this being an intermediate service which could be considered as a factor of production as it does not reach end users but is necessary for the provision of basic, value-added or broadcasting services; value-added services (VAS) to be provided under competition; and broadcasting services which are under limited competition.
- Apparatus which is designed to operate within the spectrum of radio-electric frequencies or to be connected directly or indirectly to termination points of a public telecommunications network with the purpose of sending, processing or receiving signals is subject to the attainment of a certificate of homologation and acceptance of the relevant technical standards. Certificates issued by the corresponding authorities of Member States of the E.U. are valid in Spain. 28–13

Licensing Regime including Main Licence Categories

The Spanish licensing régime distinguishes between two types of licences: the so-called concessions (*concesiones*) and authorisations (*autorizaciones*). Concessions group together various different kinds of legal transactions governed by the common principle of giving a private party part of the scope of activities originally belonging to the State Administration. In comparison, authorisations are a mere declaration of will according to which the State Administration permits private parties the exercise of rights already belonging to them subject to the prior fulfilment of certain requirements; that is, the authorisation means that the private party has a pre-existing right or faculty.[18] 28–14

Basic services, carrier services, VAS which require their own networks, those provided through leased circuits and broadcasting services are subject to the concession régime. If there is no limit as to the number of concessions which may be granted, they will be awarded by order of priority of applica-

---

[18] The distinction between concessions and authorisations is important not only for the purposes of determining the régime of telecommunications service but also for tax purposes because the granting of concessions is subject to tax payment whilst granting of authorisations is not.

tions. In other cases, the general State Contracts Legislation, the corresponding technical regulations and the rules of tender to govern each particular service, have to be complied with. At present, several of these concessions have been granted by following the procedure for inviting tenders; as has been done for paging, trunking, automatic mobile telephony, private TV and satellite TV. Meanwhile, those granted for data services were awarded by order of priority of requests.

28–15   The remaining VAS only need an administrative authorisation. In the absence of an express authorisation, requests may be deemed granted on the expiry of four months since the date of entry of the request.

The authority in charge of granting concessions and authorisations is MOPTMA.

It should also be mentioned that under the rules governing foreign investments those telecommunication services requiring their own networks and basic carrier services are considered to belong to a sector attracting specific regulation for the purposes of the rules on foreign investments in Spain, (being considered an as activity directly related to national defence). The licensee of these services must be a Spanish national. If the licensee is a corporation, capital investment by foreign individuals and corporations is permitted only up to a limit of 25 per cent; this percentage limit can be increased upon authorisation by the Government. However, these limits do not apply to the E.U.'s nationals.[19]

3. Regulation of Services

Detailed Analysis of Rights to Provide Telecommunication Services: those Subject to Monopoly and those Open to Competition.

28–16   The right to provide telecommunication services depends on their classification according to LOT. Thus, it is first necessary to establish the type of service involved in order to determine its legal régime.

The fundamental distinguishing criteria between the different types of services is whether they may be considered essentially state owned telecommunication services or whether they fall outside this classification.

(i) Essential services

28–17   The following are considered essential services:

- Those designated by LOT as final services, which are equivalent to basic services.
- Those designated by LOT as bearer services which correspond closely to carrier services.

[19] Royal Decree 671 of July 2, 1992 on Foreign Investments in Spain and LOT.

- VAS which require their own networks and those which are provided through leased circuits.
- Broadcasting services.

The provision of the above may be made directly through State owned entities or indirectly through the granting of the corresponding administrative concession.

The current régime for the provision of these essential services is the following:

Final (or basic) services: They are defined as those that supply a capacity   **28–18**
for complete communication between users, including the functioning of terminal equipment, and which generally require switching elements.

Basic telephony, telex and telegrams are defined as final services and are provided under monopoly as follows: urban, inter-urban and international telephony, mobile maritime telephony, radio telegraphy, telefax and videotext are provided by TELEFONICA in accordance with its contract with the State.[20]

Telex and telegram services correspond to OACT.[21]

Carrier (or bearer) service: Those which supply the necessary capacity for   **28–19**
the transmission of signals between the public network termination points.

They may be provided either directly through public entities or indirectly by basic services providers. Therefore, they are provided as a monopoly or with limited competition as follows:

- The carrier service for basic telephony is provided under monopoly by TELEFONICA.
- The service of lease circuits is provided by TELEFONICA and according to recent announcements made by MOPTMA will also be provided shortly by RETEVISION.
- The carrier service for trunking is provided by both TELEFONICA and RETEVISION.
- The carrier service for VSAT is directly provided by RETEVISION and OACT and indirectly by TELEFONICA.[22]
- The carrier service for broadcasting TV is provided under monopoly by RETEVISION.[23]
- The carrier service for broadcasting radio corresponds to TELEFONICA and RETEVISION and to other broadcasting stations who have

---

[20] See n. 4.
[21] OACT is entitled to provide these services according to Art. 99 of the Budget Law 1991, Law 1990/2867, developed by Royal Decree 1766 of December 13, 1991.
[22] Royal Decree 764 of May 1, 1993 grants OACT, RETEVISION and TELEFONICA legal title for the provision of this service.
[23] This fulfils that stated in Arts. 14.5 and 25.5 of LOT. In compliance with these rules RETEVISION was granted legal title to provide this service through Royal Decree 545 of May 19, 1989. It should be noted that in the regions of Catalonia, Galicia and the Basque country the Autonomous Governments have built networks which can carry broadcasting services, which for the two first cases are of doubtful legality.

been authorised to build connections where there is no available public network.

28–20   Specific VAS: VAS are defined as those which while not being broadcasting services and although using carrier or basic telecommunication services offer additional facilities to that basic service or satisfy new telecommunications needs, such as access to stored information, the sending of information or the filing and retrieval of information. These are generally considered as not being essential services with the two following exceptions:

(i) VAS requiring their own network which can only be provided through the granting of an administrative concession; the concessionaire has to meet the same requirements as those applied to the providers of basic and carrier services.[24] Automatic mobile telephony for instance, which is now one of these VAS which requires its own networks was assimilated to a basic service provided under the monopoly of TELEFONICA until July 1994. At present it is provided by two operators, TELEFONICA and AIRTEL.[25]

(ii) Value-added services provided through leased circuits: Until May 1993 they were assimilated to carrier services provided under monopoly by TELEFONICA. Now they are provided under competition by TELEFONICA and by the new concessionaires which currently number eight.[26]

28–21   Broadcasting services: Defined as those in which communication is facilitated in one direction to various points of simultaneous reception. In these services a distinction has to be made between radio broadcasting services (only sound) which are provided under competition by the State companies RNE, SA and RCE, SA (both belonging to the State owned entity RTVE) and other private entities and TV broadcasting services (sound and image) which are provided under competition by the State company TVE, SA which also belongs to the State owned entity RTVE, by the autonomous governments and private entities.

(ii) Non-essential services

28–22   Non-essential services are provided under competition through the granting of an administrative authorisation. The LOT includes in this category the following:

(i) Those services provided within a private property and those provided in a building. The administrative authorisation is generally considered as granted so no application has to be made.

(ii) Those required by companies or entities providing public services

---

[24] See n. 19.
[25] Liberalisation was effected through Royal Decree 1486 of July 1, 1994 approving the technical regulation of automatic mobile telephony.
[26] Liberalisation was effected through Royal Decree 804 of May 28, 1993 approving the technical regulation of this service.

based in a physical and continued infrastructure such as the railway company RENFE.

(iii) VAS different from those mentioned above in paragraph 28–17.

Fixed and Mobile Services

Spanish Law does not regulate services by making a distinction between those which are fixed and those which are mobile. Therefore, the régime of fixed and mobile services has to be inferred from the above mentioned rules, that is, whether a fixed or mobile service can be classified as basic, carrier, VAS or broadcasting service. In accordance with this, they are provided as follows: 28–23

(i) Fixed services

- Basic telephony is a basic service provided under monopoly by TELE-   28–24
FONICA. It is expected that liberalisation of the industry will take place by 1998.
- Telegraph and telex, both being basic services, are provided under the monopoly of OACT.
- Electronic mail is a VAS provided under competition by TELEFON-ICA and other operators.
- Data services are a VAS also provided under competition by TELE-FONICA and other operators.
- TV is a broadcasting service provided under competition by TVE and other private and State owned entities.
- Radio-broadcasting is a broadcasting service provided under competition by the State companies RNE, RCE and other private entities.
- Lease of circuits is a carrier service currently provided under a monopoly by TELEFONICA. In the near future RETEVISION will also offer a leased circuit facility.
- Carrier broadcasting TV services are provided under a monopoly by RETEVISION.
- Carrier broadcasting radio services are provided as a monopoly by TELEFONICA, RETEVISION and other private entities.
- VSAT are carrier services provided by TELEFONICA, RETEVISION and OACT.
- Cable communications are not yet regulated, although there is a projected Law (which is at the draft stage only) that will permit limited competition between TELEFONICA and another concessionaries in certain defined areas.

(ii) Mobile services

- Automatic mobile telephony is a VAS provided under competition by   28–25
TELEFONICA and AIRTEL.
- Trunking is a VAS provided under competition by TELEFONICA and various private entities.

- Paging is a VAS provided under competition by various private entities.

Satellite Services (Non-broadcasting)

28–26   The LOT foresees the use of the radioelectric public domain through the use of satellites, which shall be subject to International Law. However, for the part corresponding to Spanish sovereignty, such use is either directly or indirectly reserved for the State.

On the April 7, 1989 the Government approved the HISPASAT programme authorising the incorporation of the Company HISPASAT, SA to contract satellites and operate the system.

HISPASAT is owned by RETEVISION (25 per cent), TELEFONICA (25 per cent) and other State owned companies. It has two satellites and provides a direct broadcasting service (DBS) for five TV channels and a fixed service (FSS) for TV and data. Amongst the latter, it is used for five satellite TV channels, two of them reserved to the State TV company TVE, SA the others being awarded to private TV companies as explained below.

Regarding VSAT services, they are at present provided in a very limited way.

Radio Communications

28–27   Radio communications can be divided between broadcasting and non-broadcasting.

(i) Radio broadcasting services

28–28   Radio broadcasting services are governed by LOT, the Statute on Radio and TV[27] and the rules enacted for their implementation. They are considered as an essential service to be provided by the State.

The provision of this service either as a monopoly or competitively varies depending on the different modalities for such a provision within the frequency channels. Those provided under monopoly correspond to RNE and RCE, both of which form part of the public entity RTVE.

(ii) Non-broadcasting radio communications.

28–29   Services which can be considered non-broadcasting radio communications are those which are not provided through the use of a fixed network nor with the intervention of a satellite. In this sense, mobile communications such as trunking or maritime communications, paging and automatic mobile telephony, belong to this category of radio communications. The provisions regulating such services have already been discussed.

[27] See n. 10.

4. BROADCASTING

Basic Legislative Rules

The basic rule on broadcasting is again LOT (Articles 25, 26—5th and 6th   **28–30**
additional provisions) as developed by the Statute on radio and TV.[28]

According to these provisions, broadcasting is an essential service to be
provided by the State either directly or indirectly granting the corresponding licence. At present, broadcasting services are provided under competition by the State owned company RTVE which includes RNE, SA and
RCE, SA for radio and TVE, SA for TV. Carrier services for broadcasting
are as follows: Those for TV are under the monopoly of RETEVISION
whilst those for radio are provide by RETEVISION, TELEFONICA and
others in competition.

Terrestrial and other Delivery

Under this heading, references are only made to broadcasting TV.   **28–31**

(i) Terrestrial delivery

Until 1983 TV broadcasting through terrestrial delivery was provided   **28–32**
under monopoly by TVE, who operated two channels. However during
that year the Government was authorised to establish a third State owned
channel to be granted under a concession within the territorial scope of
each autonomous community.[29] In 1988 the indirect provision of the service by private entities was foreseen, thus ending the State monopoly.[30]
Public tenders were invited for the award of three private concessions
which were granted to GESTEVISION TELE 5, SA, ANTENA DE TV SA,
and SOCIEDAD DE TV CANAL PLUS, SA In order to be awarded a
licence an applicant must comply with the following conditions: shares can
only be held in one concessionaire, 25 per cent of the share capital is the
maximum that may be owned and foreign share capital must not exceed
25 per cent of the total.[31]

(ii) Satellite TV

Satellite TV became a reality in Spain at the end of 1992 with the enactment   **28–33**
of the Satellite TV Law.[32] The direct provision of this service lies in the

---

[28] See nn. 10 and 27.
[29] Law 46 of December 26, 1993 regulating the third channel.
[30] Indirect provisions of TV was permitted through Law 19 of May 3, 1988 on Private TV.
[31] It should also be noted that there is a Draft Law on local TV through terrestrial delivery currently being
discussed in Parliament. Its territorial scope shall be defined by the main urban center of the corresponding municipality. One licence per territory will be granted and it will be operated by the municipality
through commercial corporations or indirectly giving a concession.
[32] Law 35 of December 22, 1992 developed by Royal Decree 409 of March 18, 1993 approving the
Technical Regulation and for the provision of the service of satellite TV and of its carrier service.

hands of RTVE and indirect provision was also made by inviting public tenders for the award of three concessions which were subsequently granted to the private TV operators, GESTEVISION TELE 5, SA, ANTENA 3 DE TV SA, and SOCIEDAD DE TV CANAL PLUS, SA.[33] It should be noted that the same limits mentioned above to obtain a private territorial TV licence apply to satellite TV operators. There has been an announcement by the Government that satellite TV will be further liberalised.

(iii) Cable TV

28–34   Although not yet approved, there is a Draft Law presently being discussed in Parliament which will apply to cable telecommunications in general and will not be limited to just TV. The main rules of this Draft are the following:

- The service shall be provided per territorial areas which must include at least 20,000 inhabitants.
- The service will be provided by operators under an indirect concession granted by MOPTMA which shall invite public tenders for that concession.
- One licence shall be granted per area and TELEFONICA will also be entitled to provide the service when it so requests.
- Foreign participation in the concessionaire is not limited to E.U. capital; non E.U. capital contributions are however limited to 25 per cent, unless otherwise authorised by the Government.
- 40 per cent of the programmes have to be reserved to independent programming companies.
- There is a compulsory service to be provided, consisting in the distribution of the two State TV channels, the three private TV channels, the autonomous communities channel and those of local TV already in existence.
- Lastly and most important, it is foreseen that the licence for a cable operator includes the permission to provide any type of carrier service within its area and to provide VAS.

Regulatory Agencies

28–35   The regulatory agencies entrusted with broadcasting services are the same as for other telecommunication services, that is, basically MOPTMA, as described in Section 2 above.

Regulation of Transmission v. Content

(i) Regulation of transmission

28–36   **TV transmission:** The regulation of TV broadcasting transmission is mainly contained in LOT, which was developed when the Law for the regulation

[33] Resolutions of May 6, 1993 and October 7, 1993.

of Private TV was approved;[34] it foresaw the approval of a technical plan to cover the regulation of technical matters necessary to ensure adequate provision of the service including carrier and broadcasting systems of signals, bands, channels and frequencies reserved for the emission of programmes of the licensees of private channels. Such a technical plan governs the conditions of a technical nature necessary to ensure the adequate provision of the service and in particular the carrier and broadcasting signal systems.[35]

With regard to satellite TV, the corresponding law[36] only contains provisions relating to the obligations of operators in contracting for space capacity, terrestrials stations and up links with the entity which is licensed to provide the carrier TV broadcasting service under LOT. As explained above this licence was granted to RETEVISION.

Radio broadcasting: Regarding radio broadcasting, the rules of the LOT   **28–37** regarding transmission have also been developed through the enactment of detailed technical plans relating to different frequencies used in radio broadcasting.[37]

(ii) Regulation of content

The fundamental rules regarding the content of broadcasting in radio and   **28–38** TV are those contained in the Spanish Constitution which recognises and protects the rights of freedom of expression and of freedom to communicate or receive true information by any communication means. These freedoms are themselves limited by the obligation to respect the right to privacy, integrity and by the need to protect children and young persons.[38]

The basic legislative provisions relating to TV and radio provide that they are considered as an essential vehicle for information and for the political participation of individuals as well as of education and diffusion of Spanish culture and as a means to contribute to reality and efficiency of freedoms and equality. A significant element of the TV Statute is that it provides for the access to radio and TV spaces by the most significant social and political groups as well as providing for the limitation and control of publicity.

In all cases, the activities of the communications media must be guided by the following principles: impartiality and veracity of the information; separation between information and opinions; respect for political, religious, social, cultural and linguistic pluralism; respect for integrity, fame, the private life of individuals and for all rights and freedoms recognised by the Constitution.

---

[34] See n. 30.
[35] The National Technical Plan was approved through Royal Decree 1362 of November 11, 1988.
[36] See n. 32.
[37] Such Technical Plans are the following: the National Technical Plan of Sound Broadcasting (Royal Decree 2648 of October 27, 1978), the National Technical Plan of Sound Broadcasting in Medium Waves (Royal Decree 765 of May 21, 1993), and the National Technical Plan of Sound Broadcasting in Metrical Waves with Modulated Frequency (Royal Decree 169 of February 10, 1989).
[38] Art. 20 of the Spanish Constitution.

28–39   The legislation that contains more detailed rules concerning content is the Private TV Law.[39] As one of its purposes, private TV may be used to broaden the possibilities of information pluralism in Spain. This Law was amended taking into account the implementation in Spain of EEC Directive 89/552 on the coordination of legal rules regarding TV broadcasting activities,[40] the basic principle being the free reception within Spanish territory of TV broadcasts originating in another Member State of the E.U. As far as programming is concerned it establishes a reservation of 51 per cent of the yearly time of emission for the diffusion of European works; half of such time must be dedicated to the emission of European works in the original Spanish language. Moreover, 10 per cent of that time is reserved to the emission of European works of independent producers which have been produced in the last five years. There are also rules which limit the time dedicated to advertising.

28–40   The Law on Satellite TV[41] has only one rule relating to its content; that the programming for satellite TV must be different in its content to that of the remaining TV services.

According to the Draft Law on Cable Communications, operators must reserve a minimum 40 per cent of the audiovisual programming distributed through its network to independent programmers unless there is insufficient material of a suitable quality.

Lastly, insofar as it applies to any communication means, special attention should be paid to the Law on publicity which implements EEC Directive 84/50 of September 10, 1984 regarding misleading publicity.[42]

Monopoly and Competitive Service

(i) TV Broadcasting

28–41   In general terms, the service is provided under competition. However, as LOT does not contain the specific regulation of the legal regime corresponding to each of the TV modalities, the regulations applicable to each are detailed below.

28–42   **Terrestrial TV:** The broadcasting service is actually provided under competition by the State company TVE, SA, the autonomous communities and private entities.

28–43   **Satellite TV:** Satellite TV broadcasting service is provided under competition by the State company TVE, SA and three private entities awarded with concessions, although this will be broadened shortly.

---

[39] See n. 30.
[40] This Directive was incorporated into Spanish Law by Law 25 of July 12, 1994.
[41] See n. 32.
[42] Law 34 of November 11, 1988.

Cable TV: According to the Draft Law it will be provided under limited   **28–44**
competition by TELEFONICA and one private entity per zone.

(ii) Radio broadcasting

Whether or not competition is to be permitted in the provision of this   **28–45**
service depends on the frequency which is being used.

- Broadcasting radio through long and short waves is provided under
  monopoly by the State companies RNE, SA and RCE, SA respectively.
- Broadcasting radio through medium waves and modulated frequency
  is provided under competition by RNE, SA, RCE, SA and other State
  owned and private entities.

5. REGULATION OF APPARATUS

Regulation of Apparatus Supply and Installation

The basic regulation of apparatus supply and installation is contained in   **28–46**
LOT, which throughout its contents, and in relation to the different services
which it regulates refers expressly to terminal equipment which can be connected to the network termination points.[43]
   In this sense, LOT sets out the principle of the liberalisation of apparatus
and their free connection to the networks, on the following basis:

   (i) The corresponding technical regulations for the provision of services
       of basic and carrier telecommunication services should define the
       termination points of the network of the services where the appar-
       atus are to be connected. In the same way, entities exploiting VAS,
       are obliged to specify the termination points of the network for the
       final and carrier services used by them.
  (ii) The provision of basic and carrier services is subject to the principle
       of allowing access of all terminal equipment which could be legally
       connected to the same.
 (iii) The terminal equipment susceptible to connection to the termination   **28–47**
       points of the network may be freely acquired from the operator or
       from another entity.
  (iv) Specific regulations establish technical characteristics and operating
       conditions for equipment, apparatus and stations which use the
       spectrum of radioelectric frequencies and requirements for their
       licence holders.
   (v) However, LOT imposes the obligation to observe a series of meas-
       ures whose purpose is to guarantee that the use of the terminal
       equipment on a free market basis does not affect the security of

[43] Terminal is defined by LOT as all equipment or apparatus which transmits and receives signals via a
telecommunication network by means of network termination points defined and in accordance with
the approved specification.

users nor disturb the normal functioning of the telecommunication networks.

28–48   The general rule is that all equipment, apparatus and systems are subject to the previous acquisition of the corresponding certificates of homologation and acceptance of the relevant technical standards. The LOT lays down the procedure to homologate and accept terminal equipment and apparatus, which was further developed through a Regulation on this matter (the Regulation on terminal equipment).[44]

According to the above, the importation, manufacturing and commercialisation of terminal equipment is liberalised, providing that importer, manufacturer and distributor holds the necesary certification showing compliance with the relevant technical standards.

Finally, it should be pointed out that breach of the above principles under LOT is considered a serious or very serious violation and will therefore attract a fine of between 1 million and 10 million pesetas accordingly.

Type Approval and Testing Procedures

28–49   The LOT and the Regulation on terminal equipment contain the general provisions in relation to these points.

The Government must define and approve the technical standards in order to guarantee the efficient functioning of the telecommunication services and networks, as well as the adequate use of the radio-electrical spectrum in relation to equipment, apparatus, devices and systems.

All apparatus, equipment, devices and systems require an administrative resolution which certifies compliance with the relevant technical standards, called the "acceptance certificate" (*certificado de aceptación*). The application for this must be presented before MOPTMA by the individual or entity who wishes to obtain it for a type or model of apparatus, equipment, device or system of telecommunication.

Once the application has been submitted together with a technical report, MOPTMA will notify the interested party of the manner and testing methods to be used in order to check compliance with the applicable technical standards. The laboratory chosen will issue a technical report with the results of the tests. If this technical report is favourable, MOPTMA will issue an acceptance certificate, forwarding the same to the applicant and publishing the resolution in the Official State Bulletin. The resolution will specify the period of validity of the acceptance certificate, which can be renewed in successive periods upon request of the interested party.

28–50   The centres authorised to conduct such tests in order to check technical standards are the following:

- Laboratories of the General-Directorate of Telecommunications.

---

[44] Royal Decree 1066 of August 28, 1989 approving the Regulation of LOT regarding equipment, apparatus and systems.

- Authorised laboratories according to the procedure established by the Regulation on terminal equipment.
- Laboratories owned by the manufacturers or importers or used by them with the previous authorisation of the Telecommunications Authorities.

Regardless of the above, the certificates of acceptance or alternative pro-   28–51
cedures in compliance with the standardised norm published in the Official Journal of the European Community, issued by the competent bodies designated by Member States in accordance with E.U. legislation, have the same value as the acceptance certificate.

Terminal equipment from other E.U. Member States which conforms to national regulations having standards for such equipment which guarantee an equivalent level of the technical attainment as that established in the Spanish technical standards, will be considered to be in compliance with the enforced regulation, although they must obtain a certificate from the Spanish Administration acknowledging compliance with the essential requirements. In the same way, tests carried out in authorised laboratories of E.U. Member States which conform to the standards established for authorised laboratories in Spain, and with prior obtainment of a certificate from the Spanish Administration, will be valid in Spain.

Rights to Market

The importation, manufacture in series, sale or exposure for sale of any   28–52
apparatus, equipment, device or system is subject to the prior award of homologation and acceptance certificates of technical standards.

Independent of those sanctions set forth in LOT for cases of very serious or serious violations, the Regulation on terminal equipment establishes an additional series of obligations and responsibilities for importers, manufacturers and distributors.

Firstly, the Administration can effect controls of apparatus, equipment, devices and systems in order to verify that these continue to possess all the technical characteristics which were the basis for the granting of the acceptance certificate: if the Administration detects that certain apparatus are imported, manufactured, sold or replaced in Spain without having these characteristics, they will be considered as not covered by the acceptance certificate.

The importer, manufacturer or commercialiser should fix in an indelible   28–53
and visible manner the label or sign verifying the grant of an acceptance certificate to all those apparatus which correspond to the types or models which are covered by such certificates.

In the same way, all apparatus should have fixed to them in an indelible and visible manner the Registry number of the importer, manufacturer or commercialiser which will be granted by the Telecommunications Administration, upon prior request by the interested parties.

All commercial apparatus must be accompanied with one complete copy of the acceptance certificate together with the guarantee and user's manual.

The manufacturer, importer and, if appropriate, the distributor, must notify the Administration in writing of any modification to be introduced to a particular type or model of apparatus; if the modification affects the regulatory technical characteristics, the pertinent tests will be carried out and a new certificate will be issued.

6. TECHNICAL STANDARDS

Apparatus

28–54   As stated in the previous section, the Government upon the recommendation of MOPTMA, will approve the technical standards in order to guarantee the efficient functioning of telecommunications services and networks as well as the adequate use of the radio-electric spectrum, in relation to equipment, apparatus, devices and systems.

The Regulation on terminal equipment, in its transitional provision, establishes that the technical standards approved before the entry into force of the Regulation will be valid until the new technical standards are approved.

The Government, as of 1989, has approved the technical standards concerning additional telephone terminals used in the basic telephone service, modems for the switched telephone network and packaged switched data, amongst others.

The technical standards approved are as follows:

Royal Decree 1376, of October 27, 1989: Technical Standards of additional telephone terminal equipment used in the basic telephone service.
Royal Decree, of December 1, 1989: Technical Standards of modems for switched telephone network.
Royal Decree 116, of January 26, 1990: Technical Standards for cordless telephones.
Royal Decree 1584, of November 30, 1990: Technical Standards for facsimile terminal equipment.
Royal Decree 720, of April 22, 1991: Technical Standards for terminal equipment.
Royal Decree 1649, of November 8, 1991: Technical Standards for the connection of data terminal equipment to packaged switched data public networks.
Royal Decree 570, of May 29, 1992: Technical Standards of radio-electric terminal equipment used in the automatic mobile telephony service, in 900 MHz band.
Royal Decree 569, of May 29, 1992: Technical Standards to be complied by terminal equipment used for teletex service.
Royal Decree 986, of July 31, 1992: Technical Standards to be complied by terminal equipment used in videotex/Ibertex service.

Royal Decree 1562, of December 18, 1992: Technical Standards for user multi-line systems destined to be used as terminal equipment.
Royal Decree 80, of January 22, 1993: Technical Standards for transmission equipment of radio broadcasting sound in metric waves with frequency modulation.
Royal Decree 81, of January 22, 1993: Technical Standards for the connection of data terminal equipment to the public data network "Iberpac".
Royal Decree 376, of March 12, 1993: Technical Standards of modems for leased analogue circuits of voice bands.

Networks

The infrastructure of telecommunications in Spain can be grouped in the   28–55
following manner:

- TELEFONICA's network, which is composed of a basic transmission network which is extended throughout the entire national territory, providing interconnecting circuits amongst the nodes of switched networks; its also provides the support of carrier services for leased circuits and has a switched network which provides physical support for voice telephone services and for the range of services using voice channels requiring technical switching circuits. Moreover, TELEFONICA counts with a network for special data transmissions (called the IBERPAC network) and a network for automatic telephone mail. It is also important to point out TELEFONICA's activity in satellite communications, in which it participates through its interest in the most important entities of this sector (INTELSAT, EUTELSAT, INMERSAT, HISPASAT).
- RETEVISION's network, which consists of a broadcasting network for television and radio and a transport network for distribution, contribution and interchange of television and radio signals. RETEVISION also develops activities in satellite communications.
- OACT NETWORKS, which is composed of a network for basic transmission and two switched networks for telexes.

ONP and Similar Open Architecture/Access Principles

ONP-leased lines Directive 92/44 is still partially pending implementation   28–56
into Spanish law.
   MOPTMA has recently adopted a so called "Agreement on the Telecommunications Policy for the period 1994 to 1998"[45] where it foresees that the enhancement of VAS will be carried out by third parties through leased circuits by adopting the following measures:

- The continuation of the rebalancing of the tariffs of circuit services

[45] See n. 2.

of TELEFONICA, with a relevant decrease in prices, especially for circuits of higher capacity and lesser length.

- The lease of circuits in Spain is more expensive than in other European countries due to the fact that it is these revenues which are used to make up the deficit caused by the provision of the basic telephone service for the metropolitan area. The difficulties in reducing the tariffs are of a political nature, since it would be necessary to compensate the income reduction, and even more, the profits, at TELEFONICA's cost.

28–57
- As from January 1, 1995, resale of excess capacity of leased circuits should be authorised. This objective has just been fulfilled since February 23, 1995. Resale of capacity is now classified as a VAS defined in Section 3 above.

- To increase the offer of leased circuits as a carrier service for the data services already liberalised granting an administrative concession to RETEVISION for this purpose. In order to provide this service, RETE-VISION may use its own network or the surplus capacity of the OACT network.

CHAPTER 29

Sweden

By Olof Alffram and Susanne Themptander,
Advokatfirman Tisell & Co AB, Stockholm

1. HISTORY OF DEVELOPMENT OF TELECOMMUNICATIONS IN SWEDEN

Industry Structure

At the beginning of the century the networks in the larger Swedish cities   **29–01**
were controlled by a number of independent entities. In Stockholm there
were two networks competing to the extent that any interconnection from
time to time was made impossible. At that time some efforts were made
to introduce monopoly legislation but in the end no such regulations were
passed.

In 1918 the Telegrafverket, later Televerket, *i.e.* Swedish Telecom,
acquired the competing network in Stockholm and thereby became the sole
operator in Stockholm. This was the first step in the process of creating
the *de facto* monopoly that Swedish Telecom has enjoyed until recently.
Swedish Telecom in effect became the only operator able to offer access to
a public network. However, parts of that network, particularly in rural
areas, were owned by others for some considerable time. Not until the
1950s had Swedish Telecom acquired the entire transmission system for
telecommunications services in Sweden, including lines, switches and other
equipment.

In spite of Swedish Telecom's dominating position, Swedish Telecom has   **29–02**
never had any legally founded exclusive right to provide telecommunications services and no statutory monopoly in this field has been in force in
Sweden. Nor, until recently has there existed any licensing regime. However, as indicated above, Swedish Telecom has had a *de facto* monopoly,
the market being regulated by means of Swedish Telecom's exclusive right
to connect equipment to its network.

Beginning with a Government Bill submitted in 1980, the Government has during the subsequent decade made a number of decisions in order to facilitate competition in the telecommunications market. For instance, Swedish Telecom became obligated to keep separate accounts for parts of its operations being exposed to competition in order to prevent cross-subsidisation. Furthermore a Telecommunication Interconnect Board was formed to which Swedish Telecom's regulatory decisions regarding certain connection matters could be appealed. The connection monopoly was gradually abolished. In 1984, the connection of private telex terminals and certain computers to the telex network was permitted. Swedish Telecom's monopoly on connecting telephone sets was terminated in 1985. In 1988, the monopoly on connection of voice-band, high speed modems to the network was discontinued; the monopoly on pay telephones was abolished on January 1, 1989. As a last step, Swedish Telecom's monopoly on connection of PABX's ceased as per July 1, 1989. During this period there were also significant developments towards liberalisation of third party traffic, *i.e.* resale.

29–03   As a result of these developments towards liberalisation, consumers have been offered a growing range of equipment as well as telecommunication services. Both Swedish Telecom and other operators during the 1980s began offering services such as videotex, paging, electronic mail, telefax, packet switching and mobile telephony. A number of operators also established their own data networks, normally based on leased lines.

Presently, an increasing number of private operators are offering telecommunications services in competition with Telia (the former Swedish Telecom). This is accomplished by interconnection with Telia's switched network, by building their own infrastructure, by leasing lines from Telia and by means of leased capacity from Telia as well as from the National Rail Administration (Banverket) and others. Regarding equipment, there is a large number of suppliers present on the market.

Institutional Structure

29–04   Swedish Telecom, like most PTT's throughout Europe, has over the years had a dual role on the market, regulatory as well as operational. However, a large part of Swedish Telecom's regulatory powers have in recent years been transferred to the National Post and Telecom Agency, the PTA, formed on July 1, 1992. Through the implementation of the Swedish Telecommunications Act introduced on July 1, 1993, the PTA has assumed the remainder of Swedish Telecom's regulatory powers. At the same time Swedish Telecom was incorporated into Telia AB to become one of several competing operators in the Swedish telecommunications market. At present, all shares in Telia are owned by the State.

Background to Legislation

29–05   As mentioned above no statutory monopoly has been in force in Sweden. Neither has there been any licensing regime for telecommunications services and the provision of such services has not been governed by any specific telecommunications legislation. Historically, Governmental policies have instead been implemented through the activities of Swedish Telecom. Beyond the various instructions governing the operations of Swedish Telecom, the Swedish telecommunications market including development of infrastructure, at least from a legal point of view, essentially was liberalised in the beginning of the nineties. However, in view of the extremely rapid technical and commercial developments in the field of telecommunications, the Government announced its desire to create a regulatory system in order to guarantee fulfilment of its telecommunications policies and to ensure that efficient competition in all areas of telecommunications was maintained and promoted.

In 1990 a committee, the Commission on Frequency Regulations, was appointed for reviewing the legislation regarding allocation of radio frequencies for all applications of radio, including a reorganisation of the exercise of regulatory powers in this field. The committee submitted its report in 1991 (SOU 1991:107) and suggested for instance that the tasks of the Frequency Management within Swedish Telecom would be assumed by an independent Governmental agency, to become the PTA.

In the middle of 1990 the Swedish Parliament in view of the new situation in the telecommunications market decided that the form of association of Swedish Telecom was to be changed. In May 1991 the Government appointed a commission for the purpose of proposing a telecommunications legislation and a transfer of the rest of Swedish Telecom's authoritative powers to the PTA. The commission submitted its report (SOU 1992:70) in July 1992.   **29–06**

The above mentioned reports resulted in a Government Bill (Bill 1992/93:200) in March 1993 proposing a Telecommunications Act, a new Radio Communication Act, and finally a proposal to transfer Swedish Telecom into a limited company. Basically, the Bill proposed a new regime introducing a licensing requirement for anyone providing telephony services to fixed points of connection, mobile telecommunications services or leased circuits in a public telecommunications network. The presumption would be that a licence should be granted unless the applicant obviously would not be capable of pursuing its activities on a permanent basis, with adequate capacity and quality.

2. Existing Legislative and Regulatory Environment

Principal Legislative Instruments

The basic legislative instrument is the Telecommunications Act (SFS 1993:597), referred to below as the 1993 Act. The 1993 Act entered into force on July 1, 1993, introducing a licensing system for certain telecommunications services. The 1993 Act provides a legal framework for telecommunications on an open market. The object of the Act is, as expressed by the   **29–07**

Government, to facilitate the development of a highly competitive and efficient market by clear and stable legislation ensuring that private individuals, legal entities and public authorities shall have access to efficient telecommunications at the lowest possible cost. This implies, *inter alia*, that anyone at his or her permanent place of residence or regular business location shall have access to telephony services within a public telecommunications network. The 1993 Act has been modelled to comply with E.C. legislation and directives in the field of telecommunications. It introduces a licensing regime requiring licences for certain telecommunication services within a public telecommunications network, *i.e.* a network available to the public.

Any telecommunications activities including the use of radio transmitters also require a separate licence under the Radio Communication Act (SFS 1993:599). This Act is applicable to any possession or use of radio transmitters irrespective of the purpose thereof.

Sweden has, and as a consequence of the EEA Treaty, fully implemented the Terminal Equipment Directive (91/26) through the Swedish Act on Terminal Equipment.

Beside the legislation mentioned above, a number of minor legislative measures have been taken in connection therewith.

Agencies of Regulation: Competition Control

29–08   The regulatory authority responsible for implementing and supervising the Telecommunications Act 1993, the Radio Communication Act 1993, as well as the Act on Terminal Equipment, is as mentioned above the National Post and Telecom Agency, the PTA. The PTA has been granted a number of powers to perform its functions, for instance the authority to issue various instructions for the purpose of ensuring compliance with the telecommunications legislation as well as the authority to set forth terms and regulations to be issued by virtue of the law. The PTA is the agency responsible for issuing the licences required under the 1993 Act and under the Radio Communication Act 1993. All decisions made by the PTA may be appealed to the Administrative Court of Appeal.

As far as competition legislation is concerned, Sweden has recently adopted a new Competition Act (SFS 1993:20) that entered into force on July 1, 1993. The Competition Act 1993 is modelled on the competition rules of the European Community and the corresponding rules of the EEA Treaty. Sweden consequently has two parallel sets of rules regulating distortion of competition: one which is applicable when the distortion has effect within the European Community or an EFTA country and another which is applicable when the distortion affects the Swedish market. Should both systems be applicable simultaneously any decision under the Competition Act 1993 should conform to the E.C. rules.

The Competition Act 1993 lays down two general prohibitions; prohibition against anti-competitive co-operation and prohibition against abuse of dominant position, corresponding to the prohibitions set forth in Articles 85 and 86 of the Rome Treaty. The Act further contains provisions regarding merger control.

According to the Competition Act 1993 undertakings may notify agree-   **29–09**
ments with the Swedish Competition Authority for individual exemption
from the prohibition against anti-competitive co-operations. As a general
rule anti-competitive charges can not be levied for the period between the
notification and the decision by the Competition Authority (if any), provided that the actions of the undertaking correspond to what has been
described in the notification. Further the Competition Authority must make
a decision within three months from the notification having formally been
filed. If the Competition Authority has not taken a decision within the time
limit an exemption is considered granted for a period of five years as of
the date of the notified agreement.

The Government has issued notices on agreements of minor importance
and block exemptions for certain categories of agreements in line with what
applies within the E.C. With respect to the notice on agreements of minor
importance it can be noted that the thresholds are different from those in
the corresponding rules of the E.C.

Agreements that are prohibited pursuant to the Competition Act 1993
are void. Upon application by an undertaking concerned, the Competition
Authority may certify that an agreement or concerted practise is not prohibited under the Act (negative clearance).

The Competition Authority may order an undertaking to terminate an   **29–10**
infringement through a prohibition under penalty of a fine. An undertaking
may also be ordered to pay damages to an aggrieved party if it infringes
the prohibition against anti-competitive co-operation or abuse of dominant
position. Finally, the Stockholm City Court may, upon application of the
Competition Authority, impose anti-competitive charges which are to be
fixed according to the gravity and duration of the infringement. The maximum amount is SEK 5 million or an amount in excess thereof but not
exceeding 10 per cent of the turnover of each of the undertakings participating in the infringement.

The Competition Act 1993 is expected to have a very significant impact
on the telecommunications market. One may assume that this already has
been the case in connection with the current negotiations between Telia and
various operators regarding prices and other terms of the interconnection
agreements to be entered into.

General Rules Governing Telecommunication Installations, Services and Apparatus

As indicated above, the Telecommunications Act 1993 introduces a licensing   **29–11**
regime for certain telecommunications services within a public telecommunications network, *i.e.* a network made available to the public. The services
requiring a licence are telephony services to fixed points of connection,
mobile services and the provision of leased circuits provided, however, that
the activity in question is of such magnitude that it can be considered significant for maintaining efficient telecommunications and competition in

Sweden. The PTA is on request obligated to issue an advance decision as to whether a certain activity requires a licence. A licence may be subject to certain conditions established by the PTA ensuring that telecommunications policy objectives are met, *inter alia*, an obligation of the licensee to submit annual accounts regarding its activities under the licence. According to the 1993 Act a licence will be granted unless the applicant is obviously not able to pursue the activity for which the licence has been applied on a permanent basis and with adequate capacity and quality. Licensees are required to offer interconnection to other licensees at prices fair and reasonable in relation to costs in accordance with the ONP regulations. The 1993 Act empowers the Government with the authority to ensure that tariffs for telephony services to fixed points of connection do not exceed certain levels. For the period 1993 to 1996 a price cap is put on Telia in order to protect consumers in market segments with little or no competition. It has been decided that during this period, no other operator will be subject to any price regulations. The price cap covers basic services for households and smaller undertakings and is defined as Retail Price Index minus 1 per cent. The price cap is supervised by the PTA.

29–12   A licence is effective until further notice even though the licensing conditions that may be set forth by the PTA shall apply for a specific period of time. However, a licence to provide mobile telecommunications services may be limited in time.

As mentioned above, a licence under the 1993 Act does not eliminate the need for a licence under the Radio Communication Act 1993 in case the service contemplated includes the use of radio. Subject to suitable frequencies being available the presumption is always that an application under the Radio Communication Act 1993 shall be granted. Unlike licences under the Telecommunications Act 1993, licences for radio always apply for a fixed period of time. During the period of validity the licence cannot be revoked unless exceptional circumstances occur.

General Rules Governing Telecommunication Installations, Services, Apparatus and Licensing regime

29–13   Regarding terminal equipment, any device can be connected to a public network provided that it has been approved in accordance with the Act on Terminal Equipment. Unless certain equipment does not comply with standards adopted within the E.C., it shall be subject to a specific approval from the PTA.

3. REGULATION OF SERVICES

Analysis of Rights to Provide Telecommunication Services

29–14   The 1993 Act is applicable to any "telecommunications activities". The term "telecommunications activities" is defined as "conveyance of telecommunication messages via a telecommunications network or the provision of leased lines for such activities". A "telecommunication message" is in its turn defined as "sound, text, pictures, data or other information conveyed by means of radio transmission or light emission or electromagnetic oscillations utilizing especially devised conductor(s)". Consequently the 1993 Act also applies on mobile telephony.

Sections 2 and 3 of the 1993 Act contain a short description of the purpose of the legislation, namely to ensure that anyone shall have access to efficient telecommunications at the lowest possible cost, implying, *inter alia*, that any one shall be enabled to use, at his or her permanent place of residence or regular business location, telephony services within a public telecommunications network. The expression "telephony services" is defined as telecommunications services which consist of voice transmission and which permit transmission of telefax messages and data communication via low-speed modems.

In section 5, it is stipulated that a licence under the 1993 Act is required **29–15** in order to be entitled to provide within a public telecommunications network;

(a) telephony services (as defined above) or mobile telecommunications services (defined as telecom services with radio-aided connection of subscribers), if the activity is of an extent which is considerable with regard to the area covered, the number of users or other comparable circumstances; or
(b) leased lines offered by the owner of a network or by any other party in command of capacity in a network supplying such leased lines to a considerable extent.

A licence may relate to a specific geographical area or to the entire country.

Regarding the interpretation of the term "public", the preparatory works refer to whatever practice that may be adopted within the E.C. in the application of the ONP Directive and the Terminal Equipment Directive as well as any future national practice. It is further stated that a significant characteristic of a public network is that it is open to an unspecified number of users. The fact that an operator actively markets its services and offers connections on more or less standard terms will normally mean that the network in question will be considered public. Furthermore, the fact that an operator offers services to anyone within a specific geographic area or for customers for instance belonging to the same trade will normally also result in its network being considered public. Consequently, it is considered sufficient that the services are offered to anyone belonging to a particular category of users. Finally, the fact that a certain network interconnects with the network of Telia will no doubt have a significant impact on the judgement.

If a certain operation is insignificant from a general standpoint it is not **29–16** subject to a licence requirement: for example, operations confined to small geographic areas or to narrow fields of use. Depending on the market definition it has been indicated that market shares of 10 to 15 per cent will normally be considered significant, whereas market shares below 5 per cent normally will not. However, it is stated that all relevant factors, such as the total turnover in various sectors of a certain operator, shall be taken into consideration.

In section 11 of the 1993 Act it is explicitly provided that a licence shall be granted unless the applicant is not able to pursue the activity for which the application is made on a permanent basis and with adequate capacity and quality. In the preparatory works it is stressed that the licence requirement is not founded on any objective to restrict the possibilities of conducting telecom services, but to provide an instrument that will make it possible to exercise a reasonable degree of control over the largest operators and thereby create a market with several operators. In the preparatory works it is also stressed that a licence shall not be refused unless it is more or less obvious that the applicant is not fit to perform the service in question under the terms set forth by the PTA or otherwise applicable. In this context it is also mentioned that prospective customers can be expected to regard the licence as a kind of quality authorisation. All in all, the room for denial of an application is very limited, particularly for smaller operators that will not be subject to any far-reaching licence conditions.

29–17   Upon request the PTA is obligated to issue advance decisions as to whether a licence, according to section 5 of the 1993 Act, is required for a specific telecommunications activity. Anyone not subject to a licensing requirement in principle is at liberty to carry on his operations without any particular restraints. However, the PTA may, if required by the public interest on monitoring the development of the telecommunications sector, obligate anyone who provides within a public network services not subject to mandatory licensing according to section 5 (*i.e.* services other than telephony services, mobile telecommunications services or fixed services as well as such services not having the significance indicated in section 5) to file reports with the PTA. On the other hand, the PTA may grant individual exemptions from the mandatory licensing requirement. In such case, this requirement may be replaced by an obligation to report.

Licence Conditions

29–18   Pursuant to section 13 of the 1993 Act, a licence may be subject to various conditions to be set forth in each case by the PTA. By virtue of the 1993 Act, the PTA is furthermore empowered to issue generally applicable regulations on licence conditions. Until now, no such regulations have been issued, but it will be indicated in the licence certificates that the operator will be bound by any such future general regulations.

Section 13 contains the following examples of obligations that may be imposed on the operator:

(i) to provide on certain conditions telephony services to anyone asking for such services;

(ii) to provide, in consideration of available capacity and on certain conditions, leased lines to anyone asking for them;

(iii) to pursue the activity on the preconditions entailed by the international agreements ratified by Sweden;

(iv) to pay consideration, in the pursuance of the activity, to disabled persons' needs for special telecommunications services;

(v) to contribute to enabling telecommunications messages to be conveyed to public emergency services; **29–19**

(vi) to recognise the needs of the national defence system for telecommunications in situations of intensified military alert;

(vii) to submit annual accounts of the licensee's activities by virtue of the licence, in application of principles especially adapted to the activities and to make such accounts available to the licensing authority or any responsible party appointed by said authority;

(viii) to provide on reasonable conditions to any other licensee such published information on the subscriptions of private individuals to the extent that such information is not subject to any obligation of secrecy under the provisions of section 25 of the 1993 Act; and

(ix) to publish on reasonable conditions its own telephone directory, and make available to the public, information about the subscriptions of private individuals and legal entities to other licensees, to the extent that such information is not subject to any legal obligation of secrecy.

Regarding paragraph (i) above, it is stated in the preparatory works that **29–20** it is obvious that for the time being only Telia could be considered for this kind of obligation. It is, however, noted that the issue may become vital if one or several operators come to exploit low cost areas to the disadvantage of Telia and possibly others. It is on the other hand indicated as an alternative possibility that the Government may enter into a separate agreement with one or several operators whereby the operators for a fixed fee would undertake to connect subscribers throughout the country.

Regarding paragraph (ii), it is emphasised in the preparatory works that any obligation to offer leased lines may not impose on the operator's own needs.

Concerning paragraph (vii), the rules mentioned below regarding cost oriented tariffs for telephony services necessitate an opportunity to issue regulations on the organisation of the accounts of the licensee, in order to make it possible to excercise efficient control.

Licence conditions shall always be limited to a specific period of time. **29–21** Changes in licence conditions during such a period shall be subject either to certain reservations having been made in the licence conditions or to the consent of the licensee and after consultations with other licensees whose activities may be affected.

To this date (September, 1994) seven applications, including Telia's, have been filed with the PTA. These applications are still being processed but the first licence is expected in the very near future. The licence conditions will of course be publicly available. Regarding the licence conditions one may safely assume that smaller and less significant operators will not be subject to any particularly burdensome demands, whilst the opposite will apply to Telia, whose conditions no doubt will be very detailed and far-reaching.

Tariff Controls

29–22   Pursuant to section 18 of the 1993 Act, tariffs for telephony services delivered to a fixed termination point within a public telecommunications network and for the provision of leased lines within such a network shall be cost oriented, this term to be interpreted essentially in accordance with the ONP Directives. This section is not applicable to mobile telephony since this is considered to be subject to efficient competition already. It is specifically noted in the preparatory works that the above mentioned rule on pricing is also applicable to any price dumping that may occur. Section 18 further requires that the tariffs are to be made available to the public. It furthermore enables the Government to stipulate maximum prices for telephony services to fixed termination points within a public network. The purpose of this rule is of course to enable the Government to exercise some control pending the creation of truly efficient competition in the market. As indicated above, such a price cap is put on Telia for the period 1993 to 1996 in order to protect consumers in market segments with little or no competition. During this period, no other operator will be subject to any price regulation.

Interconnection

29–23   By virtue of section 20 of the 1993 Act a licensee is on request obligated to provide interconnection with a corresponding service operated by another licensee or by an operator having been obligated to file reports with the PTA. The licensee is in such case in return entitled to interconnection with the corresponding undertaking. Note that since the requirement to supply interconnections in principle only refers to licensees, this obligation is solely applicable to telephony services as defined in the 1993 Act, *i.e.* 1993 services in a public network consisting of voice transmission, and which permits transmission of telefax and data via low-speed modems. It is further provided that the compensation for such interconnection is to be fair and reasonable in relation to costs. It is added that when estimating the compensation, reasonable consideration shall be made for any particular obligations of the licensee granting access pursuant to its licence conditions.

Concerning telecom services other than those mentioned above, anyone supplying services to a considerable extent (with regard to area covered, number of users, etc.) is required to offer interconnection to other licensees as well as operators obligated to file reports, on competitive terms. In such case, the supplier will have a corresponding right to interconnection. It is finally provided that no operator is obligated to offer interconnection if this would impose on its own needs for network capacity.

The regulations regarding interconnection are in line with the policy formed by the Government in 1988 to ensure the maintenance of an open integrated network regarding telephony services.

The financial terms for supplying interconnection are to some extent dis- **29–24** cussed in the preparatory works. Besides general references to the ONP Directives, it has as far as telephony services are concerned been stated that an operator supplying interconnection is entitled to reasonable profits from this. The price should accordingly not only cover prime cost, depreciations, capital cost, costs for maintenance and developments but also a reasonable yield. Furthermore, the particular obligations on certain operators provided for in the licence conditions are to be taken into account. However, in the short and medium term, these special obligations can be expected to rest solely with Telia. In view of the advantages that Telia's dominant position involves this is not considered unfair. Concerning services other than telephony, it has been indicated that the expression "competitive prices" shall not enable an operator to charge excessive prices due to its dominant position. In this respect the Competition Act 1993 can certainly be applied to correct any abuse of a dominant position.

It should finally be mentioned that the cost for the activities of the PTA are to be covered by charges on licensees and applicants.

Mobile Services

It is indicated above that the Telecommunications Act 1993 is applicable **29–25** to fixed as well as mobile services, the latter defined in the 1993 Act as telecommunications services with radio-aided connection of subscribers. There are a few particular provisions in the 1993 Act for mobile services. First of all a licence may be limited to a certain period of time, which is not the case regarding fixed services. Furthermore, when it is a matter of granting a licence to provide new or substantially revised mobile services and it can be expected that the available frequency range will not be sufficient for all applicants, the applications will be subject to a procedure of general invitation to apply. The Government or the PTA may in connection therewith issue provisions as to the objective criteria to be applied when examining applications.

Under the previous radio legislation a number of licences for mobile services have been issued. The market for analogue systems is dominated by the NMT systems operated by Telia Mobile AB. Comviq AB is operating a small analogue system in addition to the NMT-systems. The GSM system was introduced in 1992. There are three operators running GSM systems; Telia Mobitel, Comviq GSM AB and Europolitan (Nordic Tel AB).

Satellite Services (Non Broadcasting)

For the time being Telia remains the signatory to INTELSAT, INMARSAT **29–26** and EUTELSAT. After the incorporation of Swedish Telecom to Telia AB and pending further legislation, Telia's signatory functions are dealt with in an agreement between Telia and the Government.

There are no national regulations particularly dealing with telecom services via satellites, other than the frequency plan for which the PTA is responsible. The regulations in the Radio Communications Act 1993 are, however, applicable also to such traffic.

Radio Communications Generally

29-27   The Swedish Radio Communications Act (SFS 1993:599) is applicable to any use of radio transmitters on Swedish territory irrespective of the field of application. With exception for Swedish national defence this Act puts a licensing requirement on anyone possessing or using a radio transmitter. Such licences are issued by the PTA. The basic purpose of the Radio Communications Act 1993 is to ensure that as many as possible shall be entitled to use radio, but that such use should be as effective as possible. The presumption when processing an application for a licence is that it shall be granted unless there are obstacles particularly mentioned in the Act, such as that the use would disturb other previously licensed users inside or outside Sweden or that the radio installation does not fulfil reasonable technical requirements. The basic issue when processing an application shall, however, not be whether the application shall be granted or not but rather on which terms the licence shall be issued.

29-28   Any licence will of course be subject to a number of circumstances such as the need to reserve frequencies for future use. In case the operation concerned requires a licence under any other regulations such as the Telecommunications Act 1993, the permit is always conditional upon such a licence having been granted. Each licence applies on a particular use of radio. Beside the frequency allocated, the licence may contain provisions with regard to technical requirements on the radio transmitter, the aerial location, traffic area and other conditions which may be of significance in order to secure an efficient use of frequencies. The licence applies for a certain period of time to be decided in each case taking into consideration factors such as expected technological developments and the financial benefit that the licensee can reasonably expect. During the period of the licence it shall normally not be possible to revoke the licence or change the licence conditions. However, intervention is possible for instance if the licensee has not complied with the licence conditions or if the licence has not been used to the extent anticipated at the time of granting the licence. Under exceptional circumstances the licence can be revoked or the conditions modified due to technological developments or changes in the use of radio abroad in accordance with international agreements.

4. Regulation of Broadcasting

Basic Legislative Rules

The former terrestrial radio and TV monopoly has been decisively abol-   29–29
ished in Sweden. Current legislation is, however, designed for the previous
situation when the former Swedish Broadcasting Corporation had the
exclusive right to broadcasting directed to the general public. When others
were granted broadcasting rights in the 1980s, special legislation was introduced permitting carefully defined broadcasting activities. Such legislation
is contained in the Cable Act 1992, the Radio Periodicals Act 1981, the
Community Radio Act 1993 and others of less importance.

Legislation introduced after 1990 is based on the assumption that anyone
may conduct broadcasting activities. Matters relating to broadcasting technology have been transferred from the Radio Act to a special Radio Communications Act 1993.

The Radio Act has two main functions; to define basic terms and preconditions relating to broadcasting and to describe the general conditions for
broadcasting to the public. The Radio Communications Act 1993 regulates
the conditions for possession and use of radio transmitters.

Separate licences are required for broadcasting to the general public and
the possession and use of radio transmitters. A radio receiver may be pos-
sessed and used by anyone.

The detailed conditions under which a licence to broadcast is granted
are contained in an agreement between the broadcaster and the Government. Such conditions may include obligations to publish replies and rectifications, prohibitions on sending commercial advertisements and an
obligation to reveal sponsors of broadcasts which are not commercial
advertising but are commissioned by someone else.

In 1985 the Radio Act Commission (the "Commission") was instructed   29–30
by the Government to review Swedish legislation in the broadcasting field.
The Commission submitted its proposal for a reform of the broadcasting
legislation during the summer 1994.

According to the proposal, the provisions which regulate radio and TV
broadcasts to the general public will be combined in two new acts: the
Radio and TV Act and a special Act concerning licence fees for radio and
TV. These two Acts are to replace the nine presently existing Acts concerning broadcasting.

In general, the proposal is based on the assumption of freedom to set
up operations in the field of radio and TV. It is, however, proposed that
a licence should be required for certain wireless transmissions at frequencies
below 1 Ghz. The reason for this licence requirement is that there must be
a possibility to control the use of transmission frequencies in areas of the
frequency spectrum which are particularly suited for radio and TV broadcasts to the general public. Licensing under the Radio Communications Act
1993 will still be required for the possession and use of transmitters.

Where a licence is not to be required according to the proposal, for instance for transmission by cable or satellite, the operator will be obliged to notify his operations for registration in order to make it possible to determine who is responsible for the operations in Sweden.

According to the proposal, the Government will have the option to introduce conditions to apply on grant of the licence. These conditions will replace the current procedure whereby public service broadcasters enter into separate agreements with the Government.

Terrestrial and other Delivery

29–31   Cable television was introduced in Sweden as a result of test operations which were governed by a special Act. This temporary legislation was replaced by three new laws in 1985; the Act on Local Cable Distribution, the Act Pertaining to Responsibility for Local Distribution and the Act Regulating Fees for Local Cable Distribution.

The technical developments in the field of satellite reception moving from low-power satellites in "stationary traffic" to medium and high-power satellites had as a result that transmissions via satellite were normally not covered by the legislation as it was regarded as direct broadcasting. Accordingly, there was no requirement to obtain a permit for satellite distribution.

The new Cable Act, effective as of January 1, 1992, was adopted in connection with new constitutional legislation. The new constitution includes a provision that any Swedish citizen or Swedish legal entity has a right to transmit radio programmes by cable.

The Cable Act 1992 pertains to cable network distribution of radio and television programmes in cable television networks that reach more than 100 households. Transmissions in conformity with such transmissions as are regulated by the Radio and Television Act are not covered by the 1992 Act. Linkage of networks and co-distribution in more than one network is permitted.

29–32   The owner or operator of a cable television network must ensure that the households in the building are able to receive, free of charge, certain transmissions specified in the Radio and Television Act. Further, they must make available, in each municipality where they have a network, one channel for television programmes originated by one or more local television operators.

An owner of a cable television network or of certain channels in a network, or anybody using such networks or channels, may distribute satellite or original programmes without a permit. The only requirement for such transmissions in Sweden is that there must be a registered "responsible publisher" resident in Sweden.

As has been mentioned above, technological developments in the area of satellite transmission had as an effect that such transmissions were generally not covered by the legislation. A new Satellite Act was therefore introduced and is effective as of January 1, 1994.

The Satellite Act 1994 regulates transmissions of TV programmes via satellite to the general public in any country party to the EES Treaty. There are no licence requirements for operating satellite activities. The operator or the entity granting satellite capacity is, however, under an obligation to register and provide certain information. The penalty for not conforming to the requirement of registration is a fine.

Regulatory Agencies

Under the present system a permit is required to broadcast transmissions which are governed by the Radio Act. The Radio Act is applicable to transmissions which are intended to be received directly by the public. The conditions under which the permit is granted are negotiated with the Government and included in a special agreement between the broadcaster and the Government. This procedure applies to wireless transmissions, excluding transmissions via satellite. As indicated above there is a proposal to replace this system, introducing a licence requirement instead.   **29–33**

With respect to broadcasting through cable there are no requirements of a licence, registration or similar whereas broadcasting via satellite requires registration.

Prior to July 1, 1994, four different regulating agencies were concerned with matters relating to broadcasting; the Radio Agency, the Cable Agency, the Local Radio Agency and the Board for Local Radio Transmissions. These have now been replaced with two new regulating agencies, *i.e.* the Radio and TV Authority and the Supervisory Agency for Radio and TV.

The Radio and TV Authority is responsible for all permissions under the special Acts in force regarding radio and TV broadcasting which are not a matter for the Government. The Radio and TV Authority is further responsible for registrations and fees and supervises the compliance with rules which do not concern the contents of transmissions. In addition the Authority takes decisions on sanctions for non-compliance with such rules.   **29–34**

The Supervisory Agency has a judge as its chairman and supervises compliance with rules concerning the content of programmes.

Regulation of Transmission v. Content

The new Constitution relating to Freedom of Speech which entered into effect on January 1, 1992 contains a constitutional protection for freedom of speech, *inter alia*, in radio and TV broadcasts.   **29–35**

According to the Constitution every Swedish citizen or legal entity has the right to transmit programmes through cable. Consequently, there may not be any requirements under Swedish law for transmissions to the public through cable. It is, however, possible to prescribe obligations for an owner or operator of a cable network to transmit certain programmes and to provide channels free of charge. As has been mentioned such rules exist in

Sweden for owners/operators of cable networks reaching more than 100 households.

The right to make wireless transmissions to the general public may be regulated by laws containing provisions on permits and conditions. Accordingly, the Government has been able to maintain the practice whereby it enters into separate agreements with public service operators including conditions for their transmission permits.

Rules regarding the contents of broadcasts are included in the Acts relating to broadcasting, the agreements between the Government and public service operators and in voluntary ethical rules for local radio transmissions.

29–36   In the Government's agreements with public service operators, the following obligations may be imposed on the broadcaster.

(a) To give a right of reply and to broadcast rectifications.
(b) To respect the privacy of individuals.
(c) To offer diverse programmes.
(d) To take into account the special impact of sound radio and TV programmes.
(e) To broadcast official announcements at the request of official authorities.
(f) To broadcast certain decisions by the Supervisory Agency.

Public service operators are further obliged to conduct their transmission rights impartially and objectively and to observe the basic principles of democracy.

The Cable Act 1992 includes provisions corresponding to the democracy rule and the obligation to take into account the special impact of sound radio and television programmes.

The Satellite Act 1994 contains provisions regarding observation of the basic principles of democracy, obligation to transmit rectifications, consideration for the special impact of radio and TV, diverse of programmes and an obligation to broadcast certain decisions by the Supervisory Agency.

5. REGULATION OF APPARATUS

Regulation of Apparatus Supply and Installation

29–37   Overall national regulation in the telecom area exists only to a limited extent. The existing regulations concern terminal equipment and the public telecommunications network and are an implementation of the E.C. Directive on Terminal Equipment.

The Terminal Equipment Act 1992 is applicable to equipment that can be connected to a network which is generally available to the public. Only equipment meeting certain requirements with respect to technical design, manufacture, labelling, control, installation and other use determined by the Government may be connected to a public network.

There are no other restrictions on the market for terminal equipment and there are a large number of suppliers.

Type Approval and Testing Procedure

The National Post and Telecom Agency, the PTA, is the authority respons- 29–38 ible for issuing regulations and the approval of equipment which may be connected to a public tele network. In the regulations reference is made to the various standards and testing procedures which have to be followed for different types of equipment. The actual testing may be conducted by any independent laboratory which meets the requirements of the PTA provided the test is conducted according to the instructions in a defined standard. At request the PTA will provide a list of approved laboratories.

6. TECHNICAL STANDARDS

The techniques in the telecommunications area are developing rapidly. 29–39 Already, telephones, telefaxes, computers and alarm equipment can be used in the conventional telecom networks. Next in turn are picture telephones and multimedia terminals for sound, text and picture.

Today the signal transmitted by a TV broadcaster to the viewer is analogue. Analogue transmission uses electrical signals altering continuously. The standard used in Sweden is Phase Alternate Link (PAL).

In the future, the digital technique will replace the analogue, first in satellite and cable networks. The terrestrial network will introduce the digital technique later due to shortage in frequencies. Digital TV transmissions will also be established in the telephony networks.

Digital transmission standards replacing PAL are under development. The transition period for complete replacement of the analogue technique with digital transmission is estimated to be fairly long, probably 10 to 20 years.

The PTA will in due course issue regulations with regard to technical requirements on terminal equipment and interfaces, based on national and international standards.

CHAPTER 30

United States

By Jeffrey Blumenfeld and Christy C. Kunin,
Blumenfeld & Cohen, Washington DC

1. OVERVIEW AND HISTORY

Overview of the United States Telecommunications Industry Today

The provision of interstate telecommunications services in the U.S. today   30–01
is largely market-driven, with regulatory oversight by the Federal
Communications Commission continuing to play an important role in
addressing those issues not adequately addressed by the marketplace. Similarly, the current telecommunications equipment industry in the U.S. is
entirely market-driven, with limited regulatory supervision related largely
to technology issues. The provision of telecommunications services within
state boundaries, which is generally subject to exclusive jurisdiction of the
state regulatory commission, is beginning to evolve to a similar market-
driven model. Most states now permit competition for intercity telephone
service, many have adopted price cap regulation (or other incentive-based
regulation) in lieu of rate-of-return regulation, and a very few permit at
least some level of competition for local exchange service.

Most other communications services, including broadcast (radio and
television), satellite, and cable services are also regulated at the federal level
pursuant to the Communications Act 1934. In addition to the Communications Act 1934, various other statutes come into play for individual categories of service, such as the Communications Satellite Act 1962
(establishing COMSAT as the U.S. signatory to INTELSAT and
INMARSAT), and the Cable Television Consumer Protection and Competition Act 1992 (amending Title VI of the Communications Act 1934).
Regulation of these services is principally governed by the Federal Communications Commission (FCC). The Commission's decisions are subject

to judicial review by the federal courts. Judicial oversight of FCC decisions has been and continues to have significant influence on regulation of the communications industries. Federal oversight of common carrier and cable television services is supplemented by state public utility commission regulation of intrastate telecommunications, and by local municipality franchise authority over cable television providers.

Although historically, telecommunications, broadcast, cable and satellite services have developed under distinctly different regulatory and policy mandates, convergence of these technologies has fostered the movement toward increasing competition within and among these services. Most segments of the communications industry are characterised by competition or emerging competition. Where competition is not yet established, policy goals of regulators, at least for the long term, tend to include fostering a competitive market.

Telecommunications

30–02   In the last 60 years, telecommunications in the U.S. has evolved from a monopoly-based system to an increasingly competitive market-based industry. The U.S. has a dynamic telecommunications industry and a well-established and pervasive telecommunications infrastructure. The U.S. is a leading exporter of equipment and services. A predicate to regulation is provision of service as a common carrier, which generally denotes "holding out" or offering service "to the public at large"; if service is private, then it is not regulated under the Communications Act 1934. Common carrier services are regulated by the FCC pursuant to Title II of the 1934 Act. The nature and extent of regulatory oversight depends on whether a particular carrier is considered "dominant" or "nondominant" under the FCC market power based classification system. All common carriers must file tariffs for provision of services within the U.S.

Satellite

30–03   Satellite services are also regulated and licensed by the FCC under Title II as common carrier services. As is also true for other common carriers, the nature and extent of regulation depends on whether a carrier is classified as dominant or nondominant. Unlike the case in many countries, COMSAT, the U.S. signatory to INTELSAT, is a private corporation, with no government ownership, but with government oversight divided among several federal agencies. In addition, since the early 1980s, the U.S. has championed competition for domestic satellite services, and has opened to competition its markets for both services and earth stations.

Broadcasting

30–04   Since its introduction, broadcasting has presented a conundrum to regulators in the U.S., requiring them to strike a balance between regulating for the public interest on the one hand and avoiding impinging on the Constitutional free speech right of broadcasters on the other. The necessity of licensing to avoid disruptive interference was apparent early in the development of broadcast services. The FCC exercises control over broadcasters through the licensing process on the basis of "spectrum scarcity". The First Amendment to the U.S. Constitution forbids any government regulation of programme content, although the FCC in fact spends a significant portion of its time on broadcast matters trying to deal with controversial content issues such as obscenity and violence in programming. Today, numerous AM and FM radio stations are licensed to operate throughout the country. Most communities are served by at least three VHF television stations, and most metropolitan areas also receive numerous VHF and UHF television signals.

Cable

Cable television has been the subject of much legislative attention in recent   30–05
years. In 1992, Congress enacted the Cable Television Consumer Protection and Competition Act 1992 to reregulate the cable television industry. Substantially reversing the deregulation provisions of the Cable Act 1984, the primary objectives of the 1992 Act are to stimulate competition in multichannel video programming services and to provide consumer protection mechanisms, including regulation of cable operators' rates. Regulatory jurisdiction is split between the Federal Communications Commission and state and local franchising authorities. Under the 1992 Act, state and local authorities are free to regulate basic service rates, grant franchises and impose customer service or other requirements on cable operators. Local licensing of cable operators varies greatly from one locality to the next, as cities and towns often impose stringent conditions and obligations on cable licensees, including dictating the number of channels that must be provided, and requiring that some channels be set aside for public, educational, and governmental purposes, as well as for access by local programme providers.

Summary of Telecommunications History and Regulatory Developments

Telecommunications Industry History

Since the invention of the telephone, telecommunications services in the   30–06
U.S. have been provided by multiple privately-owned carriers. Early in the century there were more than 5,000 telephone companies in the U.S. Most remained quite small, but others, such as Continental and GTE, grew to be multi-billion dollar companies. The largest of these was AT&T. Even today, the former Bell telephone companies provide service to just over 40 per cent of the nation's land area, but to more than 75 per cent of the nation's population.

Also called "The Bell System", prior to divestiture in 1984, AT&T was a vertically integrated company consisting of three main sets of subsidiaries: Western Electric, which manufactured telephones and other telecommunications equipment; Long Lines, which provided intercity telephone service, and; the Bell Operating Companies, which provided local exchange and short-hop intercity telephone services each within its operating territory.

The origins of the Bell System can be traced to March 7, 1876, when Alexander Graham Bell was granted a patent on the basic telephone. During the duration of its patent, Bell used patent licences to gain an interest in most local telephone companies. By 1894, when the basic telephone patent expired, the Bell System held controlling stock in most of its principal licensees. With the expiration of the Bell patents, the number of telephone companies not affiliated with the Bell group—so-called "independent" telephone companies—began to grow at a relatively rapid pace. Each of these provided only local service, and the Bell companies continued to dominate in the provision of intercity service, based on their essentially-exclusive possession of an intercity network.

30–07   In 1907, under the leadership of Theodore Vail, Bell initiated a programme of purchasing independent telephone companies. Vail established a policy goal for the Bell System of "one system under regulation", and also originated the concept of "universal service". Bell used its exclusive control of the intercity network as leverage to gain control of independent companies, by denying them the right to interconnect with that network for intercity calls. This aggressive consolidation policy based on leveraging of market power led to the first federal Department of Justice antitrust complaint against the Bell System, quickly resolved by a written commitment from AT&T Vice President Kingsbury in 1913. Under the Kingsbury Commitment, Bell agreed to stop acquiring independent companies and to permit them to interconnect with Bell's long distance (intercity) network.

Soon after the Commitment was signed, however, the Interstate Commerce Commission, which in 1910 had been given jurisdiction over some facets of telecommunications, was also given the authority to exempt AT&T from the antitrust laws in its purchase of independent telephone companies. Within several years, AT&T acquired over 200 additional local companies. By the time the Federal Communications Commission was created in 1934, the Bell System provided virtually all intercity service and the majority of local telephone service.

In 1949, the U.S. Department of Justice again filed an antitrust action against the Bell System. The focus of this complaint was the relationship between Western Electric and the Bell Operating Companies. It alleged a conspiracy between Western Electric and AT&T to monopolise the manufacture and distribution of telecommunications equipment. This lawsuit was settled in 1956 with a consent decree that required AT&T to licence its patents, restricted AT&T to providing only regulated common carrier services, and placed some restrictions on the scope of Western Electric's manufacturing and sales operations. This settlement soon came under considerable public suspicion, leading to a Congressional inquiry. The result was a report that was highly critical of the Department of Justice's handling of the settlement, essentially charging Justice with malfeasance in arranging a "back room" deal highly favorable to AT&T.

During the next two decades, new competitors arose to challenge the

Bell System's continuing monopolies in the manufacture and distribution of equipment as well as in the provision of intercity service.

The equipment challengers sought the rights to sell their equipment to   30–08
the Bell System and for their equipment to be directly interconnected to Bell System company networks. These challenges were eventually resolved through a series of private lawsuits in the federal courts and actions by the Federal Communications Commission. In the mid-1970s the Commission adopted an equipment registration programme, which is still in effect. That programme provides for private laboratories to test telephone equipment against Commission-promulgated standards, and to certify complying equipment as safe for direct interconnection with the telephone network. The registration programme began in earnest the evolution toward a competitive market for all telephone equipment, the pace of which was accelerated by the 1984 divestiture of the Bell System.

Competing intercity service providers, led by the 1963 petition of Microwave Communications Incorporated (now called MCI) to the FCC, sought the right to provide intercity telephone service in competition with AT&T. Through a series of protracted and hard-fought FCC proceedings over the next two decades, the emerging competitive carriers—termed "specialized common carriers" or "other common carriers" in regulatory parlance—gradually acquired the right to compete for all interstate services provided by the Bell System. The competing carriers were less successful in obtaining network interconnection on equal terms with AT&T's Long Lines, and the FCC enjoyed only mixed success in overcoming the Bell System's intransigence in the face of the emerging competition.

In 1974, the U.S. Department of Justice once again filed an antitrust lawsuit against the Bell System, alleging unlawful monopolisation of multiple telecommunications markets: the manufacture and distribution of customer premises equipment (such as telephones and PBXs); the manufacture and purchase of equipment used by telephone companies to provide service, and; the provision of intercity telephone service. After nearly five years of procedural delay, the lawsuit was inherited by federal judge Harold H. Greene, of the United States District Court for the District of Columbia, who re-activated proceedings in 1979. After the opening statements in January 1981 testimony began in March. The Government presented 93 witnesses in four months; AT&T presented more than 238 witnesses through December. On January 8, 1982, the parties announced that they had negotiated and signed a settlement they called the Modified Final Judgment (MFJ), based on their decision to file the settlement as a modification of the 1956 consent decree, rather than as a settlement of the 1974 complaint.

In the years since the 1956 consent decree, the laws governing antitrust   30–09
settlements had been reformed by a statute introduced by Senator John Tunney. The Tunney Act provided that in any antitrust settlement in which the Department of Justice was a party, the presiding judge was required to hold hearings on whether the proposed decree met the "public interest" standards set forth in the Act. After a six-month cycle of pleadings and

hearings, Judge Greene approved the proposed decree as being in the public interest, subject to some modifications.

As modified, the settlement required AT&T to divest itself of its 22 wholly-owned, local operating companies, as well as of its minority interests in two others (Cincinnati Bell and Southern New England Telephone Company). Thereafter, AT&T would be free to engage in any segment of the telecommunications industry, except that it was precluded for seven years from providing "electronic publishing".

The former Bell Operating Companies were subject to a series of prescriptions and proscriptions after divestiture. On the prescriptive side, the companies were required to implement "equal access"—defined as the same physical and operational network access used by AT&T itself—to all competing intercity carriers. On the proscriptive side, the companies were prohibited from entering or participating in three sets of businesses: the manufacture of both customer premises equipment and network equipment used by telephone companies to provide telephone service; the provision of information services as defined in the decree, and of intercity service outside defined zones (which have come to be called "LATAs", an acronym for "local access and transport areas", access and transport describing the services permitted by the decree, and the geographic scope of the LATA defining the area within which the activity is permitted), and; any other business unless it was a natural monopoly actually regulated by tariff. These "line of business restrictions" on Operating Company authority could be removed on a showing by the applicant company that there was "no substantial possibility that it could use its monopoly power to impede competition" in the market it sought to enter.

As is true of all consent decrees entered under antitrust actions brought by the Department of Justice (as also under other laws, such as those dealing with civil rights), the court in which the decree was originally entered retained jurisdiction to interpret and enforce the decree. Thus, Judge Greene's continued responsibility for decree administration is the rule rather than the exception in antitrust jurisprudence. What is unusual, of course, is the widespread public interest in his decisions.

30–10   The MFJ's reorganisation of AT&T became effective on September 1, 1984. Upon divestiture, and at their decision, the 22 former Bell Operating Companies formed into seven regional holding companies, sometimes referred to as "RHCs" or "RBOCs" (for Regional Bell Operating Companies). The MFJ's provisions apply only to the companies that were formerly part of the Bell System, although not to Southern New England Telephone or Cincinnati Bell.

The MFJ has been subject to increasingly intense public debate. From the first month after divestiture the RBOCs have challenged the MFJ's restrictions in every available forum, including Judge Greene's court, higher federal courts, the federal legislature, and the forums of public opinion. These efforts have been increasingly successful. As this is written, the U.S. Congress narrowly failed in its last session to significantly rewrite the Communications Act 1934, incorporating numerous MFJ modifications in the legislation. The newly-elected Congressional leadership had set a high priority on adopting such legislation in the current session.

Over the same 20 years that the Department of Justice was pursuing antitrust enforcement against the Bell System, the Federal Communications Commission was engaged in a parallel effort to deal with the issue of accommodating increasing competition in all segments of the telephone industry. Beginning in the mid-1970s, the Commission engaged in three broad sets of regulatory inquiries and rulings on these issues.

In one set, the Commission eventually instituted a scheme of tariffed, explicit charges by which local telephone companies were reimbursed by intercity carriers for use of local exchange facilities to originate and terminate interstate intercity telephone calls. These "access charges" replaced a nearly century-long enormously complicated system of "settlements" and "division of revenues" for distribution of intercity call revenues between toll carriers and local exchange carriers. The substitution of explicit and publicly-visible access charges for the negotiated and largely invisible flow of funds under the prior system was designed to help ensure that AT&T's toll services did not receive more favorable financial treatment than did toll services of the new entrants. Adjustments to the FCC's access charge scheme continue to absorb substantial Commission attention. Although access services are geographically local, the FCC has jurisdiction because the services are used to originate and terminate interstate telecommunications, and therefore are themselves considered interstate.

In a second set of proceedings, collectively called the "Computer Inquiries", the Commission promulgated successive sets of rules addressing the   **30–11** anticompetitive risks arising from vertically integrated telephone companies engaging in competitive activities (such as provision of "enhanced" or information services) that were dependent on facilities and services over which they enjoyed a monopoly. These risks—leveraging monopoly power to create potential harm to competition such as cross-subsidisation and anticompetitive pricing of competitive services, and discriminatory network access—also drove the federal lawsuit. While the MFJ resolved the potential for harm by restricting companies (at least theoretically) to the provision of either competitive or monopoly services, the Commission promulgated a set of regulatory rules that would permit the companies to provide both competitive and monopoly services.

Computer Inquiry I was soon overtaken by technological progress, and is of only historical interest. In *Computer II*, the Commission considered but rejected the adequacy of accounting practices as safeguards against anticompetitive conduct. The Commission instead required the companies to establish separate subsidiaries for their competitive and monopoly enterprises. Within two years of that decision, the Commission adopted the *Computer III* decision, which would have abolished the separate subsidiary requirements, substituting a series of accounting and other safeguards. However, *Computer III* was overturned on appeal, the court ruling that the Commission had inadequately documented and explained its change of position from *Computer II*. The case was remanded for further proceedings to the Commission, which has since adopted similar accounting and regulatory safeguards.

The third of these major regulatory efforts, called the *Competitive Carrier* docket, was initiated by the Commission in the early 1980s to streamline regulation of certain carriers, based on whether they were classified as "dominant" or "non-dominant". Dominant carriers—those with market power—continue to be subject to full regulatory oversight, including the requirement that they file for authorisation to build new facilities, and receive no presumption of lawfulness for their tariffs. However, rates for dominant carriers are subject to incentive-based price cap regulation, in lieu of the previous rate base rate-of-return approach. Dominant carriers include AT&T, the interstate operations of local telephone companies, and COMSAT. Nondominant carriers—those with no market power—are subject to relaxed regulatory oversight, including only slight entry regulation, the right to build new facilities without prior authorisation, the right to make new tariff filings effective on one day's notice, and the presumption that their tariff filings are lawful. The courts subsequently reversed the Commission's decision to eliminate tariffing for non-dominant carrier, ruling that the Communications Act 1934 requires all common carriers to file tariffs with the FCC, and that the Commission has no regulatory power to eliminate the filing requirement.

Satellite, Broadcast, and Cable Industry History

30–12   In 1962, the U.S. passed the Communications Satellite Act 1962, creating COMSAT, a privately-held corporation, to act as the U.S. signatory to the International Satellite Organisation, INTELSAT. This was the first step in a U.S. effort to spear-head creation of the international global satellite consortium, INTELSAT. Beginning in the early 1970s, the U.S. Government actively sought to introduce and promote competition in the provision of satellite services. Domestic satellite service was opened to competition in 1972. In 1981 the Commission allowed limited ancillary international services to be provided over non-INTELSAT links. In 1985, in its landmark *separate systems* decision, the FCC authorised the establishment of satellite providers to compete directly with INTELSAT in the provision of satellite services. At present, Congress and the Executive Branch of the federal Government are actively seeking to promote competition for provision of international satellite services.

The Radio Act 1912 vested authority in the Secretary of Commerce and Labour for licensing radio stations and operators. In 1922, the Secretary of Commerce developed further regulation applicable to AM radio services, resulting in a dramatic increase in the number of AM stations. By 1925, radio operators changed frequencies, power, and operating times as they pleased, creating chaos on the air. In reaction, the Radio Act 1927 created the Federal Radio Commission (FRC), a precursor to the FCC, to issue station licences, allocate frequencies, and set limits for station power. Authority to inspect radio stations and assign call signs remained with the Secretary of Commerce. After the 1927 Act was passed, the FRC found it impossible to accommodate all the then-operating stations and thereafter issued new regulations that resulted in the surrender of numerous licenses. The Communications Act 1934 centralised regulatory authority for all communications services, including broadcast services, in the newly-created Federal Communications Commission. Title III of the Communications Act 1934 continues to govern broadcast regulation.

Prior to 1962, the FCC believed that the Communications Act 1934 did not give it authority to regulate cable television. The Commission soon came under increasing pressure from broadcasters to limit the growth of cable. The FCC soon bowed to this pressure, claiming authority because of cable's "ancillary" impact on broadcasting. The Commission imposed cross-ownership rules (prohibiting common ownership of cable and television broadcast licences), and must-carry rules (requiring systems to carry the signals of local over-the-air television broadcasters) on the cable industry. Rate and entry regulation was primarily carried out through local municipalities, who granted generally-exclusive franchises to cable operators giving them access to poles and public rights of way. In 1984, Congress enacted the first comprehensive federal regulatory scheme for cable services. The thrust of the Cable Act 1984 was to deregulate cable rates and services to a large extent, at both the federal and local levels. Concern over the competitive imbalances and consumer abuses following the 1984 Act, however, led Congress to enact the Cable Television Consumer Protection and Competition Act 1992, which re-regulated cable television services.

History of State Legislation

The earliest telecommunications legislation in the U.S. was at the local   **30–13** rather than the federal level. The first oversight of telephone companies was provided by municipal statues and rules, which soon gave way to state statutes and eventually to the establishment of state regulatory commissions. In 1879, the legislatures of Missouri and Connecticut passed legislation creating the first state regulatory commissions to oversee the operations of telephone companies. Despite the push for federal legislation during the early 1900s, states did not abandon their own legislative and regulatory supervision of the industry. There are currently laws governing communications, as well as a legislatively-created regulatory agency, in each state. Cable services are subject to federal jurisdiction, but are additionally subject to state and local municipal jurisdiction over franchising and basic rates. Satellite and broadcasting are regulated solely at the federal level.

History of Federal Legislation

The early development of the Bell System, and the very inconsistent   **30–14** efforts by the federal government to supervise its activities, has been described above. Those efforts included, for example, on the one hand attempts by the Department of Justice to curb the growing power of the Bell System, as in the Kingsbury commitment limiting acquisition of independent companies, and on the other hand, Congressional authorisation to the Commerce Department to exempt such acquisitions from the antitrust laws.

These inconsistent actions led to increased pressure for federal oversight of the growing telecommunications industry during the 1920s. Among the alternatives put forth, but rejected, was federal government ownership of the telephone system. Two major investigations and reports during the early 1930s, indicating little federal regulation of telephone rates or telephone company operations, both recommended consolidating oversight of the industry into a single regulatory entity.

In 1934, Congress passed the Communications Act—which is found in Volume 47 of the U.S. Code, beginning at section 151—creating the Federal Communications Commission and vesting it with jurisdiction over interstate telecommunications activities ("interstate" activities constituting the limits of federal legislative authority). The Communications Act 1934 carried over many regulatory concepts from the Interstate Commerce Act, including the requirement that each entrant obtain a "certificate of public convenience and necessity", and the concomitant right of incumbents to challenge entry on numerous grounds, including the financial effect of entry on the incumbents. Title I, section 1 of the Communications Act 1934 set as its purpose, and the new Commission's mandate, to regulate "interstate and foreign commerce in communication by wire and radio so as to make available, so far as possible, to all the people of the United States a rapid, efficient, Nation-wide, and world-wide wire and radio communications service with adequate facilities at reasonable charges . . ." This language has generally been used as the basis for imposing universal service obligations on carriers, although the term "universal service" appears nowhere in the 1934 Act.

30–15   Title I establishes the Federal Communications Commission as an independent agency (reporting to Congress rather than to the President) charged with regulating the communications field, limits the jurisdiction of the Commission to interstate services, and reserves regulation of purely intrastate services to the states. Title II governs common carriers, and gives the Commission authority to regulate rates and other terms of service consistently with the "public interest". Title III applies to broadcasting, giving the FCC authority to issue licences consistently with the "public interest convenience and necessity," but prohibiting it from censoring content. Title VI sets forth the statutory framework, as amended by the Cable Television Consumer Protection and Competition Act 1992, for regulation of cable television services.

Finally, the Communications Satellite Act 1962—which is found in Volume 47 of U.S. Code beginning at section 701—sets forth the statutory provisions pertaining to COMSAT, INTELSAT and INMARSAT. Of course, as common carriers, satellite carriers are also subject to the provisions of Title II.

2. TODAY'S TELECOMMUNICATIONS ENVIRONMENT

Overview

The provision of telecommunications services today is largely market- **30–16** driven, at least for interstate services (subject to FCC jurisdiction), although regulatory oversight continues to play an important role in addressing issues—of market structure, inter-carrier relations, and consumer protection—not adequately addressed by the marketplace. There are several facilities-based carriers, including significantly AT&T, MCI, Sprint, LDDS, Wiltel, Cable & Wireless, and Allnet, as well as numerous resellers, some national and some regional in scope. These companies compete for business among both commercial and residential customers.

The current telecommunications equipment industry in the U.S. (which is exclusively subject to the jurisdiction of the Federal Communications Commission) is also largely market-driven, with limited regulatory oversight largely confined to technology issues. Consumers are free to choose among a wide variety of equipment from the simplest single line telephones, through multi-line home systems with sophisticated features previously found only in office systems, to the most sophisticated commercial systems. All can be sold, installed, and maintained by any commercial enterprise.

Regulatory jurisdiction over communications services is split between the federal and state authorities. Under the United States Constitution, federal legislation is limited to activities involving "interstate commerce", a term that has been liberally interpreted by the United States Supreme Court. In the course of that interpretation, the Court formulated the doctrine of federal pre-emption, holding that when Congress entirely occupies a field under the interstate commerce clause, states may not legislate inconsistently. Another legal doctrine holds that an agency created by Congress has only that authority specifically granted to it by its enabling statute. Under these doctrines, some programmes and approaches adopted by the FCC have been held to preempt state regulation, as explained under individual headings below.

The provision of telecommunications services within state boundaries, **30–17** which is largely subject to the exclusive jurisdiction of state regulatory commissions, is also beginning to evolve to a market-driven model. Most states now permit competition for intercity telephone service, many have adopted price cap or other incentive-based regulation in lieu of rate-of-return regulation, and a very few permit at least some level of competition for local exchange service.

Telephone service within state boundaries is provided by the state subsidiaries of the Regional Bell Operating Companies, as well as by more than one thousand remaining non-Bell telephone companies, of which GTE, Sprint Corp., ALLTEL, Rochester Telephone, PTI, Century, and Citizens are among the most significant.

The 22 former Bell Operating Companies are organised into seven

Regional Holding Companies: NYNEX (covering New York and the New England states), Bell Atlantic (covering the mid-Atlantic states, New Jersey to Virginia), Bell South (covering the southeastern states, Tennessee to Florida), Southwestern Bell (covering the southwestern states, Missouri, Texas, Oklahoma and Arkansas), Ameritech (covering the mid-west, from Ohio west to Illinois), US West (covering the western and northwestern states, except California and Nevada), and Pacific Telesis (covering California and Nevada).

While each of the seven regional companies operates in several states, each state regulatory commission has jurisdiction only over operations within its boundaries, posing difficulties for issues with cross-boundary effects.

Background to Legislation

30–18　Regulation, rather than legislation, has historically been the primary source of oversight of telecommunications in the U.S. At both the state and federal levels, regulatory agencies (generally called public utility or public service commissions) were created by legislation. Classically, the legislation endowed the agency with a broad mandate to regulate telecommunications companies, and other public utility companies, in the "public interest". Legislatures generally retain ongoing oversight over the operations of the regulators, through both budgetary and substantive inquiries.

Federal legislation in communications in the last 60 years has been generally limited and marginal. There has been little effort, for example, to revise the language even to keep pace with technology, so that the 1934 Act speaks of placing new facilities as "extending lines". There has also been no general revision of the 1934 Act (despite repeated attempts). Congressional action has tended to take one of two forms: Congress issues broad policy mandates with little direction, leaving the FCC to determine how to deal with conflicting technologies, claims, and even conflicting court decisions; Congress does quite the opposite, issuing detailed statutory mandates micro-managing the regulatory environment, such as with several amendments in the last decade addressing operator services (calls paid by the recipient or charged to calling cards), "900" services (calls to pre-recorded information, including adult entertainment services), cable issues, including rates and carriage of over-the-air signals, and electronic eavesdropping.

More recently, both because of controversy about the MFJ and because of the growing perception that the Communications Act 1934 is outdated and of widely-varying effectiveness, proposals to revise or replace the 1934 Act have been introduced in several recent sessions of Congress. For example, during the 1993–94 session of Congress, the Senate considered a broad bill introduced by Senator Hollings as S.1822, the "Communications Act 1994", as well as a different bill introduced by Senators Inouye and Danforth as S.1086. At the same time the House of Representatives considered a bill introduced by Representatives Markey and Fields as H.R. 3636, primarily aimed at increasing local competition and freeing the RBOCs from the MFJ's restrictions on their provision of long distance service, as well as H.R. 3626, introduced by Representatives Brooks and Dingell, also targeted at modifying the MFJ, although with a different approach.

State legislative activity has tended to be somewhat more common, and has been used both to give regulators broad policy direction—for example to accelerate the pace at which competition is introduced—as well as to bring about focused action, usually in areas in which there has been public dissatisfaction—for example requiring changes in cable television rates or terms.

Agencies of Regulation and Competition Control

Federal Communications Commission

The Federal Communications Commission, created by the Communications Act 1934, is the primary federal regulator of the communications industry. The FCC has two essential types of authority. One is judicial in nature: the Commission resolves complaints against and between providers of telecommunications services subject to its jurisdiction. The other is more classically regulatory: the Commission bears primary responsibility for promulgating rules of general industry applicability, as contrasted for example with the U.K. model, under which each carrier's DTI licence sets forth the conditions under which it operates.   30–19

All Commission actions are subject to the requirements of the Federal Administrative Procedure Act which, as the name implies, sets forth procedural requirements designed to maximise public participation and visibility, referred to as "transparency" in other countries. In addition, all final Commission actions, whether adjudications or rule makings, are subject to review by the federal courts, which have jurisdiction to reverse Commission decisions for failure to comply with the requirements of the Administrative Procedures Act. Thus, on review of Commission actions, courts are not free to substitute their own judgment on matters of substance, but can overturn Commission actions as "arbitrary and capricious" where, for example, the Commission has failed to consider alternatives to the course it has chosen, or has failed to consider evidence put before it, or has failed to articulate the basis on which it chose among available alternatives.

The FCC is an "independent agency", meaning that it is not an executive branch agency headed by a cabinet officer, but is rather a creation of Congress and is responsible to Congress. Commissioners are appointed for terms of five years, and therefore their terms are not coextensive with the President's. Thus, the FCC is somewhat more independent of presidential politics than cabinet agencies, such as the Department of Justice. On the other hand, like all such agencies, the FCC is subject to Congressional oversight, both as to the substance of its mission and as to its budget.

The Commission is made up of five commissioners, including a chairman. Commissioners are nominated by the President and confirmed by Congress. The President designates one of the commissioners to act as Chairman. The   30–20

President may change the commissioner designated as chairman, but may not remove a commissioner, except for cause. No more than three commissioners can be from the same political party.

Each Commissioner has several staff members to coordinate and supervise the work of the Commissioners' office. The Commissioners are responsible for setting policy direction for the Commission, and for making all policy decisions and all final decisions, in both adjudicative and "rulemaking" (quasi-legislative) proceedings.

The day-to-day work of the Commission is performed by the professional employees of the Commission, collectively referred to as the "staff", currently numbering nearly 1,800 employees, of whom over 1,400 are located in Washington. The FCC is currently organised into six Bureaus (Common Carrier, Cable Services, International, Mass Media, Wireless Telecommunications and Field Operations) and 10 Offices (Managing Director, General Counsel, Engineering and Technology, Legislative Affairs, Plans and Policy, Public Affairs, Administrative Law Judges, Review Board, Small Business Activities and Workplace Diversity).

As a general rule the operating Bureaus are authorised to enforce existing Commission decisions and policies directly by action of the Bureau Chief, without resort to Commissioner action. But Bureau Chiefs do not have the authority to make or interpret Commission policy without official Commission action. Of course, the line between enforcing and interpreting policy is frequently gray and indistinct.

(i) Common Carrier Bureau

30–21　The Common Carrier Bureau supervises common carriage—telecommunications services offered to the public at large—and the providers of those services. It is also the focal point of telecommunications policy debate. The chief of the Common Carrier Bureau has considerable actual authority, and is usually a trusted confidante of the Commission Chairman. Local telephone companies, because they provide the end links of interstate services, as well as the interstate providers, are subject to supervision by this Bureau. Similarly, disputes among common carriers are addressed to and by this Bureau.

(ii) Cable Services Bureau

30–22　This Bureau was established in 1994 to administer the Cable Television Consumer Protection and Competition Act 1992. The Bureau enforces regulations designed to ensure that cable rates are reasonable under the law. It is also responsible for regulations concerning "must carry" (the doctrine requiring cable operators to offer to carry over-the-air broadcast signals), retransmission consent, customer services, technical standards, home wiring, consumer electronics, cable equipment compatibility, leased access, and program access provisions. The Bureau reviews mergers, sales, and investments in cable systems. It also analyses trends and developments in the industry to assess the effectiveness of the cable regulations.

(iii) International Bureau

This new Bureau has responsibility for domestic administration of telecom-   30–23
munications provisions of treaties and international agreements to which
the U.S. is a party. The FCC participates directly in relevant international
conferences, under the auspices of Department of State. The Commission
licenses radio and cable circuits from the U.S. to foreign points and regulates the licensed operating companies. The Commission also licenses radio
stations on American planes and ships in international service, and under
international agreements and upon request, inspects the radio equipment
of foreign vessels touching U.S. ports. Further, the Bureau resolves cases
of interference between domestic and foreign stations.

   The International Facilities Division, formerly a part of the Common
Carrier Bureau, issues licences for U.S. carriers in international markets,
authorises undersea cables, and handles U.S. carrier participation in international satellite services.

   The International Policy Division, also formerly part of the Common
Carrier Bureau, handles legal proceedings on international issues and also
serves as advisor on trade issues, as representative to international conferences on telecommunications matters.

(iv) Mass Media Bureau

The Mass Media Bureau is responsible for licensing and regulating all facets   30–24
of broadcasting, including both commercial and noncommercial AM and
FM radio, traditional "over the air" television broadcasting, low power
television (LPTV) broadcasting and lottery procedures, international
broadcasting, and emerging audio and visual technologies. This Bureau also
regulates and licenses private microwave radio facilities for distribution
services such as Instructional Television Fixed Service (ITFS). Developing
and implementing enforcement and other regulatory programmes, processing applications for licences and other filings, analysing complaints,
conducting investigations, and participating in FCC hearings on mass
media issues are all responsibilities of this Bureau.

(v) Wireless Telecommunications Bureau

This newly-created Bureau consolidates responsibility for all wireless tech-   30–25
nologies, whose supervision previously had been scattered throughout several disparate Bureaus. For example, the Bureau oversees cellular service
(previously in the care of the Common Carrier Bureau), newly-authorised
(but not yet existing) "Personal Communications Services", those services
not offered to the public at large and hence called "private" radio services
(previously the charge of the Private Radio Bureau), paging services, and
other less well-known services, such as the interactive video data service
(IVDS). This Bureau is in charge of spectrum auctions for wireless services.

   The Mobile Services Division, formerly part of the Common Carrier
Bureau, issues radio licences to cellular telephone companies, operators of

paging services, operators of commercial two-way radio services, telephones aboard aircraft, and certain types of rural radio.

(vi) Office of Plans and Policy

30–26   The Office of Plans and Policy (OPP) is the primary economic and technical
policy advisor to the FCC. Responsible directly to the Commission and
supervised by the Chairman, OPP is responsible for analysing agenda items
and developing long-term policy planning. In collaboration with other FCC
Bureaus and Offices, OPP institutes policy on FCC proceedings, such as
Personal Communications Services (PCS), spectrum auctions, satellite cable
programming encryption, and the implementation of the Cable Consumer
Protection and Competition Act 1992. The OPP Chief coordinates all
policy research and development activity, both within the FCC and with
other agencies, recommends budget levels and priorities for policy research
programmes, and serves as account manager for all contract research studies funded by the Commission.

(vii) Office of the General Counsel

30–27   The responsibilities of the Office of the General Counsel are quite broad,
encompassing not only legal matters arising from the internal operations
of the Commission, but also from the Commission's mission. Thus, the
General Counsel renders advice to the Commission on procedural and substantive questions in dealing with matters within its mandate, and defends
Commission actions when challenged in the courts. (In addition, for purely
historical reasons, the Antitrust Division of the Department of Justice is
the statutory respondent in federal court challenges to FCC actions.)
Recently, a new Competition Division was formed within the Office of
General Counsel. It is charged with ensuring that the Commission's Orders
and Rulemakings promote competition in the communications industry.

(viii) Office of Engineering and Technology

30–28   The Office of Engineering and Technology (OET) renders technical advice
to the Commission on all matters within its jurisdiction. The OET has
primary responsibility for oversight of the equipment registration programme, spot-checking equipment for compliance with FCC regulations,
coordinating assignment of shared frequencies and revising the domestic
Table of Frequency Allocations. In addition, the OET monitors and analyses scientific and technical developments and information.

United States Department of Justice

30–29   The Justice Department exercises a variety of related responsibilities in
dealing with issues of regulation and competition. All of these are concentrated in the Department's Antitrust Division. The Division's work is
apportioned among a number of sections. Each section is responsible for
enforcement of the antitrust laws as applied to a particular set of industries.

Thus, all telecommunications matters are handled by a section currently titled Telecommunications, and previously known as Communications and Finance or as Special Regulated Industries (both prior names reflecting a dual jurisdiction over telecommunications and financial institutions).

One of two agencies charged with enforcing the federal competition laws (the other is the Federal Trade Commission), the Antitrust Division prosecutes both civil and criminal cases of antitrust law violations. The antitrust laws extend to activities beyond the boundaries of the U.S., where the activities will have an effect on U.S. consumers or on companies seeking to export goods or services from the U.S. The Division's antitrust enforcement extends to all commercial activities, including those of regulated industries, such as communications. The *United States v. AT&T* lawsuit was filed and tried by the Antitrust Division, and began as an investigation into AT&T's responses to emerging competition. The Division continues to examine competitive issues in the communications industry.

The Division also investigates mergers and acquisitions for compliance with federal competition laws, and has the authority to file suit to prevent or undo a transaction. Parties to a transaction must file a notice with the Federal Trade Commission based on threshold criteria concerning the size of the firms and of the transaction. This authority is exercised by the Antitrust Division on a shared basis with the Federal Trade Commission, and the agencies negotiate which of them will investigate each transaction, although generally the Division handles telecommunications matters. For example, the Division resolved an investigation of British Telecom's acquisition of 20 per cent of MCI by requiring BT to agree to certain changes in the way it operated its network within the U.K.

Finally, the Division serves as the federal government's competition advocate in regulatory proceedings, and participates in FCC proceedings involving competitive issues. The views of the Division have traditionally carried considerable weight with the FCC, so the Division can be a powerful ally or opponent to any entity's regulatory objectives.

Other Federal Agencies

Two other federal agencies are regularly involved in the communications industry: the National Telecommunications and Information Administration, and the Department of State.   30–30

(i) National Telecommunications and Information Administration

The National Telecommunications and Information Administration (NTIA) was established within the Department of Commerce in 1978. NTIA serves as the President's principal advisor on telecommunications policies. Its role is largely the formulation and articulation of broad Executive Branch policies. NTIA has no direct regulatory authority, and does not make advocacy filings before regulatory agencies.   30–31

Through its Office of Policy Analysis and Development, NTIA formulates Executive Branch policies on domestic communications issues. The

Office of International Affairs provides international policy analysis and technical guidance and represents the U.S. in international telecommunications forums to advance national strategic interests and enhance the international competitiveness of the U.S. This Office also advises the Executive Branch on bilateral and multilateral consultations and makes policy recommendations concerning regulations governing the international use of radio spectrum.

(ii) Department of State

30–32   The Department of State is the federal government's formal delegate in most international forums, although in specific instances this function may be performed by the FCC under State's auspices. Within the Department, the Bureau of International Communications and Information Policy (CIP) seeks to promote agreement among nations on key communications issues and urges private industry involvement in promoting a cooperative approach to international telecommunications. CIP also represents the U.S. in several international organisations, including the International Telecommunications Union (ITU) and INTELSAT. State Department representatives stationed abroad help promote and protect the interests of U.S. telecommunications firms.

State Regulatory Commissions

30–33   Each state has a commission with regulatory authority over communications within the state's borders. In most instances, the same commissions also supervise other regulated industries, including most commonly the energy utilities, but also including transportation companies (bus, train, boat), and often waste disposal companies and others as well. With few exceptions, the state commissions have relatively small staffs, who therefore are relatively selective in choosing the issues in which they will become heavily involved.

   In telecommunications, much of the commissions' time on a day-to-day basis is consumed dealing with rate issues of the state's telephone companies, and with consumer complaints. Over the last decade there has been considerable divergence among the states on the appropriate level of competition to be permitted. Some states, notably New York, Washington, Illinois, Maryland, and Michigan, have ruled in favor of extensive competition, including for local exchange service. Others, including California, have only very recently permitted competition for ordinary switched intra-LATA toll service, the first challenge to the local telephone companies' monopolies on such services. Licensing and operating requirements and processes vary widely among the states, and are best ascertained from the relevant Commission staffs.

Local Government

30–34   Local governments have some regulatory authority in certain industries. Notably, municipal governments have the authority to grant franchises for use of public rights of way. This authority has long been used to grant franchises for cable television service, and to impose certain service requirements on franchisees, such as the number of channels offered, the number of channels made available "without charge" to the city government and/ or for "local access".

More recently, municipal governments have imposed franchise fees on companies laying fibre optic cable in metropolitan areas to compete with local telephone companies. This has become increasingly controversial as the size of the fees has increased, and as it has become apparent that the incumbent telephone companies generally are not subject to such fees.

Agency Processes

Licensing Processes Generally

Prior to establishing operations and initiating service, prospective providers of any communications service must receive authorisation from the regulatory body or bodies with jurisdiction over the proposed service. Typically this is accomplished by obtaining a licence to operate from the relevant regulatory agencies. Common carriers seeking to offer interstate services must obtain a licence from the FCC. To the extent they also wish to provide consumers with the option of completing calls within a state they must also seek a licence from the relevant state commission. Therefore in order to offer services ubiquitously throughout the U.S., a carrier will need to obtain authorisation from the FCC and from each state commission. Cable operators must obtain a franchise from each state or local municipality in which they intend to provide service. Because broadcasting falls solely within federal jurisdiction, broadcasters need only obtain an FCC licence. **30–35**

The FCC Licensing Process

The licensing process varies with the licence sought, as discussed in the following sections.

(i) Foreign Ownership

Foreign ownership in FCC licences for common carrier, broadcast, or aeronautical service is carefully scrutinised by the FCC in the licensing process. The Communications Act 1934 specifically mandates that the Commission consider the foreign ownership of a prospective licensee when making its public interest findings on whether to award a radio licence, and establishes benchmarks applicable to foreign entities seeking to acquire an ownership interest in the parent company of a U.S. radio licensee. The Commission's inquiry focuses on both the actual entity holding the licence as well as the parent corporations controlling that entity. Section 310(b)(3) applies to the licensee. Section 310(b)(4) applies where the licence is held indirectly through a subsidiary. Specifically, section 310(b)(3) states: **30–36**

"No . . . radio license shall be granted to or held by . . . any corporation of which any officer or director is an alien or of which more than one-fifth of the capital stock is owned of record or voted by aliens or their representatives ... or by any corporation organized under the laws of a foreign country."

In addition, Section 310(b)(4) provides:

"No ... radio station license shall be granted to or held by ... any corporation directly or indirectly controlled by any other corporation of which any officer of more than one-fourth of the directors are aliens, or of which more than one-fourth of the capital stock is owned of record or voted by aliens, their representatives, or by a foreign government or representative thereof, or by any corporation organized under the laws of a foreign country, if the Commission finds that the public interest will be served by the refusal or revocation of such license."

The restrictions applicable to the actual licensee under section 310(b)(3) of the 1934 Act are considerably more stringent than those imposed on the parent of a licensee under section 310(b)(4). The section 310(b)(3) benchmark is 20 per cent, as opposed to the 25 per cent benchmark in section 310(b)(4). In addition, the actual licensee can have no foreign directors or officers, while a parent corporation may have up to 25 per cent foreign directors, but no foreign officers. The final difference is that section 310(b)(4) provides the Commission with the ability to approve licences owned and controlled by corporations above these limits if "the public interest will be served," but section 310(b)(3) provides no such "out" for licensees. Thus, as a practical matter, foreign ownership or control of a radio licensee is frequently indirect through subsidiary corporations, and subject to review under section 310(b)(4) of the 1934 Act.

30–36a   Currently, the Commission examines requests to exceed the section 310(b)(4) benchmarks on a case-by-case basis, and has generally considered the following factors;

- *national security*, including whether a country with which a prospective licensee or its parent is associated enjoys "close and friendly relations with the United States" and therefore is not a "national security concern";
- the *extent of alien participation* in the parent holding company;
- the *nature of the licence*, including whether the licensee exercises control over content (the Commission is more lenient with regard to common carrier radio licences which are "passive" in nature and confer no control over the conent of transmissions);
- *other relevant public interest factors* the Commission deems appropriate, such as increased competition or wide dissemination of licences.

The Commission has traditionally considered the following factors when examining the extent of alien participation in the parent corporation of a Title III licensee:

- *where* the parent corporation is *incorporated* (the U.S. or elsewhere),

- the *citizenship* or the *stockholders, officers* and *directors* of the parent corporation,
- whether there are *intermediate* corporations between the licensee and the parent corporation that are incorporated in the U.S., are owned by U.S. citizens or interests, and have U.S. officers and directors.

A review of selected Commission cases involving application in practice **30–36b** of the foreign ownership rules in section 310(b) of the 1934 Act, both for highly visible transactions and for those transactions involving Canadian ownership of common carrier licences indicates that each transaction is carefully reviewed on a case-by-case basis. The Commission clearly permits foreign ownership above the statutory limits, but frequently imposes additional safeguards.

Thus, in *GRC Cablevision*, the Commission allowed 60 per cent foreign ownership of a parent corporation, where the parent corporation was incorporated in the U.S. and the majority of directors were U.S. citizens. Similarly, in *Millicom*, the Commission allowed more than 25 per cent of the directors to be non-U.S. citizens, where the majority were U.S. citizens and where 90 per cent of the shareholders were U.S. citizens. In *Teleport Transmission Holdings*, 65 per cent foreign ownership in a parent corporation was allowed where 75 per cent of the directors and officers were U.S. citizens.

Most recently, in the highly publicised British Telecom acquisition of MCI stock, the Commission introduced a new element into its public interest inquiry: reciprocal market access. Such inquiry into the foreign market opportunities for U.S. firms has routinely been the focus of the Commission's public interest determination in the award of international section 214 authority.

In response to considerable pressure from foreign governments and providers, both the Commission and the United States Congress are considering revisions to the foreign ownership rules. If Congress amends the Communications Act 1934 to either eliminate or modify the foreign ownership restrictions, the Commission could implement the changes quite soon. Absent such legislation, however, the Commission is moving on its own accord to revise its "public interest" review of foreign ownership.

On February 17, 1995, the Commission issued a Notice of Proposed **30–36c** Rulemaking requesting comments on its tentative conclusions to revise the foreign ownership rules applicable to Title III common carrier radio licensees as well as to Title II section 214 authorisations. (*Re Market Entry and Regulation of Foreign-affiliated Entities*, Notice of Proposed Rulemaking, IB Docket No. 95–22, (Rel. February 17, 1995). The Commission proposes to review "effective market access" for U.S. carriers in the foreign country on a service-by-service basis, but the inquiry into foreign market opportunities for U.S. firms would not be dispositive of the Commission's public interest determination. In addition, the Commission also requested comment on whether, in addition to a review of the "effective market access" in the foreign jurisdiction, it should also consider other factors such as the state of liberalisation in the foreign country's other radio-based service markets, national security, or the competitiveness of the applicant's target market in the U.S.

Thus, the Commission could continue to authorise foreign ownership interests higher than the amounts set forth in section 310 of the 1934 Act even in those situations where it does not find that there is "effective market access" abroad. In practice, this means that the Commission would be adding an additional item to its lit of public interest concerns. But, since the Commission can and does make such determinations now (as it did in the case of the BT/MCI merger), the rulemaking may not result in a drastic change in policy or practice. Until the rulemaking is complete, it is difficult to ascertain exactly what the focus of the Commission's inquiry will be.

As this is being written, Congress is considering several legislative initiatives that address the foreign ownership issues, primarily in an effort to liberalise the requirements. One Bill, H.R. 514 proposes to eliminate the foreign ownership restrictions entirely. The major telecommunications legislation, proposed by Senator Pressler, however, would impose an "equivalent market opportunities" test similar to that proposed by the Commission and includes a "snapback" provision that would permit the Commission to withdraw licensing authority if market conditions significantly worsened in the foreign jurisdiction. The most contentious issues in the two Bills involve whether the revised ownership limits should apply to broadcast licences as well as common carrier licences. As passed by the Senate subcommittee, the revised rules would *not* apply to broadcast licences and would be limited to common carrier applications only.

(ii) Common Carrier—Landline

30–37   So long as the applicant is not a "dominant" carrier (currently limited to local exchange companies and AT&T), obtaining a licence for domestic intercity authority is simple and rapid. FCC licensing for interstate common carriage is governed by section 214 of the Communications Act 1934, which provides that carriers must apply for and obtain a "certificate [of] public convenience and necessity" for the construction of any facilities to be used in providing service. The applicant submits a section 214 application along with the appropriate filing fee, the application is reviewed by the FCC, and if found acceptable, it is placed on public notice. Competing parties are given an opportunity to file oppositions to the application. Absent any opposition, the applications are granted within 30 to 60 days after the public notice date.

(iii) Common Carrier—Wireless (Cellular and PCS)

30–38   Dissatisfied with the lottery experience, and particularly with having awarded valuable licences for no payment, the Commission sought and received Congressional authorisation to award new wireless licences by auction. Licences for PCS services—a designation with little current mean-ing other than wireless, and hopefully innovative, services for both voice and data (paging) applications—are being awarded by auction. Applicants pay a fairly steep fee to participate in the auction, calculated on the basis of the number of potential customers within the geographic scope of the licence sought, as well as the amount of spectrum authorised for the exclus-ive use of the licensee. The applicants then bid, by a set of currently-evolving rules, until one emerges the high bidder. The winning bidder for each licence receives the exclusive right to use the allocated spectrum for 10 years for the purposes set forth in the licence.

(iv) Common Carrier—International

In order to provide international resale or facilities-based services, carriers   30–39
must request section 214 authority by formal application, accompanied by a statement which describes the proposed construction and that it will serve the public interest, convenience and necessity. The carrier must also submit the appropriate filing fee with the application. Applications for authority to resell switched services, to resell private lines for private line service, and to resell private lines for switched services or for interconnected private line services are eligible for "streamlined" processing, that is, uncontested applications are usually granted by the FCC 45 days after the date of public notice listing the application as accepted for filing. While facilities-based applications are not eligible for "streamlined" processing, the majority of these applications are granted within 90 days after public notice, provided no competing party has filed a petition to deny the application.

(v) Common Carrier—Satellite

The licensing process for satellite service is set out in Part 25 of the FCC's   30–40
rules, and covers constructing and operating a transmit/receive earth station or a transmit-only earth station, registration of a domestic receive-only earth station, and licensing of an international receive-only earth station. The application process is similar to that for domestic facilities and interna-tional licensing, except the satellite applicant submits FCC form 493 along with appropriate filing fee. If successfully reviewed by the FCC, the applica-tion is placed on public notice and absent any filed oppositions, the applica-tion is usually granted within 60 days after the public notice date.

(vi) Broadcast

Applicants seeking to build and operate a broadcast station, whether AM,   30–41
FM, or television, must file a form 301 application for a "construction permit" with the Commission. The Commission reviews the application and, if the basic legal, technical, and financial qualifications to operate a station have been met, puts the application out for public notice. If no competing application for the same licence is filed the application is

granted. Most frequently, competing applications are filed on every application for a new station. The Commission must conduct a "comparative hearing" among competing applicants to award the licence to the "most qualified" applicant, since applications to use the same frequency allocation are mutually exclusive.

The FCC streamlined the comparative hearing process in 1990 to expedite the award of construction permits. Under its new procedures, the Commission compares the qualifications of competing applicants based solely on the papers they have filed with the Commission, and awards the licence to the applicant it decides will best promote the twin goals of diversity and integration of station ownership and management. When an application is granted for a new station, the applicant receives a construction permit from the Commission.

This licensing process has been heavily criticised by station applicants, who argue that the process unnecessarily delays licence awards until exhaustion of the review and appeals, and results in largely arbitrary awards that do not promote the Commission's goals. To that end, many industry critics have suggested that the Commission exercise its power to award licenses through lotteries or auctions.

The Rulemaking Process

30–42   The FCC has promulgated an extensive set of rules and regulations governing the matters within its jurisdiction. These rules are codified in Volume 47 of the "Code of Federal Regulations." Suggestions for changing these can come from the Commission itself (either the Commissioners or the staff), or from the affected parties, or from the general public, or as a result of action by Congress or a court. No matter the source, the suggestion is either filed formally as, or treated as, a Petition for Rulemaking. Each petition goes first to the relevant bureau or office for staff review. If the Petition meets certain simple threshold requirements, it is put out for public comment by publication in the FCC's weekly notices, and the public at large is invited to comment within 30 days.

After reviewing the comments, the Commission can deny the petition, or investigate it further by issuing either a Notice of Inquiry, which seeks information and suggestions on a course to follow, or a Notice of Proposed Rulemaking, which seeks more focused comments on specific courses of action, including where the Commission is considering adopting specific rule changes. In either case, the proceeding is formally designated with a docket number, and the public is invited to file initial comments and also to reply to comments filed by others. After considering the comments, the Commission issues a report and order, announcing its decision, which can range from adopting rules to a decision that no change is required. There are frequently petitions for reconsideration, leading to further reports and orders in the same docket.

Most states have similar processes for rulemaking, although the terminology varies widely, subject to requirements of openness and fairness imposed by commission rule, or legislation, or court decision. In some states, participants have a right to court review of unfavorable commission actions. In other states, the courts have the discretion to agree or refuse to hear appeals from state agency action.

The Complaint Process

Complaints at the FCC can take a number of forms, and follow a number   30–43
of processes. Each of the Bureaus has an Enforcement Division, which is charged with handling formal complaints filed either by members of the public or by other regulated entities. These are handled according to rules established by the relevant Bureau. Formal complaints are filed in writing, with the subject entity being given a reasonable time for filing a written response. There may also be opportunities for some discovery and for hearings. Formal complaints are ultimately decided by the Bureau Chief, and if necessary thereafter by the full Commission. More frequently (at least on the common carrier side), disputes are handled informally. The aggrieved party speaks with the relevant Bureau, whose staff will then meet with the parties in an attempt to resolve the issue without resort to the more time-consuming formal complaint process.

Most states have some combination of formal and informal complaint processes. As is true of other matters also, the complaint process at the state level can be considerably more informal than at the FCC, permitting the participants considerable latitude and creativity. In addition, the power of particular staff positions can vary widely, so that in some states certain staff members effectively have the power to decide complaint issues with a considerable degree of assurance that the commission will agree.

Working with Other Agencies

The Department of Justice is a law enforcement, not a regulatory, agency.   30–44
As such, there are no procedural rules for dealing with the Department. Traditionally, the Division's attorneys responsible for telecommunications matters are quite willing to meet with industry participants. Information from third parties is an important source of the Division's active investigations. The views of the Division should always be taken into account in formulating strategy on communications matters involving competitive issues.

As with most other facets of dealing with state commissions, the procedures for dealing with the commissions and their staff vary greatly, and are governed solely by state law, regulation, and practice. Thus it is very important to be familiar with the state's practices and, if possible, the individuals who comprise the staff and the commission.

3. REGULATION OF COMMON CARRIERS

As discussed in previous sections, regulation of common carriers and their   30–45
services is divided between federal and state authority, depending on

whether the service is provided across or within state boundaries. At both the federal and state levels, the day-to-day regulatory issues of concern to common carriers include rules governing entry and licensing, any limits on ownership (such as a ban on owning both a telephone company and a cable company, or a limit on ownership by other than a U.S. citizen), the requirement of rate regulation, the rules governing pricing practices (for example, whether prices must be cost-based), and the requirements for making prices known (whether by tariff or otherwise).

Federal regulation is entirely within the jurisdiction of the Federal Communications Commission. As the technology and market for common carrier services has evolved, the Commission has determined that competition has often proven more effective than traditional regulation as a means of meeting its statutory objective to regulate common carriers consistent with the "public interest". Consequently, according to the Commission, it seeks to eliminate or avoid imposing unnecessary regulatory burdens on carriers and consumers.

Common carriers, including wireline, nonwireline, and satellite carriers, are governed by Title II of the Communications Act 1934. The definition of "common carrier", found in common law (decisions of court) rather than in the 1934 Act, generally denotes "holding out" or offering service nondiscriminatorily "to the public at large", that is to anyone who is willing and able to meet the terms and conditions of the offering, as compared with private arrangements.

In section 201 of the Communications Act 1934, the Commission is charged with regulating common carriers consistently with the "public interest", and common carriers are required to make their services available upon reasonable request.

Reasonable and Nondiscriminatory Obligations

30–46   Under section 201(b) of the 1934 Act, common carrier rates and charges must be "just and reasonable". Moreover, under section 202(a), common carriers may not "make any unjust or unreasonable discrimination [or preference] in charges, practices, classifications, regulations, facilities or services." This requirement has been interpreted by the Commission to require common carriers to make their tariffed services available on the same terms and conditions to all "similarly situated" customers. Title II also authorises the Commission to prescribe rates, require tariffs and other information to be filed by carriers, conduct hearings, accept complaints, and award damages for violations of the Act.

Tariff Regulation

30–46a   The Commission may only reject a filed tariff upon a finding that it is "patently unlawful". The Commission construes this legal standard very narrowly, requiring an overt violation of the Communications Act or Commission rules to be readily apparent on the face of the tariff. Such a finding is rare, and thus tariffs are rarely rejected. If the Commission or another interested party suspects that the tariff may violate the law, the Commission has authority to suspend the tariff and investigate. The maximum suspension period is specified in the Communications Act, and changes from time to time. At the conclusion of the suspension period, the tariff goes into effect, even if the Commission has not concluded its investigation.

Satellite Services

The U.S. does not require that all international satellite transmission be 30–47 carried over the facilities of the INTELSAT signatory. The FCC has authorised "separate systems" to provide services in competition with INTELSAT. These separate systems, therefore, provide users a choice among various satellite providers. In addition, COMSAT Corporation, the U.S. signatory to INTELSAT, is required by the FCC to allow resale of its services. Thus, a carrier can purchase a block of capacity from COMSAT and resell that capacity in smaller pieces to other end users. Finally, end users are authorised in the U.S. to purchase services directly from COMSAT, which cannot insist on dealing only with other carriers. This gives users the option of dealing directly with COMSAT for the space segment (satellite transponder capacity) and making arrangements for the domestic links, earth stations and international links independently.

Cross-ownership rules

A recent development of interest to common carriers in the US is a federal 30–48 court ruling holding that the FCC's long-standing ban on "cable-telco cross ownership"—the common ownership of both a telephone company and a cable television company—is invalid because it violates the free speech clause of the First Amendment to the U.S. Constitution. In addition to the stated impact of permitting telephone companies to own cable television companies, the ruling will have the perhaps more significant effect of permitting telephone companies to provide video programming directly, over their existing facilities, or over additional facilities they add to their networks. Some telephone companies, notably Bell Atlantic, have been experimenting with technologies to permit the provision of video over ordinary copper twisted pair.

State Regulations of Common Carriers

State regulation is too varied to be treated in this chapter. Each state 30–48a commission is established and governed by specific state statutory language. There is considerable similarity among the state statutes on overall goals and procedures. There is also divergence on some important policy issues, such as the weight to be accorded competitive considerations in making regulatory decisions, and the authorisation of intrastate competition at various levels of the market (intercity interLATA, intercity intraLATA, and local exchange). Similarly, while State commissions tend to share certain concerns—notably the preservation and extension of universal service, and the protection of those customers with the fewest telecommunications choices—they differ widely in opening their state's telecommunications markets to competition. State regulatory issues must be approached on a state-specific basis.

4. REGULATION OF BROADCAST

30–49   Broadcast—meaning over-the-air broadcast of radio and television signals—is subject solely to federal regulatory jurisdiction and is covered by Title III of the Communications Act 1934. The Commission is authorised, under section 302, to issue licences for use of the radio spectrum and to regulate the broadcasters consistently with the "public interest convenience and necessity." Title III establishes the terms for radio and television licences, and gives the Commission authority to use comparative hearings or random selection (lotteries) for award of construction permits and licences.

Because Title III specifies that broadcasting is not a common carrier operation, broadcasters, unlike common carriers, are not required to sell or give time to all who seek to go on the air, nor are they subject to regulation of rates and business affairs. Programming is primarily the responsibility of the broadcast licensees, and the Commission does not ordinarily monitor individual programs, management, or rates, and similarly does not require the filing of scripts. Indeed, section 326 of Title III specifically prohibits the Commission from censoring broadcast programming. Section 315 establishes special rules applicable to political broadcasting, and section 312(a)(7) gives legally qualified political candidates limited rights of access to broadcast time.

Regulation of broadcasting by the Commission begins with the allocation of frequencies. In accordance with international allocation parameters, and in coordination with the ITU, the Commission allocates radio spectrum for broadcast services. Next the Commission assigns frequencies within the allocated frequency bands to each station or licensee.

30–50   Licensees receive "renewal expectancy", as a result of which the licence is generally renewed to the initial holder after expiration of each licence term. Although competing applications can be filed upon a renewal application, competitors rarely succeed in overcoming the "renewal expectancy" afforded the incumbent. The Commission also oversees assignments (in which assets are transferred to a new licensee) and transfers of control (in which the licensee remains the same, but control of the entity changes) to ensure that the public interest will be served. The Commission inspects licensees' operations for compliance with the Commission's rules. The Commission is particularly concerned that licensees comply with the ownership requirements set forth in section 310 of the 1934 Act.

Several Commission rules address the concentration of ownership of broadcast stations and are under reconsideration by the Commission. The Commission's "multiple ownership rules", revised in 1992, allow ownership of up to 20 AM and 20 FM stations, on a national basis, as of

September 1, 1994. Common ownership of up to two AM and two FM stations is permitted in markets with 15 or more stations, provided that the commonly owned stations do not have a combined audience share of 25 per cent or more. In markets with fewer than 15 stations, the Commission's rules permit ownership of up to three stations, so long as no more than two stations offer the same service. These limitations are somewhat relaxed for stations owned or controlled by members of specified ethnic groups.

Under Commission rules limiting the number and audience reach of television stations, a single entity may hold an attributable interest in up to 12 stations reaching 25 per cent of households. The "duopoly" and the "one-to-a-market" rules prohibit any entity from owning two or more radio stations in the same service if there is a prohibited amount of signal coverage contour overlap between or among the stations, or more than one television station within a local service area. In addition, a broadcast licence holder is prohibited from owning a daily newspaper if the broadcast service area encompasses the entire community in which the paper is published. Similarly, the Commission prohibits ownership of a reportable interest in both a cable system and a television station with overlapping service areas.

The Commission's Financial Interest and Syndication (Fin-Syn) rules 30–51 restrict a network's financial interest in "first run" syndicated non-network-produced programming. As revised by the FCC's 1992 decision, these rules only apply to "prime time" programming. The Prime Time Access rule requires television stations (in the 50 largest markets) that are owned or affiliated with a national network to broadcast no more than three hours of network entertainment programming during the "prime time" hours of 7 pm to 11 pm eastern standard time.

The First Amendment of the U.S. Constitution flatly prohibits the government from making any law interfering with the exercise of free speech. This restriction forms the basis for the Communications Act's prohibition on censorship. Despite the Communications Act's prohibition on content regulation, however, historically the Commission, and the courts, have expended as much energy on this topic as on technical and licensing regulations. Thus, the Commission mandates that political candidates be given "equal time" during political campaigns, that "indecent" programming be broadcast only at certain hours, placing restrictions on the proximity of children's programming and advertising of related products and prohibiting "obscene" broadcasts altogether. The tension between the First Amendment prohibition on government control of broadcast content and the public's demand for "decency" in programming has been litigated since the inception of the broadcast industry.

Industry Structure

Broadcast licensees are regulated as either commercial or noncommercial 30–52 stations. Commercial stations are funded primarily by revenue from advertising broadcast during programming. In contrast, noncommercial stations obtain their revenues primarily from two sources: government funding through the Corporation for Public Broadcasting, and private funding through viewer and corporate donations.

Commercial Television Broadcasting

30–53   The commercial television industry consists of individual broadcast stations located throughout the country, which operate on VHF and UHF frequencies authorised by the FCC. Generally, VHF frequencies are preferable to UHF frequencies because signal propagation is considerably more favorable in the VHF portion of the spectrum.

Most commercial station licences are held by firms or individuals who own more than one station. Approximately half of all commercial stations are affiliated with one of the four national programming networks: ABC, CBS, NBC and, the newest, FOX. In most cases, network affiliates are VHF licensees. Under the affiliation agreements, the affiliates are typically required to carry a certain amount of network programming, and may be required to carry exclusively network programming during certain hours. The networks not only compensate their affiliates for carrying network programmes (and accompanying commercial advertisements), but also commonly provide slots within network "feeds" where affiliates can insert locally generated advertising.

While many commercial stations are network affiliates, a rapidly growing number of "independent" stations, typically operating in the UHF range, have developed. Unlike affiliates, independent stations must purchase virtually all of their programming on the open market, and offer largely syndicated programming, sports and second-run movies.

"Over the air" television—also called "free" television, because the viewer does not pay the programme provider—is funded by the sale of advertising time by networks and local stations. Commercial spots for a particular programme are sold in 30 or 15 second increments with the rates based on the size and demographic characteristics of the audience. No domestic commercial stations are owned or controlled by the Government.

Commercial Radio Broadcasting

30–54   Broadcast radio licensees are authorised by the FCC to provide either AM or FM radio broadcasts. Both AM and FM broadcasters develop programming narrowly targeted for specific audiences with certain demographic characteristics, broadcasting a specific type of music, as well as news, weather, traffic, sports, and talk segments. Most stations are authorised by the FCC to operate on a 24-hour basis, although AM radio "daytimers" are limited to broadcast between sunrise and sunset to avoid interference with other AM stations.

Radio networks sell advertising time to national advertisers covering all markets in the country. Individual stations primarily sell time to local advertisers. Third-party firms (called "reps") represent individual stations to advertisers placing "spot buys", that is individual advertisement appearances, rather than contracts for multiple appearances.

Public Broadcasting

Noncommercial broadcasting includes "public," educational, and instructional services. The Public Broadcasting Act 1967 amended Title III of the Communications Act 1934 to create the Corporation for Public Broadcasting (CPB), a private, nonprofit organisation, with a board of directors appointed by the President of the United States. This hybrid nongovernmental structure was designed to prevent direct influence on programming decisions by the Government. Congressional appropriations for public broadcasting are given to the CPB to fund and support public television. CPB established two sources of public programming, the Public Broadcast System (PBS) for television programming, and National Public Radio (NPR) for radio programming. PBS and NPR distribute programmes simultaneously to all public television and public radio stations, respectively.   30–55

5. REGULATION OF CABLE TELEVISION SERVICES

Regulation of cable televisions services, like common carrier telecommunications, is divided between federal and local authorities. Unlike telephony, however, the vast majority of cable system rate and entry (franchise) regulation is administered by local governments (cities, towns and counties), rather than by state public utility commissions. This unusual treatment stems from cable television's unique and erratic regulatory history. Cable regulation continues to change in unanticipated ways as a result of frequent legislative and judicial modifications to the First Amendment rights of cable operators and their potential competitors and the relative powers of the various governmental bodies.   30–56

The fundamental source for local regulatory power over cable television services stems from a "police power" rationale for protecting health, safety, and welfare. The municipal or other local government conditions the cable company's use of public rights of way in a franchise agreement. Until the mid-1960s, such franchise-based regulations—which typically would specify channel carriage, rate and public access requirements, as well as payment of fees to the franchising authorities—were the regulatory mechanism for cable television. In some states, this power was exercised through special state cable boards or commissions, but typically cable regulation was administered solely by municipal governments.

In 1965, however, the FCC initiated supplemental regulation of cable television systems, principally to protect against a perceived threat by cable television against "free" over-the-air broadcast television stations. Based on its so-called "ancillary jurisdiction," subsequently affirmed by the U.S.   30–57

Supreme Court in 1968, the FCC promulgated rules that, among other things, limited the number of broadcast signals a cable system could carry, required carriage of certain "significantly viewed" local broadcast stations, and granted local broadcast stations exclusivity protection against cable system retransmission of syndicated programming. (Some residual FCC power over cable systems also stems from Title III of the Communications Act 1934, as many cable systems use microwave radio for the transmission and receipt of programming at head-end facilities).

Beginning in 1980, the FCC and later Congress began reconsidering this regulatory scheme and expanding the scope of the Commission's regulatory authority. First, the Commission repealed the "distant signal" and "syndicated exclusivity" rules. They found that broadcast television had continued to flourish despite the rapid growth of cable systems from geographically limited retransmitters of hard-to-receive signals (Community Antenna Television, or CATV), to more robust providers of a broader mix of local broadcast, satellite-delivered and Pay TV programming, such as Time Inc.'s (now Time-Warner) revolutionary Home Box Office service. (The "must-carry" rules, discussed below, remained in effect.) Second, the Commission preempted local rate regulation of "pay" cable programming, effectively limiting the scope of municipal rate regulation to "basic" cable programming. Finally, in 1984, Congress enacted the first major modifications to the Communications Act 1934 (Title VI) specifically addressing cable services. The Cable Act 1984, among other things, codified municipal franchising power and the Commission's federal jurisdiction, but most significantly deregulated basic cable rates for the vast majority of cable systems. The 1984 Act specified that basic cable rates could only be regulated in the absence of "effective competition," which the Commission subsequently interpreted to mean any market served by three or more unduplicated broadcast stations.

30–58   The 1984 Act's policy of fostering cable television's growth through deregulation was reversed, in large part, a mere eight years later by the Cable Television Consumer Protection and Competition Act 1992. While the 1984 Act assumed that video distribution alternatives, such as microwave-delivery systems (MMDS or wireless cable) and Direct Broadcast Satellite (DBS) services provided competitive alternatives to cable systems, the 1992 Act is based on a diametrically different set of policies. These included, for instance, congressional findings that most cable systems are *de facto* monopolies, that rate regulation is essential to protect consumers in the absence of real competition, and that vertical integration of large multiple system operators (MSOs), such as Tele-Communications, Inc., has created anticompetitive incentives for cable programmers to withhold programming from competing delivery systems in order to stifle potential competition. Thus, while the 1984 Act *deregulated* cable services, the 1992 Act substantially *reregulated* cable services, as well as the structure of the cable industry itself. Neither the 1984 nor 1992 Acts, however, set standards for competition by cable systems with local telephone companies for common carrier telecommunications, currently permitted by several states and under consideration by a number of others.

The Cable Act 1992 required the formulation of 22 new sets of FCC regulations and numerous other reports to Congress. Interested parties are urged to study this massive and complicated statute and its companion regulatory record in detail, as this overview only summarises the more significant of the Act's provisions and effects. These fall into seven major categories:

(i) Effective Competition

Like the 1984 Act, the Cable Act 1992 conditions municipal rate regulation   **30–59** on the absence of "effective competition" for cable services. However, contrary to the 1984 Act's presumption, the 1992 Act provides that, absent a showing to the contrary, cable systems do not face "effective competition". It establishes a three-part test which cable systems must meet to demonstrate their qualification for exemption from rate regulation, This includes factors such as penetration (less than 30 per cent of households subscribe), competitive alternatives (at least two multichannel video programming distributors serving 50 per cent of households with at least a 15 per cent market share) or municipal-owned cable system competition (at least 50 per cent of households served).

(ii) Rate Regulation.

The 1992 Act confines municipal rate regulation authority to the "basic"   **30–60** tier of cable services and associated equipment. It adds specific authority for the FCC to regulate cable television services, including all other service tiers *except* programming sold on a per-channel or per-programme basis (pay or "pay-per-view" programming). The 1992 Act also directs the Commission to establish standards for ensuring "reasonable" basic cable rates for systems not subject to effective competition, based on seven specific factors, including competitive rate levels, programming and operating costs, franchise fees, and reasonable profit margins. Municipalities may regulate basic rates in accordance with these standards once they certify to the FCC that minimum procedural safeguards are in place, although these certifications are subject to challenge by cable operators.

These rate provisions serve as the basis for two extensive and complicated sets of Commission regulations. First, the FCC defined rate "benchmarks" for basic cable prices to function as a zone of reasonableness, using an average per-channel pricing methodology. These benchmarks were applied by requiring cable systems, in a series of 1992 orders, to roll-back rates for basic cable services between 10 and 17 per cent. Cable systems are permitted, however, to make cost-of-service showings to the FCC if they believe their own costs justify departure from the rate benchmarks. Second, changes from the rate benchmarks are governed by a set of "price cap" requirements, under which annual inflation and other economic indices, along with productivity offsets, are used as a surrogate for direct price or cost review. "Transition relief" mechanisms for smaller cable systems mitigated the impact of the mandated rate reductions. In mid-1994, however, the Commission modified the rate benchmarks to provide financial incentives for cable systems to add additional basic cable channels, and entered sanctions against numerous cable systems that were found to have reconfigured their service tiers or prices in an effort to evade the Commission's rate regulations.

(iii) Must-Carry and Retransmission Consent.

30–61   The 1992 Act codified the Commission's long-standing rules under which cable systems are required to carry significantly-viewed local broadcast stations, reversing a federal appeals court decision of 1985 that overturned must-carry requirements on constitutional grounds. The 1992 Act ties must-carry authority to retransmission consent, in that broadcast stations can opt either for required mandatory carriage on a cable system or for carriage on the system only with their express consent (and negotiated compensation in lieu of statutory copyright licence payments). However, in 1994, the U.S. Supreme Court overturned substantial portions of the must-carry requirements on constitutional grounds, ruling essentially that the 1992 Act's provisions were content-based regulation of video "speech" and that the economic findings of Congress were insufficient to override cable systems' First Amendment rights as programming distributors. Related provisions of the Cable Act 1992, dealing with carriage of public (non-commercial television) stations and channel positioning of broadcast stations, were unaffected by this ruling.

(iv) Cross-Ownership Restrictions.

30–62   The Cable Act 1992 also codified the Commission's long-standing rule prohibiting local telephone companies from providing cable television service in their telephone service areas. These cross-ownership rules, however, are presently under challenge from all of the RBOCs and many of the independent telephone companies on First Amendment grounds, under which courts have held that restrictions on telephone company entry into cable television violate the First Amendment as a prior restraint on speech. The U.S. Supreme Court is expected to consider the first appeal in one of these cases in late 1995. (Just as the telephone industry has attacked the 1992 Act's cross-ownership restrictions, the cable industry has challenged the FCC's decision to permit telephone companies to offer "video dial tone" services under the Administrative Procedure Act. These cases were pending as this chapter was written.)

(v) Programming Competition.

30–63   The 1992 Act permits the FCC to regulate the structure of the cable television industry, and the relationship between cable systems, cable programming services, and their competitors in order to encourage the development of a competitive video programming delivery market. Cross-ownership restrictions against cable system control of competing local video delivery media were added to the Communications Act 1934. The 1992 Act also directs the Commission to set standards for horizontal concentration of cable MSOs, under which the FCC rules require that no MSO can own or control systems reaching more than 30 per cent of cable subscribers nationwide. Finally, the 1992 Act established an enforceable right of cable programming providers for nondiscriminatory access to cable systems, and of competing multichannel video distributors for nondiscriminatory access to cable programming. These provisions, arising from the Congressional view that vertical integration in the cable industry was stifling potential competition, prohibit policies and practices determined by the FCC to constitute "unfair competition" under the Act's standards.

(vi) Public Educational Governmental and Commercial Access.

The 1984 Act included standards requiring cable systems, in certain circum-   **30–64**
stances, to make channel capacity available for public, educational, and governmental (PEG) purposes and for commercial leased access by programming producers. These provisions of the 1984 Act went largely unused, however, and were modified in the Cable Act 1992 in an effort to encourage more diverse access to monopoly cable systems. The 1992 Act regulates the amount cable systems can charge for commercial access, permits franchising authorities to include mandatory, enforceable PEG requirements in franchise agreements, and requires a proportional amount of system channel capacity to be set-aside for commercial leased access. A complaint procedure for resolution of commercial access disputes is also provided for in the 1992 Act.

(vii) Consumer Protection.

Finally, the 1992 Cable Act includes numerous provision designed to pro-   **30–65**
vide consumers with greater information on and recourse against cable systems, such as customer service requirements, subscriber billing standards, equipment compatibility standards and inside wiring protection, all of which are implemented in detail in the FCC's cable regulations.

6. REGULATION OF EQUIPMENT

The FCC has sole authority in the field of communications equipment regu-   **30–66**
lation. The Commission's adoption of an equipment registration programme, set forth in Part 68 of the Code of Federal Regulations, has entirely preempted the field of equipment (apparatus) regulation. Thus, states are not permitted to impose any requirements or limitations on telecommunications equipment that are inconsistent with, or would frustrate, the specifics and purposes of the federal programme.

For a full century from the invention of the telephone, there was no conceptual or regulatory distinction between "service" and "equipment".

Telephone companies provided "end to end service", as part of which they placed telephone equipment on the customers' premises for the customer to use. All terminal equipment was owned by the telephone company, and none was available for sale to or private ownership by customers. While the equipment was provided under tariff, it was not tariffed distinctly as equipment, but rather subsumed in the service tariffs.

The first challenges to this approach began during World War II. Thirty years of court challenges and regulatory proceedings at the FCC finally resulted in the Commission's adoption in 1975 of the equipment registration programme described in paragraph 30–08. That programme provides for private laboratories to test telephone equipment against Commission-promulgated standards, which are set forth in detail in Part 68 of the Commission's Rules, and to certify complying equipment as safe for direct interconnection with the telephone network. That action marked the beginning of meaningful regulation in the market for telecommunications equipment. In 1982, the Commission "detariffed" both terminal equipment and inside wire, prohibiting telephone companies from providing these under tariff, but permitting them to sell or lease the equipment and inside wire to end users. This action, combined with the 1984 divestiture of the Bell System, greatly accelerated the pace of that evolution.

30–67   Today, terminal equipment ranging from single line telephones to complex electronic switching equipment is marketed like any other consumer appliance or electronic device, by a wide variety of wholesale and retail entities, including retail store-fronts, consultants, and direct from manufacturers (particularly as to the most complicated devices).

Today, obtaining type approval for telecommunications equipment is quite straightforward, assuming the equipment meets the requirements of Part 68 of the Code. The applicant notifies the Office of Engineering and Technology (OET) of the desire to have a device certified. OET provides a set of forms to be completed and filed, and the names of independent laboratories that have been accredited to certify compliance. The applicant then provides samples of the device to the laboratory, which tests the device, and certifies that it either complies or does not. A compliant device is then issued an FCC registration number, which must be displayed on every device sold. It is illegal to sell devices before they have been certified, or without the certification number displayed, even if the device has been certified.

In addition to Part 68 type-approval, many varieties of modern electronic devices include computer processor-based technologies must also comply with the Commission's Part 15 rules governing devices that emit radio frequency (RF) energy, including computers and peripheral devices. Part 15 requires prior testing and/or certification of any device that emits RF energy either intentionally or unintentionally, to ensure conformity with restrictions on power, interference, and other technical characteristics. In general, the rules permit the operation by end users of Part 15 RF-emitting devices without Commission licensing, so long as the equipment complies with the applicable technical specifications, which are more stringent for residential equipment (Class B digital devices) than equipment marketed for commercial or business uses (Class A digital devices). This is an exception to the Commission's general requirement that end users obtain licences to operate devices that intentionally emit RF energy.

7. Technical Standards

As discussed in the previous section, certification of the technical standards   30–68
of terminal and other telecommunications equipment is solely the province of the FCC. States are not permitted any regulation of equipment regulated by the FCC. Technical standards for telecommunications services, on the other hand, are subject to both federal and state jurisdiction. The defining boundary of the jurisdictional split is the ability to identify segregable services as "intrastate" as compared with "interstate".

For intrastate communications, state commissions routinely regulate technical issues such as quality of service—for example, probability of blocking and echo/loss—as well as customer service, billing, subscription termination, deposit, and other commercial practices.

In contrast, the FCC has generally not promulgated by regulation specific technical or customer service standards applicable to telecommunications, leaving the development of standards to consensus-based industry forums and standards-setting bodies. Many *de facto* network interface standards in U.S. telecommunications—for instance the SONET standard for fiber optic interconnection—have been developed under the auspices of the T.1 Committee, a certified standards developing organisation affiliated with the American National Standards Institute (ANSI). Other technical standards have been developed by the Alliance for Telecommunications Industry Solutions (ATIS), the Industry Carriers Compatibility Forum (ICCF), and its subcommittees, including the Industry Numbering Committee (INC). Frequently, the FCC has deferred to such consensual standards without promulgating them as a matter of federal regulatory law.

Telephone numbering issues present particular technical and regulatory   30–69
problems in the U.S. From a technical perspective, the dramatic growth in telephone number usage and the intricate technical requirements of new services such as "800" toll free services have resulted in the "exhaustion" of numbering resources, such as number plan areas (NPAs), commonly referred to in the U.S. as "area codes". From a regulatory perspective, the North American Numbering Plan (NANP) has been administered by Bell Communications Research (Bellcore) since the 1984 Bell System divestiture. Bellcore is a joint venture commonly owned by all seven of the RBOCs. After years of industry wrangling, the competitive issues associated with Bellcore's supervision of number assignment for RBOC competitors and potential competitors, persuaded Bellcore in 1993 to request the FCC to select a neutral third party to take over NANP administration responsibility. Unlike regulatory authorities in other nations, for instance the U.K.'s

OFTEL, the FCC apparently is unwilling to directly administer or regulate numbering, and has not yet acted on this unopposed petition.

A considerably more controversial subject has been the effort by regulators, at the urging of new entrants, to open up the incumbents' networks, both technically and operationally, to allow for efficient interconnection and interoperability of otherwise separate networks. This effort is generically referred to as "open network" or "ONA", the latter a term describing particular efforts in this direction by the FCC.

Incumbent carriers traditionally resist "open network" initiatives, based on their perception that the results would be contrary to their business interests. Entrants press for such initiatives for the same reasons. Therefore, open network initiatives are largely a matter of regulatory proceedings rather than of industry collaborative processes.

30–70   The FCC's ONA efforts began in 1986 and continued through final decisions in 1991. The result was a set of detailed rules requiring local exchange carriers to offer, under tariff, discrete network functions that together comprised existing services offered by the carriers. While the proceeding was successful in the sense that final rules were adopted, there has been little if any innovative use of the discrete network functions. Explanations for this vary, but tend to focus on two criticisms. One is that the FCC did not require a sufficient level of disaggregation, with the result that the local exchange carriers can offer the discrete functions in pacakages that are relatively unattractive in price or performance, rather than as individual functions. The other is that the FCC required those seeking unbundling to submit requests to the local exchange carrier, and permitted the local exchange carriers to refuse the request if the local exchange carrier determined the unbundling was not "feasible". At the outset of the ONA programme, numerous requests for unbundling were submitted to the LECs, the vast majority of which were deemed not feasible. None of those requests has since been granted, and the requests soon slowed to a trickle. The FCC's ONA programme is now essentially defunct.

Open network, or network disaggregation, is one of the most controversial, and arguably one of the most important technical and operational issues in the achievement of fully competitive and fully interoperable networks. It continues to be a central issue at both the federal and state levels. Activity in this area is currently concentrated at the state levels, as new entrants push state regulators for the authority, as well as the technical and operational terms needed, to compete in the provision of local exchange service, currently the last stronghold of telecommunications monopoly.

Index